AF412405

Atomic and Molecular Photoabsorption

Absolute Total Cross Sections

Atomic and Molecular Photoabsorption

Absolute Total Cross Sections

Joseph Berkowitz

*Argonne National Laboratory,
Argonne, IL 60439, USA*

ACADEMIC PRESS

A Division of Harcourt, Inc.

San Diego   San Francisco   New York   Boston
London   Sydney   Tokyo

Academic Press
A Division of Harcourt, Inc.
Harcourt Place, 32 Jamestown Road, London NW1 7BY, UK
http://www.academicpress.com

Academic Press
A Division of Harcourt, Inc.
525 B Street, Suite 1900, San Diego, California 92101-4495, USA
http://www.academicpress.com

ISBN 0-12-091841-2

Library of Congress Catalog Number: 2001088743

A catalogue record for this book is available from the British Library

Typeset by Laser Words Pvt. Ltd., Chennai, India
Printed and bound in Great Britain by MPG Books, Bodmin, Cornwall

01 02 03 04 05 06 07 MP 9 8 7 6 5 4 3 2 1

Contents

Acknowledgment .... **vii**

1 Introduction **1**

  1.1   Introduction 1
  1.2   Reference Table: Sum Rules (Rydberg Units) 6
        Definitions 6
        Conversion Factors 7

2 Atoms **8**

  2.1   Atomic Hydrogen 8
  2.2   Helium 8
  2.3   Lithium 18
  2.4   Atomic Nitrogen 27
  2.5   Atomic Oxygen 35
  2.6   Neon 43
  2.7   Sodium 55
  2.8   Atomic Chlorine 66
  2.9   Argon 82

3 An Aside: The Quantum Yield of Ionization **95**

  3.1   Introduction 95
  3.2   Detailed Studies in the Domain of Competitive Processes 98
  3.3   Involvement of Quantum Yield (η_i) with M_i^2 121

4 Diatomic Molecules **125**

  4.1   Molecular Hydrogen (H_2) 125
  4.2   Molecular Nitrogen (N_2) 140
  4.3   Molecular Oxygen (O_2) 147
  4.4   Carbon Monoxide (CO) 156
  4.5   Nitric Oxide (NO) 166
  4.6   Hydrogen Chloride (HCl) 175

5 Triatomic Molecules **181**

    5.1   Water (H_2O) 181
    5.2   Carbon Dioxide (CO_2) 189
    5.3   Nitrous Oxide (N_2O) 197
    5.4   Nitrogen Dioxide (NO_2) 206
    5.5   Hydrogen Sulfide (H_2S) 214
    5.6   Sulfur Dioxide (SO_2) 221
    5.7   Ozone (O_3) 228

6 Polyatomic Molecules **237**

    6.1   Ammonia (NH_3) 237
    6.2   Methane (CH_4) 246
    6.3   Acetylene (C_2H_2) 252
    6.4   Ethylene (C_2H_4) 260
    6.5   Ethane (C_2H_6) 267
    6.6   Methanol (CH_3OH) 274
    6.7   Benzene (C_6H_6) 281
    6.8   Buckminsterfullerene (C_{60}) 287
    6.9   Sulfur Hexafluoride (SF_6) 300
    6.10 Silane (SiH_4) 309

7 Aspirations for the Future **317**

References **318**

Index **343**

Acknowledgment

The author wishes to express his gratitude to Dr. Mitio Inokuti of Argonne National Laboratory, who has been unstinting of his time while sharing his extensive knowledge of several topics in this monograph. He is also thankful to Dr. Inokuti for accepting the onerous task of reading and commenting on the manuscript.

1

Introduction

1.1 Introduction

Is there ever an optimum time to review an active field of research? For the present topic, a critical evaluation requires a significant investment of time. During this interval, additional information becomes available. One tries to compensate, but it is a series without a convergence limit.

Two decades ago, this author made such an effort (Berkowitz, 1979) at a time when electron synchrotrons were mostly first generation. Even though the field was developing rapidly, general principles could be presented, and one could broadly map the landscape. In the intervening time, new technology and improved calculational capability (and also more experimental sites, more researchers) have vastly increased the available information. It was predictable that increased light intensity from second- and third-generation synchrotrons would enable more difficult, differential cross sections to be measured. But improvements have also been realized in the determination of absolute total cross sections. In the high-energy region, prior data were largely confined to x-ray lines, but currently the smooth continuum emanating from synchrotrons can map out the structure in the vicinity of K- and L-edges of atoms and molecules. In the vacuum ultraviolet, early synchrotron data were encumbered by scattered light and second-order radiation, but more recent experiments, particularly by Holland and collaborators, appear to have overcome these problems. Samson and collaborators have largely avoided such potential uncertainties by using a many-line spark source, and with improved measurement techniques have reduced their combined statistical and systematic error to 1–3%.

Very sharp structure occurs in the photoabsorption spectra of small (usually di- and tri-atomic) molecules in the sub-ionization region. This can lead to saturation at peaks in photoabsorption measurements, if the instrumental resolution is broader than the inherent line width, in experiments based on the Beer–Lambert law. Inelastic energy loss experiments, employing a thin target, effectively avoid saturation, although their resolution is typically poorer than optical studies. Hence, they can provide a check on optical measurements where saturation is possible. Brion and collaborators have recently published a summary of their (e,e) data. Our practice is to utilize photoabsorption measurements when problems such as saturation can be avoided, since the electron energy loss data require an auxiliary normalization (such as the Thomas–Reiche–Kuhn sum rule (TRK) for $S(0)$), which implies a redundancy, since we are utilizing the sum rules to make selections from available data. In some instances (He, H_2), accurate calculations of sub-ionization oscillator strengths are available.

In the earlier volume (Berkowitz, 1979), attention was focused on three sum rules, $S(-2)$, $S(-1)$ and $S(0)$. See 1.2 Reference Table for definitions of $S(p)$. Here, we have extended the study to include $S(+1)$ and $S(+2)$, which emphasize energies typically beyond K-edges. At these high energies, the electric dipole photoabsorption cross section continues to decline precipitously (as $\sim E^{-3.5}$) and higher-order processes (particularly Rayleigh and Compton scattering) begin to dominate. The very low cross sections make measurements difficult. The conventional experiment in the past has measured attenuation, which is the sum of photoabsorption and scattering processes. The photoabsorption component has been inferred by subtracting calculated scattering cross sections from the photoattenuation cross section. The uncertainty can be large, when the subtracted component is $>90\%$ of photoattenuation (see, e.g., the helium section). Experiments with third-generation synchrotrons are currently being undertaken to directly determine the individual contributions. The expectation value of $S(+2)$ can play an important role, since it is related to the photoabsorption component alone.

The expectation value for $S(+2)$ is proportional to the electron density at the nucleus. Atomic Hartree–Fock calculations provide this quantity rather accurately. Equivalent calculations for N_2 and O_2 demonstrate that atomic additivity works quite well. We shall find empirically that it is generally a good approximation for molecules. It is not as good for $S(+1)$, where a term dependent upon correlation enters into the expectation value equation (see Reference Table for sum rules).

As implied above, direct measurements of photoabsorption cross sections above $\sim 30\,\mathrm{keV}$ are scarce and uncertain. Of necessity, we invoke calculated values. Between $10^4–10^5\,\mathrm{eV}$, we avail ourselves of the recently tabulated, calculated atomic photoabsorption cross sections of Chantler (1995). At these high energies, molecular cross sections can be accurately approximated as atomic sums for the relatively low Z atoms considered. For $h\nu > 10^5\,\mathrm{eV}$, we have calculated atomic cross sections using a hydrogenic equation with screening given by Bethe and Salpeter (1977). It is non-relativistic, assumes only electric dipole processes, and applies only to K shells. At $10^5\,\mathrm{eV}$, we have compared the atomic cross sections provided by this equation with the calculated values of Chantler and the 'photoelectric' cross sections given by Hubbell (1969) for selected atoms, and find very good agreement. However, at higher energies Hubbell's cross sections rapidly exceed the hydrogenic formula, presumably because they contain relativistic and non-dipole effects. They may be more physically realistic, but their spectral sum would be inconsistent with the expectation values (especially for $S(+2)$), since the derivation of these sum rules implies non-relativistic, electric dipole photoabsorption. Specifically, the equation we use for all cases except He and H_2 (which require special consideration) is

$$\sigma = 6.8 \times 10^{-16}(Z - 0.3)^6 (Ry/\nu)^4 f(\chi),$$

where σ is in cm^2, Z the atomic number, Ry = Rydberg, $\chi = \sqrt{\nu_1/(\nu - \nu_1)}$ and $f(\chi) = \exp(-4\chi)\arctan(1/\chi)/1 - \exp(-2\pi\chi)$. Here, ν is the running photon frequency, and ν_1 corresponds to the K-edge. At energies well above the K-edge, the experimental value of ν_1 can be used.

After evaluating these cross sections, we compute $S(p)$, $0 \leq p \leq 2$, for successive decades of energy between 10^5–10^9 eV, and record these values for each system. Only at $h\nu \sim 10^9$ eV does asymptotic behavior $\sigma \propto E^{-3.5}$ manifest itself. We demonstrate this by fitting these very high-energy cross sections to an appropriate expansion, $\sigma \propto AE^{-3.5} + BE^{-4.0} + CE^{-4.5}$, and find that the first term dominates. This expansion permits analytical integration from $10^9 < h\nu < \infty$. The resulting values of $S(+2)$ are usually within 2% of the 'expectation' values.

Some may object to using the term 'expectation value' in this context, since it is usually reserved for describing the eigenvalue of an operator, or equivalently, a matrix element. We justify its use here because (apart from $S(-2)$) the $S(p)$ are matrix elements, multiplied by constants. Other terms, such as 'expected value' or 'anticipated value' fall short of conveying the intended meaning.

The expectation value for $S(-2)$ is proportional to the electric dipole polarizability, more specifically the static polarizability. Compared to Berkowitz (1979), more careful attention has been given to data sources, and to the occasionally important infrared contribution, which should not be included in the current comparisons. Bishop and Cheung (1982) summarized available information on this topic for a number of molecules considered in this monograph. Among the alkali atoms, the resonance transition typically accounts for approximately 99% of the polarizability. Recently, some impressive experiments have provided accurate oscillator strengths for these transitions. At least as noteworthy is a measurement of the electric dipole polarizability of sodium by Pritchard and collaborators (Ekstrom *et al.*, 1995), in which two interfering atomic beams are used as an interferometer.

In addition to lithium and sodium, we include atomic nitrogen, oxygen, chlorine, ozone and C_{60} in this monograph. Berkowitz (1979) was largely confined to permanent gases and vapors with sufficient vapor pressure at room temperature to enable Beer's law measurements. This excluded most atoms, and the large class of transient molecular species. New techniques of calibration have been reported recently for calibrating the absolute cross sections of high-temperature vapors. For others, typified by N, O and Cl, experimentalists have tried to exploit specific techniques related to the method of generation. Needless to say, the uncertainties here are much larger than with permanent gases. For even the simplest, relatively stable transient molecular species, the available data were judged too fragmentary for inclusion.

Modern ab initio calculations (many-body perturbation theory, random phase approximation with exchange, R-matrix) have become increasingly useful for atoms, supplementing and in some cases correcting experimental cross sections. Recent applications to open-shell atoms are particularly welcome, because of the aforementioned calibration problems. Thus far, calculations for molecules other than H_2 have been less successful. Special mention should be made of the highly correlated wave functions which have been used to calculate $S(p)$ for He, Li and H_2.

In the past 10–15 years, W. J. Meath and collaborators have presented their distribution of oscillator strengths for many of the atoms and molecules presented here. Their approach is a constrained optimization procedure based on Lagrange multiplier techniques. Various sources of input data are used, subject to satisfying experimental molar refractivities (related to $S(-2)$) and the TRK sum rule. The results are presented as integrated oscillator strengths encompassing various energy intervals. The procedure will generally alter the input data, sometimes beyond the stated experimental uncertainty. It is not unique, since it depends upon the input data available. They stress that their dipole oscillator strength distributions (DOSDs) are not totally reliable in local detail, since the constraint procedures 'cannot completely offset the errors that are inherent in the input information'.

Our goal is somewhat different. We wish to find the best local cross sections, using the sum rules as a guide. This can be very important in regions of sharp structure, where the experimental resolution can influence the maximum and minimum cross section. Some initial filtering of older and/or less precise data is performed, and where necessary, subjective judgments are made, based on the track record of the experimental group. Numerous graphical comparisons are presented. Atomic additivity is employed where related experiments have established its validity, typically in regions devoid of structure. Where possible, direct measurements rather than mixture rules are used in regions displaying structure near K- or L-edges. In many instances newer, and usually more precise data have become available and are incorporated. The presentation is intended to enable the reader to find the best choice of photoabsorption cross sections for the specified system at any given energy.

This information can be utilized to evaluate properties other than the moments of the oscillator strength distribution $S(p)$. These include four differently weighted averages of $\ln E_n$, which we represent compactly (Fano and Cooper, 1968)

$$\ln I(p) = \sum_n E_n^p (\ln E_n) f_n / S(p)$$

They concern the total inelastic scattering cross section for grazing collisions of fast-charged particles with the target species ($p = -1$), the average energy loss, or stopping power in these collisions ($p = 0$), its mean fluctuation ($p = 1$) and the Lamb shift ($p = 2$). The lower energy range of the oscillator strength distribution may be used to estimate the C_6 constant for intermolecular van der Waals interactions.

A related goal is to determine absolute partial cross sections, which can involve states of the ion, stages of ionization, or (with molecules) the abundance of different fragments. The latter measurements are usually presented as branching ratios, e.g., in photoelectron spectroscopy or photoionization mass spectrometry, but can be placed on an absolute scale using absolute photoionization cross sections. For atoms, there is practically little distinction between photoabsorption and photoionization cross sections, except in isolated cases where autoionization may be restricted by selection rules. With molecules, there is usually a region between the ionization potential and roughly $20\,\mathrm{eV}$ where other mechanisms, typically direct dissociation and predissociation, compete with direct ionization and autoionization. Hence, auxiliary measurements are required to determine the fractional ionization, referred to as the quantum yield of ionization η_i. In this monograph, we devote an entire chapter to a survey of the mechanisms underlying η_i for various molecules, and we attempt to rationalize why η_i approaches unity at $\sim 20\,\mathrm{eV}$ for all molecules studied, regardless of size. The totality of ionizations at all energies is related to the absolute ionization cross section for electron impact at very high energies (see Berkowitz (1979)), and comparisons between those two different experiments are made here.

In recent years, a technique called ZEKE (zero electron kinetic energy) has been introduced in photoionization studies. Pulsed laser radiation, involving one or more photons, excites atoms or molecules to within $<10^{-3}\,\mathrm{eV}$ of the first (or higher) ionization potential. After a short delay, a weak electric field induces ionization by the Stark effect. In the present study, our primary interest in ZEKE is that it can provide precise ionization potentials, which mark a convenient separation of photoabsorption/photoionization processes. It has been found that the states probed by ZEKE are longer-lived than expected for high n, low ℓ excited states. Two conflicting views of this long lifetime were presented, one involving dispersion of initially low ℓ states to the entire ℓ (and m_ℓ) manifold by the presence of weak external fields and ions, the other invoking an interaction between the high Rydberg electron and the core. Currently, experiments favor the former as an explanation for the long ZEKE lifetimes, but the latter is almost certainly involved in determining the quantum yield of ionization. It is an intrinsic property of the molecule, and potentially holds greater interest than environmental (electric field, point charge) effects.

This monograph terminates at $Z = 18$, both because detailed absolute photoabsorption cross sections are less complete for atoms and molecules of heavier atoms, and because the sum rules may begin to depart from the formulas given in the Reference Table, due to relativistic effects and higher multipole effects. These sum rules are based on electric-dipole-allowed transitions. Recently, considerable attention has been directed at non-dipolar effects, at photon energies as low as several hundred eV. They are observed as backward-forward asymmetries in photoelectron angular distributions, and arise from cross terms (electric dipole-electric quadrupole, E1-E2, and electric dipole-magnetic dipole, E1-M1) in the multipole expansion of the incident electromagnetic wave. In photoabsorption, only even multiples occur in the expansion; the influence on the sum rules is negligible for low Z, and for incident wavelengths much longer than orbital dimensions.

1.2   Reference Table: Sum Rules (Rydberg Units)[†]

	Spectral sum	Expectation value
$S(-2)$	$\displaystyle\sum_n f_n/E_n^2 + \int_{IP}^{\infty} (1/E^2)(\mathrm{d}f/\mathrm{d}E)\,\mathrm{d}E$	$\alpha/4a_0^3$
$S(-1)$	$\displaystyle\sum_n f_n/E_n + \int_{IP}^{\infty} (1/E)(\mathrm{d}f/\mathrm{d}E)\,\mathrm{d}E$	$\displaystyle\left\langle 0 \left\| \left(\sum_i^N r_i\right)^2 \right\| 0 \right\rangle \Big/ 3$
$S(0)$	$\displaystyle\sum_n f_n + \int_{IP}^{\infty} (\mathrm{d}f/\mathrm{d}E)\,\mathrm{d}E$	N
$S(+1)$	$\displaystyle\sum_n f_n \cdot E_n + \int_{IP}^{\infty} E(\mathrm{d}f/\mathrm{d}E)\,\mathrm{d}E$	$\displaystyle(4/3)\left\{ E_0 + 1/2 \sum_{i\neq j}(0\|p_i \cdot p_j\|0)\right\}$ $\displaystyle= 4/3\left\langle 0 \left\|\left(\sum_{i=l}^N p_i\right)^2\right\| 0 \right\rangle$
$S(+2)$	$\displaystyle\sum_n f_n \cdot E_n^2 + \int_{IP}^{\infty} E^2(\mathrm{d}f/\mathrm{d}E)\,\mathrm{d}E$	$\displaystyle(16\pi/3)\sum_a Z_a \left\langle 0 \left\| \sum_{i=l}^N \delta(r_{ia}) \right\| 0\right\rangle$

In most molecules and some atoms, individual transitions below the IP may not be resolved. The summation must then be replaced by an integral.

The expectation values are for neutral atoms and molecules. They are derived with a non-relativistic Hamiltonian, with transitions confined to the dipole approximation. The nuclei are considered point charges at fixed positions. For non-spherical species, the expectation value is an average over orientations. The polarizability α related to $S(-2)$ is the static electric dipole polarizability, determined from electronic transitions, and sometimes referred to as α^{uv}. Some experiments yield $\alpha^{tot} = \alpha^{ir} + \alpha^{uv}$, where α^{ir} is the contribution of vibrational excitations in the infrared.

Hirschfelder *et al.* (1964) provide a more detailed description of these sum rules, applicable to molecules as well as atoms, and incorporating the finite mass of the nuclei. For $S(0)$, the correction is of order m/m_α, and for $S(+2)$, of order $(m/m_\alpha)^2$, where $m =$ electron mass, $m_\alpha =$ nuclear mass. When applied to H_2, $S(0)$ becomes 2.001 (0.05% correction), and proportionately much less for $S(+2)$. For $S(-1)$ and $S(+1)$ the corrections are less transparent, but we shall rarely encounter sufficiently accurate ab initio calculations of these expectation values to warrant concern. On the other hand, we use $S(0) = N$, the Thomas–Reiche–Kuhn sum rule, which should be accurate to $\sim0.1\%$.

Definitions

$$f = \text{oscillator strength (dimensionless)}$$
$$E = \text{energy (Rydbergs)}$$
$$\alpha = \text{polarizability (cm}^3)$$
$$a_0 = \text{Bohr radius (cm}^3)$$
$$r_i = \text{vector position of } i^{\text{th}} \text{ electron}$$
$$N = \text{total no. of electrons}$$
$$E_0 = \text{total electron kinetic energy}$$
$$p_i, \ p_j = \text{momentum vector of } i^{\text{th}}, \ j^{\text{th}} \text{ electron}$$
$$Z_a = \text{nuclear charge of } a^{\text{th}} \text{ nucleus}$$
$$\delta(r_{ia}) = \text{charge density of } i^{\text{th}} \text{ electron at nucleus } a$$

[†]See, e.g. Dalgarno and Lynn (1957); Bethe and Salpeter (1977) p. 358

Conversion Factors

$$\sigma(\text{Mb}) = 109.760\,97\,(41)\,\text{Mb eV}\,(\mathrm{d}f/\mathrm{d}E)$$
$$\sigma(\text{Mb}) = 8.067\,283\,(64)\,\text{Mb Ryd}\,(\mathrm{d}f/\mathrm{d}(E/R))$$
$$\text{eV} \cdot \text{Å} = 12\,398.418\,57\,(49)$$

2
Atoms

2.1 Atomic Hydrogen

These absolute cross sections are known more accurately from theory than experiment, and are included here for completeness. The values of $S(+2)$ and $S(+1)$ are utilized in the estimation of hydrogen-containing molecules (except H_2) by summing atomic quantities.

The oscillator strengths for the Lyman series (1s $\to$ np) are given by Bethe and Salpeter (1971, p. 263).

$$f_n = \frac{2^8 n^5 (n-1)^{2n-4}}{3(n+1)^{2n+4}}$$

These oscillator strengths, and the corresponding transition energies, are given explicitly by Morton (1991) and also Verner *et al.* (1994) to $n = 30$.

The ionization potential of atomic hydrogen is 13.598 44 eV. The oscillator strength distribution in the continuum is given by

$$\mathrm{d}f/\mathrm{d}\varepsilon = (2^7/3)(1+k^2)^{-4} \exp\left(\frac{4}{k}\arctan k\right)[1 - \exp(-2\pi/k)]^{-1},$$

where $\varepsilon = k^2$ is the electron kinetic energy, and the incident photon energy $E = 1 + \varepsilon$. Dillon and Inokuti (1981) have derived a 5-term series expansion for the above expression, accurate to 1% for $0 < \varepsilon < 5$, or hv up to 81.63 eV. Figure 2.1 is a histogram showing how $\Delta f/\Delta E$ smoothly merges with $\mathrm{d}f/\mathrm{d}E$ at the onset of the continuum.

The values of $S(p)$, in Rydberg units, are: $S(-2) = 9/8$, $S(-1) = 1.0$, $S(0) = 1.0$, $S(+1) = 4/3$ and $S(+2) = 16/3$.

2.2 Helium

Helium warrants special consideration, because one anticipates a closer concordance between the spectral distribution of oscillator strengths with sum rules than with all other systems except atomic hydrogen. Here, calculated values surpass experimental results in accuracy in almost all spectral regions. Indeed, helium has been a veritable proving ground for various theories incorporating correlation. At

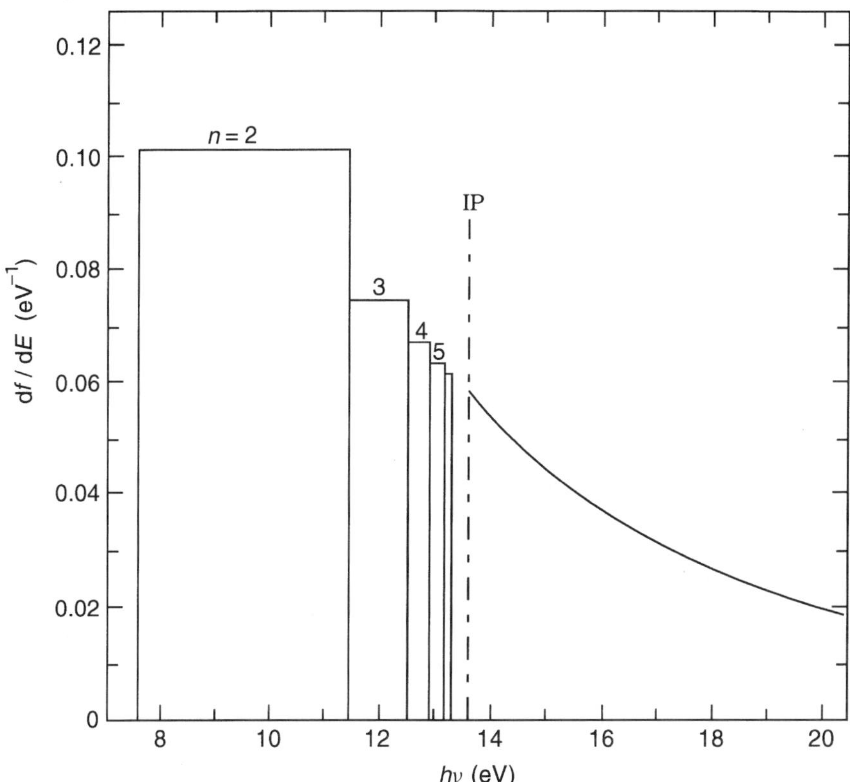

Fig. 2.1 A histogram displaying the convergence of the average oscillator strength in the discrete spectrum to the continuum cross section at the IP, for atomic hydrogen

least three sum rule analyses (Cooper, 1996; Berkowitz, 1997b; Yan *et al.*, 1998) have been reported in recent years. We shall focus on the areas of agreement and disagreement, and try to elicit the best current values.

2.2.1 The data

The ionization potential of helium, based on spectral analysis, is $198\,310.7745(40)\,\mathrm{cm}^{-1}$, or $24.587\,399\,89(50)\,\mathrm{eV}$ (Martin, 1984). Recent Lamb shift experiments obtained $198\,310.6672(15)\,\mathrm{cm}^{-1}$ (Eikema *et al.*, 1997) and $198\,310.6711(16)\,\mathrm{cm}^{-1}$ (Bergeson *et al.*, 1998).

a The discrete spectrum

Already in 1971, Schiff *et al.* (1971) calculated the oscillator strengths of the lowest four transitions ($1\mathrm{s}^2 \rightarrow 1\mathrm{snp}$, $n = 2$–5) to high precision. For the first two, they obtained 0.2762 and 0.0734. Recently, Drake (1996) calculated $0.276\,164\,7$ and $0.073\,434\,9$. Experimentally, Gibson and Risley (1995) determined 0.2700

± 0.0076 and 0.0737 ± 0.0023, respectively. Berkowitz (1997b) lists other calculations up to $n = 9$, and references to other experiments. The three sum rule analyses cited above agree on the contribution of the discrete spectrum to all five sum rules, to the third significant figure. From quantum defect extrapolation, it may be concluded that $\sigma = 7.40$ Mb at the ionization threshold.

b   The continuum

Before embarking on this domain, some general observations are in order. The onset for double ionization is 79.005 eV. Two-electron excitations, most often resulting in autoionization, are observed beginning at $E_r = 60.150$ eV. The anomalous profile of this first resonance may significantly influence the underlying continuum down to $\sim$57 eV. We shall treat the oscillator strength contributions of these resonances separately, as perturbations on the smooth continuum. Above 79 eV, the various theories applied to photoabsorption may be calculating single ionization only, or the sum of single and double ionization. According to Cooper (1996), the random phase approximation (RPA) calculations of Amusia *et al.* (1976a) include single and double ionization, those of Bell and Kingston (1971) are estimates of single ionization, while the more recent calculations of Hino *et al.* (1993) and Kornberg and Miraglia (1993) explicitly provide single and double photoionization cross sections. To the latter, we may add the subsequent convergent close-coupling calculations of Kheifets and Bray (1998a,b). An ambiguity may exist in the high-energy asymptotic behavior.

b.1 IP – 120 eV   Figure 2.2 presents the experimental data of Samson *et al.* (1994a). They assert an accuracy of ± 1–1.5% from IP – 60 eV, and $\pm 2\%$ from 60–120 eV. Their cross section at threshold (7.40 Mb) is precisely that evaluated from the discrete spectrum. Also shown are length (upper cap) and velocity (lower cap) calculations by Bell and Kingston (1971), and RPA calculations by Amusia *et al.* (1976a). Not shown are calculations by Stewart (1978) from IP – 58.6 eV, which fall very close to the experimental curve. The gap $\sim$60 eV is the region of prominent resonances. This is the only spectral region where experimental accuracy appears to exceed that of calculations, although the results of Stewart are very slightly higher. At lower energies, the length form of Bell and Kingston agrees better with experiment, as they anticipated. The RPA values are larger than the experimental results below 60 eV, but are in good agreement above the double ionization threshold. The contributions to the sum rules, based on the experimental data, have been reported previously (Berkowitz, 1997b) and are included in Table 2.1.

b.2 Resonances, 60–72 eV   The 'excess oscillator strength' in each resonance is represented by the expression

$$f_{xs} = \frac{mc^2}{2e^2}\sigma\rho^2\Gamma(q^2 - 1)$$

given by Codling *et al.* (1967). Here m and e are the mass and charge of the electron, c the velocity of light, σ is the continuum cross section and ρ is the

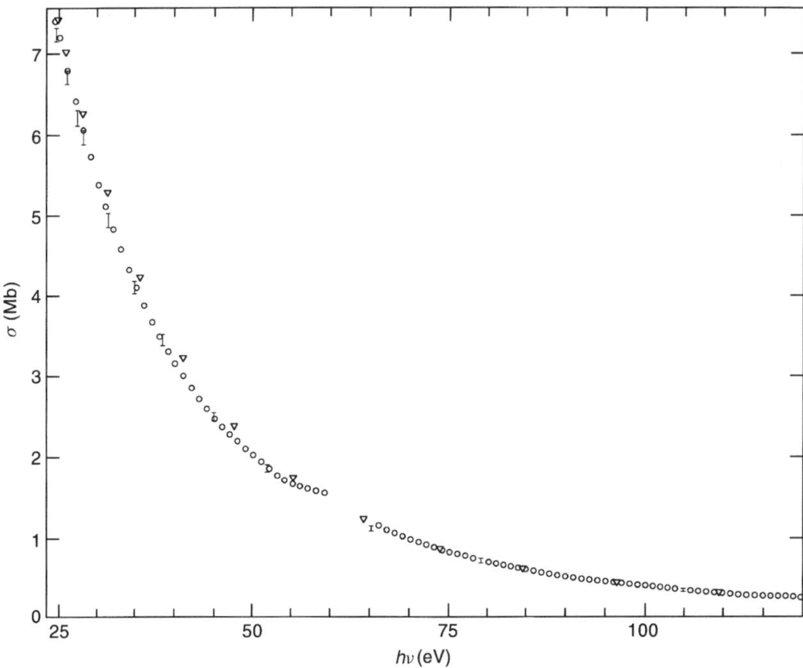

Fig. 2.2 Absolute photoabsorption spectrum of helium. 25–120 eV. ○ Samson *et al.* (1994a); I Bell and Kingston (1971); ▽ Amusia *et al.* (1976a)

Table 2.1 Spectral sums, and comparison with expectation values for helium[a]

Energy, eV	$S(-2)$	$S(-1)$	$S(0)$	$S(+1)$	$S(+2)$
21.2–IP[b]	0.1632 (0.1624)[c]	0.2629 (0.2617)	0.4247 (0.4229)	0.688 (0.685)	1.12 (1.12)
IP–120[d]	0.1809 (0.1825)	0.4747 (0.4778)	1.4143 (1.4188)	4.984 (4.989)	21.25 (21.27)
60–72 (Reson.)[e]	0.0003 (0.0005)	0.0015 (0.0032)	0.0066 (0.0147)	0.030 (0.068)	0.13 (0.32)
120–280[f]	0.0009 (0.0009)	0.0099 (0.0098)	0.1159 (0.1154)	1.438 (1.433)	18.89 (18.84)
280–1000[f]	– (–)	0.0008 (0.0008)	0.0240 (0.0237)	0.762 (0.750)	27.28 (26.80)
1000–13 600[g]	– (–)	– (–)	0.0015 (0.0015)	0.196 (0.194)	34.86 (34.50)
13 600–∞[h]	– (–)	– (–)	– (–)	0.006 (0.006)	17.73 (17.75)
Total	0.3453 (0.3463)	0.7498 (0.7533)	1.9870 (1.9970)	8.104 (8.125)	121.26 (120.60)
Expectation value[i,j]	0.3458[k]	0.7525	2.0	8.167 45	121.336

[a]In Rydberg units.
[b]As evaluated by Berkowitz (1997b).
[c]Quantities in parentheses from Yan *et al.* (1998a). Discrete spectrum and resonances as given, continuum values integrated from their eq. 14.
[d]As evaluated in b, based on Samson *et al.* (1994a).
[e]Calculated from experimental parameters of Schulz *et al.* (1996).
[f]Based on calculated ($\sigma^+ + \sigma^{++}$) from Kheifets and Bray (1998a), velocity form.
[g]Based on calculated ($\sigma^+ + \sigma^{++}$) from Hino *et al.* (1993) and Ishihara *et al.* (1991).
[h]From asymptotic formula of Salpeter and Zaidi (1962).
[i]Pekeris (1959).
[j]Drake (1996).
[k]Bishop and Pipin (1995).

autoionization to dipole correlation coefficient (Fano and Cooper, 1965). The values of Γ and q are taken from recent experimental data of Schulz *et al.* (1996). The value of ρ is unity for the dominant first transition, and is used throughout, though it may be lower for the higher transitions. The appropriate continuum cross section is interpolated from Fig. 2.2. The computation yields an excess oscillator strength for the resonances given by Schulz *et al.* of 0.0066, slightly higher than given earlier by Berkowitz (1997b), 0.0058, but lower than that of Yan *et al.* (1998a), 0.0147.

b.3 120–280 eV Earlier sum rule analyses had led Berkowitz (1997b) to conclude that the selected cross sections of Samson *et al.* (1994a) in this region were low, and an alternative compilation (Bizau and Wuilleumier, 1995) was interposed. Yan *et al.* (1998a) came to a similar conclusion and they interpolated the region between 170 eV and 2 keV on a plot of $E^{7/2}\sigma(E)$ versus E. They subsequently generated a 7-term expansion for $\sigma(E)$, from IP to infinity. The function $E^{7/2} \cdot \sigma(E)$ is still rising rather rapidly between 120 eV and 2 keV, while $\sigma(E)$ is steeply declining. Figure 2.3, a plot of $E^{5/2}\sigma(E)$ versus E, enables us to distinguish among various calculations and experiment in greater detail. The dashed line follows the polynomial fit of Yan *et al.* (1998a), which we utilize provisionally as a basis of comparison. The calculations of Bell and Kingston (1971) are in better agreement in the length form at the lower energies, but at

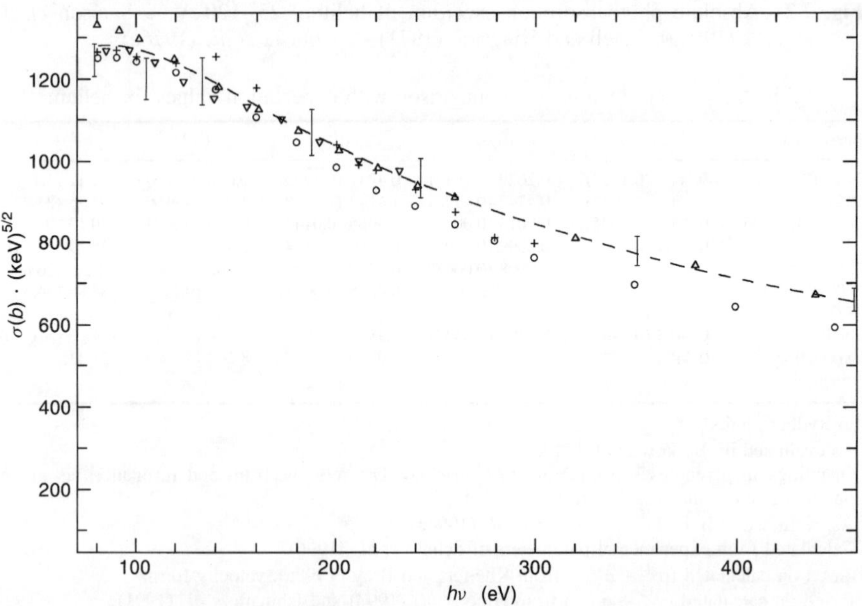

Fig. 2.3 $\sigma(b)^* (\text{keV})^{5/2}$ versus $h\nu$, 80–460 eV, for helium. - - - Yan *et al.* (1998a); o Samson *et al.* (1994a); I Bell and Kingston (1971); + Bizau and Wuilleumier (1995); △ Kheifets and Bray (1998a); ∇ Amusia *et al.* (1976a)

higher energies the velocity form is preferable, as expected. This conclusion is valid even though double ionization occurs in this range, and their calculation is limited to single ionization. At 120 eV, the experimental value of Samson *et al.* (1994a) is in good agreement with Yan *et al.*, but falls below this reference with increasing energy, as already noted by Berkowitz (1997b) and Yan *et al.* The compilation of Bizau and Wuilleumier (1995) is more erratic, hovering near the reference (80–120 eV), exceeding it by 6.5% (140 eV), then declining more rapidly and approaching Samson *et al.* at 280 eV. The RPA calculations of Amusia *et al.* (1976a) follow the reference fairly well, with no deviant trend, from 84.6–231.5 eV. The convergent close-coupling calculations $(\sigma^+ + \sigma^{++})$ of Kheifets and Bray (1998a), velocity form, tend to be high for $h\nu < 120$ eV, but above this value they follow the reference. Tang and Shimamura (1995) used hyperspherical coordinates with close-coupling (HSCC) and pseudostates for discretizing the continuum He^+ states. They calculated σ^+ and σ^{++} between 79–280 eV, and present their data in compressed figures. Our manual extraction of their results $(\sigma^+ + \sigma^{++})$ shows excellent agreement with the reference from 120–280 eV, with somewhat higher values at lower energies, i.e., similar to the results of Kheifets and Bray. They also find excellent agreement with the R-matrix calculations of Meyer and Greene (1994). To avoid confusion, the data of Tang and Shimamura and Meyer and Greene are not shown in Fig. 2.3. In summary, very good agreement exists among several types of calculations, which are preferred over the experimental compilations in this range, as indicated by earlier sum rule analyses. To evaluate the $S(p)$, we have used the data of Kheifets and Bray (1998a), made available to us in digital form, rather than the polynomial of Yan *et al.* (1998a) which required semiempirical interpolation in this region.

b.4 280–1000 eV Figure 2.4 extends the plot of $E^{5/2}\sigma(E)$ versus E to 2300 eV. In this expanded view, the velocity form of the calculations of Kheifets and Bray (1998a) is seen to be slightly higher than given by the polynomial of Yan *et al.* The experimental compilation of Samson *et al.* continues to be distinctly lower. The MBPT calculations of Hino *et al.* (1993) now appear, and are displayed as σ^+ and $(\sigma^+ + \sigma^{++})$. Their results fall below Yan *et al.* for $h\nu < 1$ keV, but at 1 and 2 keV they are in very good agreement. We tentatively continue with the data of Kheifets and Bray (1998a) in this interval, fitted to a 4-term polynomial, although the Bell and Kingston (1971) calculations favor the interpolation of Yan *et al.*

b.5 1000–3000 eV; 3000–13 600 eV Experimental photoabsorption cross sections become increasingly difficult to measure here. The photoelectric effect wanes asymptotically as $E^{-7/2}$, and scattering begins to dominate. Large errors can result by subtracting calculated Rayleigh and Compton scattering cross sections from total attenuation measurements. This becomes eminently clear when comparing experimental and calculated cross sections on a plot of $E^{7/2} \cdot \sigma(E)$ versus E, as is done in Fig. 2.5. Several calculations generally agree that this

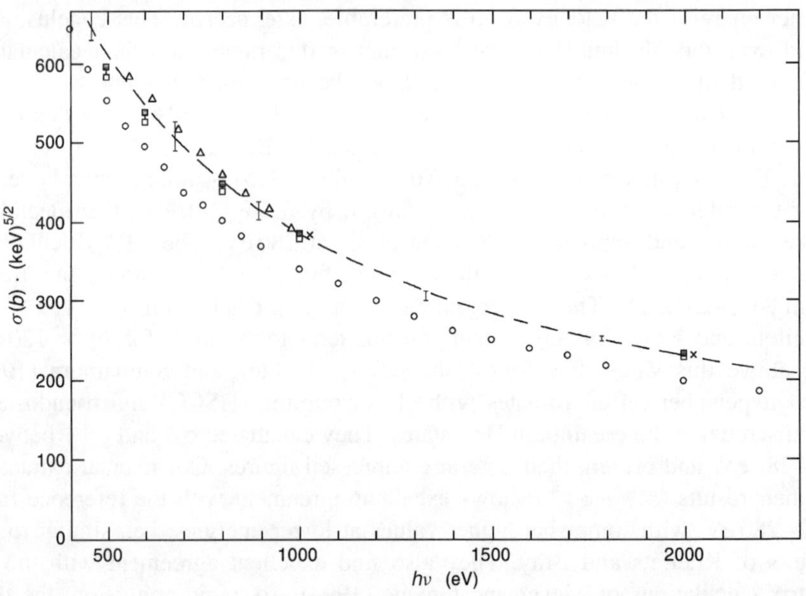

Fig. 2.4 $\sigma(b)^*(\text{keV})^{5/2}$ versus $h\nu$, 450–2000 eV, for helium. - - - Yan *et al*. (1998a); I Bell and Kingston (1971); △ Kheifets and Bray (1998a) vel.; × Kheifets and Bray (1998b) accel.; □, ■ Hino *et al*. (1993) σ^+ or $(\sigma^+ + \sigma^{++})$; o Samson *et al*. (1994a)

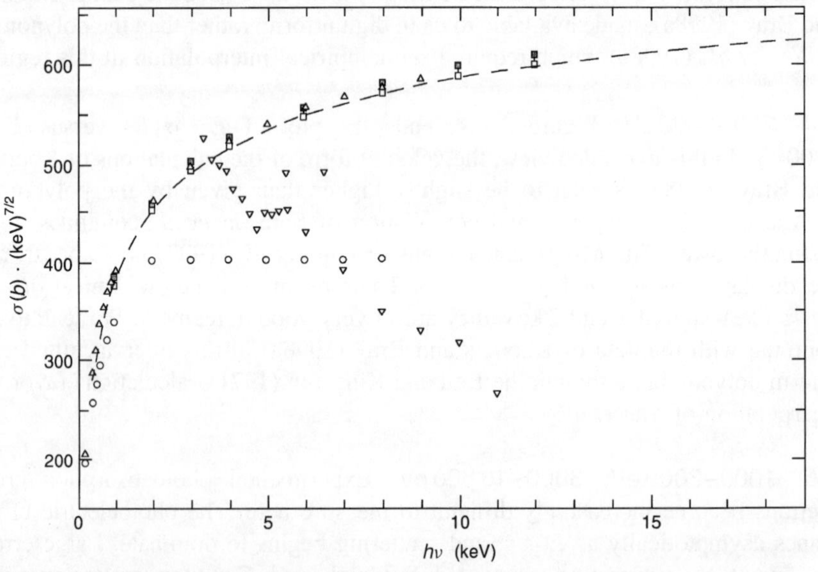

Fig. 2.5 $\sigma(b)^*(\text{keV})^{7/2}$ versus $h\nu$, 0.1–18 keV, for helium. - - - Yan *et al*. (1998a); △ Kheifets and Bray (1998a) (vel) to 980 eV, Kheifets and Bray (1998b) (accel) to 9025 eV; □, ■ Hino *et al*. (1993) σ^+ or $(\sigma^+ + \sigma^{++})$; o Samson *et al*. (1994a); ▽ Azuma *et al*. (1995)

function increases monotonically while experimental inferences by Samson *et al.* (1994a) and Azuma *et al.* (1995) either manifest an abrupt constancy far short of the asymptote or an erratic, dramatic diminution.

The calculated cross sections plotted include the previously encountered MBPT results of Hino *et al.* (1993), the convergent close coupling data of Kheifets and Bray (1998a,b), the fitted polynomial of Yan *et al.* (1998a) and the asymptotic equation of Salpeter and Zaidi (1962).

$$\frac{\mathrm{d}f}{\mathrm{d}E\,(\mathrm{Ry})} \cong 309 E_{\mathrm{Ry}}^{-7/2}(1 - 2\pi/E_{\mathrm{Ry}}^{1/2} + \cdots),$$

which they recommend for $E \gtrsim 1000$ Ry units (13.6 keV). We defer discussion of this equation, but note here that the leading term is the same as that of Yan *et al.*, and the second term nearly so, accounting for the smooth merger of the two. It is noteworthy that the single ionization cross section (σ^+) of Hino *et al.* also merges with this asymptotic form. At 12 keV, $\sigma^+(\mathrm{Hino}) = 0.0996$ b, $\sigma(\mathrm{Yan}) = 0.0996$ b and $\sigma(S-Z) = 0.0997$ b. If we accept Cooper's (1996) assertion that the calculations of Hino *et al.* are indeed partial cross sections for σ^+ and σ^{++}, then these must be summed for the present purposes. We make this heuristic assumption, and fit the ($\sigma^+ + \sigma^{++}$) by regression to two 4-term polynomials (1–3 keV; 3–13.6 keV). We prefer the values of Hino *et al.* in this interval to those of Kheifets and Bray (1998b), since the latter display disparity in length, velocity and acceleration gauges. The coefficients of the various fitted polynomials are collected in Table 2.2, while the contributions to $S(p)$, obtained by analytical integration, are given in Table 2.1.

b.6   13 600 eV $\rightarrow \infty$ The asymptotic form of the oscillator strength's dependence on energy is given by Salpeter and Zaidi (1962) as

$$\frac{\mathrm{d}f}{\mathrm{d}E} = \left(\frac{512}{3}\right) \langle \delta(r_1) \rangle (1 - 2\pi/E^{1/2} + \cdots) E^{-7/2},$$

Table 2.2   Coefficients of the polynomial $\mathrm{d}f/\mathrm{d}E = ay^2 + by^3 + cy^4 + dy^5$ fitted to data at various energies[a]

Energy range, eV	a	b	c	d
IP–120	0.307 867	3.025 638	−3.860 86	1.447 231
120–280	−0.068 16	5.149 822	−6.421 27	1.921 883
280–1000	−0.035 76	4.878 073	−11.384 3	49.633 03
1000–3000	−0.007 651	3.112 359	21.601 08	−204.666
3000–13 600	−0.000 34	1.207 182	222.051 7	−8 701.494

[a]$\mathrm{d}f/\mathrm{d}E$ in Ry units, $y = B/E$, $B = \mathrm{IP} = 24.5874$ eV.

with E in Ry units. With $\langle \delta(r_1) \rangle$, the charge density at the nucleus in the ground state given by Drake (1996) as $1.810\,429\,318\,49$, we obtain

$$\frac{\mathrm{d}f}{\mathrm{d}E} = 308.98(1 - 2\pi/E^{1/2} + \cdots)E^{-7/2}.$$

The coefficient is shown by Salpeter and Zaidi to be the sum of 287.6 $[(1s)^2 \rightarrow (1s\varepsilon p),\ ^1P]$, 13.045 $[(1s)^2 \rightarrow 2s\varepsilon p\ ^1P]$ and 8.45 $[(1s)^2 \rightarrow (ms,\ \varepsilon'p)$ and $(\varepsilon's,\ \varepsilon p)$, $^1P]$. Berkowitz (1997b) used only the first component, which resulted in a slightly lower value of $S(+2)$. The third component includes $\varepsilon's$ states of low excitation energy, i.e. the εp electron contains most of the excitation energy. Salpeter and Zaidi note that the third component comprises roughly 3% of the dominant $(1s)^2 \rightarrow (1s, \varepsilon p)$, ^1P transition, and suggest that inclusion of $\varepsilon's$ with higher excitation energy may increase the contribution of this component to 4–5%. The implication is that the asymptotic form given is not quite complete. The difference between 3% and 4–5% is tantalizingly close to 1.6%, the asymptotic ratio of σ^{++}/σ^{+} which is currently favored.

Dalgarno and Sadeghpour (1994) describe a similar equation derived by Dalgarno and Stewart (1960) which is said to include single ionizations to the ground and all excited states of He^+ and the double ionization process. However, they qualify this observation by stating 'As the hard photon 'collides' with the atom, the full two-electron wavefunction collapses onto the nucleus, releasing a photoelectron with almost all of the photon energy and angular momentum, <u>leaving the other photoelectron with little energy</u> and zero angular momentum'. They do not offer an estimate of the oscillator strength neglected in this approximation.

Subsequently, Yan $et\ al.$ constructed an analytical formula joining the asymptotic form ($E > 13\,\mathrm{keV}$) with theoretically calculated values (2–$13\,\mathrm{keV}$) and a semi-empirical interpolation (0.17–$2\,\mathrm{keV}$) that merges with the data of Samson $et\ al.$ at still lower energy, with an eye toward satisfying the five sum rules. Their formula is

$$\sigma(E) = \frac{733.0}{E(\mathrm{keV})^{7/2}} \left(1 + \sum_{n=1}^{6} \frac{a_n}{x^{n/2}} \right) \text{barns},$$

with $x = E/24.58\,\mathrm{eV}$. Below, we compare the first three terms with corresponding ones from Salpeter and Zaidi, in Rydberg units.

	Salpeter and Zaidi (1962)	Yan $et\ al.$ (1998)
1st term:	$308.98\ E^{-7/2}$	$309.28\ E^{-7/2}$
2nd term:	$-1941.38\ E^{-4}$	$-1971.10\ E^{-4}$
3rd term:	$7082.8\ \ E^{-9/2}$ (incomplete)	$8280.62\ E^{-9/2}$

We note that the leading term is virtually identical, but the second term may already reflect the semi-empirical adjustments made by Yan $et\ al.$ The third term cannot be directly compared, because Salpeter and Zaidi do not include the contributions of ms, εp and ε's, εp to this term. For sum rule analysis in the interval 13.6 keV to infinity, the differences are inconsequential. Integration of the full seven-term expansion of Yan *et al.* yields $\Delta S(+2) = 17.75_3$ Ry units, while the Salpeter–Zaidi formula gives $\Delta S(+2) = 17.73_0$ Ry units. For $\Delta S(+1)$, both give 0.005_6 Ry units, and contributions to the other $S(p)$ may be neglected.

2.2.2 The analysis

The expectation values of $S(p)$, $-1 \leq p \leq 2$, originally given to high accuracy by Pekeris (1959) and verified to even higher accuracy by Drake (1996) are given in Table 2.1. For $S(-2)$, we utilize the recent calculation by Bishop and Pipin (1995).

Table 2.1 lists the contributions to $S(p)$ in the energy intervals discussed, and in parentheses, the corresponding values from Yan *et al.* Their values for the discrete spectrum and resonance region are shown, together with continuum contributions evaluated from their 7-term expansion.

For both the current selections of data and the analytical formula of Yan *et al.*, the spectral sums agree with the expectation values to better than 1% for all $S(p)$. For $S(-1)$, $S(0)$ and $S(+1)$, the analytical formula appears to be slightly better, but this can be traced primarily to the resonance contributions. Yan *et al.* apparently based their resonance contributions on integration over calculated resonance profiles, whereas our method involved experimental parameters and an equation for excess oscillator strength. For $S(+2)$, the present selection is closer to expectation, and can be traced to our use of $(\sigma^+ + \sigma^{++})$ between 1.0–13.6 keV from Hino *et al.* The observations in 2.2.1.b.5, 2.2.1.b.6 and Fig. 2.5 suggested that the asymptotic formula derived by Salpeter and Zaidi (1962) and implicitly used by Yan *et al.* (1998a) may not fully account for σ^{++}. If the asymptotic formula is increased by 1.6% (the currently favored σ^{++}/σ^+), then $S(+2)$ would be enhanced by 0.28, with no significant effect on the other $S(p)$. This is probably an excessive increase, but it is certainly within the tolerance of sum rule analysis.

Numerically, these are fine points. The main conclusions from this analysis are that the analytical formula of Yan *et al.* is a very good approximation to the photoabsorption cross section of helium from IP$-\infty$, and that recent calculations support the semi-empirical interpolation used by Yan *et al.* in arriving at their formula. This follows not only from the excellent agreement of the total spectrum with expectation values, but also from the very good agreement in the various energy intervals shown in Table 2.1. In the course of arriving at this level of concordance, we have seen the limitations of existing experimental data. Also, the various calculations appear to achieve their highest accuracy in different energy domains, and we have used graphical representations to make our selections. Finally, it must be kept in mind that the cross sections and expectation values refer to electric dipole selection rules in the non-relativistic range.

2.3   Lithium

2.3.1   The data

The ionization potential of atomic lithium is $43\,487.19 \pm 0.02\,\mathrm{cm}^{-1}$, or $5.391\,724 \pm 0.000\,003\,\mathrm{eV}$ (Moore, 1971).

a   *The discrete spectrum*

The experimental oscillator strengths in the discrete spectrum were analyzed by Martin and Wiese (1976). For the transitions $2^2\mathrm{S} \to n^2\mathrm{P}$, $n = 2-7$, they chose the relative oscillator strengths of Filippov (1932), normalized to an f value (0.753) for the resonance transition $2^2\mathrm{S} \to 2^2\mathrm{P}$ calculated by Weiss (1963). The precise f value for the resonance transition is important since (as will be shown later) it contributes $\sim$99% to the total $S(-2)$, or static polarizability. Gaupp *et al*. (1982) presented what appeared to be a very precise experimental value, $f = 0.7416 \pm 0.0012$, using the decay in flight of a fast, laser-excited beam. This triggered a re-examination by both theorists and experimentalists. Weiss (1992) obtained $f = 0.7478$ from an extensive CI calculation, and pointed out his value and other independent calculations differed from Gaupp *et al*. by 4 experimental standard deviations. Quite recently, Yan and Drake (1995), using variational wave functions in Hylleraas coordinates, obtained $f = 0.746\,957\,2$ (10). Almost concurrently, two experimental results appeared on the radiative lifetime of the $2^2\mathrm{P}$ state, which could be converted to f values. Volz and Schmoranzer (1996), using in principle the same experimental technique as that of Gaupp *et al*., obtained $\tau = 27.11$ (6) ns, or $f = 0.7469$ (16). Alexander *et al*. (1996) used association of cold Li atoms to determine the long range vibrational energies of the $A^1\Sigma_u{}^+$ state of Li_2. The result could be expressed in terms of the $2^2\mathrm{P}_{1/2}$ lifetime of $^7\mathrm{Li}$, $\tau = 27.102$ (7) ns or $f = 0.7471$ (2). These latter experimental results, in excellent agreement with the calculated value of Yan and Drake, appear to settle the matter.

After the analysis of experimental data by Martin and Wiese (1976), the oscillator strengths of the higher transitions $2^2\mathrm{S} \to n^2\mathrm{P}$, $n \geq 3$, have been calculated, rather than measured. (Nagourney *et al*. (1978) reported the lifetime of $3^2\mathrm{P}$ to be 203 (8) ns, but the decay of this state is primarily to $3^2\mathrm{S}$, and an accurate branching to $2^2\mathrm{S}$ would be necessary to infer the desired f value.) Some of these calculated f values are shown in Table 2.3, together with the experimental inferences of Martin and Wiese. We choose the values compiled by Verner *et al*. (1994), which agree well with the close-coupling calculations of Peach *et al*. (1988), and are included in the Opacity Project data.

One of the noteworthy features of the oscillator strength distribution in lithium is that a Cooper minimum occurs in the discrete spectrum, between the $2^2\mathrm{P}$ and $3^2\mathrm{P}$ upper states. This can be seen in Martin and Wiese (1976), Fig. 4 and in Barrientos and Martin (1987), Fig. 1, where a histogram is depicted. Although these are not the best choices of oscillator strength, the location of the Cooper minimum is validated by all the calculations and the experimental values.

Table 2.3 Oscillator strengths in the discrete spectrum of lithium

n(upper state)	(Martin and Wiese, 1976)	(Lindgård and Nielsen, 1977)	Peach *et al.* (1988) Model pot.	Close coupling	Verner *et al.* (1994)
2p	0.753	0.7412	0.7435	0.7475	0.748
3p	0.0055	0.004225	0.00488	0.00481	0.00481
4p	0.0045	0.003949	0.00435	0.00430	0.00431
5p	0.0027	0.002377	0.00260	0.00257	0.00258
6p	0.0017	0.001463	–	–	0.00158
7p	0.0011	0.0009496	–	–	0.00102
8p	–	0.0006676	–	–	0.000697
9p	–	0.0004571	–	–	0.000495
10p	–	0.0003447	–	–	0.000363
11p	–	0.0003511	–	–	–
12p	–	0.0001956	–	–	–

Table 2.4 Contributions from the discrete spectrum to the $S(p)$ sums (in Ry units)

np	E_n (cm^{-1})	$S(-2)$	$S(-1)$	$S(0)$	$S(+1)$	$S(+2)$
2	14903.88	40.495399	5.499847	0.746957	0.101447	0.013778
3	30925.63	0.060564	0.017068	0.00480	0.001356	0.000382
4	36469.80	0.039023	0.012969	0.00431	0.001432	0.000476
5	39015.56	0.020410	0.007257	0.00258	0.000917	0.000326
6	40390.84	0.011663	0.004293	0.00158	0.000582	0.000214
7	41217.35	0.007230	0.002716	0.00102	0.000383	0.000144
8	41751.63	0.004815	0.001832	0.000697	0.000265	0.000101
9	42118.26	0.003360	0.001290	0.000495	0.000190	0.000073
10	42379.16	0.002434	0.000940	0.000363	0.000140	0.000054
11	42569.1					
$\sum_{n=10.5}^{\infty}$		0.010865	0.004256	0.001667	0.000653	0.000256
Total discrete		40.655763	5.552468	0.764479	0.107365	0.015804

The contributions of the discrete oscillator strengths to the $S(p)$ sums are gathered in Table 2.4. The short portion between $n = 10$ and the IP is bridged by linear extrapolation of (df/dE) and the corresponding integrations for $S(p)$. For this purpose (see below), the photoionization cross section at the IP is taken as 1.49 Mb.

b   The ionization continuum

b.1   IP to 500 Å (24.797 eV)   In the near photoionization continuum, absolute cross sections were measured by Hudson and Carter more than 30 years ago (Hudson and Carter, 1965; 1967). In such experiments there is the usual difficulty of establishing an accurate pressure for a non-permanent gas. An additional complication is the concomitant presence of dimer (Li_2), which can exacerbate the error when the atomic cross section is low, as it is here. Peach *et al*. (1988) have calculated the atomic cross section in this region, using the R-matrix method. The shape of the σ (Mb) versus λ (Å) curve is similar to the experimental curve, but the absolute value is slightly lower at the IP, and the discrepancy increases with energy to 60% at 575 Å. For the reasons mentioned, the calculation appears more reliable in this region. A more recent calculation by Chung (1997) is in excellent agreement with that of Peach *et al*. Consequently, the graph of Fig. 3 from Peach *et al*. was digitized and fitted to a 4-term polynomial. The coefficients are listed in Table 2.5, and the contribution to $S(0)$ in Table 2.6.

b.2 24.797–65.0 eV continuum This region consists of a $2s \to \varepsilon p$ continuum, declining in intensity, and autoionizing structure attributable to K-shell excitation, with limits $1s2s^3S$ (IP $= 64.41$ eV) and $1s2s^1S$ (IP $= 66.15$ eV). Data for the underlying continuum are taken from Peach *et al*. (1988), Fig. 2, $h\nu = 24.44 - 32.60$ eV, and continued to 65.0 eV from Lisini (1992). Both calculations used the R-matrix method. The underlying continuum was fitted to another 4-term polynomial. The declining continuum approaches zero in cross section before the onset of inner-shell continua. The combined oscillator strength of the discrete spectrum and continuum to 65 eV amounts to $f = 0.990$, which is attributable to the essentially separable excitation and ionization of the 2s electron.

b.3 The resonances 1s 2s np These single-electron transitions are listed in Table 2.7. The energies and oscillator strengths are experimentally determined, but receive strong support from calculations by Lisini (1992).

b.3.1 Two-electron excitations and unassigned bands These transitions, primarily having lower oscillator strengths, are listed separately in Table 2.8. They are based on experimental data of Mehlman *et al*. (1978a; 1978b; 1982).

Table 2.5 Coefficients of the polynomial $(df/dE) = ay^2 + by^3 + cy^4 + dy^5$ fitted to data at various energies[a]

Energy range, eV	a	b	c	d
IP (5.3917)–24.797	1.404 752	−1.980 46	0.860 344	−0.100 32
24.797–65.0	1.565 44	0.134 004	−34.773 4	99.333 65
75.0–200.0	−26.733 2	4013.147	−57 295	267 916.6
200.0–851.5	−1.741 31	2425.186	−10 513.8	−346 844
851.5–2000	−0.752 29	2081.107	26 561.62	−1 752 736
2000–10 000	−0.344 17	1547.128	172 584.8	−24 462 221

[a] df/dE in Rydbergs, $y = B/E$, $B = IP = 5.391\,724$ eV.

Table 2.6 Spectral sums, and comparison with expectation values for lithium (in Ry units)

Energy range, eV	$S(-2)$	$S(-1)$	$S(0)$	$S(+1)$	$S(+2)$
0–IP ⎱ [a] 0–5.3917 ⎰	40.655 763	5.552 468	0.764 479	0.107 365	0.015 804
Continuum					
5.3917–24.797[b]	0.313 709	0.210 421	0.164 341	0.150 241	0.165 849
24.797–65.0[b,c]	0.008 913	0.022 614	0.061 435	0.179 861	0.567 670
Resonances					
58.91–65.65[d]	0.017 61	0.077 57	0.342 23	1.511 1	6.678 8
64.9–73.82[e]	0.002 09	0.010 16	0.048 96	0.236 3	1.139 9
Continuum					
65.0–75.0[f]	0.008 355	0.043 205	0.223 668	1.159 2	6.015 0
75.0–200.0[f]	0.018 771	0.138 558	1.086 950	9.153 2	83.197 3
200.0–851.5[g]	0.000 646	0.012 666	0.270 782	6.541 4	184.540 0
851.5–2000[g]	0.000 002	0.000 157	0.013 004	1.134 4	104.824 0
2000–10 000[h]	–	–	0.002 160	0.508 2	142.372 7
10^4–10^{5}[h]	–	–	0.000 044	0.066 7	109.512 8
10^5–10^{6}[i]	–	–	–	0.002 3	36.164 3
10^6–10^{7}[i]	–	–	–	0.000 1	11.847 3
10^7–10^{8}[i]	–	–	–	–	3.789 0
10^8–10^{9}[i]	–	–	–	–	1.202 5
10^9–∞[i]	–	–	–	–	0.556 7
Total	41.025 9	6.067 8	2.978 1	20.750 4	692.59
Expectation values	41.03 ±0.05[j]	6.072 043[k]	(3.0)	20.746 42[k]	695.765 8[k]
		6.065[j]	2.999[j]		
		6.210 9[l]	(3.0)	19.820[l]	694.37[l]
Other values[m]	40.9	6.07	(3.0)	20.88	720.0

[a]See Table 2.4 and text.
[b]Peach *et al*. (1988).
[c]Lisini (1992); Chung (1997).
[d]See Table 2.7.
[e]See Table 2.8.
[f]See Fig. 2.6.
[g]Henke *et al*. (1993).
[h]Chantler (1995).
[i]Assuming hydrogen-like behavior, K-shell only.
[j]Hylleraas coordinates, configuration interaction, Pipin and Bishop (1992). See Table 2.8 and text for other recent calculations corroborating this result for $S(-2)$.
[k]Hylleraas-type functions, variational calculation, King (1989).
[l]Hartree–Fock calculations, presented to compare with results from more highly correlated wave functions. From Fraga *et al*. (1976).
[m]Zeiss *et al*. (1977).

b.4 **The 65.0–200.0 eV continuum** The results of several authors, including experimental data of Mehlman *et al*. (1978a; 1982), calculations by Lisini (1992) and Amusia *et al*. (1976b) and the compilation of Henke *et al*. (1993) are displayed in Fig. 2.6. From a minimum at ∼65 eV, the cross section rapidly rises to a maximum at ∼70 eV, then declines monotonically to 200 eV. There

Table 2.7 Energies and oscillator strengths for the 1s2snp Rydberg series in atomic lithium

Assignment	Energy, eV	f	Reference
1s(2s2p^{3}P) ^{2}P	58.910	0.24	a
1s(2s2p^{1}P) ^{2}P	60.396	0.007	a
(1s2s^{3}S)3p ^{2}P	62.419	0.053	a
(1s2s^{3}S)4p ^{2}P	63.356	0.019	a
(1s2s^{3}S)5p ^{2}P	63.753	(0.009)	b
(1s2s^{3}S)10p ^{2}P	64.260	(0.001)	b
(1s2s^{1}S)3p ^{2}P	64.046	0.001	c
(1s2s^{1}S)4p ^{2}P	65.29	(0.0005)	d
(1s2s^{1}S)5p ^{2}P	65.65	0.0003$_3$	e

[a]Assignment and energy from Ederer *et al.* (1970). Oscillator strength from P. Gerard, Ph.D. thesis, Université de Paris-Sud (1984), cited by Lisini (1992).

[b]Ederer *et al.* (1970) give assignments and energies. Oscillator strengths estimated by $(n^*)^3$ dependence.

[c]Ederer *et al.* (1970) give assignment and energy. Oscillator strength calculated by Lisini (1992).

[d]Assignment and energy from Gerard (1984), cited by Lisini (1992). Oscillator strength estimated by $(n^*)^3$ dependence.

[e]Assignment and energy from Gerard (1984), cited by Lisini (1992). Relative oscillator strength from Gerard (1984).

Table 2.8 Energies and oscillator strengths for 2-electron transitions and unassigned bands in atomic lithium

Assignment	Energy, eV	f	Reference
Broad, unassigned	64.9	0.035	a,b
Unassigned	64.5	(0.003)	a
(1s2p^{1}P)3s ^{2}P(?)	65.3	(0.003)	c
(1s2p^{1}P)4s ^{2}P(?)	66.5	(0.002)	c
(1s2p^{1}P)5s ^{2}P(?)	66.96	(0.001)	c
(1s2p^{1}P)6s ^{2}P(?)	67.18	(0.0005)	c
1s(3s3p ^{3}P) ^{2}P	71.14	0.003 4	d
1s(3s3p ^{1}P) ^{2}P	71.47	0.000 72	d
1s(3s3p ^{3}P) ^{2}P	72.71	0.000 104	d
1s(3,4 ^{1}P) ^{2}P	73.12	0.000 107	d
1s(3,3 ^{1}P) ^{2}P	73.35	0.000 058	d
1s(3,4 ^{3}P) ^{2}P	73.44	0.000 009 4	d
1s(3,4 ^{1}P) ^{2}P	73.67	0.000 026	d
1s(3,4 ^{1}P) ^{2}P	73.82	0.000 034	d

[a]Mehlman *et al.* (1978).

[b]Mehlman *et al.* (1982).

[c]Mehlman *et al.* (1978a), Fig. 1. May include nd, as well as ns resonances. Oscillator strengths are estimated from figure.

[d]Mehlman *et al.* (1982), Table III.

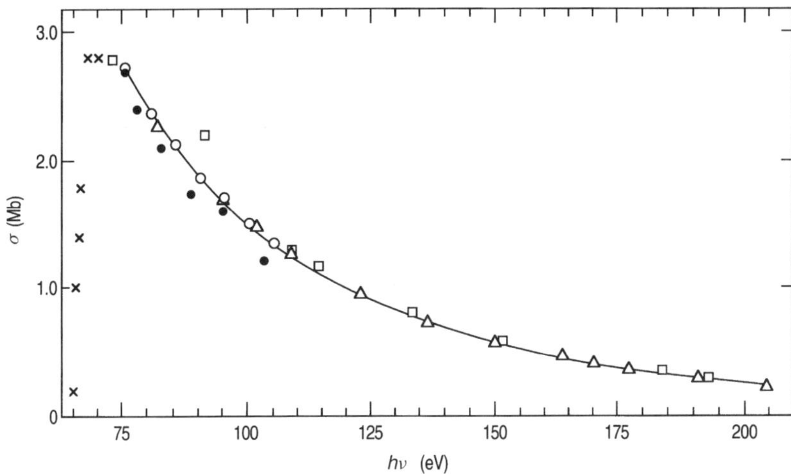

Fig. 2.6 Absolute photoabsorption spectrum of lithium, 65–200 eV. ● Mehlman *et al.* (1982); ○ Lisini (1992); △ Amusia *et al.* (1976b); □ Henke *et al.* (1993); × Mehlman *et al.* (1978a)

is fairly good agreement among the various data sets, although the experimental data of Mehlman *et al.* (1982) are ~20% lower. This has been corroborated by a recent accurate calculation by Fang and Chung (2001). We have treated the data by trapezoidal rule integration between 65–75 eV, then choosing a consensus of calculated data points between 75–200 eV, which is fitted by a polynomial function. The coefficients of this function are included in Table 2.5, the contributions to $S(p)$ in Table 2.6.

b.5 The continuum: 200–10 000 eV In Fig. 2.7, we compare data from the experimentally based compilation of Henke *et al.* (1993) with the calculated cross sections of Chantler (1995). The Henke values are slightly higher below 2 keV, but become increasingly more so above 2 keV. The cross sections of Henke *et al.* (1993) are unlikely to be based on gas phase data. We adopt their values only to 2 keV, then transfer to the calculated Chantler values between 2–10 keV. The cross sections in this energy region primarily affect $S(+2)$, with a small influence on $S(+1)$. Our choice leads to a reasonable value of $S(+2)$, vide infra, whereas use of the Henke values throughout this range would significantly overestimate $S(+2)$. The data in each domain are fitted by regression to 4-term polynomials. These functions are analytically integrated to yield the $S(p)$. The coefficients of the polynomials are given in Table 2.5, the values of $S(p)$ in Table 2.6.

b.6 The continuum: 10^4–10^5 eV We continue using the calculated cross sections of Chantler in this interval.

2.3.2 The analysis

The measured static electric dipole polarizability of atomic lithium is 24.3 $\pm0.5 \times 10^{-24}$ cm^3 (Molof *et al.*, 1974). This corresponds to $S(-2) = 41.0 \pm0.8$

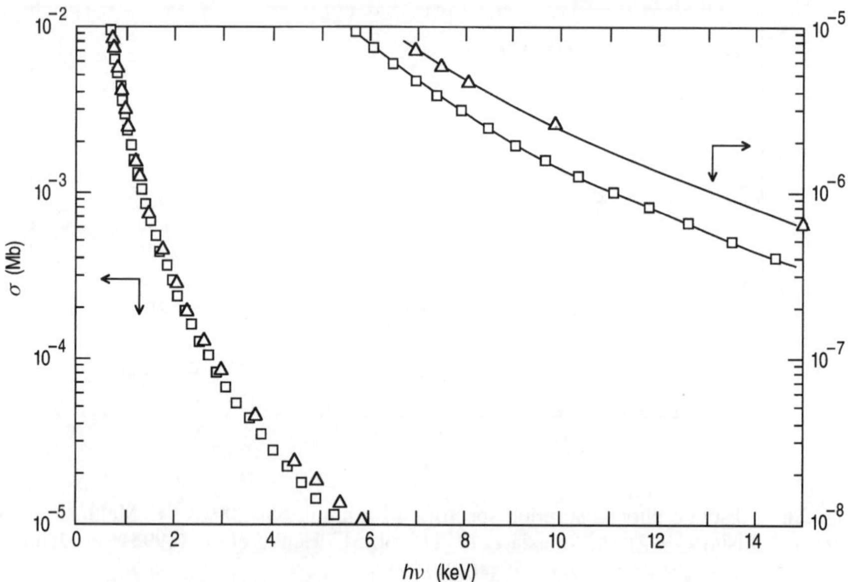

Fig. 2.7 Absolute photoabsorption spectrum of lithium, 1–15 keV. △ Henke *et al.* (1993); □ Chantler (1995)

Table 2.9 Recent calculations of the static electric dipole polarizability (α) of atomic lithium. [α(a.u.) = 4 × $S(-2)$, Ry units]

α (a.u.)	Method	Reference
164.1	Configuration interaction, Hylleraas coordinates	Pipin and Bishop (1992)
164.2 ±0.1	Møller–Plesset 4, and also coupled cluster	Kassimi and Thakkar (1994)
164.08	Full-core plus correlation, plus relativistic correction, 393-term basis set	Wang and Chung (1994)
163.91 164.01 ±0.22	1-electron model, with valence-K shell correlation effects	Laughlin (1995)

in Ry units. A contemporary calculation, also with an estimated uncertainty of 2%, arrived at $S(-2) = 41.1$, using the coupled-electron-pair approximation (Werner and Meyer, 1976). In the 1990s, more extensive calculations were performed, which reduced the uncertainty substantially. Some of these calculations are summarized in Table 2.9, where α is in atomic units (α(a.u.) = $4S(-2)$ Ry units). There is now excellent agreement among at least four high quality calculations that α(Li) = 164.1 ±0.2 a.u., or $S(-2) = 41.03$ ±0.05. This is almost exactly the value resulting from the spectral analysis. Of course, the contribution of the resonance line (40.6558 Ry units) represents 99.1% of the total $S(-2)$, but the final result attests to the accuracy and consistency of both calculations. From the standpoint of sensitivity analysis, it also demonstrates that the remainder of the spectrum cannot be tested by $S(-2)$. To a lesser extent, this is also true of $S(-1)$, where the resonance transition contributes 90.5% to the total. Our spectral sum for $S(-1) = 6.0678$ lies midway between the Hylleraas coordinate-based calculations of King (1989) (see Appendix), and Pipin and Bishop (1992), and within 0.1% of both (see Table 2.6). The spectral sum for $S(0)$ acquires only $\sim 25\%$ from the resonance transition, but the total is nevertheless 99.3% of the required Thomas–Reiche–Kuhn value. The major contributions occur between threshold and 200 eV, where several sources of data exist.

The spectral sum for $S(+1)$ is most sensitive to the 65.0–2000 eV region. It is fortuitously close to the expectation value ($<0.02\%$ difference).

The spectral sum for $S(+2)$ is 0.5% lower than King's predicted value. Had we chosen the cross sections of Henke *et al.* (1993) between 2–10 keV, the spectral sum would have been $\sim 8\%$ higher than the expectation value. The sum rule analysis clearly favors the calculated Chantler cross sections in this case, where gas phase experimental values are very difficult to deduce.

The Hartree–Fock calculations of Fraga *et al.* (1976) are included in Table 2.6 because they are available for every atom, unlike the highly correlated calculations which can be performed for Li. The Hartree–Fock results exceed King's by 2.3% for $S(-1)$, but are lower by 4.7% for $S(+1)$. Both $S(-1)$ and $S(+1)$ are sensitive to correlations. The value of $S(+2)$, which is essentially the electron density at the nucleus, is much less dependent on correlation, and is manifested by the good agreement (within 0.2%) between the Hartree–Fock and King values.

2.3.3 Appendix: determination of S(−1), S(+1) and S(+2) from calculations of King (1989)

King (1989) does not give the values of $S(-1)$, $S(+1)$ and $S(+2)$ directly. Instead, he gives certain expectation values for the ^{2}S ground state, from which these sums can be deduced. King's values in his Table III are in atomic units, which we shall retain until conversion to Ry units at the end.

1. $S(-1)$

From the derivation (see Reference Table),

$$S(-1)_{\text{au}} = \frac{2}{3}\left\langle (\vec{r}_1 + \vec{r}_2 + \vec{r}_3)^2 \right\rangle$$

$$= \frac{2}{3}\left\langle \left(r_1^2 + r_2^2 + r_3^2 + 2\vec{r}_1 \cdot \vec{r}_2 + 2\vec{r}_2 \cdot \vec{r}_3 + 2\vec{r}_3 \cdot \vec{r}_1\right)\right\rangle$$

$$\left\langle r_1^2 + r_2^2 + r_3^2 \right\rangle = \left\langle r_i^2 \right\rangle = 18.354\,74$$

$$\text{(Col. 7, Table III, King)}$$

$$\left\langle r_{12}^2 \right\rangle = \left\langle (\vec{r}_1 - \vec{r}_2)^2 \right\rangle = \left\langle \left(r_1^2 - 2\vec{r}_1 \cdot \vec{r}_2 + r_2^2\right)\right\rangle,$$

and analogously for $\vec{r}_2 - \vec{r}_3, \vec{r}_3 - \vec{r}_1$.

$$\left\langle r_{ij}^2 \right\rangle = \left\langle (\vec{r}_1 - \vec{r}_2)^2 \right\rangle + \left\langle (\vec{r}_2 - \vec{r}_3)^2 \right\rangle + \left\langle (\vec{r}_3 - \vec{r}_1)^2 \right\rangle = 36.848\,09$$

(Col. 7, King)

$$\left\langle r_{ij}^2 \right\rangle = 2\left\langle r_i^2 \right\rangle - 2\left\langle \vec{r}_i \cdot \vec{r}_j \right\rangle$$

$$2\left\langle \vec{r}_i \cdot \vec{r}_j \right\rangle = 2(18.354\,74) - 36.848\,09 = -0.138\,61$$

$$S(-1)_{\text{a.u.}} = \frac{2}{3}(18.354\,74 - 0.138\,61) = 12.144\,087$$

$$S(-1)_{\text{Ry}} = \frac{1}{2}(12.144\,087) = 6.072\,043$$

2. $S(+1)$

From the derived equation (Reference Table),

$$S(+1)_{\text{au}} = \frac{4}{3}\left\langle \frac{1}{2m}\left(\sum_{i=1}^{3} p_i\right)^2 \right\rangle$$

$$= \frac{4}{3} \cdot \frac{1}{2m}\left\langle (p_1^2 + p_2^2 + p_3^2 + 2\vec{p}_1 \cdot \vec{p}_2 + 2\vec{p}_2 \cdot \vec{p}_3 + 2\vec{p}_3 \cdot \vec{p}_1) \right\rangle$$

Now $p = -i\hbar\nabla$, $p_i^2/2m = -\nabla_i^2/2m = -\dfrac{1}{2}\nabla_i^2$ (in a.u.)

$$\varepsilon_0 = -\frac{1}{2m}(p_1^2 + p_2^2 + p_3^2) = -\frac{1}{2}\nabla_i^2 = 7.478\,059$$

(King, Table III, Col. 7)

$$\varepsilon_1 = \frac{1}{m}\left\langle (\vec{p}_1 \cdot \vec{p}_2 + \vec{p}_2 \cdot \vec{p}_3 + \vec{p}_3 \cdot \vec{p}_1) \right\rangle = -\vec{\nabla}_i \cdot \vec{\nabla}_j = 0.301\,846\,7$$

(King, Col. 7)

$$S(+1) = -\frac{4}{3}(7.478\,059 + 0.301\,846\,7) = 10.373\,208 \text{ a.u.}$$

or $S(+1)_{\text{Ry}} = 2S(+1)_{\text{au}} = 20.746\,42$

3. $S(+2)$

From the Reference Table (in atomic units),

$$S(+2) = \frac{4\pi Z}{3}\left\langle \delta(\vec{r}_i) \right\rangle$$

$$= \frac{4\pi}{3} \cdot 3(13.841\,82) \text{ (King, Table III, Col. 7)}$$

$$S(+2)_{\text{a.u.}} = 173.941\,44$$

or $S(+2)_{\text{Ry}} = 4S(+2)_{\text{au}} = 695.7658$

2.4 Atomic Nitrogen

2.4.1 The data

The ionization potential of atomic nitrogen, forming the $^3P_0$ state of N^+, is $117\,225.66 \pm 0.11\,\mathrm{cm}^{-1} \equiv 14.534\,128 \pm 0.000\,013\,\mathrm{eV}$ (Eriksson, 1986). The spin-orbit split states $^3P_1$ and $^3P_2$ are excited by $0.006\,034$ and $0.016\,217\,\mathrm{eV}$, respectively (Eriksson, 1986).

a   The discrete spectrum

The electronic ground state has the configuration $1s^2 2s^2 2p^3$, $^4S^o_{3/2}$. Electric dipole-allowed transitions converging to the ^3P ground state of N^+ can be anticipated, having the structure $1s^2 2s^2 2p^2 (^3P_{0,1,2})$ ns, nd. L–S coupling and spin preservation are fairly good approximations here, so that the more strongly allowed transitions have a ^4P upper state. A recent compilation of transition probabilities for atomic nitrogen by Wiese *et al.* (1996) demonstrates that this is generally true for ns ($n = 3$–9) and nd ($n = 3$–6), but weaker transition probabilities are observed for (^3P) 3d, ^2F and ^4D. In addition, the transition $\ldots 2s^2 2p^3 (^4S^o) \rightarrow \ldots 2s 2p^4 (^4P)$ occurs below the IP, and has a significant oscillator strength. The contributions of these transitions to the $S(p)$ are summarized in Table 2.10.

A histogram of the oscillator strength distribution of the ns series is shown in Fig. 2.8. The average oscillator strength declines sharply between $n = 3$ and $n = 4$, then remains essentially constant up to $n = 9$. We make a short extrapolation

Table 2.10 Contributions of the discrete spectrum to $S(p)$ sums in atomic nitrogen
[$S(p)$ in Ry units]

	$S(-2)$	$S(-1)$	$S(0)$	$S(+1)$	$S(+2)$
1. ns(^4P) series					
3	0.4491	0.3411	0.259	0.1967	0.1494
4	0.0316	0.0298	0.028 2	0.0267	0.0252
5	0.0109	0.0109	0.010 9	0.0109	0.0109
6	0.0051	0.0052	0.005 35	0.0055	0.0056
7	0.0028	0.0029	0.003 04	0.0032	0.0033
8	0.0017	0.0018	0.001 89	0.0020	0.0021
9	0.0011	0.0012	0.001 26	0.0013	0.0014
10–∞	0.0036	0.0038	0.004 1	0.0043	0.0046
2. nd (^4P) series					
3	0.0800	0.0764	0.073 0	0.0698	0.0666
4	0.0365	0.0367	0.036 9	0.0371	0.0373
5	0.0186	0.0192	0.019 7	0.0203	0.0208
6	0.0108	0.0112	0.011 7	0.0122	0.0127
7–∞	0.0262	0.0276	0.029 2	0.0309	0.0326
3. 3d (^2F)	0.0074	0.0071	0.006 75	0.0065	0.0062
3d (^4D)	0.0021	0.0020	0.001 88	0.0018	0.0017
4. 2s2p^4 (^4P)	0.1368	0.1098	0.088 2	0.0708	0.0569

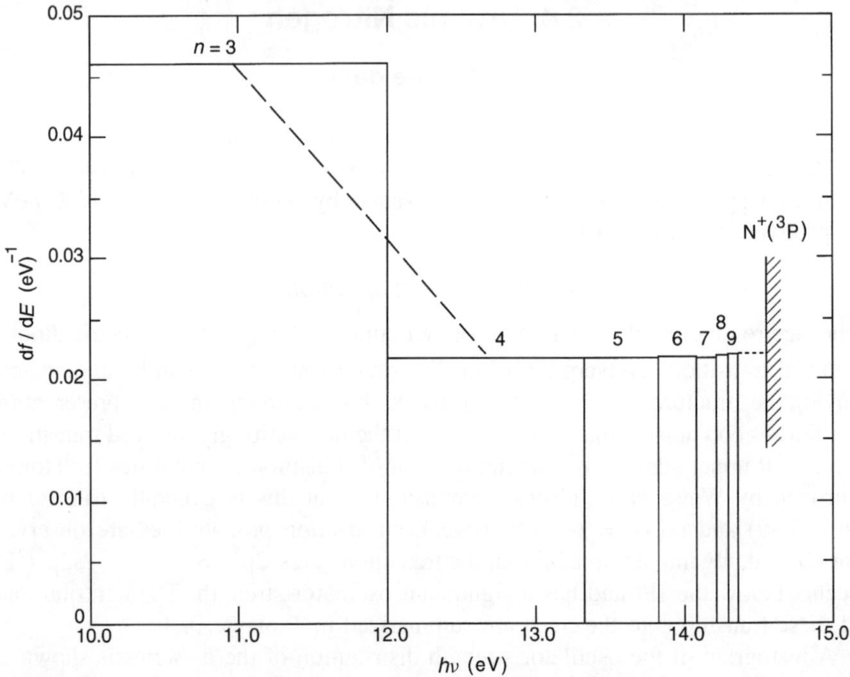

Fig. 2.8 Histogram for the series $\ldots 2p^3$, $^4S_{3/2} \to \ldots 2p^2$ (^{3}P)ns, ^{4}P in atomic nitrogen

of this behavior to the IP. (For the present purposes, we take IP as the weighted average of $^3P_{0,1,2}$ and sum the fine-structure oscillator strengths within each Rydberg member.) The extrapolated value of $\mathrm{d}f/\mathrm{d}E$ for the ns series is 0.022/eV.

The histogram of the oscillator strength distribution of the nd (^{4}P) series (Fig. 2.9) displays a monotonic, essentially linear increase between $n = 3$–6. The extrapolated value of $\mathrm{d}f/\mathrm{d}E$ for the nd (^{4}P) series is 0.095/eV. From the sum of $\mathrm{d}f/\mathrm{d}E$ for the two series, we obtain a photoionization (photoabsorption) cross section of 12.8 Mb at the IP, or perhaps slightly more, allowing for the weak nd ^{2}F and nd ^{4}D series. Assuming the linear extrapolations to the IP in Figs. 2.8 and 2.9, we compute $S(p)$ for the extrapolated regions and list them in Table 2.10.

b The continuum

b.1   14.534–30.0 eV; 30.0–49.6 eV   In addition to ^{3}P, the $2s^2 2p^2$ N$^+$ configuration also gives rise to ^{1}D (16.433 eV) and ^{1}S (18.587 eV). However, the transition probability from the neutral ground state (^{4}S) to these doublet continua appears to be insignificant. The $2s2p^3$ configuration gives rise to ^{5}S$^\circ$ (20.3347 eV) (Eriksson, 1958), ^{3}D$^\circ$ (25.97 eV), ^{3}P$^\circ$ (28.08 eV) and ^{3}S$^\circ$ (33.77 eV) (Moore, 1971). A prominent np Rydberg series converging to ^{5}S$^\circ$, with characteristic

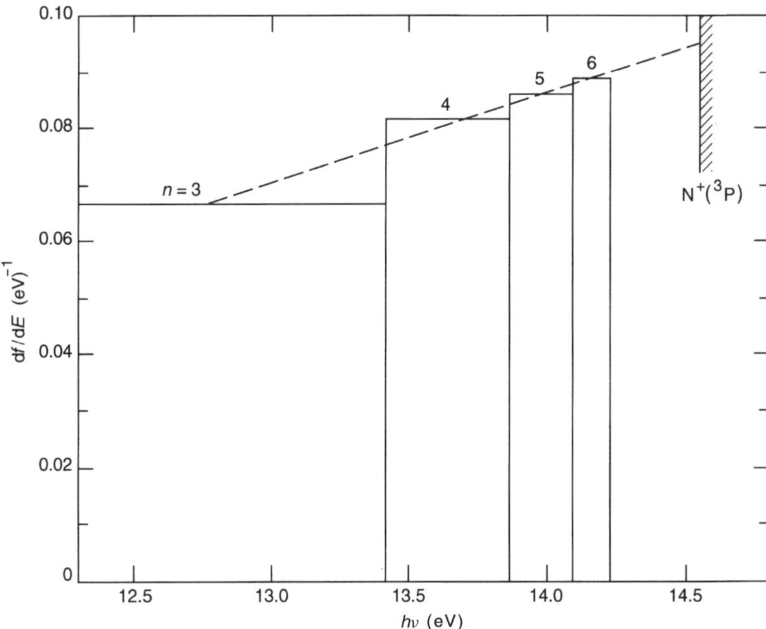

Fig. 2.9 Histogram for the series $\ldots 2p^3$, $^4S_{3/2} \rightarrow \ldots 2p^2$ $(^3P)nd$, ^4P in atomic nitrogen

asymmetric autoionization resonances has been observed (Carroll *et al.*, 1966; Dehmer *et al.*, 1974; Samson and Angel, 1990; Schaphorst *et al.*, 1993); and calculated (Le Dourneuf *et al.*, 1979; Bell and Berrington, 1991; Nahar and Pradhan, 1997; Burke and Lennon, 1996). Much weaker resonances converging to $^3D^o$, $^3P^o$ and $^3S^o$ have been calculated by these authors, but to our knowledge have not yet been observed experimentally.

The absolute photoionization cross section of nitrogen atom between the IP and $\sim$40 eV reported by three groups is depicted in Fig. 2.10. Two are calculations, one (Samson and Angel, 1990) is based on experiment, with some assumptions to be discussed. A more recent calculation (Anderson and Veseth, 1994) has been omitted from Fig. 2.10 to minimize confusion, but will be considered later. The two calculations shown in Fig. 2.10 utilize the random phase approximation with exchange (RPAE), (Cherepkov *et al.*, 1974) and the R-matrix method, (Le Dourneuf *et al.*, 1979). The RPAE calculation did not include the channels involving 2s excitation and ionization, and hence no resonances were found. The R-matrix calculation included these channels, and calculated resonances converging to $^5S^o$, with parameters in good agreement with experiment. These have been excised from Fig. 2.10, but some evidence can be seen for resonances converging to the excited states of the sp^3 configuration. The RPAE calculation has a similar shape to the R-matrix calculation, but is uniformly lower in cross section. Le Dourneuf *et al.* (1979) attribute the lower cross sections in the RPAE calculation to their neglect of the 2s photoejection channels.

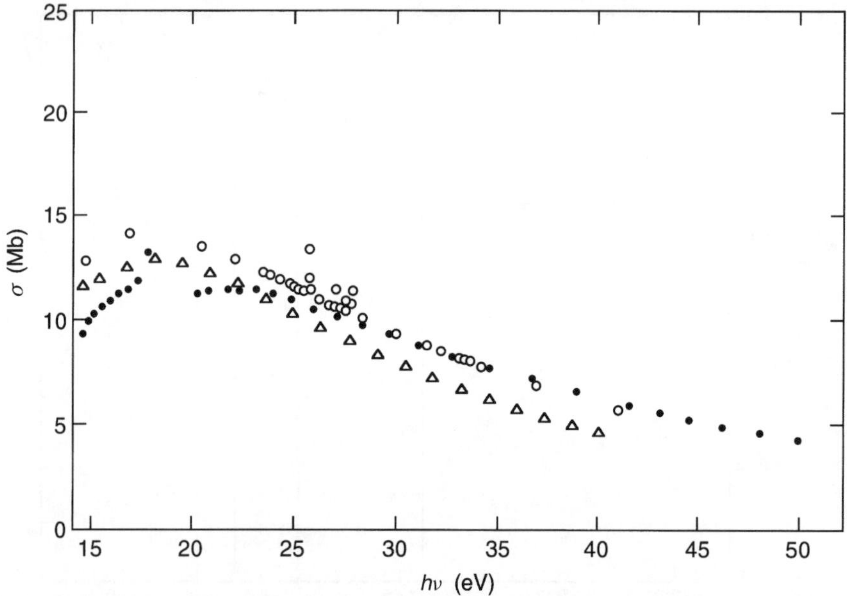

Fig. 2.10   Absolute photoabsorption spectrum of atomic nitrogen, 15–50 eV. • Samson and Angel (1990); ○ Le Dourneuf *et al*. (1979); △ Cherepkov *et al*. (1974)

Samson and Angel (1990) measured the relative photoionization cross section of N from 44.3–850 Å (14.59–280 eV). Then, they normalized the high energy region ($E > 49.6$ eV) to 1/2 the cross section of N_2, and also σ (NH_3)–3σ (H). This is a reasonable assumption for sufficiently high energy. In Fig. 2.10, the calculated values of Le Dourneuf *et al*. (1979) merge with the normalized experimental values of Samson and Angel (1990) above 30 eV.

Samson and Angel now encountered another problem. The long wavelength data (850–300 Å) and short wavelength data (500–44.3 Å) were taken with two different monochromators. Compounding this problem, 'in the overlapping region between 300 and 500 Å no reliable data could be obtained because of the presence of higher order spectra and/or weak light intensity'. Their strategy was to extrapolate the higher energy curve to the lower energy region, but also to adjust the lower energy region so that the total oscillator strength was 7.0 (satisfying the TRK sum rule). This required an estimate of all the other contributions to $S(0)$, including the discrete portion. For the discrete contribution, they chose 0.96 ±50% from Wiese *et al*. (1966). As we have seen in the previous section and Table 2.10, the current value of $S(0)$, discrete, is 0.581. By choosing a higher value for $S(0)$, discrete, they force their continuum data between IP–30 eV to have lower cross sections. This can readily be seen in Fig. 2.10, where the data of Samson and Angel drop below the cross sections of Le Dourneuf *et al*. (1979), and even below those of Cherepkov *et al*. (1974). Samson and Angel apparently recognized the uncertainty of their lower energy cross sections, noting that they could get good agreement with Le Dourneuf *et al.* if they pivoted the lower energy data at 250 Å (49.6 eV) and decreased their discrete oscillator strength by ~50%.

Our goal here is to select the best oscillator strength distribution from existing data. In addition to the TRK sum rule, we shall ultimately test alternative data sets with the $S(-2)$ sum rule, which is particularly sensitive to low energy data, and is based on a fairly well-known polarizability (see Sect. 2.4). Furthermore, we expect continuity in df/dE across the IP boundary. In Sect. 2.4.1.a, we found $\sigma = 12.8\,\mathrm{Mb}$ at the IP. This is close to the value from Le Dourneuf *et al.*, as shown in Fig. 2.10, whereas the Samson and Angel data give ~9.5 Mb at the IP. The other R-matrix calculations give values of σ at the IP of 12.3 Mb (Bell and Berrington, 1991); 11.4 Mb (Nahar and Pradhan, 1997); and 12.8 Mb (Burke and Lennon, 1996), consistent with the value from Le Dourneuf *et al.*

For evaluation of the $S(p)$, we have fitted separate polynomials to the data of Samson and Angel (1990) and Le Dourneuf *et al.* (1979) between IP–30.0 eV. Between 30.0–49.6 eV, only the data of Samson and Angel were used. The contributions to $S(p)$ are recorded in Table 2.11. The coefficients of the polynomials are given in Table 2.12.

b.2 Resonances, 17.898–20.335 eV These resonances, discussed in Sect. 2.4.1.b.1., involve the transitions $\ldots 2s^2 2p^3(^4S) \rightarrow \ldots 2s^2 2p^3(^5S)np$, (^4P). The equation for 'excess oscillator strength' f_{xs} given in Sect. 2.1.b.2 is used. Dehmer *et al.* (1974) give experimentally deduced values for ρ^2, Γ and q for the first two transitions, while Le Dourneuf *et al.* (1979) provide corresponding values based on their calculations which are in good agreement with experiment. Taking $\sigma \approx 14\,\mathrm{Mb}$ (Le Dourneuf *et al.*, 1979) we obtain $f_{xs} = 0.0053$ and 0.0022 for $n = 3$ and 4, respectively. Assuming ρ^2 and q remain constant in the series, and taking Γ for the higher resonances from Carroll *et al.* (1966), modified by the correction given in Dehmer *et al.* (1974), we can compute f_{xs} for the entire series. The contribution of the entire series to $S(p)$, recorded in Table 2.11, is very small, and justifies the neglect of the still weaker series converging to higher limits as found by Le Dourneuf *et al.* (1979).

b.3 49.6–180 eV In this energy region, Samson and Angel (1990) show convincingly in their Fig. 1 that the photoabsorption cross section of atomic nitrogen tracks $1/2\,\sigma$ (N_2), and also σ (NH_3)-3σ (H). The compilation of Henke *et al.* (1993), based on similar premises, may also be used, but the points are sparser. We utilize the data from Table I of Samson and Angel, fitted to a 4-term polynomial $\sum_{n=2}^{5} a_n y_n$ by regression analysis. The coefficients of this polynomial are given in Table 2.12, and the evaluated $S(p)$ are listed in Table 2.11.

b.4 180–409.9 eV Here, the cross section attributable to photoejection from 2p and 2s orbitals declines monotonically until it approaches the K-edge. We take the K-edge value of 409.9 eV to be the same as that for N_2 (Jolly *et al.*, 1984). We anticipate some pre-edge structure, as in N_2, which is considered separately below. To estimate the continuum contribution to $S(p)$, we can utilize the

Table 2.11 Spectral sums, and comparison with expectation values for atomic nitrogen[a]

Energy, eV	$S(-2)$	$S(-1)$	$S(0)$	$S(+1)$	$S(+2)$
Discrete → IP_{avg}[b]	0.8243	0.6867	0.5811	0.5000	0.4373
IP–30.0	0.7828[c]	1.1533[c]	1.7705[c]	2.8322[c]	4.7123[c]
	(0.6753)[d]	(1.0095)[d]	(1.5730)[d]	(2.5562)[d]	(4.3092)[d]
30.0–49.6[d]	0.1595	0.4348	1.2096	3.4358	9.9628
Resonances 17.898–20.335[e]	0.0069	0.0096	0.0134	0.0187	0.0262
49.6–180[d]	0.0499	0.2487	1.3607	8.3683	59.3357
180–409.9[d]	0.0006	0.0109	0.1976	3.7751	76.2433
180–409.9[f]	(0.0007)	(0.0112)	(0.1990)	(3.7130)	(78.2429)
Resonances Pre-K edge[g]	0.0001	0.0041	0.12	3.5350	104.1346
409.9–2042.4[f]	0.0009	0.0356	1.6417	88.2175	5 756.54
2 042.4–10 000[f]	–	0.0004	0.0931	23.0813	6 834.63
10^4–10^{5}[h]	–	–	0.0032	4.0058	6 809.03
10^5–10^{6}[i]	–	–	–	0.1965	3 124.57
10^6–10^{7}[i]	–	–	–	0.0067	1 077.18
10^7–10^{8}[i]	–	–	–	0.0002	350.34
10^8–10^{9}[i]	–	–	–	–	111.79
10^9–∞[i]	–	–	–	–	51.86
Total	1.8250	2.5841	6.9909	137.973	24 370.8
Expectation values	1.91 ±0.10[j]	2.5471[k]	7.		24 156.3[l]
	2.05 ±0.04[m]				
	1.84 ±0.02[n]				24 156.8[o]
Other values[p]	(1.8193)	2.567	(7.)	137.54	24 600

[a] $S(p)$ in Ry units.
[b] See Table 2.10 and text.
[c] From Le Dourneuf *et al.* (1979).
[d] From Samson and Angel (1990).
[e] See text.
[f] From Henke *et al.* (1993).
[g] From Akimov *et al.* (1988).
[h] From Chantler (1995).
[i] Using the hydrogenic equation of Bethe and Salpeter (1977).
[j] From Alpher and White (1959).
[k] From Thomas Müller, private communication.
[l] From Fraga *et al.* (1976).
[m] Wettlaufer and Glass (1972).
[n] From selection of data in Table 2.13, taking $\alpha(a_0^3) = 7.36$ ±0.07.
[o] From Bunge *et al.* (1993).
[p] Zeiss *et al.* (1977).

cross sections of Samson and Angel (1990), or alternatively the compilation of Henke *et al.* (1993). They have been separately fitted, and both values of $S(p)$ are included in Table 2.11. The good agreement provides internal support for the cross sections used here.

Table 2.12 Coefficients of the polynomial $df/dE = ay^2 + by^3 + cy^4 + dy^5$ fitted to data at various energies[a]

Energy range, eV	a	b	c	d
14.534–30.0[b]	9.877016	−11.8001	4.255618	−0.74703
30.0–49.6	6.954833	7.861052	−36.2889	24.07203
49.6–180.0	1.579575	30.49315	−1.26292	−142.487
180.0–409.9[c]	2.709094	23.20204	67.99257	−943.426
409.9–2042.4	1.376059	3899.576	−65606.8	284359.2
2042.4–10000	−1.87659	4570.961	−132009	4130131

[a] df/dE in Ry units, $y = B/E$, $B = $ IP $(^3P_0) = 14.5341$ eV.
[b] From Le Dourneuf *et al.* (1979).
[c] From Samson and Angel (1990).

b.5 Pre-K edge resonances Although these have not been observed for atomic nitrogen, Zhadenov *et al.* (1987) argue that they should appear more analogous to corresponding structures in N_2 than those in NH_3. In N_2, they are dominated by a sharp peak at ~401 eV, which has an oscillator strength $f = 0.21 \pm 0.02$. They conclude that, taking other transitions into account, a reasonable estimate of the pre-edge oscillator strength per N atom is $f = 0.12$ (see Akimov *et al.* (1988)).

b.6 409.9–2042.4 eV; 2042.4–10000 eV The data in these sections are taken from Henke *et al.* (1993). The data are partitioned and individually fitted to two 4-term polynomials for improved accuracy. As before, the contributions to $S(p)$ are recorded in Table 2.11, the polynomial coefficients given in Table 2.12.

b.7 10^4–10^5 eV The calculated cross sections of Chantler (1995) are utilized in this high-energy domain. They have been compared with the compilation of experimental data of Henke *et al.* (1993) between 1–21 keV, and the agreement is excellent.

2.4.2 The analysis

Before we can assess the relative merits of the data entering the spectral sums in Table 2.11, we turn to some theoretical expectations. The total $S(0)$ should, of course, equal 7.00, given by the Thomas–Reiche–Kuhn sum rule. The value of $S(-2)$ is related to the static electric dipole polarizability (α) by $\alpha = 4a_0^3 S(-2)$. Alpher and White (1959), and later Wettlaufer and Glass (1972) measured the specific refractivity of atomic nitrogen in shocked N_2 at several wavelengths. The results, though plausible, had stated uncertainties of 2–15%. Zeiss and Meath (1997) attempted a Cauchy expansion, and obtained a static value of 7.277 a_0^3. The experimental values of dynamic polarizability with the lowest error bars were 8.19 ± 0.14 a_0^3 at 6943 Å (Wettlaufer and Glass, 1972) and 7.74 ± 0.50 a_0^3 at 5446 Å (Alpher and White, 1959). Various calculations have been listed in Table 2.13, in reverse chronological order. A reasonable choice that encompasses most of the

Table 2.13 Various determinations of the static electric dipole polarizability (α) of atomic nitrogen

$\alpha(a_0^3)$	Method	Reference
$7.6_3 \pm 0.4$	Expt. – specific refractivity in shocked N_2 gas	Alpher and White (1959)
7.3581	Restricted Hartree–Fock, finite perturbation method	Stiehler and Hinze (1995)
6.259–6.575	Many-body perturbation theory (MBPT)	Andersen and Veseth (1994)
7.3	Second-order perturbation theory (CASSCF)	Anderson and Sadlej (1992)
$7.36_2 \pm 0.05$	MBPT – 4th order, polarized basis set	Sosa and Ferris (1990)
7.33	Polarized pseudo-state, superposition of configuration, using 'short-range correlation'	Hibbert et al. (1977)
7.49	Variational calculation	Nesbet (1977)
7.43 ± 0.15	Pseudo-natural-orbital, coupled-electron pair approximation	Werner and Meyer (1976)

calculations is $\alpha = 7.36 \pm 0.07 \ a_0^3$. A glaring exception is the result of Andersen and Veseth (1994), $\alpha = 6.259$–$6.575 \ a_0^3$. However, these authors note that 'CI and RHF methods are particularly suited for calculating static polarizabilities' whereas their method was constructed for obtaining dynamic polarizabilities, and is not expected to yield 'static polarizabilities… as good as those obtained by other methods'. The values of $S(-2)$ equivalent to α cited above are listed at the bottom of Table 2.11.

Let us return briefly to the calculations of Andersen and Veseth. They calculated the total photoionization cross section $\sigma(\omega)$ by inverting the integral equation relating the dynamic polarizability $\alpha(i\eta)$ to $\sigma(\omega)$. Although their static polarizability was lower than given by experiment and other calculations, their total photoionization cross section was significantly higher than that of Le Dourneuf et al. (1979), which in turn is higher than the experimental choice of Samson and Angel (1990) between IP and 30 eV. For this reason, we have not plotted the cross section of Andersen and Veseth in Fig. 2.10.

A value of $S(+2)$ at the Hartree–Fock level has been taken from Fraga et al. (1976) A confirmatory calculation of $S(+2)$ is given by Bunge et al. (1993). This $S(+2)$ value is expected to be fairly accurate, since it is minimally dependent on correlation. However, $S(-1)$ and $S(+1)$ are sensitive to correlation effects.

If we now compare our spectral $S(p)$ (using the calculated cross sections of Le Dourneuf et al. (1979) in the sensitive region, IP–30 eV) with predicted values, we find excellent agreement for $S(0)$, 6.99 cf. 7.00, and very good agreement for $S(-2)$, 1.825 cf. 1.91 ± 0.10 (expt.), 1.84 ± 0.02 (calc.). For $S(-1)$, the spectral sum is lower than the Hartree–Fock value (2.8084), as expected, but close to the value obtained by Müller (1996) using a correlated wave function. Had we used the cross sections of Samson and Angel (1990) in the region IP–30 eV, we would have obtained $S(0) = 6.79$ (3% low) and $S(-2) = 1.718$ (about 6–7% low). The $S(-1)$ value would be lower than Müller's calculated number by 4.2%.

For $S(+1)$, the energy domain above the K-edge amounts to 83.7%, most of which (80.7%) is contained in the 409.9–10 000 eV region which is based on the Henke *et al.* (1993) compilation. The spectral sum is 1–2% higher than the Hartree–Fock value (136.59), but this is the direction often found when correlation is added. The evidence supports the conclusion that the Henke compilation in the energy range used is accurate to 1–2%. The spectral sum for $S(+2)$ derives about 1/2 of its total from the experimentally based Henke data and 1/2 from the calculated cross sections at higher energies. This quantity is <1% higher than the presumably reliable Hartree–Fock calculated value, providing additional support for the Henke data and in this case, the calculations as well.

Overall, the level of agreement is astonishingly good for a non-permanent gas, where absolute calibrations and Beer–Lambert type measurements could not be used. One simplification which seems to be borne out by this analysis is that additivity, i.e., $\sigma(N) = 1/2\sigma(N_2)$ works well for sufficiently high energy, which in this case appears to occur for $h\nu > 30$ eV. Another important factor is the importance of high quality calculations. We have already seen their effect in the data of Le Dourneuf *et al.* (1979) for the IP–30 eV region. In addition, Wiese *et al.* (1996) note that their principal data sources for oscillator strength in the discrete spectrum come from advanced atomic structure calculations, although some experimental emission data were utilized. Also, the calculated values of polarizability not only support the experimental value, but in all likelihood improve upon it. Although much depends on calculations in this analysis, they are quite different calculations for different regions of the spectrum. The sum rule analysis indicates that they mesh rather well, since the very good concordance with predicted sums has been obtained without forcing agreement with any of the sum rules.

Zeiss *et al.* (1977) also found good agreement with the sum rules, based on earlier data. Their method requires $S(-2)$ and $S(0)$ to have the correct sums, but cannot assure that local oscillator strengths or cross sections are accurate.

2.5 Atomic Oxygen

2.5.1 The data

The ionization potential of atomic oxygen, from the $^3P_2$ ground state of O to the $^4S_{3/2}$ state of O^+, is 109 837.03 $\pm$0.06 cm^{-1} $\equiv$ 13.618 055 $\pm$0.000 007 eV (Eriksson and Isberg, 1963). The lowest-lying ionic states ^2D and ^2P have average excitation energies above $^4S_{3/2}$ of 3.3251 and 5.0175 eV, respectively.

a The discrete spectrum

The electronic ground state of the atom has the configuration $1s^2 2s^2 2p^4$, $^3P_{2,1,0}$. For the ion, we have $1s^2 2s^2 2p^3$, $^4S^\circ_{3/2}$, with excited states $^2D^\circ_{5/2,3/2}$ and $^2P^\circ_{3/2,1/2}$. Electric dipole allowed transitions ns and nd are observed converging to these various ionization limits. Wiese *et al.* (1996) have recently published a critical compilation of oscillator strengths for many of these transitions. All but one [$(^2D)3s$] of the transitions appearing below the IP converge to the $^4S_{3/2}$ ground state of the ion. Since the fine structure splitting of the ^3P neutral ground state is comparable to kT, the relative populations of $^3P_2$, $^3P_1$, $^3P_0$ depend upon the experimental temperature. Although Wiese *et al.* report oscillator strengths involving individual multiplets, they also provide the oscillator strength from ^3P, which we pragmatically adopt here. In Table 2.14, we list the oscillator strengths for transitions appearing below the IP, and their contributions to the sum rules. Histograms of the individual ns and nd series analogous to Fig. 2.1 display a smooth decline for $n \geq 4$. At the ^4S convergence limit, the nd series corresponds to $\sigma \sim 2.74$–$2.96\,$Mb, the ns series to $\sigma \sim 0.67\,$Mb. Their sum is somewhat larger than the observed continuum cross section at the IP, $\sigma \cong 2.75\,$Mb, which we attribute to the acknowledged 10% uncertainties in the nd oscillator strengths and the extrapolation. However, Bell and Kingston (1994) point out that several R-matrix calculations agree on a threshold value of $\sim 4\,$Mb. See also Nahar (1998).

b The continuum

Atomic oxygen is a transient species, typically generated in the laboratory by electric discharge. This circumstance precludes utilization of methods based on the Beer–Lambert law for the determination of absolute cross sections. As an alternative, the number density of atomic oxygen can be estimated by knowing the number density (or pressure) of molecular oxygen prior to discharge, and its reduction during the discharge. Corrections can be made for the concomitant production of an excited state of molecular oxygen, $a^1\Delta$. With known values of the (stable) molecular photoionization cross section, the atomic cross section

Table 2.14 Contributions to the sum rules of discrete transitions in atomic oxygen

Transition $2p^4\ ^3P\rightarrow$	λ(Å)	E (eV)	$S(-2)$	$S(-1)$	$S(0)^+$	$S(+1)$	$S(+2)$
$(^4S)3s$	1304	9.51	0.106 3	0.074 3	0.051 9	0.036 3	0.025 4
$(^4S)4s$	1040	11.92	0.011 93	0.010 45	0.009 16	0.008 03	0.007 03
$(^4S)3d$	1026	12.08	0.025 5	0.022 6	0.020 1	0.017 9	0.015 9
$(^2D)3s$	989	12.54	0.065 1	0.060 0	0.055 3	0.050 95	0.046 95
$(^4S)5s$	977	12.69	0.003 80	0.003 55	0.003 31	0.003 09	0.002 88
$(^4S)4d$	972	12.76	0.015 7	0.014 7	0.013 8	0.012 9	0.012 1
$(^4S)6s$	951	13.04	0.001 71	0.001 64	0.001 57	0.001 50	0.001 44
$(^4S)5d$	949	13.06	0.006 84	0.006 57	0.006 31	0.006 06	0.005 82
$(^4S)7s$	938	13.22	0.009 3	0.000 90	0.000 877	0.000 85	0.000 83
$(^4S)6d$	937	13.23	0.003 84	0.003 73	0.003 63	0.003 53	0.003 43
$(^4S)8s$	931	13.32	0.000 56	0.000 55	0.000 537	0.000 53	0.000 51
$(^4S)7d$	930	13.33	0.002 40	0.002 35	0.002 30	0.002 25	0.002 21
$(^4S)8d$	926	13.39	0.001 59	0.001 56	0.001 54	0.001 52	0.001 49
$n \rightarrow \infty$			0.002 81	0.002 81	0.002 81	0.002 81	0.002 81
Total			0.249 0	0.205 7	0.173 1	0.148 2	0.128 8

can be inferred. This approach was used by Samson and Pareek (1985) using photoionization mass spectrometric (PIMS) detection, and later by van der Meer *et al.* (1988) using photoelectron spectroscopy (PES). The two results differed significantly. At 584 Å, Samson and Pareek found $\sigma(O) = 13.2$ Mb, while van der Meer *et al.* obtained a value of 8.3 Mb. Subsequently, Berkowitz (1997b) showed by sum rule analysis that the higher value was clearly preferred. With some slight changes, this analysis is presented below.

b.1 Autoionization peaks (878–676 Å) Relative photoionization cross sections have been reported in several studies. Dehmer *et al.* (1973), using PIMS, obtained a spectrum of atomic oxygen between threshold and 650 Å, and later (Dehmer *et al.*, 1977) between threshold and 731 Å, at a higher resolution. This region is dominated by autoionization peaks, mostly attributable to ns and nd series converging to O^+ (^2D) and (^2P). Angel and Samson (1988) concentrated on the underlying continuum, and obtained relative cross sections between threshold and 260 Å, with additional measurements of multiple ionization which enabled them to extend their study to 44.3 Å. The relative cross sections of Angel and Samson were placed on an absolute scale by utilizing the calibration of Samson and Pareek (1985).

There are two approaches we can adopt to the calibration of the relative photoionization spectra of Dehmer *et al.* (1973; 1977).

(i) Usually, autoionization rates are at least three orders of magnitude faster than radiative rates. Hence, for atomic systems, the photoabsorption cross section is usually virtually identical to the photoionization cross section. However, exceptions occur when autoionization is forbidden by L–S selection rules, but can nonetheless proceed through spin-orbit interactions. In such cases, autoionization and fluorescence may be competitive. There are two such regions in atomic oxygen, the transitions $2p^4\ {}^3P \rightarrow 2p^3(^2P)$ 3s at 878–879 Å and $(2s)^2(2p)^4\ {}^3P \rightarrow 2s2p^5$ at 792 Å. Both of these regions are split into multiplets. Both Wiese *et al.* (1996) and Doering *et al.* (1985) provide oscillator strengths for these transitions. Dehmer *et al.* (1977) report the branching ratios into autoionization and fluorescence. Hence, it is possible to deduce the oscillator strengths attributable to autoionization for these transitions. Since these autoionization peaks have very high q values (see Fano (1961)), they can be treated as triangular functions whose areas are the oscillator strengths. (This is a valid procedure for thin targets.) In this fashion, the ordinate in the photoion yield curve can be converted into an absolute cross section scale. The peak heights will vary with the resolution of the experiment (0.42 Å FWHM, Dehmer *et al.*, 1973); (0.16 Å FWHM, Dehmer *et al.*, 1977), but the areas should be invariant. In this analysis, we have concentrated on the 792 Å region, since it is given in both spectra.

(ii) If, for the moment, we accept the calibration used by Angel and Samson (1988), we note that $\sigma(720.0$ Å$) = 9.00$ Mb. In the photoion yield curve of Dehmer *et al.* (1973), Fig. 2, the continuum intensity measures 1.9 mm, with a background of ~ 0.1 mm. Therefore, we conclude that the ordinate corresponds to ~ 5.0 Mb mm^{-1}. Similarly, Angel and Samson give $\sigma(732.2$ Å$) = 8.03$ Mb, which corresponds to ~ 1.0 mm in the continuum intensity of Dehmer *et al.* (1977), Fig. 1. Hence, for this figure, the ordinate is ~ 8.0 Mb mm^{-1}. Upon measurement of the area of the 792 Å region in both figures, we obtain oscillator strengths that are within $\pm 10\%$ of those inferred from the first procedure outlined above.

Having established an absolute scale for the ordinates, we extract areas and convert to oscillator strengths, as well as the corresponding $S(p)$. The oscillator strengths deduced from Dehmer *et al.* (1973; 1977) are in fair agreement with one another. Wiese *et al.* offer oscillator strengths for some of these autoionization peaks, based on calculations of Butler and Zeippen (1991) and Hibbert *et al.* (1991). The agreement between the 'experimental' and calculated values is patchy, but typically poorer than between the experimental calibrations, though within a factor of 2.

Taking into account the extracted oscillator strengths based on normalization (Dehmer *et al.*, 1973; 1977), the compilation of Wiese *et al.* (1996) and the two regions studied by Doering *et al.* (1985), we arrive at a selection of oscillator strengths of the autoionizing levels, which is listed in Table 2.15. The sum of

Table 2.15 Contributions to the sum rules from autoionizing transitions in atomic oxygen

Transition $2p^4 \, ^3P \rightarrow$	λ(Å)	E (eV)	$S(-2)$	$S(-1)$	$S(0)^a$	$S(+1)$	$S(+2)$
(^2P) 3s	878	14.12	0.074	0.077	0.080	0.083	0.086
(^2D) 4s	817–818	15.18	0.007	0.008	0.009	0.010	0.011
(^2D) 3d	811	15.29	0.0030	0.0034	0.0038	0.0043	0.0048
(^2D) 3d^1	805	15.40	0.0073	0.0083	0.0093	0.0106	0.0120
$2s2p^5 \, ^3P$	792	15.65	0.050	0.057	0.066	0.076	0.087
(^2D) 5s	775–776	16.00	0.0021	0.0025	0.0029	0.0034	0.0040
(^2D) 4d	770–771	16.10	0.0128	0.0152	0.0179	0.0212	0.0251
(^2D) 6s, 5d	756–759	16.40	0.0059	0.0071	0.0085	0.0102	0.0103
(^2D) 6d	748–750	16.57	0.0028	0.0034	0.0041	0.0050	0.0061
(^2D) 7d	744–745	16.66	0.0020	0.0024	0.0030	0.0036	0.0044
(^2D) 8d	741	16.73	0.0013	0.0016	0.0020	0.0025	0.0030
(^2D) 9d	739	–	0.0007	0.0008	0.0010	0.0012	0.0015
(^2D) 10d	738	16.80	0.0007	0.0008	0.0010	0.0012	0.0015
(^2D) 11d–15d	735–737	16.84	0.0016	0.0020	0.0025	0.0031	0.0038
(^2P) 4s, (^2P) 3d	725	17.10	0.0047	0.0059	0.0074	0.0093	0.0117
(^2P) 5s	701	17.69	0.0010	0.0013	0.0017	0.0022	0.0029
(^2P) 4d	697	17.79	0.0020	0.0026	0.0034	0.0045	0.0059
(^2P) 6s, 5d	686	18.07	0.0015	0.0020	0.0026	0.0035	0.0046
(^2P) 6d	680	18.23	0.0007	0.0009	0.0012	0.0016	0.0021
(^2P) 7d	676	18.34	0.0003	0.0005	0.0006	0.0008	0.0011
Total	–	–	0.1814	0.2027	0.2279	0.2572	0.2908

[a]Selected oscillator strengths based on Wiese *et al.* (1996), Doering *et al.* (1985) and normalized, integrated autoionization peaks from Dehmer *et al.* (1973; 1977) as described in text. The 878 Å and 792 Å clusters include non-autoionizing components.

the oscillator strengths for these peaks, including the 878 and 792 Å clusters as absorption (not just ionization) is 0.2288.

b.2 Continuum (910.5–490 Å; 430–260 Å) As shown in Fig. 3 of Angel and Samson (1988), the underlying continuum between 910.2–490 Å has a step-like structure corresponding to the formation of the ^{4}S, ^{2}D and ^{2}P states of O$^+$. Integration of such a pattern can be negotiated more accurately by graphical or trapezoidal methods, rather than processing via fitted function. Thus, the tabulated data of Angel and Samson have been graphically integrated in each of the three steps (915–732 Å, 732–665 Å, 665–490 Å) and are recorded separately in Table 2.16. Between 490–430 Å, autoionizing features appear, corresponding to transitions converging on the 2s2p^4 edge. This region is considered in the following section. The tabulated data of Samson and Angel recommence at 430 Å and continue to 260 Å. This smoothly declining domain is fitted by regression to a 4-term polynomial, whose coefficients are given in Table 2.17, and integrated to provide the corresponding $S(p)$.

b.3 The region 490–430 Å; structure and continuum This region contains resonances having characteristic asymmetric profiles (Fano parameter $q \sim 1$) which are typical of inner s valence shell excitations in first row atoms (Berkowitz et al., 1992). In Fig. 6 of their review article, Bell and Kingston (1994) display a calculated (Bell et al., 1989) and an experimental (Angel and Samson, 1988) spectrum of this region. The asymmetric features track, but the experimental spectrum does not reveal some of the fine structure, due to limited resolution. The calculated spectrum has an underlying continuum varying slightly from $\sigma(490\,\text{Å}) = 11.5\,\text{Mb}$, to $\sigma(435\,\text{Å}) = 10.6\,\text{Mb}$. The experimental spectrum has apparently been arbitrarily displaced, with $\sigma(490\,\text{Å}) = 6.44\,\text{Mb}$ and $\sigma(440\,\text{Å}) \cong 5.94\,\text{Mb}$. Actually, Angel and Samson (1988) present this spectrum without an ordinate scale, but elsewhere tabulate $\sigma(490\,\text{Å}) = 12.0\,\text{Mb}$ and $\sigma(430\,\text{Å}) = 11.5\,\text{Mb}$. Hence, there is very good agreement between calculated and experimental values for the underlying continuum. In fact, several calculations (Taylor and Burke, 1976; Pradhan, 1978; Bell et al., 1989) are in substantial agreement regarding the magnitude of the underlying continuum and the shape of the resonances. One of these (Taylor and Burke, 1976) has analyzed the resonances in terms of the Fano parameters ρ, Γ and q (see Sect. 2.2.1.b.2). Using their values, we compute $f_{\text{xs}} \cong 0.0007$ as the 'excess oscillator strength' for all the resonances in this region, compared to $f \cong 0.378$ for the underlying continuum. In this approach, the contribution of the resonances to $S(p)$ is inconsequential. Alternatively, we have normalized the experimental spectrum of Angel and Samson by requiring that the cross section at its extremities (490 Å and 430 Å) matches their tabulated values. The procedure used here is not unique – we have merely displaced the spectrum shown in Fig. 6 of Bell and Kingston so that $\sigma(490\,\text{Å}) = 12.0\,\text{Mb}$. Graphical integration now includes the resonances, and yields $f = 0.376$. This approach assumes thin target conditions, i.e. the area

Table 2.16 Spectral sums, and comparison with expectation values for atomic oxygen[a]

Energy range, eV	$S(-2)$	$S(-1)$	$S(0)$	$S(+1)$	$S(+2)$
$0–13.618(IP)$[b]	0.2490	0.2057	0.1731	0.1482	0.1288
$IP–18.64$[c]	0.1814	0.2027	0.2279	0.2572	0.2908
(autoionization only)					
(continuum)					
$IP–16.93$[d]	0.0879	0.0941	0.1056	0.1212	0.1373
$16.93–18.64$[d]	0.0869	0.1119	0.1494	0.2237	0.2571
$18.64–25.30$[d]	0.3034	0.4810	0.7683	1.2373	2.0052
$490–430$ Å,					
$25.30–28.83$[d]	0.0954	0.891	0.3756	0.7468	1.4871
$28.83–47.69$[e]	0.2099	0.5513	1.4777	4.0451	11.3037
$47.69–280$[e]	0.0810	0.4181	2.4917	18.1305	166.65
$(47.69–280)$[f]	(0.0829)	(0.4296)	(2.5791)	(19.1712)	(179.73)
$280–552.5$[f]	0.0002	0.0063	0.1694	4.7071	135.87
K edge$–572.8$[g]	0.0001	0.0045	0.1837	7.4275	300.5
$572.8–2622.4$[f]	0.0004	0.0240	1.5150	110.8101	9 672.0
$2622.4–10\,000$[f]	–	0.0004	0.0976	29.7146	10 328.5
$10^4–10^5$ [h]	–	–	0.0058	7.3793	12 641.6
$10^5–10^6$	–	–	–	0.3826	6 101.0
$10^6–10^7$	–	–	–	0.0136	2 130.7
$10^7–10^8$	–	–	–	0.0004	696.0
$10^8–10^9$	–	–	–	–	222.4
$10^9–\infty$	–	–	–	–	103.2
Total[i]	1.2956	2.2908	7.7408	185.35	42 487.0
Total[j]	(1.2975)	(2.3023)	(7.8282)	(186.39)	(42 500.1)
Expectation values	$1.31_6 \pm 0.01_7^k$	2.520^p	8.0	183.9^p	$41\,776.9^p$
	1.3344^l	2.326^q		181.8^q	$41\,775.4^r$
	1.353^m				
	1.3014^n				
	1.332^o				
Other values[s]	1.238_8	2.324	(8)	186.64	42 440

[a]In Ry units.
[b]Wiese *et al*. (1996); See Table 2.14.
[c]Details in Table 2.15.
[d]Angel and Samson (1988), graphical integration. See text.
[e]Angel and Samson (1988), polynomial fit.
[f]Henke *et al*. (1993), polynomial fit.
[g]Stolte *et al*. (1997), graphical integration.
[h]Chantler (1995).
[i]Using Angel and Samson (1988).
[j]Using Henke *et al*. (1993).
[k]Wettlaufer and Glass (1972).
[l]Saha (1993).
[m]Werner and Meyer (1976).
[n]Allison *et al*. (1972).
[o]Kelly (1969).
[p]Fraga *et al*. (1976).
[q]Müller (1996).
[r]Bunge *et al*. (1993).
[s]Zeiss *et al*. (1977).

Table 2.17 Coefficients of the polynomial $\mathrm{d}f/\mathrm{d}E = ay^2 + by^3 + cy^4 + dy^5$ fitted to data at various energies[a]

Energy range, eV	a	b	c	d
18.64–25.30	134.6053	−605.259	946.0393	−498.886
28.83–47.69	37.02097	−201.608	489.3173	−422.79
47.69–280[b]	4.16384	105.2181	−467.293	569.8476
(47.69–280)[c]	2.607738	169.0597	−960.076	1614.156
280–K edge	2.378651	130.4815	582.0583	−16781
572.8–2622.4	−4.92454	9896.02	−315578	4274390
2622.4–10000	−2.63618	9253.657	−110742	19425324

[a]$\mathrm{d}f/\mathrm{d}E$ in Rydberg units, $y = B/E$, $B = IP = 13.618\,\mathrm{eV}$.
[b]Fitted to data of Angel and Samson (1988).
[c]Fitted to data of Henke *et al.* (1993).

spanned by the resonances should yield the correct oscillator strength, despite experimental broadening. We record this result in Table 2.16.

b.4   260 Å (47.687 eV)–280 eV   Alternative values of the photoabsorption cross section in this continuum region are given by Angel and Samson (1988) and Henke *et al.* (1993). They are in rather good agreement between 47.687 and 150 eV. At higher energies, the cross sections of Angel and Samson are ~15% lower, but they merge at 270–280 eV. Their partial sums are listed separately in Table 2.16. Their differences are quite small compared to the total sums.

b.5   280 eV–K-edge   Stolte *et al.* (1997) have recently presented absolute photoionization cross sections for atomic oxygen in the K-edge region. Their data explicitly exclude valence shell contributions. To compensate, this section carries the valence shell contribution through the K-shell region, i.e. from 280–552.5 eV, the upper limit of the data of Stolte *et al.* We traverse the continuum by fitting the compiled points of Henke *et al.* (1993) to a 4-term polynomial.

b.6   K-edge to 572.8 eV   Figure 1 of Stolte *et al.* (1997) displays absolute cross sections for formation of O^+ and O^{2+}, constituting Rydberg series converging to $1s2s^2 2p^4$ (^{4}P) at 544.03 eV and $1s2s^2 2p^4$ (^{2}P) at 548.85 eV, and a continuum extending to 552.5 eV. We numerically integrate this structure, combining the O^+ and O^{2+} contributions (mention is made in Stolte *et al.* (1997) of $O^{3+}/O^{2+} \approx 1/30$). For the prominent leading peak ($1s2s^2 2p^5$) at 527 eV, we find $f \cong 0.045$, compared to the previously estimated $f \cong 0.064$ (Berkowitz, 1997a), but there is compensation from the higher transitions. At 552.5 eV, the Stolte data give $\sigma \cong 0.49\,\mathrm{Mb}$, and the valence shell contributes ~0.03 Mb; at 572.8 eV, $\sigma \cong 0.497\,\mathrm{Mb}$ (Henke *et al.*, 1993). We assume a linear decline in this interval. The contributions to $S(p)$ appear in Table 2.16.

b.7   572.8–10 000 eV   Data from the compilation of Henke *et al.* (1993) are fitted to two 4-term polynomials, 572.8–2622.4 eV and 2622.4–10 000 eV.

b.8 10^4–10^5 eV   We use the calculated cross sections of Chantler (1995).

2.5.2 The analysis

The experimental static electric dipole polarizability of atomic oxygen is not well known. Several groups (Alpher and White, 1959; Anderson *et al.*, 1967; Wettlaufer and Glass, 1972) have used optical interferometry in shocked O_2 gas to obtain specific refractivities, from which the dynamic refractivity and polarizability can be deduced. The available data display scatter with wavelength, making extrapolation to infinite wavelength hazardous. Alpher and White (1959) report $\alpha = 0.77 \pm 0.06 \times 10^{-24}$ cm^3, which is an average of three wavelengths. From Anderson *et al.* (1967) we deduce $\alpha = 0.87 \pm 0.01 \times 10^{-24}$ cm^3 at 5200 Å. However, from Wettlaufer and Glass (1972) we obtain $\alpha = 0.81 \pm 0.02 \times 10^{-24}$ cm^3 at 5300 Å, and $0.78 \pm 0.01 \times 10^{-24}$ cm^3 at 6943 Å. The latter, which claims good precision, originates from the same laboratory as Anderson *et al.*, is more recent, was obtained at longer wavelength, and is probably the closest one can get to an experimental value. It corresponds to $S(-2) = 1.31_6 \pm 0.01_7$. Several high quality calculations exist. Expressed as $S(-2)$, they include 1.3344 (Saha, 1993), 1.353 (Werner and Meyer, 1976), 1.3014 (Allison *et al.*, 1972) and 1.332 (Kelly, 1969). Our spectral sum is lower, but by <3% of the average theoretical value and the selected experimental result. Similarly, the spectral sum for $S(0)$ is ∼3% lower than the TRK sum, and even closer using the values of Henke *et al.* between 47.7–280 eV. The Hartee–Fock value for $S(-1)$, 2.520 Ry units (Fraga *et al.*, 1976) is characteristically too high. With inclusion of correlation (Müller, private communication, 1996), this expectation value diminishes to within 1.5% of the spectral sum. Thus, there is remarkable consistency among the three lower $S(p)$ – each is lower by ≤3% of expectation. This suggests that the slight shortfall is not localized in one spectral region. The spectral sum for $S(-2)$ depends primarily upon the absolute calibration used by Angel and Samson (1988), and to a lesser extent on the oscillator strengths in the discrete region (Wiese *et al.*, 1996). The calibration used by Angel and Samson was that determined earlier by Samson and Pareek (1985), which had an estimated error of $\pm 9\%$. The present analysis implies that the uncertainty may be closer to 3%. Since this calibration was used implicitly in evaluating the oscillator strengths of the autoionization peaks listed in Table 2.15, the level of agreement provides some confidence in these values.

Bell and Kingston (1994) have reviewed several ab initio calculations, which are generally in good agreement with the experimental values used here. In their view, 'theory suggests that the cross section for photoionization of ground state oxygen is known to better than 5%'. Two major discrepancies exist. At the photoionization threshold, theory predicts an abrupt step-like behavior, whereas the experimental data indicate an abrupt onset, followed by a gradual increase. Also, above the ^{2}P threshold (18.64 eV) the theoretical cross section is rather flat, whereas the experimental data lie higher, and display a broad maximum.

The Hartree–Fock value for $S(+1)$, 183.9 Ry (Fraga *et al.*, 1976), is typically too low. Use of correlated wave functions generally increases this value, and the spectral sum, $S(+1) \approx 185\text{--}186$ Ry units, supports this view. Müller's (1996) calculation yields a lower value. He notes that his wave function, designed for optimizing inelastic scattering factors, tends to display erratic behavior with respect to electron correlation effects in the core region. By contrast, the Hartree–Fock value for $S(+2)$ should be adequate. Two almost identical values [41 776.9 (Fraga *et al.*, 1976); 41 775.4 (Bunge *et al.*, 1993)] have been reported, which are $\sim$1.7% lower than the spectral sum.

Zeiss *et al.* (1977) were dependent on older experimental values in both the discrete and continuum regions, and also on an older calculation by Henry (1967). The K-shell region was mimicked by mixture rules. The existing data were adjusted to satisfy the TRK sum rule, and also their chosen value of $S(-2)$. The latter is based on a Cauchy plot and analysis by Zeiss and Meath (1977) with input from the aforementioned experiments and the calculation by Kelly (1969). Unfortunately, their Cauchy expansion does not fit these data well. Their extrapolated refractivity falls 7% below Kelly's calculation (which subsequent calculations support), and also below most of the experimental points. The good agreement they obtain with our inferred $S(+1)$ and $S(+2)$ is attributable to the high-energy photoabsorption cross sections, which have not changed appreciably in the intervening years. However, large differences can be found in localized cross sections, such as the autoionization region and the detailed structure of K-shell excitation.

2.6 Neon

2.6.1 The data

Neon has the ground state configuration $1s^2 2s^2 2p^6$. The first ionization potential, corresponding to the ionic state $1s^2 2s^2 2p^5\,^2P_{3/2}$, is $173\,929.75 \pm 0.06$ cm^{-1} $\equiv$ $21.564\,538 \pm 0.000\,007$ eV (Kaufman and Minnhagen, 1972). The excited spin-orbit state $^2P_{1/2}$ lies 780.4240 cm^{-1} higher, with IP $= 21.661\,298 \pm 0.000\,007$ eV (Yamada *et al.*, 1985), see also Harth *et al.* (1987).

a The discrete spectrum

Electric dipole allowed excitation in the valence region, $1s^2 2s^2 2p^6 \rightarrow 1s^2 2s^2 2p^5 ns$, nd gives rise to five series. Three of these, ns(3/2)$^o_1$, nd(1/2)$^o_1$ and nd(3/2)$^o_1$ converge to the ground state of Ne$^+$, $^2P_{3/2}$, where the notation is J_cK, as generally used. The other two, ns$'$(1/2)$^o_1$ and nd$'$ (3/2)$^o_1$, converge to the $^2P_{1/2}$ excited state. Above the $^2P_{3/2}$ threshold, these latter two series are degenerate with a continuum, and display autoionization features. All five series have been observed in single-photon excitation (photoabsorption), four of them to high quantum numbers (Baig *et al.*, 1984; Ito *et al.*, 1988). The nd $(1/2)_1^o$ and nd$(3/2)_1^o$ series differ by the spin-orbit splitting of the Rydberg electron, which is only $112\,\mathrm{cm}^{-1}$ ($0.0139\,\mathrm{eV}$) for $n = 3$ and diminishes by $\sim(n*)^{-3}$ thereafter. As a result, Baig and Connerade (1984) have been able to follow the weak nd$(1/2)_1^o$ series only to $n = 11$, Ito *et al.* (1988) to $n = 15$, whereas the strong nd$(3/2)_1^o$ series could be observed to $n = 44$ and 65, respectively. In determinations of oscillator strength, these two series may not be resolved, and then the measured oscillator strength represents their sum. In that case, we have effectively two Rydberg series (ns and nd) whose oscillator strengths must be assessed up to the $^2P_{3/2}$ limit, and another two (ns$'$ and nd$'$) which will be followed to the $^2P_{1/2}$ limit.

Direct measurement of these oscillator strengths by photoabsorption and the use of the Beer–Lambert law is extremely difficult, because the lines are very sharp (even in the autoionization region). Saturation at the line center is almost unavoidable. Chan *et al.* (1992) discuss this problem in considerable detail, and consequently advocate their method of inelastic electron scattering in the forward direction at high incidence energies, which is a non-resonant process applicable to thin targets. Even though their resolution ($0.048\,\mathrm{eV}$) is poor, by optical standards, the area under the peak is proportional to the oscillator strength (f), and absolute values can be inferred with appropriate normalization. These authors have used a modified Thomas–Reiche–Kuhn sum rule, and more recently (Olney *et al.* (1977)) experimentally known polarizabilities and the $S(-2)$ sum rule for their normalizations.

A variety of techniques have been devised to circumvent the saturation problem (see Chan *et al.*, 1992a). Two of them have seen frequent application to the neon problem, and are briefly described here. The first, initiated by Westerveld *et al.* (1979) uses an electron beam to excite resonance radiation in the atoms. This radiation is partly absorbed in the gas between the beam and a vacuum ultraviolet spectrometer. The intensity of the resonance radiation is recorded as a function of gas pressure, and a formalism is used to derive the oscillator strength. Recent applications of this principle for neon have been reported by Gibson and Risley (1995) and Tsurubuchi *et al.* (1990). The second, described by Aleksandrov *et al.* (1983) uses a method of total absorption in an optically thick layer. The incident light was monochromatized synchrotron radiation. The 'equivalent width' of an absorption line is plotted as a function of $(Nl)^{1/2}$, where N is the number density of atoms and l is the column length. The slope is shown to be proportional to oscillator strength.

The Beer–Lambert law, in differential form, is

$$-\mathrm{d}I = I_o \sigma N \, \mathrm{d}l,$$

where I_o is the incident light flux, $\mathrm{d}I$ is its diminution upon passing through a path $\mathrm{d}l$ and σ is the absorption cross section. Above the ionization potential, we can identify the loss of photons, $\mathrm{d}I$, with the gain in ion signal in a thin target photoionization experiment, when properly normalized. Hsu *et al.* (1996) have obtained a photoionization spectrum of neon which displays the ns′ and nd′ autoionizing resonances between the $^2P_{3/2}$ and $^2P_{1/2}$ thresholds. Their relative photoionization yield can be normalized at the onset of the $^2P_{1/2}$ continuum, using the accurate absolute photoionization cross section reported by Samson *et al.* (1991). The area of each peak is now equal to the corresponding oscillator strength, even though the instrumental width is substantially larger than the inherent line width, provided that the target is sufficiently thin to avoid saturation. This appears to be the case for the data of Hsu *et al.* We have utilized this strategy to estimate ns′ and nd′ oscillator strengths near the $^2P_{1/2}$ threshold.

For the lower-lying excitations, recent determinations of oscillator strength (experimental and theoretical) are summarized in Table 2.18. (Earlier values can be found in Chan *et al.* (1992a).) Aleksandrov *et al.* (1983) succeeded in resolving the closely spaced d doublets (see above); we record their sum, since they are not resolved at high *n*, and it facilitates comparison with other evaluations. It is also convenient to couple ns and ns′ for a given *n* (and also nd and

Table 2.18 Oscillator strengths for the ns, ns′ ($n = 3-6$), nd, nd′ ($n = 3-5$) discrete transitions in neon

E, eV	16.670 833	16.848 059		
	f, 3s ($^2P_{3/2}$)	f, 3s′ ($^2P_{1/2}$)	Sum	Ratio
Expt.	0.010 95 (32)[a]	0.1432 (38)[a]	0.1542	13.08
	0.0118 (6)[b]	0.159 (8)[b]	0.171	13.47
	0.0122 (6)[c]	0.123 (6)[c]	0.135	10.08
	0.012 (3)[d]	0.144 (24)[d]	0.156	12.0
	0.0109 (9)[e]	0.147 (12)[e]	0.158	13.49
	0.0121 (8)[f]	0.148 (13)[f]	0.160	12.23
Theory		0.159[g]	0.163[h]	13.27[i]
	0.0106[d]	0.141[d]	0.152[d]	13.3[d]
Selection	0.011	0.145	0.156	13.18
E, eV	19.688 201	19.779 778		
	f, 4s ($^2P_{3/2}$)	f, 4s′ ($^2P_{1/2}$)		
Expt.	0.0128 (10)[e]	0.0153 (12)[e]	0.0281	1.195
	0.0129 (6)[b]	0.0165 (8)[b]	0.0294	1.279
	0.0145 (35)[d]	0.0185 (60)[d]	0.033	1.276
Theory			0.028[h]	1.282[i]
	0.0124[d]	0.0160[d]	0.0284[d]	1.29[d]
Selection	0.012 85	0.0159	0.0288	1.237
E, eV	20.570 571	20.662 780		
	f, 5s ($^2P_{3/2}$)	f, 5s′ ($^2P_{1/2}$)	Sum	Ratio
Expt.	0.0061 (5)[e]	0.0042 (3)[e]	0.0103	0.688
	0.006 37 (32)[b]	0.004 61 (23)[b]	0.0110	0.724
	0.0083 (31)[d]	0.0049 (17)[d]	0.0132	0.590
Theory				0.726[i]
	0.0060[d]	0.0043[d]	0.0103[d]	0.717[d]
Selection	0.0063	0.0044	0.0107	0.698

Table 2.18   (*Continued*)

E, eV	20.949 289	21.043 553		
	f, 6s (^{2}P$_{3/2}$)	f, 6s′ (^{2}P$_{1/2}$)		
Expt.	0.003 30 (30)[b]	0.001 56 (16)[b]	0.004 86	0.473
	0.0045 (19)[d]	0.003 (1)[d]	0.0075	0.667
Theory				0.641[i]
	0.0031[d]	0.0018[d]	0.0049[d]	0.581[d]
Selection	0.0033	0.0016	0.0049	0.485

E, eV	20.033 422 (avg)	20.139 464		
	f, 3d (^{2}P$_{3/2}$)	f, 3d′ (^{2}P$_{1/2}$)		
Expt.	0.0186 (9)[b]	0.006 65 (33)[b]	0.0253	0.358
	0.0222 (46)$_\text{sum}^\text{d}$	0.0082 (29)[d]	0.0304	0.369
		0.0064 (5)[e]		
Theory			0.021[h]	
	0.0176$_\text{sum}^\text{d}$	0.0064[d]	0.024[d]	0.364[d]
Selection	0.019 (1)	0.0065	0.0255	0.342

E, eV	20.705 514 (avg)	20.805 518	Sum	Ratio
	f, 4d (^{2}P$_{3/2}$)	f, 4d′ (^{2}P$_{1/2}$)		
Expt.	0.009 44 (47)	0.004 39 (22)[b]	0.0138	0.465
	0.0147 (36)$_\text{sum}^\text{d}$	0.005 (2)[d]	0.0197	0.340
Theory	0.0091$_\text{sum}^\text{d}$	0.0041[d]	0.0132[d]	0.451[d]
Selection	0.0094	0.0044	0.0138	0.468

E, eV	21.015 664	21.114 022		
	f, 5d (^{2}P$_{3/2}$)	f, 5d′ (^{2}P$_{1/2}$)		
Expt.	0.005 43 (54)[b]	0.002 29 (23)[b]	0.007 72	0.422
Theory	0.0050[d]	0.0024[d]	0.0074[d]	0.480[d]
Selection	0.0054	0.0024	0.0078	0.444

[a]Gibson and Risley (1995).
[b]Chan *et al*. (1992a).
[c]Tsurubuchi *et al*. (1990).
[d]Aleksandrov *et al*. (1983).
[e]Westerveld *et al*. (1979).
[f]Bhaskar and Lurio (1976).
[g]Stewart (1975).
[h]Amusia (1990).
[i]Semenov and Strugach (1968).

nd′) since theoretical values are available for their sum and ratio, which can be compared with experiment. In making a selection, primary attention is given to error bars reported, but comparison with other values, and plausible concordance with predicted sum and ratio is also considered.

From the selected values, histograms have been constructed for each of the series ns, nd, n′s and n′d, and are displayed in Figs. 2.11–2.14. Extrapolation of the ns and nd series to the ^{2}P$_{3/2}$ threshold is relatively straightforward, and gives $(\mathrm{d}f/\mathrm{d}E) = 0.0137$/eV for the ns series, 0.027/eV for the nd series, summing to 0.0407/eV, or $\sigma = 4.46_7$ Mb. For the ns′ and nd′ series, additional values

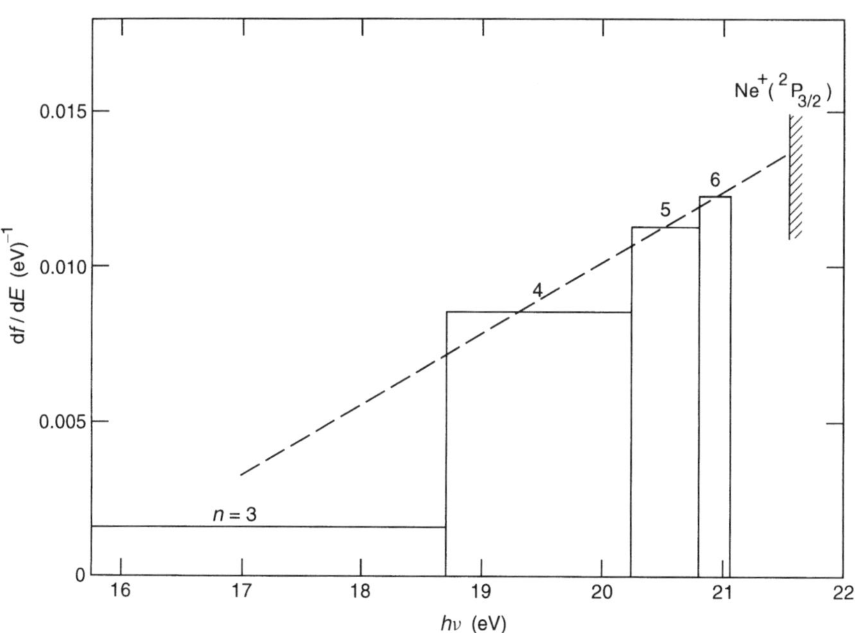

Fig. 2.11 Histogram for the series $\ldots 2p^6$, $^1S_0 \rightarrow \ldots 2p^5$ ($^2P_{3/2}$)ns, ^{1}P in neon

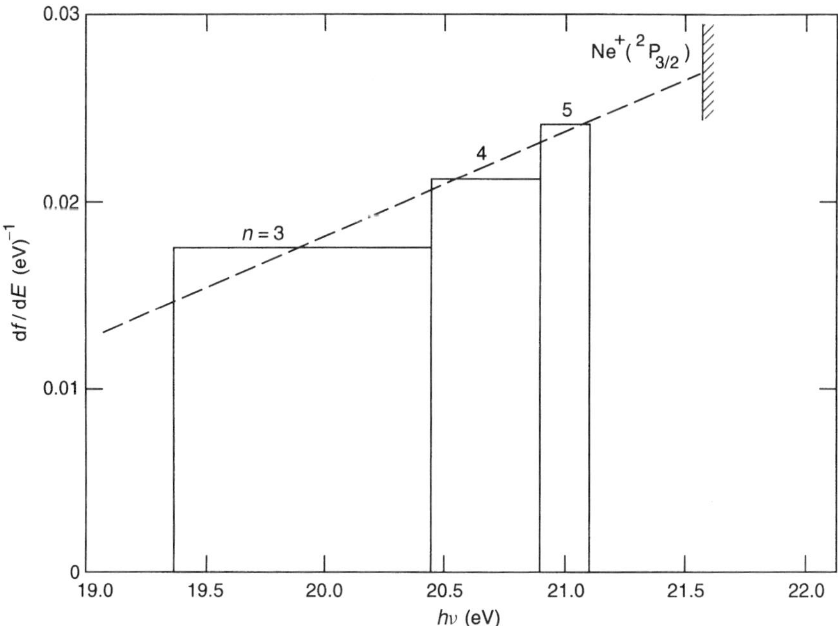

Fig. 2.12 Histogram for the series $\ldots 2p^6$, $^1S_0 \rightarrow \ldots 2p^5$ ($^2P_{3/2}$)nd, ^{1}P in neon

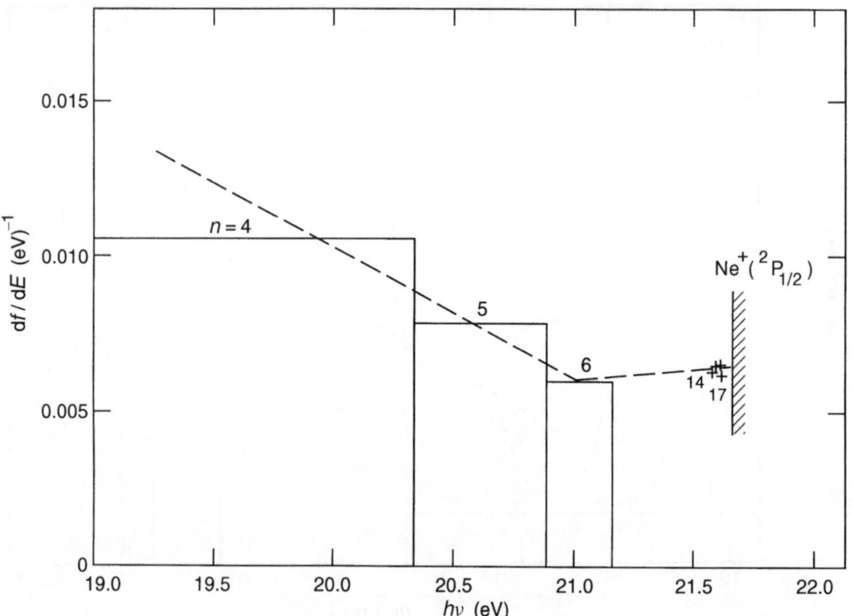

Fig. 2.13   Histogram for the series $\ldots 2p^6$, $^1S_0 \rightarrow \ldots 2p^5$ ($^2P_{1/2}$)ns, ^1P in neon

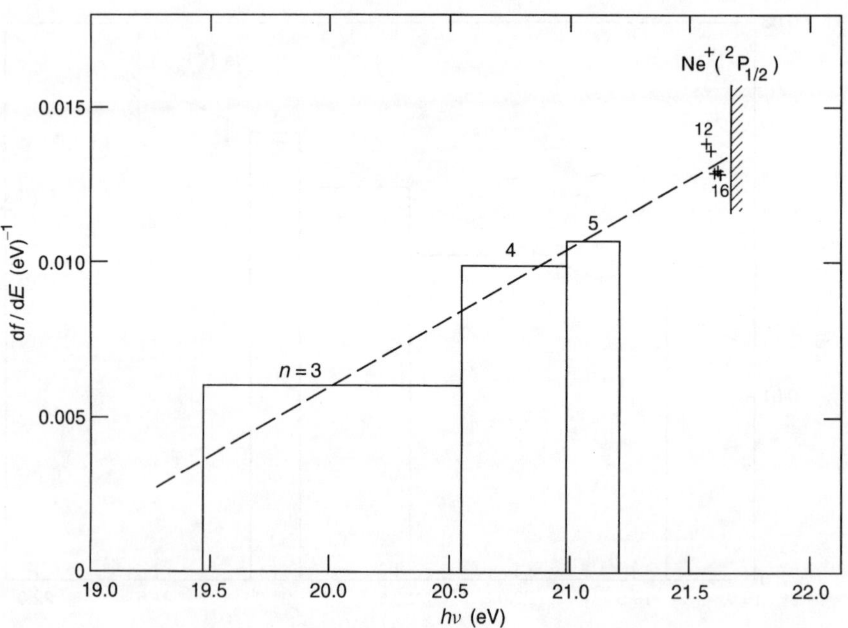

Fig. 2.14   Histogram for the series $\ldots 2p^6$, $^1S_0 \rightarrow \ldots 2p^5$ ($^2P_{1/2}$)nd, ^1P in neon

at high n have been extracted from the data of Hsu *et al*., as described earlier. They corroborate a more-or-less linear extrapolation to the $^2P_{1/2}$ threshold for the nd' series, but reveal a change of sign in the slope for the ns' series, which Aleksandrov *et al*. (1983) had already seen at $n' = 6$. The extrapolated value of df/dE for the nd' series is 0.0135/eV; that for the ns' series of 0.0065/eV. Their sum (0.0200/eV) corresponds to $\sigma = 2.19_5$ Mb. The ratio of $\sigma(^2P_{3/2}) : \sigma(^2P_{1/2})$ is 2.03, close to the non-relativistic statistical weight (L–S coupling) of 2.0. Samson *et al*. (1975) have measured a ratio of 2.18 ±0.17 over a broad energy range 0–19 eV above the $^2P_{1/2}$ threshold. The sum of $\sigma(^2P_{3/2})$ and $\sigma(^2P_{1/2})$ is 6.66 Mb, close to the accurate determination of Samson *et al*. (1991) at the $^2P_{1/2}$ threshold, 6.39 Mb.

With this degree of concordance, it is now possible to estimate the contributions of higher ns, ns' series ($n > 6$) and nd, nd' series ($n > 5$) to the total oscillator strength, and to the other $S(p)$ sums, making use of Figs. 2.11–2.14. This information is included in Table 2.19, together with the contributions of

Table 2.19 Contributions from the discrete spectrum to the $S(p)$ sums in neon (in Ry units)

a. Series converging to Ne$^+$($^2P_{3/2}$)

nl	$S(-2)$	$S(-1)$	$S(0)$	$S(+1)$	$S(+2)$
3s	0.0073_3	0.0089_8	0.0110	0.0134_8	0.0165
4s	0.0061_4	0.0088_8	0.0128_5	0.0186	0.0269
5s	0.0027_6	0.0041_7	0.0063	0.0095_3	0.0144
6s	0.0013_9	0.0021_4	0.0033	0.0050_8	0.0078
$\sum_{7s}^{\infty}$	0.0026_3	0.0041_3	0.0064_7	0.0101_4	0.0159
3d (sum)	0.0087_6	0.0129	0.0190	0.0280	0.0412
4d (sum)	0.0040_6	0.0061_8	0.0094	0.0143	0.0218
5d (sum)	0.0022_6	0.0035	0.0054	0.0083_4	0.0129
$\sum_{6d}^{\infty}$	0.0045_7	0.0071_7	0.0112_5	0.0176_5	0.0277
Continuum, IP (3/2)–IP (1/2)	0.0014_9	0.0023_6	0.0038	0.0059_6	0.0095
Sub-total	0.0413_9	0.0604_1	0.0887_7	0.1310_8	0.1946

b. Series converging to Ne$^+$($^2P_{1/2}$)

	$S(-2)$	$S(-1)$	$S(0)$	$S(+1)$	$S(+2)$
3s'	0.0946	0.1171	0.1450	0.1796	0.2223
4s'	0.0075	0.0109	0.0159	0.0231	0.0336
5s'	0.0019	0.0029	0.0044	0.0067	0.0101
6s'	0.0007	0.0010	0.0016	0.0025	0.0038
$\sum_{6s'}^{\infty}$	0.0012_7	0.0020	0.0031_5	0.0049	0.0077
3d'	0.0030	0.0044	0.0065	0.0096	0.0142
4d'	0.0019	0.0029	0.0044	0.0067	0.0103
5d'	0.0010	0.0015	0.0024	0.0037	0.0058
$\sum_{5d'}^{\infty}$	0.0023	0.0036	0.0057	0.0090	0.0142
Sub-total	0.1141_7	0.1463	0.1890_5	0.2458	0.3220
Total	0.1556	0.2067	0.2778	0.3769	0.5166

the lower n transitions and the underlying continuum between the $^2P_{3/2}-^2P_{1/2}$ thresholds.

b The continuum

b.1 $(2p)^{-1}$ to $(2s)^{-1}$, 21.6613–48.4750 eV   For enhanced accuracy and subsequent analysis, the continuum is partitioned. Initially, it is convenient to consider the domain between the 2p and 2s edge. In this range, the absorption cross section first increases to a maximum at $\sim$32 eV, and then declines. The $(2s)^{-1}$ ionization potential is established by the resonance transition in Ne$^+$, $2s^2 2p^5\,^2P_{3/2} \rightarrow$ 2s $2p^6\,^2S_{1/2}$, given as 460.7284 Å by Persson (1971). The corresponding energy, 26.910 48 eV, added to IP($^2P_{3/2}$) gives 48.4750 eV for the L$_1$ edge. The cross section in this energy range is mostly smooth, but is punctuated by 2s $\rightarrow$ np resonances having a characteristic asymmetric shape (Codling et al., 1967). These will be examined separately in the following section.

Among recent sources, Samson et al. (1991) have presented accurate data ($\pm$3%), but the figures shown cover limited wavelength regions. Much earlier determinations by Samson (1966) cover the requisite range with $\pm$5% accuracy. Chan et al. (1992a) provide extensive data from the ionization threshold to 250 eV, using high-energy electron inelastic scattering normalized by a modified Thomas–Reiche–Kuhn sum rule. Since the present work involves sum rule testing of experimental results, utilizing the results of Chan et al. introduces an element of circular reasoning, but it is nevertheless useful to compare with other measurements. Finally, Bizau and Wuilleumier (1995) have presented their recommended cross sections from threshold to 280 eV. The latter appear to be a rough mean of the Chan and Samson values, which differ only slightly. We have fitted the Chan and Samson data sets individually by regression analysis with a 4-term polynomial. From these fitted functions, we obtain contributions to $S(0)$ of 2.0416 from the Samson data, 2.0197 from Chan et al., in the interval 21.6613–48.4750 eV. (Samson (1966) also obtained 2.04.) The values of $S(p)$ in this range are listed in Table 2.20. The coefficients of the fitted function for the Samson data, which should tend to reduce statistical scatter, are recorded in Table 2.21.

b.2 Resonances, 45.55–48.83 eV   Codling et al. (1967), and later Aleksandrov et al. (1983) and Langer et al. (1997) recorded and analyzed these resonances. Four are early members of the series $2s^2 2p^6(^1S_o) \rightarrow 2s2p^6 np(^1P_1)$; the other two are two-electron excitations with upper states $2s^2 2p^4 3s3p$. All three groups have fitted the asymmetric shapes to Fano parameters; Codling et al. (1967) and Aleksandrov et al. (1983) calculated the 'excess oscillator strength', f_{xs}, for each transition. Here, we take an average of their results, and extrapolate the $2s2p^6$np values to the series limit. Langer et al. (1997) report values for q, Γ and ρ^2, but not σ. If we borrow σ from Codling et al. (1967), the values of f_{xs} deduced from their parameters agree with those of Codling, within experimental error, for $n = 3-5$. (For one of the two-electron excitations, the Langer values

Table 2.20 Spectral sums, and comparison with expectation values for neon ($S(p)$ in Ry units)

Energy, eV Discrete $\to {}^2P_{1/2}$	$S(-2)$	$S(-1)$	$S(0)$	$S(+1)$	$S(+2)$
0–21.661[a]	0.1556	0.2067	0.2778	0.3769	0.5166
IP $\to$ 2s edge					
21.661–48.475	0.3557[b]	0.8307[b]	2.0416[b]	5.2726[b]	14.2540[b]
	(0.3521)[c]	(0.8223)[c]	(2.0197)[c]	(5.2076)[c]	(14.0495)[c]
resonances					
45.5–48.8[d]	0.0003	0.0011	0.0037	0.0127	0.0572
48.475–250.0	0.1282[b]	0.7276[b]	4.8088[b]	37.7118[b]	361.4259[b]
	(0.1239)[c]	(0.6956)[c]	(4.5671)[c]	(36.1651)[c]	(352.2703)[c]
250.0–280.0[e]	0.0004	0.0070	0.1366	2.6551	51.6431
$\to$ K edge					
280.0–870.25[f]	0.0007	0.0203	0.5943	19.1649	669.5803
resonances					
867.25–870.25[g]	–	0.0001	0.0069	0.4390	27.754
870.25–2984.3[f]	0.0002	0.0168	1.5628	160.992	18 691.1
2984.3–10 000[f]	–	0.0006	0.1762	59.8076	22 700.8
10^4–10^{5h}	–	–	0.0157	20.0611	34 974.5
10^5–10^{6i}	–	–	0.0001	1.1568	18 559.4
10^6–10^{7i}	–	–	–	0.0423	6 644.9
10^7–10^{8i}	–	–	–	0.0014	2 189.0
10^8–10^{9i}	–	–	–	–	701.4
10^9–∞^i	–	–	–	–	325.9
Total	0.6411	1.8109	9.6245	307.694	105 912.2
Expectation values	0.6656(7)[j]	1.924[k]	10.0		103 870.5[l]
		1.8806[m]			103 868.5[n]
Other values	(0.6673)[o]	1.9005[o]	(10.0)[o]	304.0[o]	105 120[o]
	0.6428[c]				
	0.6658[p]	1.8005[p]			

[a] See Table 2.19.
[b] Samson *et al.* (1991); Samson (1966).
[c] Chan *et al.* (1992a).
[d] Codling *et al.* (1967); Aleksandrov *et al.* (1983).
[e] Bizau and Wuilleumier (1995).
[f] Henke *et al.* (1993).
[g] Esteva *et al.* (1983); Wuilleumier (1971).
[h] Chantler (1995).
[i] Hydrogenic calculation, K-shell only, from Bethe and Salpeter (1977).
[j] See text.
[k] Kim *et al.* (1973).
[l] Fraga *et al.* (1976).
[m] Saxon (1973).
[n] Bunge *et al.* (1993).
[o] Kumar and Meath (1985a).
[p] Olney *et al.* (1997).

Table 2.21 Coefficients of the polynomial $\mathrm{d}f/\mathrm{d}E = ay^2 + by^3 + cy^4 + dy^5$ fitted to data at various energies[a]

Energy range, eV	a	b	c	d
21.661–48.475	16.136 68	−38.9941	35.960 58	−12.319
48.475–250.0[b]	5.806 197	73.98 605	−328.517	353.6014
250.0–280.0	−2071.35	76 715.55	−942 469	3 857 740
280.0–870.25	0.766 927	170.951	−971.918	2 191.973
870.25–2984.3	3.401 404	4 655.156	−92 346.7	683 347
2984.3–10 000	−2.139 04	5 956.894	−185 028	2 306 776

[a]$(\mathrm{d}f/\mathrm{d}E)$ per Ry unit, $y = B/E$, $B = \mathrm{IP}(^2\mathrm{P}_{1/2}) = 21.6613\,\mathrm{eV}$.
[b]These cross sections may be low by ∼6%. See Sect. 2.6.2, Analysis.

are about 50% higher than f_{xs} from Codling.) Aleksandrov *et al.* mention that this increases the total cross section by 0.5% at the series limit. The two-electron excitations are treated as isolated members, the upper one occurring beyond the 2s edge. The contribution of all these resonances to the $S(p)$, as recorded in Table 2.20, is not very significant.

b.3   48.475–280 eV   In this declining continuum region, the recommended values of Bizau and Wuilleumier (1995) closely follow the data of Samson *et al.* (1991) and Samson (1996), supplemented by measurements of Watson (1972). Earlier values reported by West and Marr (1976) are shown to be higher, while the more recent numbers from Chan *et al.* (1992a) are somewhat lower. For purposes of comparison, we have fitted the data of Samson *et al.* (1991) supplemented by Samson (1966) with values from Watson (1972), which mesh well with Samson's data in the region of overlap and extend to 230 eV. Higher-energy points, taken from Bizau and Wuilleumier (1995), are used to complete the region 48.475–250 eV. The data of Chan *et al.* (1992a) are similarly fitted to a 4-parameter polynomial over the same energy interval and the respective contributions to $S(p)$ are evaluated, and recorded in Table 2.20. As expected from the earlier observations, the contribution to $S(0)$ is lower using the Chan *et al.* data by about 0.24 units. Since our total spectral sum for $S(0)$ is about 0.38 units lower than required by the TRK sum rule (see below), even when using the Samson–Watson data, the lower values from Chan *et al.* are disfavored. Consequently, Table 2.21 lists the parameters for the function fitted to the Samson–Watson data. Cross sections for the 250–280 eV interval are taken from Bizau and Wuilleumier (1995), although the provenance of these values is unclear. They do provide a smooth transition to the compilation of Henke *et al.* (1993), which is utilized in the next section.

b.4   280–870.25 eV (K-edge)   The photoabsorption cross section of neon declines smoothly, by more than an order of magnitude, between 280 eV and the K-edge, according to Henke *et al.* (1993). There are some resonances due to Rydberg excitation just prior to this edge, which are considered separately in the succeeding section. Values for the K-edge varying from 870.1(2) eV (Hitchcock and Brion (1980)) to 870.31 eV (Thomas and Shaw, 1974) have been reported, the most recent being 870.28 eV (Esteva *et al.*, 1980). We choose a weighted average of 870.25 eV. The cross sections in the stated interval listed by Henke *et al.* have been fitted in the manner previously described, to a 4-term polynomial. Values of $S(p)$ have been computed from this function, and are recorded in Table 20.20. The coefficients of the function are given in Table 20.21.

b.5   Rydberg resonances approaching K-edge (867.25–870.25 eV)   Esteva *et al.* (1983) and Wuilleumier (1971) display figures on an absolute cross section scale for these resonances, while Hitchcock and Brion (1980) provide a relative scale. We estimate $f \cong 0.0041$ for the 1s $\rightarrow$ 3p resonance (0.9 Mb×0.50 eV) from Fig. 1 of Esteva *et al.*. Higher members have been scaled accordingly. The contribution of the series to $S(p)$ is included in Table 2.20.

b.6   870.25–10 000 eV   The cross sections given in the compilation of Henke *et al.* (1993) extrapolate to $\sigma = 0.36$ Mb just above the K-edge, in very good agreement with Fig. 1 of Esteva *et al.* (1983). Above 6 keV, the Henke cross sections merge smoothly with the calculated values of Chantler (1995). In the stated interval, these cross sections have been partitioned, and fitted with two 4-term polynomials, as previously described. Values of $S(p)$ computed from this function are listed in Table 2.20, and the coefficients are given in Table 2.21.

b.7 10^4–10^5 eV   The calculated cross sections of Chantler have been used here.

2.6.2   The analysis

Experimental determinations of the polarizability (α) of neon, and hence $S(-2)$, are based on measurements of dielectric constant or refractive index. For many years, the value $\alpha = 0.3956(4) \times 10^{-24}$ cm^3 obtained by Orcutt and Cole (1967) from dielectric constant measurements has been the accepted value. More recent dielectric constant measurements which displayed a temperature dependence for α were reported by Lehmann *et al.* (1987), but were discredited by Hohm and Kerl (1990), who measured the refractive index at one wavelength (6329.9 Å), but over a wide temperature range. The refractive index was measured at four wavelengths by Burns *et al.* (1986). One of these wavelengths essentially coincided with the wavelength employed by Hohm and Kerl, and the agreement on refractive index was very good. By fitting the dynamic polarizability at these four wavelengths, we deduce a static dipole polarizability of 0.3938×10^{-24} cm^3. Scaling this value to the slight difference between the measurements of Hohm and Karl and Burns *et al.* leads to $\alpha = 0.3946(4) \times 10^{-24}$ cm^3, or $S(-2)= 0.6656(7)$ Ry units, slightly lower than the value of Orcutt and Cole, but almost within the combined error limits.

There have been a number of recent calculations of $S(-2)$, or α, at higher levels of theory. The values obtained, in equivalent α (Å)^3, are 0.3998 (Maroulis and Thakkar, 1989); 0.3965 (Saha and Caldwell, 1991); 0.3897 (Rice *et al.*, 1991); 0.3905–0.3994, (Kobayashi *et al.*, 1993) and 0.3971–0.4029 (Woon and Dunning, 1994). They are generally in reasonable agreement with our derived experimental value.

A glance at Table 2.20 reveals that the spectral sum for $S(-2)$ is about 3.7% lower than the value determined from α. The spectral sum for $S(0)$ is also shy of the Thomas–Reiche–Kuhn value (10.0), by about the same relative amount. Since both spectral sums are low, it is clear that where a choice existed between the cross sections of Samson *et al.* (1991) and Chan *et al.* (1992a), the Samson values are preferred, since the Chan values are lower still. It is also apparent that cross sections for $E > 250\,\text{eV}$ are too small to explain the discrepancy between spectral sums and anticipated sums. Three energy regions may be suspect – the discrete spectrum, 21.66–48.48 eV and 48.48–250 eV. Since these regions contribute different proportions to $S(-2)$ and $S(0)$, the discrepancy cannot be localized to one domain. The most precise measurements encompass the 21.66–48.48 eV region, but even here the accuracy claimed is $\pm3\%$ (Samson *et al.*, 1991). The upper limit of the error bar here could halve the deviation for $S(-2)$, but only accounts for 1/6 of the discrepancy for $S(0)$. Similarly, the uncertainty in the discrete spectrum could account for part of the difference for $S(-2)$, but it would play an insignificant role for $S(0)$. Consequently, the most likely culprit to explain the discrepancy in $S(0)$ is the 48.48–250 eV region, which would require an increase in cross sections of $\sim6\%$.

Kumar and Meath (1985a) evaluated the oscillator strength distribution in neon using their fitting technique, which assures conformity to the $S(0)$ and $S(-2)$ (their selection) sum rules. Upon comparing their distribution with the present one, a major difference can be seen between 48.48–250.0 eV, where Kumar and Meath infer $\Delta S(0) = 5.2912$ and the present value is 4.8088. The difference (0.48) is slightly more than the current shortfall in $S(0)$, about 0.38. Kumar and Meath rely primarily on the older data of Ederer and Tomboulian (1964), modified by their fitting procedure. The Ederer cross sections are about 10% higher than those of Samson (1966), Samson *et al.* (1991) and Watson (1972), which agree well with one another (see Fig. 2.15). Ederer and Tomboulian assert an accuracy of 5% or better, Samson *et al.* (1991) and Watson (1972) claim $\pm3\%$ accuracy. In light of these observations, it is surprising that the present analysis favors the Ederer and Tomboulian data. Despite all the prior studies on neon, it appears that further work is necessary to resolve this discrepancy.

For $S(-1)$, the Hartree–Fock value is 10% higher than the spectral sum (Fraga *et al.*, 1976). Inclusion of correlation typically reduces this quantity. Kim *et al.* (1973), using a Bethe–Goldstone correlated wave function, obtained $S(-1) = 1.924$, while Saxon (1973) calculated $S(-1) = 1.8806$ with a 50-configuration wave function. Saxon's value is 3.7% higher than the spectral sum, the same discrepancy as found for $S(-2)$ and $S(0)$, and hence more consistent with the

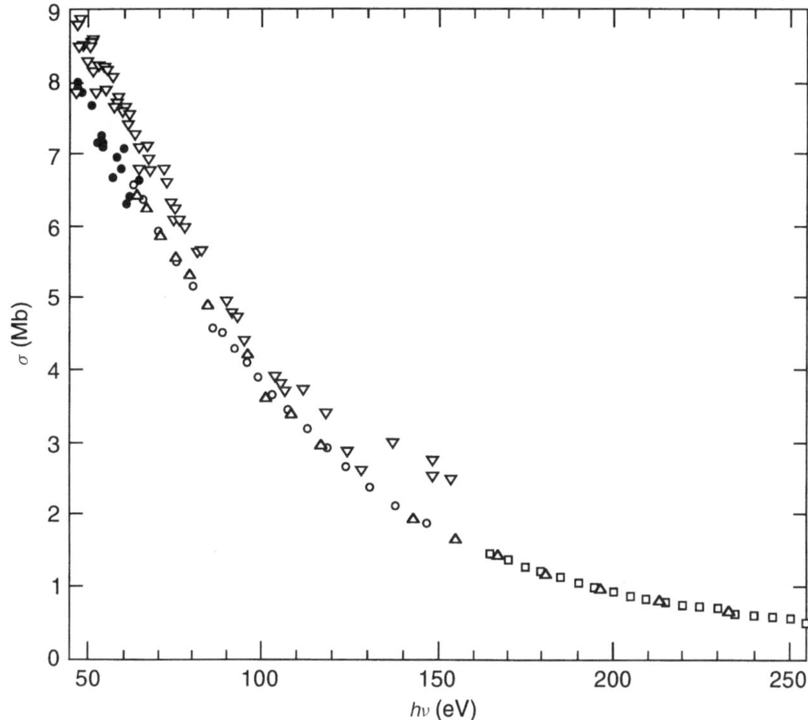

Fig. 2.15 Absolute photoabsorption spectrum of neon, 50–250 eV. • Samson (1966); ∘ Samson *et al.* (1991); △ Watson (1972); ▽ Ederer and Tomboulian (1964); □ Bizau and Wuilleumier (1995)

present analysis. Kumar and Meath (1985a) arrive at a slightly higher value, $S(-1) = 1.9005$.

The Hartree–Fock value for $S(+1)$, 302.84, is 1.6% lower than our spectral sum. Correlation effects are expected to increase this value, and indeed Saxon's 50-configuration wave function yielded $S(+1) = 308.80$ Ry units, just 0.36% larger than our spectral sum. For $S(+2)$, the Hartree–Fock value should be fairly reliable, since the influence of correlation is minimal. In this case, the Hartree–Fock value is about 2% lower than the spectral sum. The $S(+2)$ value given by Kumar and Meath (1985a) is even closer to the 'expectation' value, but their $S(+1)$ determination, 304 Ry units, is probably too low.

2.7 Sodium

2.7.1 The data

The ionization potential of atomic sodium is 41 449.44 ±0.03 cm^{-1} = 5.139 075 (4) eV (Martin 1980).

a The discrete spectrum

a.1 The 3s ^{2}s $\rightarrow$ 3p ^{2}P resonance transition As with all the alkali elements, the intravalence resonance transition predominates the discrete spectrum in sodium. A precise value is required, because (see Sect. 1, Introduction) it contributes $\sim$99% to $S(-2)$ or α. In 1982, Gaupp *et al.* (1982) reported a rather precise value for the 3p $^2P_{1/2}$ state of 16.40 $\pm$0.03 ns, and hence an oscillator strength ($^2P_{1/2} + {}^2P_{3/2}$) of 0.9536 $\pm$0.0016. Theorists soon began to question the precision of this experiment (see Brage *et al.* 1994), just as was done for the Gaupp *et al.* results for lithium. Several recent experiments, with comparable or higher precision, are shown in Table 2.22, together with contemporary ab initio calculations. The agreement among the recent experiments and with ab initio calculations is excellent, and can be summarized as $f_{1/2} = 0.310(1)$, $f_{3/2} = 0.641(1)$ and $f_{tot}(3s \rightarrow 3p) = 0.961(1)$, which is not greatly different from the value of Gaupp *et al.*, but the discrepancy is 4–5 times their error bar.

a.2 The 3s ^{2}S $\rightarrow$ np ^{2}P (n $\geq$ 4) transitions Wiese *et al.* (1969) list oscillator strengths for many higher transitions, based largely on early calculations and experimental, relative oscillator strengths. All but 3s $\rightarrow$ 4p have been dropped by Wiese and Martin (1980). Surprisingly, the oscillator strengths for 3s $\rightarrow$ np ($n = 6$–9) have been retained in recent compilations (Morton, 1991; Verner *et al.* 1994). The f values for $n = 4, 5$ are in good agreement with the relative oscillator strengths obtained by Filippov and Prokofjew (1928), when normalized to

Table 2.22 Recent determinations of oscillator strength for the $3s^2S \rightarrow 3p^2P$ resonance transition in atomic sodium

a. Experiment

Method	$\tau_{1/2}$(ns)[a]	$\tau_{3/2}$(ns)[a]	$f_{1/2}^{b}$	$f_{3/2}^{b}$	f_{total}
BGLS[c]	16.299(21)	16.254(22)	0.319 92(41)	0.640 32(87)	0.960 24(132)
Linewidth[d]		16.237(35)		0.640 98(138)	0.961 18(207)
C$_3$ analysis[e]	16.280(16)	16.230(16)	0.320 29(32)	0.641 26(64)	0.961 55(96)

b. Ab initio theory

MCHF-CCP[f]	0.9603
MCHF-CI[g]	0.9614
Consensus:	0.961(1)

[a]Lifetime of $^2P_{1/2}$ and $^2P_{3/2}$ states.
[b]Oscillator strength of 3s ^{2}S $\rightarrow$ 3p $^2P_{1/2,3/2}$.
[c]Beam-gas laser spectroscopy, decay time of laser-excited fast beam, from Volz *et al.* (1996).
[d]Natural linewidth of 3s ^{2}S $\rightarrow$ 3p $^2P_{3/2}$ transition, from Oates *et al.* (1996).
[e]Contribution of C$_3$ (i.e., $1/r^3$ term) to Na$_2$ potential at large r, from Jönsson *et al.* (1996)
[f]Multi-configuration Hartree–Fock, core polarization, from Brage *et al.* (1994). Relativistic correction given in g, below.
[g]Multi-configuration Hartree–Fock, configuration interaction, from Jönsson *et al.* (1996). From given line strength S, $2f = 303.8\lambda^{-1}S$.

Table 2.23   Oscillator strengths for the 3s ^{2}S$\to$ np ^{2}P ($n > 3$) transitions in atomic sodium

np	Compilations[a]	LN (1977)[b]	McC (1983)[c]	MB (1986)[d]	EBS (1974)[e]
4	0.013 45	0.015 35	0.013 05	0.014 05	0.016(3)
5	0.001 905	0.002 564	0.001 905	0.002 273	0.0025(5)
6	0.000 730 7	0.000 841 4	0.000 6	0.000 723 8	
7	0.000 363 4	0.000 380 5		0.000 318 0	
8	0.000 192 2	0.000 205 2		0.000 167 9	
9	0.000 115 0	0.000 127 6		0.000 101 0	
10	0.000 077[f]	0.000 084 34		0.000 065 94	

[a]Verner *et al.* (1994), taken from Morton (1991).
[b]Lindgåird and Nielsen (1977). Calculation using numerical Coulomb approximation.
[c]McEachran and Cohen (1983). Calculation, frozen-core Hartree–Fock, non-empirical polarization potential.
[d]Martin and Barrientos (1986). Calculation using their preferred semi-empirical dipole operator to take polarization into account.
[e]Erman *et al.* (1974). Experimental values.
[f]Wiese *et al.* (1969). This is the source of $n = 6-9$ in this column.

the current value for the resonance transition, but for higher n, the Filippov values fall below the compilation values. In Table 2.23, we list these oscillator strengths, as well as some more recent calculated values, and limited experimental data. As expected from the dominance of the 3s $\to$ 3p transition, these oscillator strengths are very small. The more recent calculations follow the same pattern as the compilations, based on earlier calculations. The $n = 4$ and $n = 5$ transitions in the compilation have been up-dated, making use of the presumably more accurate calculations of McEachran and Cohen (1983). The oscillator strengths for $n = 6-10$ are within 10% of the calculated values of Lindgård and Nielsen (1977) and lie between the values of Lindgård and Nielsen and Martin and Barrientos (1986). Since the total contribution to f is $\sim$0.017 between $n = 4-10$, and anticipated uncertainty in this quantity is $\sim$10%, we shall retain the compilation values for subsequent computations.

The average value of f_n/E_n declines monotonically with increasing n, and merges smoothly with the value of $\mathrm{d}f/\mathrm{d}E$ at the onset of the continuum. Thereafter, it wanes to a Cooper minimum at $\sim 6.3\,$eV (see, for example, Fig. 2, Barrientos and Martin 1987). This continuum will be considered in more detail in the next section. For the present purposes, we take $\sigma = 0.13\,$Mb at the IP (see below) and linearly interpolate between this value and $\mathrm{d}f/\mathrm{d}E$ at $n = 10$. The contributions in this interval to $S(-2)$, $S(-1)$ and $S(0)$ are, respectively: 0.001 770, 0.000 658 and 0.000 245. The additions to $S(+1)$ and $S(+2)$ are negligible. The contributions of the discrete spectrum to $S(p)$ are summarized in Table 2.24.

b   The continuum

b.1   IP–14.0 eV   Figure 2.16 displays experimental cross sections between the IP and 22 eV obtained by Hudson and Carter (1967; 1968) and four subsequently calculated curves by Chang and Kelly (1975), Butler and Mendoza

Table 2.24 Spectral sums, and comparison with expectation values for atomic sodium. ($S(p)$ in Ry units)

Energy, eV	$S(-2)$	$S(-1)$	$S(0)$	$S(+1)$	$S(+2)$
Discrete					
$2.1037(n = 3)$[a]	40.197	6.215	0.961(1)	0.1486	0.0230
$3.7531-4.9764$[b]	0.2078	0.05891	0.01683	0.00479	0.00133
$(n = 4-10)$					
$4.9764-5.1391$[c]	0.001770	0.000658	0.000245	0.000091	0.000034
$(n = 10 \to$ IP)					
Σ Discrete	40.4066	6.2746	0.9781	0.1535	0.0244
continuum					
IP-6.359[d]	0.0029_6	0.0011_8	0.0004_7	0.0001_9	0.0000_7
$6.359-14.0$[d]	0.0109_5	0.0084_4	0.0067_3	0.0055_2	0.0046_5
$14.0-36.5$[d]	0.0088_2	0.0133_8	0.0218_5	0.0386_3	0.0726_4
2p Resonances					
$30.77-38.71$[e]	0.0024	0.0064	0.0172	0.0460	0.1202
Continuum					
$36.5-53.7$[d]	0.0535	0.1882	0.6659	2.3723	8.4997
2s Resonances					
$66.4, 69.4$[f]	0.0009	0.0044	0.0215	0.1058	0.5205
Continuum					
$53.7-311.7$[d]	0.1416	0.8826	6.5153	58.8100	671.0789
$311.7-1079.1$[d]	0.0008	0.0239	0.7958	29.5225	1237.9348
1s Resonances					
$1075-1088$[g]	–	0.0001	0.0128	1.0139	80.278
Continuum					
$1079.1-3691.7$[h]	0.0001	0.0133	1.5285	195.1164	27976.602
$3691.7-10000$[h]	–	0.0004	0.1576	62.6656	26940.60
10^4-10^{5}[i]	–	–	0.0237	30.4689	53584.9
10^5-10^6	–	–	–	1.8289	29437.0
10^6-10^7	–	–	–	0.0678	10679.5
10^7-10^8	–	–	–	0.0023	3534.1
10^8-10^9	–	–	–	0.0001	1134.0
$10^9-\infty$	–	–	–	–	526.5
Total	40.6286	7.4169	10.7454	382.22	155811.7
Expectation	40.68(14)[j]	7.57^k_8	11.0	389.15[k]	153660.4[l]
values					153764.9[d]
Other values	40.68[n]	7.55[n]	11.0[n]		

[a]Table 2.22.
[b]Table 2.23.
[c]See text.
[d]Table 2.25.
[e]Table 2.26.
[f]Sect. 2.7.1.b.5.
[g]Table 2.28.
[h]Henke *et al.* (1993).
[i]Chantler (1995).
[j]From $\alpha = 24.11(8) \times 10^{-24}$ cm^3, Ekstrom *et al.* (1995).
[k]Fischer *et al.* (1998).
[l]Fraga *et al.* (1976).
[m]Bunge *et al.* (1993).
[n]Kharchenko *et al.* (1997).

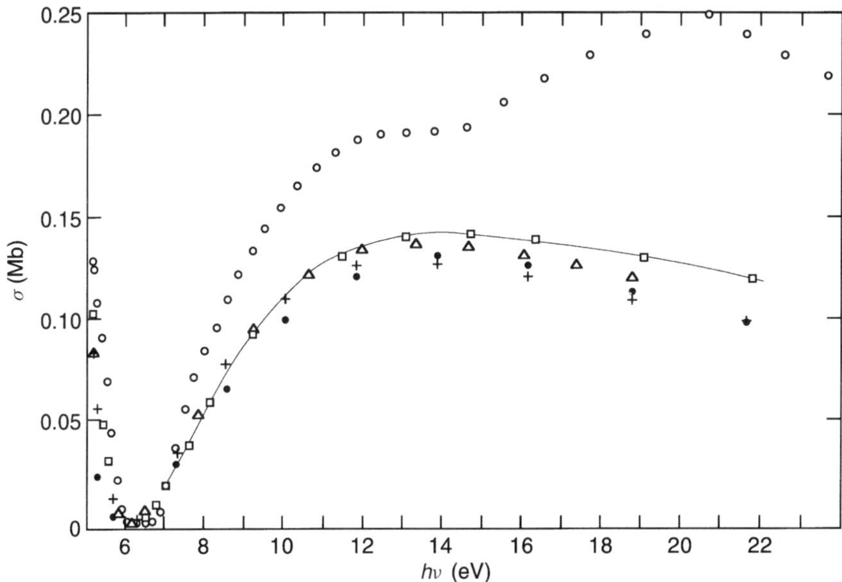

Fig. 2.16   Absolute photoabsorption spectrum of sodium, 5–22 eV. ○ Hudson and Carter (1967); ● Chang and Kelly (1975); + Dasgupta and Bhatia (1985); △ Butler and Mendoza (1983); □ Saha *et al.* (1988)

(1983), Dasgupta and Bhatia (1985) and Saha *et al.* (1988). Although all the curves agree on the approximate location of the Cooper minimum, the calculated curves differ from experiment both below and above this minimum. (A later calculation by the Kelly group (Isenberg *et al.*, 1985) focused on higher energy resonances and noted that the earlier (1975) calculation by Chang and Kelly was more reliable here.) For the descending portion between IP and the Cooper minimum, support for the experimental data of Hudson and Carter can be found in an earlier calculation of Boyd (1964) which arrived at almost the same cross section at the ionization threshold (0.136 Mb length form, 0.126 Mb velocity form) as the experiment (0.13 Mb). An earlier experiment by Ditchburn *et al.* (1953) also supports this value. The most recent of the calculations cited, by Saha *et al.* (1988), extrapolates to ∼0.11 Mb at the IP. However, on the high energy side of the Cooper minimum all calculations (including Boyd's) are in fair agreement with one another, but are substantially lower than the data of Hudson and Carter. Various authors (Marr and Creek, 1968; Chang, 1975) have speculated that the experimental values may be influenced by much larger dimer cross sections. From the available evidence, we choose to follow the experimental curve from the IP to the Cooper minimum, and the calculated values of Saha *et al.* (1988) to higher energy. The selected values are given in Table 2.25. The values of $S(p)$ calculated from these fitted functions are recorded in Table 2.24.

Table 2.25 Selected values of photoabsorption cross sections for atomic sodium, IP-K edge, smooth continuum

hν, eV	σ, Mb	hν, eV	σ, Mb	hν, eV	σ, Mb
5.140	0.130[a]	13.0	0.1406[b]	53.7	9.10[e]
5.167	0.126[a]	14.0	0.1416[b]	61.96	9.0[e]
5.277	0.110[a]	15.0	0.141[b]	72.4	6.949[f]
5.391	0.092[a]	16.0	0.1397[b]	91.5	5.727[f]
5.511	0.070[a]	17.0	0.137[b]	108.5	4.505[f]
5.636	0.045[a]	18.0	0.1335[b]	114.0	4.16[f]
5.767	0.022[a]	19.0	0.1305[b]	132.8	3.22[f]
5.905	0.008[a]	20.0	0.127[b]	151.1	2.57[f]
6.049	0.001[a]	21.0	0.1232[b]	183.3	1.733[f]
6.359	0.000[a]	21.769	0.1203[b]	192.6	1.569[f]
6.529	0.0026[b]	24.73	0.104[c]	220.1	1.164[f]
6.799	0.0091[b]	28.14	0.093[c]	277.0	0.683[f]
7.079	0.0184[b]	31.81	0.083[c]	311.7	0.531[f]
7.619	0.0389[b]	35.76	0.074[c]	392.4	0.284[f]
8.159	0.0595[b]	38.1	0.41[d]	452.2	0.197[f]
8.5	0.071[b]	40.0	0.95[d]	511.3	0.144[f]
9.0	0.086[b]	42.0	2.04[d]	524.9	0.134[f]
9.249	0.0938[b]	44.3	3.41[d]	572.8	0.1077[f]
9.5	0.10[b]	45.0	4.57[e]	637.4	0.0825[f]
10.0	0.11[b]	45.5	4.93[e]	676.8	0.0699[f]
10.5	0.118[b]	48.0	6.13[e]	705.0	0.0626[f]
11.0	0.126[b]	49.3	7.18[f]	776.2	0.0485[f]
11.5	0.132[b]	50.0	6.71[e]	851.5	0.0376[f]
12.0	0.1365[b]	51.3	7.60[e]	929.7	0.0296[f]
				1011.7	0.0233[f]
				1041.0	0.0214[f]
				1079.1	0.0196 (K-edge)

[a]Hudson and Carter (1967).
[b]Saha *et al.* (1988).
[c]Dasgupta and Bhatia (1985), normalized to b.
[d]Baig *et al.* (1994), normalized at 45.0 eV to e.
[e]Codling *et al.* (1977).
[f]Henke *et al.* (1993).

b.2 14.0–36.5 eV As can be seen partly in Fig. 2.16 and more extensively in Table 2.25, the photoabsorption cross section declines monotonically from a plateau at 14 eV to a minimum at $\sim$36.5 eV. Data between 24.73–35.76 eV are taken from Dasgupta and Bhatia (1985), Table VII, col. 5, normalized to the calculated values of Saha *et al.* (1988) at 21 eV. Autoionizing resonances begin to appear at $\sim$31 eV and $\sim$36–38 eV, signaling the onset of excitation from the inner (2p) shell. The oscillator strengths of these resonances are treated separately below. The underlying continuum due to 2p ionization begins to increase at about 36.5 eV, as seen in Baig *et al.* (1994), Fig. 1. See also Wolff *et al.* (1972), Fig. 2.

b.3 Autoionizing resonances, $2p^6 3s \rightarrow 2p^5 3s$ nl Beginning at 30.768 eV, corresponding to the excitation $2p^6 3s \rightarrow 2p^5 3s^2$, and proceeding to the series limits ($2p^5 3s\ ^3P$, ~38 eV; $2p^5 3s\ ^1P$, 38.46 eV) a plethora of autoionizing peaks have been observed (Wolff $et\ al.$, 1972; Baig $et\ al.$, 1994). In addition, there is a prominent doublet beyond this limit at 38.556/38.707 eV, which has been assigned to the double excitation $2p^5 4s^2$, $^2P_{3/2} - {}^2P_{1/2}$. Here, we attempt to estimate the oscillator strengths of these transitions. Most of the resonances appear to be sharp. In those cases, we assume a triangular peak shape, with instrumental line width. One exception is the $2p^5 3s$ ($^1P_1$) 7d resonance at 324.823 Å $\equiv$ 38.170 eV, which has a broad, asymmetric shape. Baig $et\ al.$ (1994) have fitted this shape and extracted the Fano parameters q and Γ. In this case, we estimate the 'excess' oscillator strength f_{xs} using the equation given in Sect. 2.2.1.b.2.

Wolff $et\ al.$ provide an overview spectrum (their Fig. 1) with an absolute cross section scale. They give a more detailed spectrum (their Fig. 2) in arbitrary units which seems to match the absolute scale in Fig. 1. Baig $et\ al.$ present densitometer traces in arbitrary units. These have been converted to absolute cross sections by choosing regions of the underlying continuum, e.g. 45.0 eV in their Fig. 1, 38.5 eV in Fig. 4, and normalizing to the corresponding continuum cross section given in Table 2.25. The oscillator strengths estimated by these procedures are listed in Table 2.26. The domain from 30.77–37.0 eV, and the double excitations at ~38.6 eV, are based on the data of Wolff $et\ al.$, while the 37.9–38.5 eV data are from Baig $et\ al.$.

The sum of the contributions of these resonances is given in Table 2.24. The estimated oscillator strengths may be uncertain by a factor 2, but their total contribution to each of the $S(p)$ in Table 2.24 is not large enough to significantly affect the final sums. We have neglected higher two-electron excited states between ~39–46 eV (see Wolff $et\ al.$ (1972), Fig. 3: Baig $et\ al.$ (1994), Figs. 5 and 6). These features either have asymmetric peak shapes ($q \approx |0.5|$) or are window resonances, which implies very small f_{xs}.

b.4 36.5–53.7 eV Beginning at about the $2p^5 3s$ edge, the photoabsorption cross section increases (see Baig $et\ al.$, 1994, Fig. 1) to a maximum at about 60 eV (Codling $et\ al.$, 1977). The densitometer trace of Baig $et\ al.$ (1994) is normalized at 45 eV to the (digitized) absolute cross section of Codling $et\ al.$ (1977). The selected values appear in Table 2.25. The corresponding contributions to $S(p)$ are listed in Table 2.24.

b.5 Autoionizing resonances, $2s^2 2p^6 3s \rightarrow 2s 2p^6 3s$ np This resonance series signals the excitation and ultimate ionization of an electron from the 2s shell. Early spectra are given by Wolff $et\ al.$ (1972), Fig. 4 and Codling $et\ al.$ (1977), Fig. 6. A more recent scan can be seen in LaVilla $et\ al.$ (1981), Fig. 1. The dominant peak here is the first one, $2s 2p^6 (3s3p\ ^3P)$, ^2P at 66.37 ± 0.03 eV. From $q = -2.6$, $\Gamma = 0.20$ eV and estimates of $\sigma \approx 8.5$ Mb, $\rho^2 \sim 1/8$,

Table 2.26   Estimated[a] oscillator strengths of autoionizing resonances, 30.77–38.5 eV

Energy, eV	Upper state	f
30.768	$2p^5 3s^2$, $^2P_{3/2}$	0.001 75
30.934	$2p^5 3s^2$, $^2P_{1/2}$	0.001 32
35.768	$2p^5 3s(^3P)4s$, $^2P_{3/2}$	0.0006_8
35.790	4s, $^2P_{1/2}$	0.0002_8
35.985	$2p^5 3s(^3P)3d$	0.0014
36.018	$2p^5 3s(^3P)3d$	0.0004
36.056	$2p^5 3s(^3P)3d$	0.0008_7
36.129	$2p^5 3s(^3P)3d$	0.0005_3
36.217	?	0.0007_9
36.906	$2p^5 3s(^3P)4d$	0.0008_1
36.929	$2p^5 3s(^3P)4d$	0.0006_5
37.070	$2p^5 3s(^3P)4d$	0.0002_7
37.273	$2p^5 3s(^3P)5d$	0.0002_5
37.292	$2p^5 3s(^3P)5d$	0.0002_1
37.497	?	0.0002_5
37.699	?	0.0002_1
37.994	$(^1P_1)7s$	0.0002
38.128	$(^1P_1)8s$	0.0001_4

Energy, eV	Upper state	f
38.212	$(^1P_1)9s$	0.0000_9
38.267	10s	0.0000_6
38.305	11s	0.0000_5
38.333	12s	0.0000_3
38.354	13s	0.0000_3
38.370	14s	0.0000_2
38.384	15s	0.0000_2
37.865	$2p^5 3s(^1P_1)5d$	0.0002
38.051	6d	0.00049_5
38.170	7d	0.0003
38.234	8d	0.0002_6
38.283	9d	0.0001_4
38.317	10d	0.0001
38.342	11d	0.0000_8
38.361	12d	0.0000_6
38.376	13d	0.0000_5
38.387	14d	0.0000_4
38.397	15d	0.0000_3
38.405	16d	0.0000_3
38.410	17d	0.0000_2
38.415	18d	0.0000_2
38.421	19d	0.0000_2
38.425	20d	0.0000_1
38.556	$2p^5 4s^2$, $^2P_{3/2}$	0.0026
38.707	$^2P_{1/2}$	0.0013

[a]Based on data of Wolff *et al.* (1972) for 30.77–37.7 eV and 38.6 eV doublet, Baig *et al.* (1994) for 37.9–38.4 eV. See text for details of estimation.

we calculate $f \approx 0.0175$. The only other significant peak, $2s2p^6(3s4p\ ^3P)$, ^2P occurs at 69.4 eV, with an estimated oscillator strength of 0.004. All other features in this region are weaker and more difficult to estimate.

b.6   53.7–1079.1 eV   In Fig. 2.17, we compare the data of Codling *et al.* (1977) with the compilation of Henke *et al.* (1993). The agreement is only fair in the region of overlap, 50–245 eV. The Henke data fall on a smooth curve, about which the Codling data oscillate. Codling *et al.* describe their overall error to be 20–25%, largest at the lowest and highest energies. There is also fluctuation in the region of the peak. At 50 eV, there is good agreement. Our choice here is to transfer from the Codling data at 62 eV to the Henke data at 72.4 eV, which merges the data sets without an abrupt discontinuity. The data of Henke *et al.* can then be utilized up to the K-edge, at 1079.1 eV (Banna *et al.*, 1978; Tuilier *et al.*, 1982). For enhanced accuracy, this extended energy domain is fitted in two segments, 53.7–311.7 eV and 311.7–1079.1 eV, each with a 4-term polynomial

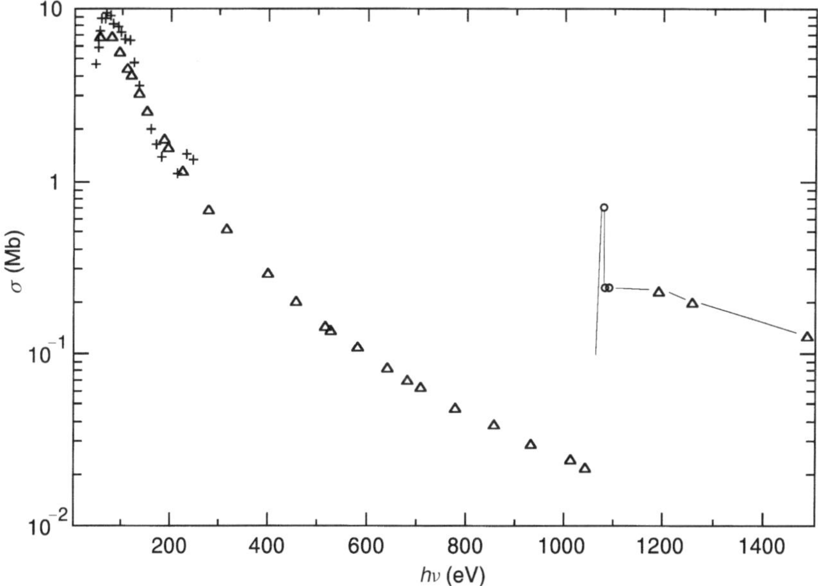

Fig. 2.17 Absolute photoabsorption spectrum of sodium, 120–1500 eV. + Codling *et al.* (1977); △ Henke *et al.* (1993); ○ Yavna *et al.* (1986a)

Table 2.27 Coefficients of the polynomial $df/dE = ay^2 + by^3 + cy^4 + dy^5$ fitted to data at various energies[a]

Energy range, eV	a	b	c	d
53.7–311.7	307.4663	−191.325	−32 305.7	149 599.7
311.7–1079.1	−17.4	35 158.05	−2 197 604	61 866 490
1079.1–3691.7	2494.821	−2 164 279	884 106 572	−101 454 776 511
3691.7–10 000	−35.4965	613 534.4	−88 069 064	11 161 381 369

[a]df/dE in Rydberg units, $y = B/E$, $B = \mathrm{IP} = 5.139\,077$ eV.

in $y = B/E$. The coefficients of the polynomials appear in Table 2.27. Integration of the df/dE function and its moments provides the contributions to $S(p)$, which are recorded for each domain in Table 2.24.

b.7 Resonances around K-edge, 1075–1088 eV These resonances can be classified as single excitations $[\rightarrow$ 1s $2s^2$ $2p^6$ 3s (^{1}S,^{3}S) np] and double excitations $[\rightarrow$ 1s $2s^2$ $2p^6$ nl,n′l′] with l = 0, 1 and l′ = 1, 2 in the latter case. LaVilla (1979) obtained an early photoabsorption spectrum with a resolution of 0.8 eV; a spectrum with better resolution (0.4 eV) and improved signal-to-noise was subsequently presented by Tuilier *et al.* (1982). A calculated spectrum that appears to fit the latter data very well for single excitations, and in addition provides an absolute cross section scale lacking in the experimental data was obtained by Yavna *et al.* (1986a). The energies, assignments and oscillator strengths for the single excitations are listed in Table 2.28a.

The double excitation spectrum has been calculated by Sukhorukov *et al.* (1987). See also Yavna *et al.* (1986b). The calculations appear to agree well with the spectrum of Tuilier *et al.* (1982), and provide absolute cross sections. The energies, assignments and oscillator strengths for these double excitations are listed in Table 2.28b. The total contributions of these resonances to $S(p)$ are recorded in Table 2.24.

b.8 Post K-edge continuum, 1079.1–10 000 eV In Fig. 2.17, some data points from the calculation of Yavna *et al.* (1986a) are compared with data from the compilation of Henke *et al.* (1993) between 1079.1–1500 eV. The agreement is fairly good, and provides some justification for the use of the Henke data to 10 000 eV. The data have been fitted by two 4-term polynomials, one spanning the range 1079.1–3691.7 eV, the other 3691.7–10 000 eV, for enhanced accuracy. The coefficients of the polynomials appear in Table 2.27, and the contribution of each portion to $S(p)$ is recorded in Table 2.24.

b.9 $10^4 - 10^5$ eV We use the calculated cross sections of Chantler (1995).

Table 2.28 Near K-edge resonances in atomic sodium

a. Single excitations[a]

Energy, eV	Assignment	Oscillator strength
1074.95	$1s2s^2\,2p^6 3s(^1S)3p$	0.006 67
1076.47	$(^3S)3p$	0.000 70
1078.17	$(^3S)4p$	0.001 45
1078.41	$(^1S)4p$	0.000 59
1078.95	$(^3S)5p$	0.000 55
1079.05	$(^1S)5p$	0.000 20

b. Double excitations[b]

Energy, eV	Assignment	Oscillator strength
1081.4	$1s\ldots4s\ (^3S)3p$	0.000 84
1081.5	$1s\ldots4s\ (^1S)3p$	0.000 58
1083.1	$1s\ldots5s\ (^3S)3p$	0.000 22
1083.5	$1s\ldots6s\ (^3S)3p$	0.000 06
1083.6	$1s\ldots6s\ (^1S)3p$	0.000 02
1083.8	$\ldots4p\ 3d$	0.000 12
1085.4	$\ldots4p\ 3d$	0.000 33
1085.7	$\ldots4s\ 4p$	0.000 24
1087.1	$\ldots4s\ 5p$	0.000 12
1088.0	$\ldots4s\ 6p$	0.000 11

[a]Yavna *et al.* (1986a).
[b]From Sukhorukov *et al.* (1987).

2.7.2 The analysis

The static electric dipole polarizability (α) of sodium has been measured recently to 0.3% accuracy by Ekstrom *et al.* (1995) using an elegant technique, atom beam interferometry influenced by an electric field. The value they report, $\alpha = 24.11(6)(6) \times 10^{-24}$ cm^3, is equivalent to $S(-2) = 40.68(14)$, where the experimental uncertainties, statistical and systematic, have been summed in quadrature. This value is in excellent agreement with the spectral sum, $S(-2) = 40.6286$. The resonance transition accounts for 98.9% of the total. Hence, the oscillator strength for the resonance transition (see Table 2.22) and the current value of α are consistent to a high degree, but conversely the $S(-2)$ sum rule is relatively insensitive to the remainder of the spectrum.

The spectral sum for $S(0)$, 10.746, lies 2.3% below the Thomas–Reiche–Kuhn requirement of 11. This deficit cannot be attributed to the discrete spectrum, because of the excellent agreement with $S(-2)$, and also because the total oscillator strength up to the onset of inner shell excitations is essentially unity, as expected for 3s excitation plus ionization. Nor is this deficit to be attributed to K-shell ionization, since the value from Table 2.24 is $\sim$1.69. Kharchenko *et al.* (1997) show that subtraction of the 'forbidden' 1s–2p oscillator strength (2×0.18) from 2 yields an expected oscillator strength beyond the K-shell of 1.64, slightly lower than our value. Clearly the major source of the deficit lies in the continuum between 36.5–1079.1 eV, the bulk of which exists between 53.7–311.7 eV. The data sources here are Codling *et al.* (1977) and Henke *et al.* (1993), seen in Fig. 2.17. The contribution to $S(0)$ is very sensitive to the cross section values near the peak ($\sim$60 eV). The values of Codling *et al.* lie higher than those of Henke *et al.* in this region, by nearly 2 Mb. We chose the Henke values above 62 eV for reasons based on smoothness and continuity. An increase in our chosen values by 2 Mb over a 20 eV width would increase $S(0)$ by 0.36, greater than our deficit, attesting to the sensitivity of $S(0)$ to the 60–80 eV region.

This sensitivity carries over to the analysis of $S(-1)$. The matrix elements determining its value in an ab initio calculation (see Reference Table) depend upon the degree of correlation in the wavefunction. In the case of lithium, where a highly correlated Hylleraas type wavefunction was available, the resulting $S(-1)$ was 2.3% lower than that from a Hartree–Fock wavefunction. The corresponding Hartree–Fock wave function for sodium yields $S(-1) = 8.206$ (Fraga *et al.*, 1976). Recent results from a highly correlated wavefunction for sodium give $S(-1) = 7.57_8$, 8.3% lower (Fischer *et al.*, 1998). The current spectral analysis arrives at $S(-1) = 7.41_7$, 2.1% lower than the result of Fischer *et al.* (1998). Applying the same assumptions used in discussing the deficit in $S(0)$, i.e., 2 Mb increase over 20 eV at $h\nu = 60$ eV, would increase $S(-1)$ to 7.5, accounting for half the deficit relative to Fischer *et al.*.

Kharchenko *et al.* (1997) have performed an analysis similar to the present one for $S(-2)$, $S(-1)$ and $S(0)$. They relied more heavily on calculated values of cross sections, in most cases providing shell-wise partial cross sections, which were summed. They achieved perfect agreement for $S(-2)$ and $S(0)$, and argued that $S(-1) = 15.1$ a.u., or 7.55 Ry units. The present results, when corrected for the presumed 2 Mb deficit at 60–80 eV, are also in good agreement, although the uncertainty is sufficient to encompass the value of Fischer *et al*. (The value attributed to Fischer *et al*. by Kharchenko *et al*. as 'Note added in Proof', $S(-1) = 15.47$ a.u. $= 7.73_5$ Ry units, was an early draft. The value cited here is the correct one.)

As can readily be seen from a glance at Table 2.24, the $S(+2)$ spectral sum is predominantly dependent upon values above the K-edge, while $S(+1)$ acquires $\sim 3/4$ of its value in that domain. The presumed deficit in cross section at 60–80 eV would increase $S(+1)$ by ~ 1.6 above its spectral sum of 382.2. The Hartree–Fock value for $S(+1)$ is 376.15 (Fraga *et al*., 1976). The calculated value of $S(+1)$ is also dependent upon the correlatedness of the wavefunction. Recent multiconfiguration Hartree–Fock calculations by Fischer *et al*. (1998) can be used to calculate $S(+1) = 389.15$, which exceeds the spectral sum by $\sim 1.3\%$.

The value of $S(+2)$, which is essentially the electron density at the nucleus, is much less dependent on correlation. For lithium, Hartree–Fock and Hylleraas-type calculations agreed to within 0.2%. The present spectral sum for sodium exceeds the Hartree–Fock value by 1.3%, which lends credence to the Henke *et al*., Chantler and Bethe–Salpeter cross sections in this instance.

2.8 Atomic Chlorine

2.8.1 The data

The electronic ground state of atomic chlorine may be written as $1s^2 2s^2 2p^6 3s^2 3p^5$, $^2P_{3/2}$. Its spin-orbit partner, $^2P_{1/2}$, is excited by 882.36 cm$^{-1} \equiv 0.109$ eV, and is not significantly populated in a quasi-thermal, room-temperature experiment. The ionization potential corresponds to $\ldots 3s^2 3p^4$ ($^3P_2$), and occurs at 104 591.0 ± 0.3 cm$^{-1} \equiv 12.96763$ ± 0.00004 eV (Radziemski and Kaufman, 1969). The accompanying spin-orbit states are $^3P_1$ (13.05392 ± 0.00005 eV) and $^3P_0$ (13.09118 ± 0.00005 eV). The same ionic configuration gives rise to the excited states $^1D_2$ (14.41249 ± 0.00004 eV) and $^1S_0$ (16.42406 ± 0.00004 eV) (Radziemski and Kaufman, 1974).

a *The valence shell spectrum*

Valence shell excitation can be expected to give rise to various Rydberg series of the type $\ldots 3s^2 3p^5$, $^2P_{3/2} \rightarrow \ldots 3s^2 3p^4$ ($^3P, ^1D, ^1S$)ns, nd. Those with the strongest transitions conserve spin. Most of the transitions involving ^1D and ^1S cores occur above the adiabatic ionization potential, and appear as autoionization features (Ruščić and Berkowitz, 1983). These appear to be relatively 'pure' states with little mixing, since regular series with nearly constant quantum defects and shapes within a series are observed. Before undertaking the analysis of the contributions of these higher-energy series to the oscillator strength, we turn to the transitions converging on the ground state, ^3P.

a.1   ns series approaching ^{3}P   Here, the information is sparse, confusing and sometimes contradictory. Only the first transition doublet, $3s^2 3p^5$, $^2P_{3/2} \rightarrow$ $3s^2 3p^4$ $(^3P)4s$, $^2P_{3/2,1/2}$ appears to be relatively pure. For the $^2P_{3/2} \rightarrow ^2P_{3/2}$ transition at 1347.24 Å, experimental oscillator strengths of 0.10 ±0.03 (Clyne and Nip, 1977), 0.109 ±0.010 (Schwab and Anderson, 1982) and 0.153 ±0.011 (Schectman et al., 1993) have been reported. Some support for the recent (and higher) value is provided by contemporary configuration interaction calculations. Ojha and Hibbert (1990) used a large multiconfiguration expansion and the CIV3 code to obtain $f = 0.132$, while Biémont et al. (1994) used the SUPERSTRUCTURE code, with the most important configuration interaction and incorporated relativistic effects, yielding $f = 0.147$. (Here, only the length gauge results are given; Biémont et al. note that their length/velocity values are closer than those of Ojha and Hibbert.) A more recent, but less extensive calculation by Lavin et al. (1997) gives $f \sim 0.13$. We adopt $f \sim 0.14$, roughly within the error limits of the highest experimental value, giving some weight to the other experiments and calculations.

For the companion $^2P_{3/2} \rightarrow ^2P_{1/2}$ transition, there appears to be only one experimental value, 0.028 ±0.006 (Clyne and Nip, 1977), but the calculated values (0.0264, Ojha and Hibbert (1990); 0.0299, Biémont et al. (1994)) are close. We shall combine the f values for the $^2P_{3/2} \rightarrow ^2P_{3/2,1/2}$ transitions as the 4s value in attempting to construct a histogram.

The $3p^4(^3P)5s$ configuration mixes with the $3p^4(^3P)3d$, as discussed by Schectman et al. and Biémont et al. Radziemski and Kaufman (1969) locate five energy levels with the $(^3P)5s$ configuration, which they describe in J_cK notation as $2[2]_{5/2}$, $2[2]_{3/2}$, $1[1]_{3/2}$, $1[1]_{1/2}$ and $0[0]_{1/2}$. Ojha and Hibbert describe the same levels in LS notation as $^4P_{5/2}$, $^2P_{3/2}$, $^4P_{3/2}$, $^4P_{1/2}$, and $^2P_{1/2}$, respectively. Neither J_cK nor LS, but rather intermediate coupling appears to be necessary, according to Radziemski and Kaufman (1969) and Biémont et al. Both Ojha and Hibbert and Biémont et al. calculate oscillator strengths to these five upper levels. The sums of the respective f values differ by almost an order of magnitude between Biémont et al. (0.031) and Ojha and Hibbert (0.0044). This is reflected primarily in transitions to the $J = 5/2$ state, where the respective values are 0.0166 and 0.001 12. There are no experimental measurements available to distinguish between these calculated quantities. In Fig. 2.18 (an abbreviated histogram of the $(^3P)ns$ transitions), both sums (Biémont et al.; Ojha and Hibbert) are sketched in for $n = 5$. From this limited information, we surmise that the slope of df/dE is negative and approaches the IP with a value of about 0.010 ±0.005$(eV)^{-1}$, or about 1 Mb. We can infer from continuum cross sections (see below) that the cross section at the ^3P threshold is about 20 Mb. (Here, as in Fig. 2.18, we take a weighted average of $^3P_{2,1,0}$ for the ^3P threshold and sum the contributions to the cross section.) Thus, the bulk of the oscillator strength approaching the ^3P threshold should derive from $3p \rightarrow$ nd-like transitions.

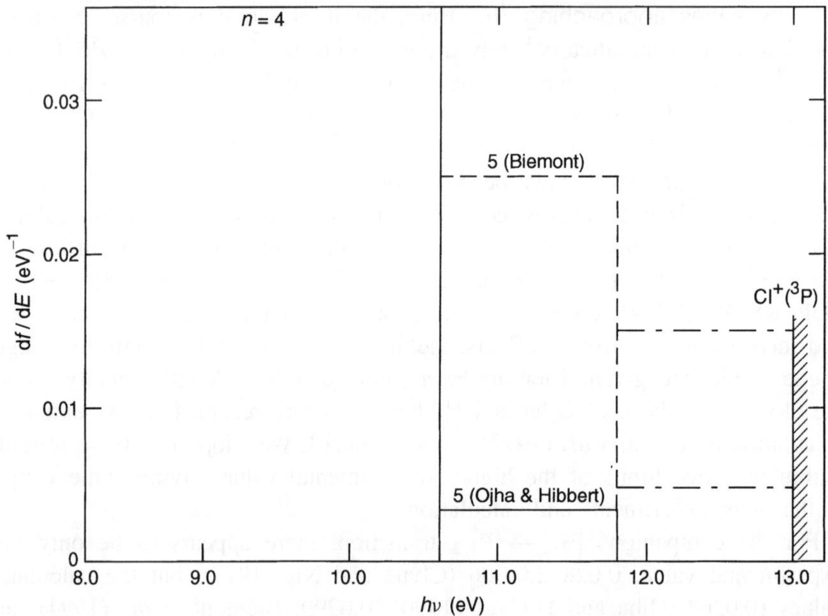

Fig. 2.18   Histogram for the series $\dots 3p^5$, $^2P_{3/2} \rightarrow \dots 3p^4$ (^3P)ns in atomic chlorine

a.2 nd series approaching ^3P As the prior section has forewarned, the (^3P)3d configuration mixes with (^3P)5s and is split. Thirteen states ranging from $^4F_{5/2}$ to $^2P_{1/2}$ have been calculated to be accessible from the ground state. Of these, seven may be considered to have significant oscillator strength. Two have been measured by Schectman *et al.* (1993): $^2F_{5/2}$, at 1088.062 Å, with $f = 0.081 \pm 0.007$, and $^2D_{5/2}$ at 1097.369 Å, with $f = 0.0088 \pm 0.0013$. Biémont *et al.* (1994) believe that their calculations support the experimental findings of Schectman *et al.* at the wavelengths indicated, but question the labels of the states, which hark back to the assignments of Radziemski and Kaufman (1969). We are concerned here with the mapping of the oscillator strength distribution, rather than the assignments of individual transitions. Toward this end, it is satisfying to note that the summed oscillator strength of all thirteen transitions is 0.156 (Biémont *et al.*, 1994) and 0.142 (Ojha and Hibbert, 1990) while the sum of only two experimental transitions is 0.0898. We adopt $f_{avg} = 0.149$ for the sum of all transitions to (^3P)3d.

Transitions to the split levels of the (^3P)4d, 5d and 6d configurations can be found in the compilation of Verner *et al.* (1994), which are taken from the compilation of Morton (1991), which in turn are based on calculations by Kurucz and Peytremann (1975). Their reliability is dubious, but they have been included as alternatives in the abbreviated histogram of Fig. 2.19. We can, however, establish approximately the value of df/dE for the nd series at the ^3P continuum, since we have previously concluded that the observed cross section is predominantly attributable to this series. With this limited information, we have estimated plausible upper and lower limits to the $S(p)$ for the 'ns' and 'nd' series converging on ^{3}P, and record them in Table 2.29. The range of uncertainty is modest for the 'ns' series, but substantial for the 'nd' series.

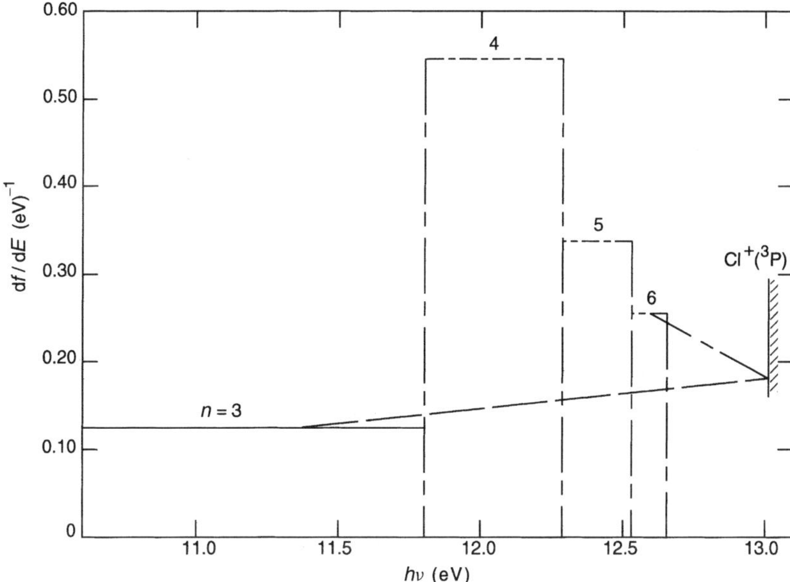

Fig. 2.19 Histogram for the series $\ldots 3p^5$, $^2P_{3/2} \rightarrow \ldots 3p^4$ (^{3}P)nd in atomic chlorine

Table 2.29 Contributions of the $(3p)^{-1}$ spectrum to $S(p)$ sums in atomic chlorine[a]

	$S(-2)$	$S(-1)$	$S(0)$	$S(+1)$	$S(+2)$
1. $3p^4(^3P)$ns					
4[b]	0.364	0.248	0.168	0.114	0.007
5[c]	0.0062–0.0443	0.0052–0.371	0.0044–0.0310	0.0037–0.0259	0.0032–0.0216
6–∞[c]	0.0138–0.0413	0.0120–0.0361	0.0106–0.0317	0.0093–0.0278	0.0082–0.0245
Σ	0.3840–0.4496	0.2652–0.3212	0.1830–0.2307	0.1270–0.1677	0.0884–0.1231
2. $3p^4(^3P)$nd					
3[d]	0.2152	0.179	0.149	0.1239	0.1031
4–∞[e]	0.3281–0.6420	0.2935–0.5775	0.2630–0.5198	0.2360–0.4682	0.2121–0.4219
Σ	0.5433–0.8572	0.4725–0.7565	0.4120–0.6688	0.3599–0.5921	0.3152–0.5250
3. $3p^4(^1D)$ns					
4[f]	0.133 15	0.102 06	0.078 24	0.0600	0.0460
5[g]	0.007 47	0.007 02	0.0066	0.0062	0.0058
6[h]	0.0027	0.0027	0.0027	0.0027	0.0027
7[h]	0.0014	0.0014	0.0014	0.0014	0.0014
8[h]	0.000 75	0.000 77	0.000 795	0.0008	0.0008
9–∞	0.0018	0.0019	0.0020	0.0021	0.0022
Σ	0.1473	0.1159	0.0917	0.0732	0.0589

Table 2.29 (*Continued*)

	$S(-2)$	$S(-1)$	$S(0)$	$S(+1)$	$S(+2)$
4. $3p^4(^1D)nd$ $(^2P+^2D)$					
3^i	(0.1006)	(0.094)	(0.087)	(0.081)	(0.075)
4^h	0.0372	0.0367	0.0362	0.0357	0.0353
5^h	0.0200	0.0203	0.0206	0.0209	0.0212
6^h	0.0114	0.0118	0.0121	0.0124	0.0128
$7-\infty$	0.0282	0.0295	0.0308	0.0323	0.0337
Σ	0.1974	0.1923	0.1867	0.1823	0.1780
5. $3p^4(^1D)nd(^2S)$					
3^j	0.0104–0.1917	0.0097–0.1739	0.009–0.1577	0.0084–0.1430	0.0078–0.1297
4^h	0.003 81	0.003 80	0.003 79	0.003 78	0.003 77
5^h	0.002 68	0.002 74	0.002 79	0.002 85	0.002 90
6^h	0.001 72	0.001 77	0.001 83	0.001 89	0.001 95
$7-\infty$	0.005 41	0.005 66	0.005 93	0.006 20	0.006 49
Σ	0.0240–0.2053	0.0237–0.1879	0.0233–0.1720	0.0231–0.1577	0.0229–0.1448
6. Continuum, $^3P_{avg} - {}^1D$, 13.010 13–14.412 50 eV[b]					
	0.3253	0.3286	0.3324	0.3360	0.3403
7. $3p^4(^1S)ns(^2S)$					
3^k	0.0285	0.0264	0.0244	0.0226	0.0209
4^h	0.0047	0.0051	0.0055	0.0060	0.0065
5^h	0.0022	0.0025	0.00287	0.0033	0.0037
$6-\infty$	0.0034	0.0040	0.0047	0.0056	0.0066
Σ	0.0388	0.0380	0.0375	0.0375	0.0377
8. $3p^4(^1S)nd(^2D)$					
3^h	0.012	0.013	0.014	0.015	0.016
4^h	0.0097	0.0110	0.0125	0.0142	0.0161
5^h	0.0041	0.0047	0.0055	0.0064	0.0074
6^h	0.0020	0.0023	0.00273	0.0032	0.0038
$7-\infty$	0.0040	0.0048	0.0057	0.0068	0.0082
Σ	0.0318	0.0358	0.0404	0.0456	0.0515
9. Continuum, $^1D-{}^1S$, 14.412 50–16.424 07 eV					
	0.4750	0.5372	0.6084	0.6900	0.7836
10. $3s3p^6(^2S)^l$					
	0.0230	0.0179	0.014	0.0109	0.0085

[a]$S(p)$ in Ry units.

[b]Several experimental sources; see text for choice.

[c]Lower limit, Ojha and Hibbert (1990), upper limit, Biémont *et al.* (1994).

[d]Combination of experiments and calculations; see text.

[e]Lower limit, extrapolation; upper limit, calculation of Kurucz and Peytremann (1975), cited by Morton (1991).

[f]Morton (1991).

[g]Estimated from relative intensities in Radziemski and Kaufman (1969).

[h]From Fig. 1, Ruščić and Berkowitz (1983).

[i]Estimated from $(n^*)^{-3}$ dependence.

[j]Lower limit estimated from $(n^*)^{-3}$ behavior; upper limit, calculation of Kurucz and Peytremann (1975), cited by Morton (1991).

[k]Calculation of Kurucz and Peytremann (1975), cited by Morton (1991).

[l]From Ojha and Hibbert (1990), geometric mean of length and velocity formulations.

a.3 ns series approaching ^{1}D The $3s^2 3p^4(^1D)4s$ configuration gives rise to $^2D_{5/2}$ (1188.7742 Å) and $^2D_{3/2}$ (1188.7515 Å). Wiese and Martin (1980) provide data for $^2D_{5/2}$ ($f = 0.074$) and $^2D_{3/2}$ ($f = 0.0057$). Verner *et al.* (1994) take their values from Morton (1991), who weighs experimental data of Clyne and Nip (1977) and Schwab and Anderson (1982) as well as calculated values from Ojha and Hibbert (1990) and selects $f = 0.07277$ ($^2D_{5/2}$) and $f = 0.005469$ ($^2D_{3/2}$). The later calculations of Biémont *et al.* (1994) give $f_{avg} = 0.0714$ ($^2D_{5/2}$) and $f = 0.0055$ ($^2D_{3/2}$). The agreement of several sources is fairly good. We accept the Morton selection, which gives $f = 0.07824$ for the sum of $^2D_{5/2}$ and $^2D_{3/2}$ transitions.

We are unable to find experimental or calculated oscillator strengths to $(^1D)5s$. Radziemski and Kaufman (1969) list these transitions at 969.919 Å ($^2D_{5/2}$) and 969.912 Å ($^2D_{3/2}$). We make the crude assumption that the ratio of oscillator strengths to $(^1D)4s$ and $(^1D)5s$ are proportional to the intensities given by Radziemski and Kaufman, and thereby estimate $f \cong 0.0066$ for $(^1D)5s$. For $(^1D)6s$, 7s, 8s, we utilize Fig. 1 of Ruščić and Berkowitz (1983), which has approximately the correct normalization. The sharp peaks are assumed to be triangular, with a half-width equal to the experimental resolutions, 0.28 Å. The areas yield $f = 0.0027$, 0.0014 and 0.000795 for $(^1D)6s$, 7s and 8s, respectively. The corresponding histogram appears in Fig. 2.20, and the contributions of this series to $S(p)$ are given in Table 2.29.

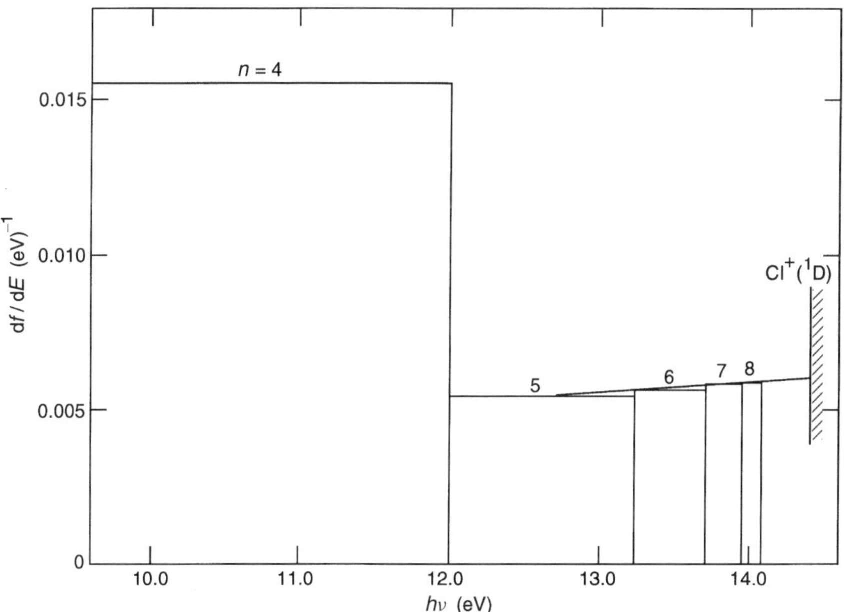

Fig. 2.20 Histogram for the series $\ldots 3p^5$, $^2P_{3/2} \to \ldots 3p^4$ $(^1D)ns$ in atomic chlorine

a.4   nd (^{2}P,^{2}D) series approaching ^{1}D   The dominant features in the spectrum (Fig. 1) of Ruščić and Berkowitz (1983) constitute a broad, dispersion-like series converging to ^{1}D, having a shape similar to the nd series in Ar converging to ^{2}P$_{1/2}$. It was originally thought to be a (^{1}D) nd ^{2}P series, but shortly thereafter was recognized as the unresolved (^{1}D)nd (^{2}P,^{2}D) series (Ruščić *et al.*, 1984). The first three members, (^{1}D)4d,5d,6d had previously been fitted to the Fano line-profile equation (Fano, 1961) from which the line-profile index q and spectral width Γ had been extracted (Ruščić and Berkowitz, 1983). We now estimate the 'excess' oscillator strength f_{xs} for each member by the relation given in Sect. 2.2.1.b.2, with ρ^2 taken as unity. The resulting values of f_{xs} are 0.0362, 0.0206 and 0.0121 for $n = 4$, 5 and 6, respectively. The (^{1}D)3d(^{2}P) and (^{1}D)3d(^{2}D) states are tentatively identified by Radziemski and Kaufman (1969) at 980.92 Å and 978.59 Å, i.e. below the adiabatic IP. Neither experimental nor calculational oscillator strengths are available for these transitions. We crudely estimate the sum of (^{1}D)3d(^{2}P $+^2$D) by assuming (n*)$^{-3}$ dependence, thereby obtaining $f_{3d} \sim 0.087$. The corresponding histogram appears in Fig. 2.21. The extrapolated value of df/dE at the ^{1}D threshold is reasonable, as will be shown below.

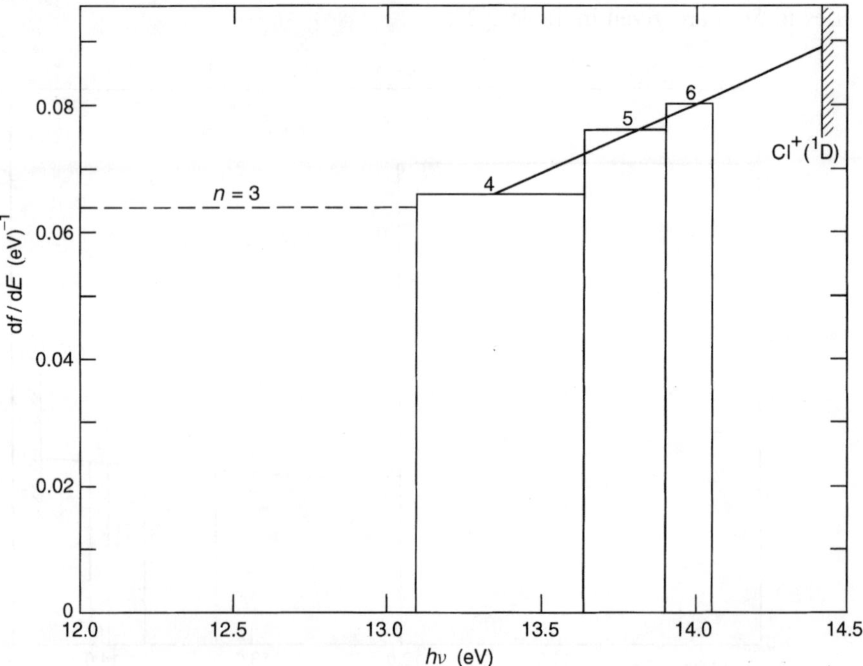

Fig. 2.21   Histogram for the series …3p^5, ^{2}P$_{3/2}$ → …3p^4 (^{1}D)nd, ^{2}D + ^{2}P in atomic chlorine

a.5 nd(^{2}S) series approaching ^{1}D In the photoionization spectrum (Fig. 1 of Ruščić and Berkowitz, 1983) there is a second sharp series converging to ^{1}D, which does not appear in the calculated spectrum of Brown *et al.* (1980). Although originally assigned as the 'missing' nd(^{2}D) series, it was soon recognized to be an nd(^{2}S) series. In L–S coupling, this series should not autoionize, since the continuum (^{3}P,εs,εd) does not have an S component. Hence, it is not seen in the calculated spectrum of Brown *et al.* However, it does occur through weak spin-orbit interaction. The oscillator strengths of $n = 4$, 5 and 6 are estimated to be $f = 0.003\,79$, $0.002\,79$ and $0.001\,83$, respectively from triangular areas having the instrumental band width.

Both the location and oscillator strength of the antecedent (^{1}D)3d(^{2}S) are uncertain. Radziemski and Kaufman (1969) assign the transition to (^{1}D)3d(^{2}S) to a line at 1004.6776 Å. This corresponds to an effective quantum number $n^* = 2.563$, or $\delta = n - n^* = 0.437$. Such a quantum defect is closer to ...(^{1}D)nd, ^{2}P/^{2}D than ...(^{1}D)nd, ^{2}S, where we find $\delta \approx -0.03$. Both Verner *et al.* (1994) and Morton (1991) cite the same assignment. They list a rather large oscillator strength, $f = 0.1577$, based on a calculation by Kurucz and Peytremann (1975). Using our measured value for $n = 4$, and $(n^*)^{-3}$ behavior, we would estimate $f \approx 0.009$ for this transition, to be about 18 times weaker. We retain both f values, as upper and lower limits, in Table 2.29. A histogram of the ...(^{1}D)nd ^{2}S series is shown in Fig. 2.22.

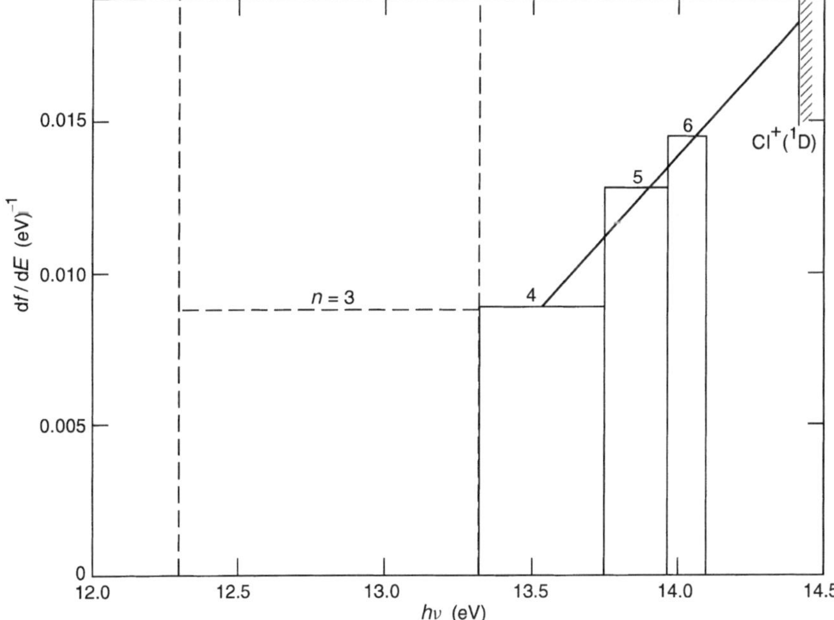

Fig. 2.22 Histogram for the series ...3p^5, ^{2}P$_{3/2}$ → ...3p^4 (^{1}D)nd, ^{2}S in atomic chlorine

The extrapolated values of df/dE for the three prominent series converging to ^{1}D are 0.0061 (ns ^{2}D), 0.089 (nd ^{2}D/^{2}P) and 0.0182 (nd ^{2}S). The sum, $0.1133(\text{eV})^{-1}$, or 12.4 Mb, corresponds to the partial cross section of ^{1}D at its onset. van der Meulen *et al.* (1992) report 11.7 $\pm$2.5 Mb for this cross section at 21.2 eV. Using the total cross section of Samson *et al.* (1986) and the intermediate coupling geometric ratios of ^{3}P, ^{1}D and ^{1}S from Berkowitz and Goodman (1979), we obtain 12.9 Mb for ^{1}D. Hence, we conclude that the oscillator strengths of the three series for $n \geq 4$ are reasonable, as are the slopes of df/dE. An additional ...(^{1}D)nd ^{2}F series has been predicted by Robicheaux and Greene (1992) from their R-matrix calculation. The ^{2}P $\rightarrow$ ^{2}F photoabsorption is forbidden in L–S coupling, but can occur by spin-orbit interaction. It is lost in the noise of the photoionization spectrum of Ruščić and Berkowitz (1983), but has been observed as a weak feature in the differential cross section of Benzaid *et al.* (1996). It is of negligible importance for the present analysis.

a.6 **The ^{3}P$_{\text{avg}}$–^{1}D continuum, 13.010 13–14.412 50 eV** We take the continuum cross section just at the ^{3}P threshold to be $\sim$20 Mb. Determinations must be indirect because autoionization immediately causes the cross section to vary, but measurements at $h\nu = 21.2$ eV hover about this value. van der Meulen *et al.* (1992) find a somewhat lower cross section, Samson *et al.* (1986) combined with branching ratios from Berkowitz and Goodman (1979) yield a higher value, while van der Meer *et al.* (1986) obtain 19.7 $\pm$2.5 Mb. Calculated values (see Table III of van der Meulen *et al.* (1992)) also are in general agreement. The incremental cross section attributable to ^{1}D is about 12 Mb, as noted in Sect. 2.8.1.a.5 and also in the references just cited. For this relatively short energy region between ^{3}P$_{\text{avg}}$ and ^{1}D, we assume an underlying continuum increasing linearly from 20–32 Mb, and calculate the contributions to $S(p)$ accordingly. They are recorded in Table 2.29.

a.7 **ns series approaching ^{1}S** From Fig. 1 of Ruščić and Berkowitz (1983) we compute the areas (and hence, oscillator strengths) for $n = 5$ ($f \cong 0.0055$) and $n = 6$ ($f \cong 0.002\,87$) with some deconvolution. Radziemski and Kaufman (1969) list ...(^{1}S) 4s at 984.2865 Å. In this case, the quantum defects for $n = 4$, 5 and 6 match rather well. The oscillator strength to ...(^{1}S) 4s is given by both Verner *et al.* (1994) and Morton (1991) as 0.0244, and can be traced to a calculation by Kurucz and Peytremann (1975). From the histogram constructed in Fig. 2.23, we note that this value is at least plausible. The extrapolation to df/dE at the ^{1}S threshold is uncertain, but the absolute quantities here are very small. The contribution of this series to the $S(p)$ is duly recorded in Table 2.29.

a.8 **nd series approaching ^{1}S** This series appears in its entirety in Fig. 1 of Ruščić and Berkowitz (1983). From the areas of the first four members, we compute $f = 0.014$, 0.0125, 0.0055 and 0.0027 for $n = 3$, 4, 5 and 6, respectively. The histogram (Fig. 2.24) displays a negative slope approaching the ^{1}S

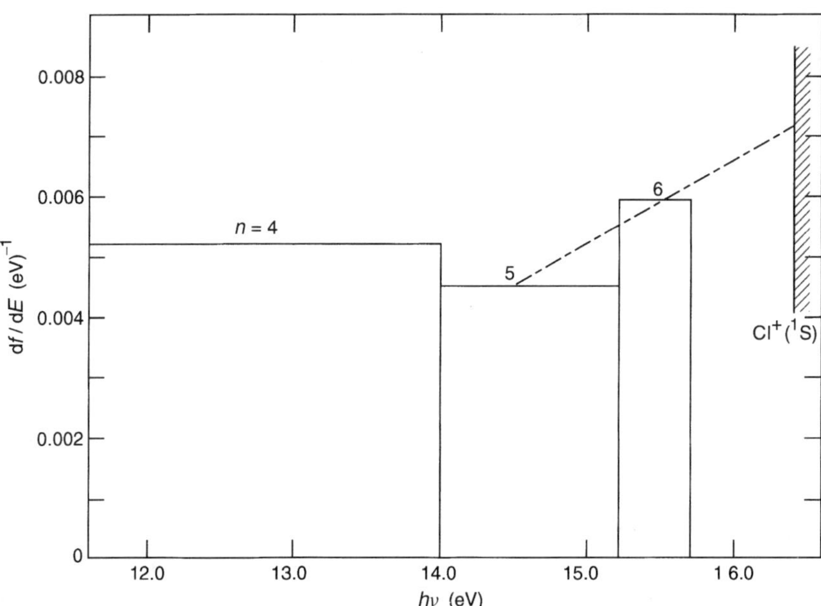

Fig. 2.23   Histogram for the series $\ldots 3p^5, {}^2P_{3/2} \to \ldots 3p^4\,({}^1S)ns, {}^2S$ in atomic chlorine

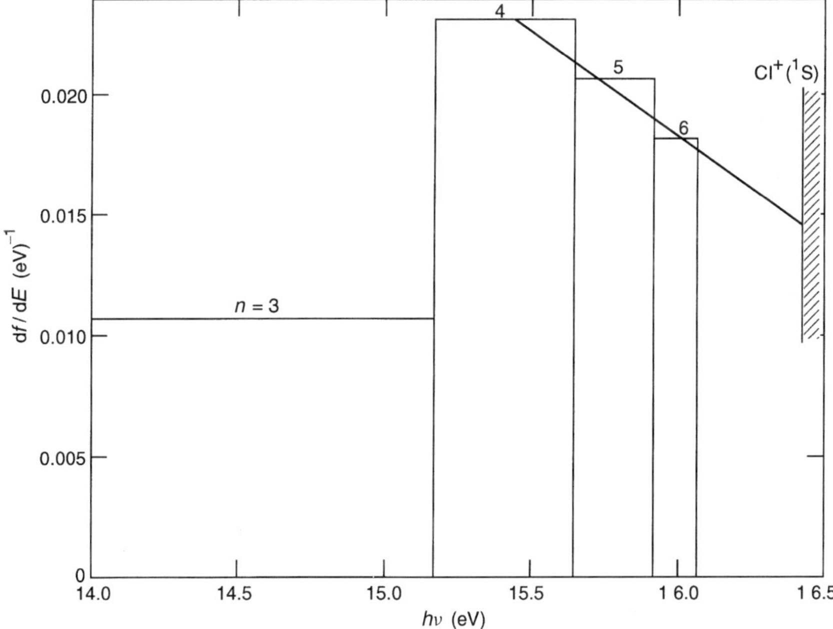

Fig. 2.24   Histogram for the series $\ldots 3p^5, {}^2P_{3/2} \to \ldots 3p^4\,({}^1S)nd, {}^2D$ in atomic chlorine

threshold, though this could result from experimental uncertainty, deconvolution and measurement of areas. The absolute values are small, and partially compensatory, since an artificially low oscillator strength for the nd series may appear as a larger value for the adjacent ns series. The contributing oscillator strengths are shown in Table 2.29. The sum of the extrapolated values of df/dE at the ^1S threshold corresponds to 2.4 Mb, which is just the partial cross section of ^1S at 21.2 eV given by van der Meulen $et\ al.$ (1992). Using the total cross section of Samson $et\ al.$ (1986) at 21.2 eV, together with the intermediate coupling branching ratios of Berkowitz and Goodman (1979), one obtains $\sigma(^1S) = 2.6$ Mb.

a.9 The $^1D-^1S$ continuum, 14.412 50–16.424 07 eV Assuming linear behavior for this underlying continuum, the cross section will increase from 32 to 32.4 Mb. The contribution of this continuum to $S(p)$ is listed in Table 2.29.

a.10 An interloper–3s3p^6 The contributions listed above refer to $(3p)^{-1}$ excitation or ionization. One inner shell transition, to 3s3p^6, occurs in this domain, at 1167.1479 Å (Radziemski and Kaufman, 1969). Verner $et\ al.$ (1994) and Morton (1991) cite a calculated oscillator strength of 0.009 57, obtained by Ojha and Hibbert (1990). However, this is only a 'length' value, and differs substantially from their 'velocity' gauge number, 0.020 55. Here, we choose the geometric mean, $f = 0.014$.

b The continuum

b.1 16.424 07–43 eV (Cooper minimum) Samson $et\ al.$ (1986) have obtained experimental photoionization cross sections for atomic chlorine from the onset of the $3s^23p^4(^1S)$ continuum to 78.5 eV, normalized to 1/2 the absolute Cl_2 cross section between 27.6–41.3 eV. At the ^1S onset, they give $\sigma(Cl) = 43.6 \pm 3.5$ Mb. van der Meulen $et\ al.$ (1992) note that their cross sections are about 25% lower. Support for a lower value can be extracted from the figures of calculations by Brown $et\ al.$ (1980) and Robicheaux and Greene (1992), which successfully reproduce the resonances with an underlying continuum at the ^1S threshold of $\sim$36 Mb. Our deliberations arrive at $\sim$34.2 Mb. Since Samson $et\ al.$ provide a convenient, continuous data set over an extended range, we shall utilize their cross sections, recognizing that a discontinuity with our earlier results exists at 16.42 eV.

In the interval 16.42–43 eV, a Rydberg series of window resonances is observed, corresponding to the transitions $\dots3s^23p^5(^2P_{3/2}) \rightarrow 3s3p^5np$ (van der Meulen $et\ al.$, 1992) which converge to $\dots3s3p^5(^3P_{2,1,0})$ at 24.5437, 24.6221 and 24.6635 eV, respectively (Radziemski and Kaufman, 1974) and $\dots3s3p^5(^1P_1)$ at 27.307 eV (van der Meulen $et\ al.$, 1992). The 'excess' oscillator strength due to these resonances is obviously negative. van der Meulen $et\ al.$ (1992) have computed $f = -0.011$ for the $n = 5$ resonance, and estimate that the entire Rydberg series 'is unlikely to exceed -0.1'. Rather than explicitly evaluating this negative contribution, we note here that there is almost certainly a Rydberg series preceding the $L_{II,III}$ edge (i.e. $<208\,\mathrm{eV}$) whose magnitude can be estimated from absolute cross sections of Cl_2 and HCl given by Ninomiya *et al.* (1981). It is of comparable magnitude, and of opposite sign, to the window resonance series. The contributions of both series are assumed to cancel for $S(0)$, and are also negligible for the other $S(p)$.

The data of Samson *et al.* (1986) have been fitted by regression to a 4-term polynomial in inverse energy, $\sum_2^5 a_n y^n$, where $y = B/E$, and $B = 16.42407\,\mathrm{eV}$ is the ^1S ionization potential, in the stated interval. The derived polynomial is integrated to determine the contributions to $S(p)$, which are listed in Table 2.30. The coefficients of this polynomial are given in Table 2.31.

b.2 43–91.5 eV; 91.5–208 eV

In this domain, the photoabsorption cross section recovers a little from the Cooper minimum to a plateau at $\sim70\,\mathrm{eV}$, then declines monotonically until it approaches the $L_{II,III}$ edge. The data of Samson *et al.* (1986) merge smoothly with the compiled points of Henke *et al.* (1993) at $\sim80\,\mathrm{eV}$ (see Fig. 2.25). The data have been partitioned into two regions, 43–91.5

Table 2.30 Spectral sums, and comparison with theoretical sums for atomic chlorine[a]

Energy, eV	$S(-2)$	$S(-1)$	$S(0)$	$S(+1)$	$S(+2)$
Valence spectrum,[b] $\rightarrow\ ^1S$ (16.42)	2.1899–2.7507	2.0271–2.5313	1.9294–2.3826	1.8855–2.2930	1.8850–2.2514
16.42–43[c]	1.6258	2.5372	4.0105	6.8974	12.2453
43–91.5[c]	0.0242	0.1108	0.5314	2.6628	13.8393
91.5–208 (L_{III})[d]	0.0079	0.0726	0.7005	7.1388	76.7270
208.0–705.0[d]	0.0118	0.2610	6.3792	174.2109	5330.8087
705.0–2830 (K)[d]	0.0003	0.0189	1.4574	127.1684	12 908.667
270[e]	–	0.0004	0.008	0.16	3.15
2821[f]	–	–	0.0066	1.37	283.8
2830–10 000[d]	–	0.0046	1.4010	469.450	178 017.8
10^4–10^5[g]	–	0.0001	0.1479	195.9397	360 475.0
10^5–10^6[h]	–	–	0.0012	14.2517	233 696.0
10^6–10^7[h]	–	–	–	0.5778	91 551.6
10^7–10^8[h]	–	–	–	0.0199	31 131.6
10^8–10^9[h]	–	–	–	0.0006	10 078.9
10^9–∞[h]	–	–	–	–	4685.1
Total	3.86–4.42	5.03–5.54	16.57–17.03	1001.7–1002.1	928 267.0
Expectation values	3.678 ±0.073[i]	5.9638[j] 5.0350[k]	17.0	1009.389[k]	916 850.8[j] 916 944.9[l]

[a] $S(p)$ in Ry units.
[b] See Table 2.29 and text.
[c] Samson *et al.* (1986).
[d] Henke *et al.* (1993).
[e] Ninomiya *et al.* (1981).
[f] Bodeur *et al.* (1990).
[g] Chantler (1995).
[h] Hydrogenic calculation, K-shell only, from Bethe and Salpeter (1977).
[i] Reinsch and Meyer (1976).
[j] Fraga *et al.* (1976).
[k] Müller (1996).
[l] Bunge *et al.* (1993).

Table 2.31 Coefficients of the polynomial $\mathrm{d}f/\mathrm{d}E = ay^2 + by^3 + cy^4 + dy^5$ fitted to data at various energies[a]

Energy range, eV	a	b	c	d
16.424–43.0	−35.06	127.3753	−114.244	27.094 18
43.0–91.5	8.219 563	−11.3732	−86.6082	170.7537
91.5–208.0	8.329 238	−3.755 01	−175.934	408.3603
208.0–705.0	−335.036	29 627.42	−611 141.0	3 870 240.0
705.0–2830	8.522 174	5829.088	−123 326.0	1 177 086.0
2830–10 000	1.148 096	101 785.3	−6 984 793.0	84 375 767.0

[a]$\mathrm{d}f/\mathrm{d}E$ in Ry units, $y = B/E$, $B = 16.424\,07\,\mathrm{eV}$.

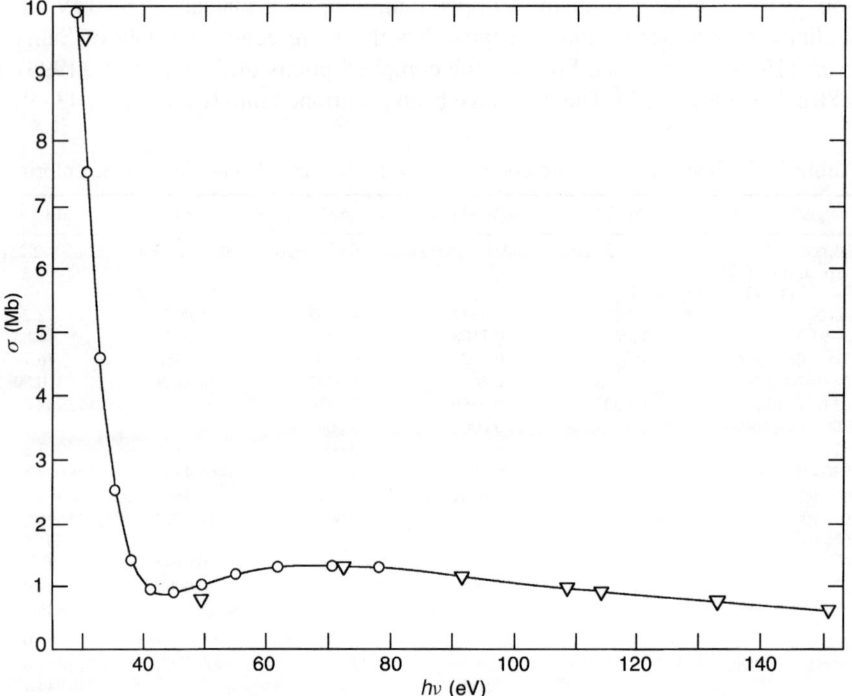

Fig. 2.25 Absolute photoabsorption spectrum of atomic chlorine, 30–150 eV. ○ Samson *et al.* (1986); ▽ Henke *et al.* (1993)

and 91.5–208 eV, for improved fitting. The aforementioned 4-term polynomials have been matched to each segment. The values of $S(p)$ for each region are recorded in Table 2.30, the coefficients of the polynomials in Table 2.31.

b.3 **208–705 eV; 705–2830 eV** The L_{III} ($^2p_{3/2}$) edge occurs at 208 eV. Gluskin *et al.* (1977) obtain 208.0 ±0.3, Ninomiya *et al.* (1981) find 208.26

±0.4, while Jolly *et al.* (1984) select 207.8 eV, all for Cl_2. The L_{II} edge is about 1.6 eV higher. (Actually, with two open shells, there are many more fine-structure limits. Caldwell *et al.* (1999), vide infra, calculate ten ionization thresholds in a span of $\sim$5 eV.) Ninomiya *et al.* (1981) have measured the pre- and post-edge photoabsorption cross section for HCl and Cl_2. They observe excitation to valence anti-bonding orbitals (which have no analog in atomic chlorine) and then a Rydberg series, which should bear some relation to that in atomic chlorine. The summed oscillator strength for this Rydberg series is about 0.02 in HCl, and less in Cl_2, which is the basis for our earlier conclusion that it tends to compensate for the negative oscillator strength in the $3s^{-1}$ window series.

Recently, Caldwell *et al.* (1999) have measured photoion yield spectra for atomic chlorine between 198–212 eV. They find a number of sharp lines, some superposed on the continuum which increases roughly linearly between 206–210 eV. These sharp lines do not fit a simple Rydberg series. Calculations show that no pure-coupling representation (LS or JJ) is very satisfactory. No absolute calibration of intensities was attempted, but values of gf were calculated for the most prominent lines. Their sum is 0.0739. If g refers to the degeneracy of the lower state, $^2P_{3/2}$, then $f \cong 0.0185$, comparable to that found for HCl and Cl_2. If we normalize the continuum in their experimental spectrum at 210 eV to $\sim$2 Mb (see below), the summed oscillator strength in their spectrum is about 1/3 this value. However, this does not include the pseudo-continuum between 206–208 eV, which they attribute to 'line broadening due to increased interactions with more and more continua'.

At the $L_{II,III}$ edge, the cross section increases abruptly to about 2 Mb in HCl and Cl_2 (per Cl atom), then more gradually to $\sim$2.5 Mb at 220 eV, according to Ninomiya *et al.* (1981). The sparse points of Henke *et al.* (1993) offer 3.65 Mb at 220.1 eV. The experimentally based compilation of Henke *et al.* agrees quite well with the calculated values of Chantler (1995) above $\sim$300 eV. Chantler's L_{III} edge (200 cV) is lower than the experimental value (208 eV), and his maximum cross section above the edge (5.31 Mb) is higher than the Henke value and about twice that deduced from the data of Ninomiya *et al.* from HCl and Cl_2. We choose the Henke data in the post L_{II} edge region as a suitable compromise. No significant increase is observed at the L_I ($2s^{-1}$) edge at 278.74 eV (Jolly *et al.*, 1984), although Ninomiya *et al.* observe some structure prior to this edge, which they attribute to transitions to anti-bonding orbitals. We can anticipate an analog in atomic chlorine for the allowed $\dots 2s^2 2p^6 3s^2 3p^5 \rightarrow \dots 2s 2p^6 3s^2 3p^6$ transition, for which we estimate $f \sim 0.008$, i.e. half the oscillator strength of the Cl_2 structure in this region. With increasing energy, the photoabsorption cross section continues to decline until it approaches the K-edge, given as 2830.2 eV by Bodeur *et al.* (1990) for Cl_2. The most prominent pre-K edge feature for Cl_2 is assigned to a transition to an anti-bonding orbital with 3p character, followed by a Rydberg excitation. Here again, there is an analog in atomic chlorine, the nominally allowed $1s \rightarrow 3p$ transition, whose oscillator strength we estimate as half that of the anti-bonding feature in Cl_2, or $f \sim 0.0066$. The contributions of these isolated features are not significant; however, they are included in Table 2.30. The continuum between the $L_{II,III}$ and K-edges is partitioned into two segments, 208–705 eV and 705–2830 eV, and fitted to separate 4-term polynomials, using the data of Henke *et al.* (1993). The derived values of $S(p)$ for each domain are given in Table 2.30, and the coefficients of the polynomials in Table 2.31.

b.4   Post K-edge: 2830–10 000 eV   The photoabsorption cross section of Cl_2 increases by about one order of magnitude in a span of ~ 3 eV at the K-edge (Bodeur *et al.*, 1990), and one anticipates similar behavior for atomic chlorine. Indeed, the experimental compilation of Henke *et al.* and the calculated values of Chantler show just this behavior (see Fig. 2.26). Note the good agreement between Henke and Chantler, alluded to earlier. Also, the magnitude of the increase in the atomic cross section at the K-edge, 0.10 Mb, is 1/2 the Cl_2 cross section observed by Bodeur *et al.* (1990). Hence, the contributions to the $S(p)$ should be reliable in the stated interval. Accordingly, the cross sections given by Henke *et al.* have been fitted to the usual 4-term polynomial, with coefficients listed in Table 2.31, and corresponding $S(p)$ values in Table 2.30.

b.5 10^4–10^5 eV   The calculated cross sections of Chantler (1995) are utilized in this high energy domain.

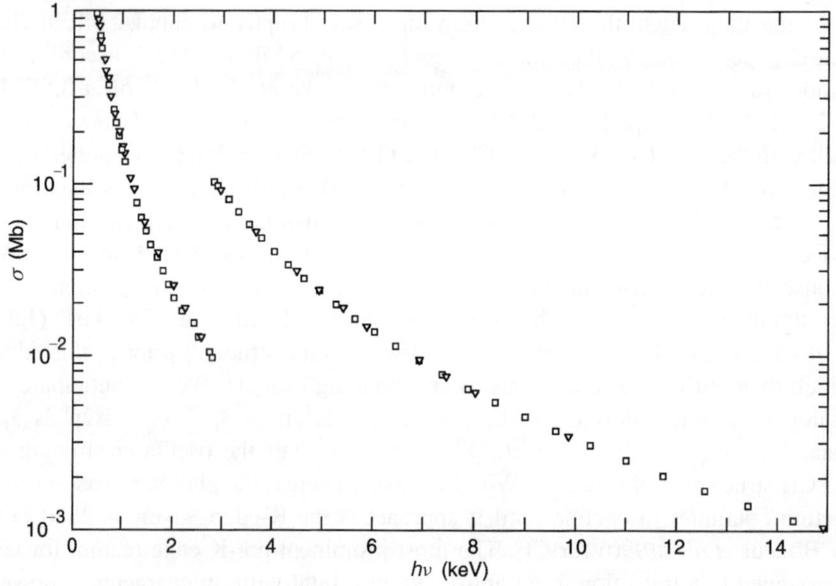

Fig. 2.26   Absolute photoabsorption spectrum of atomic chlorine, 0.5–15 keV. □ Chantler (1995); Henke *et al.* (1993)

2.8.2 The analysis

The purpose of this exercise is to utilize the sum rules as a guide in selecting among differing sources of data, to narrow the uncertainty in oscillator strengths and cross sections. The sum rules also localize the region of uncertainty. A glance at Table 2.30 reveals that contributions to $S(-2)$ are very small beyond $\sim 50\,$eV, whereas $S(0)$ has significant accretion to at least $5\,$keV. The largest relative uncertainty exists for $S(-2)$. Hence, it is important to establish the most reliable polarizability (α) for atomic chlorine. Unfortunately, we are unaware of any experimental determinations of this quantity. In Table 2.32, we have collected various calculated values. The oldest calculated value, $14.71\ a_0^3$ is based on CEPA (coupled-electron-pair approximation) by Reinsch and Meyer (1976), and has a good track record. It was originally stated to be accurate to $\pm 2\%$ for atoms through the second row, and seems to have held up. The many-body perturbation theory calculations of Kutzner *et al.* (1988) have a larger predicted error bar, but the mean of the length and velocity results is only slightly larger than the CEPA value. The CASPT2 value, $14.63\ a_0^3$, is also believed by the authors (Andersson and Sadlej, 1992) to be quite accurate, and is about as much lower than the CEPA value as the mean MBPT result is higher. The restricted Hartree–Fock results are all significantly lower. Hence, we adopt the CEPA value here. We note that α (Cl) $< 1/2\alpha(\mathrm{Cl}_2)$. This behavior is also true of oxygen, but the reverse is the case for hydrogen and nitrogen. A rationale can be found in the corresponding ionization potentials, which are higher for O and Cl than for O_2 and Cl_2, but lower for H and N than for H_2 and N_2.

With $\alpha(\mathrm{Cl}) = 14.71\ \pm 0.29\ a_0^3$, $S(-2) = 3.678\ \pm 0.073\,$Ry units. Even the lower limit of our spectral sum for $S(-2)$ exceeds the required sum. The lower limit was based on linear extrapolations to the anticipated values at the limits (the partial ionization cross sections) whereas the upper limits stemmed primarily from old calculations by Kurucz and Peytremann (1975). The $S(-2)$ sum rule clearly disfavors the latter for the $3\mathrm{p}^4(^3\mathrm{P})n\mathrm{d}$ series.

This conclusion forces us to select the lower value of the spectral sum for $S(0)$, 16.5732, although the higher value comes very close to 17.00, the result required by the Thomas–Reiche–Kuhn sum rule. Consequently, we look to the continuum regions to understand the 2.5% deficit. There are four major

Table 2.32 Calculated values of the static electric dipole polarizability (α) of atomic chlorine, in atomic units

14.20 ±0.28	Restricted Hartree–Fock	Reinsch and Meyer (1976)
14.71 ±0.29	Coupled-electron-pair approx.	
14.98 ±1.5	MBPT – length	Kutzner *et al.* (1988)
14.58 ±1.5	MBPT – velocity	
14.13	Restricted Hartree-Fock	Andersson and Sadlej (1992)
14.63	CASPT2[a]	
14.238	Numerical RHF	Stiehler and Hinze (1995)

[a]Complete-active-space self-consistent-field, second-order perturbation treatment.

contributing segments. The post K-edge region is expected to contribute about 0.3 less than 2.0. In argon, we found 1.63, in atomic chlorine we obtain 1.56. The difference, 0.07, is small compared to the overall deficit, 0.43. The two major contributors to $S(0)$ are 16.42–43 eV (4.01) and 208–705 eV (6.38). In the former region, we chose the calibration of Samson *et al.* (1986) over that of van der Meulen *et al.* (1992), which was 25% lower. The current results tend to support the higher values. In the 208–705 eV region (post $L_{II,III}$ edge), we opted for the compilation of Henke *et al.* (1993), rather than the lower values that would be forthcoming from 1/2 $\sigma(Cl_2)$ or $\sigma(HCl)$ obtained by Ninomiya *et al.* (1981). Here again, the higher values used in the computation are favored. Thus, no particular region is suspect. We must consider the 2.5% deficit quite satisfactory for this difficult case, where even the measurements of Samson *et al.* are given $\pm 8\%$ uncertainty.

Hartree–Fock sums are available for $S(-1)$, $S(+1)$ and $S(+2)$ (Fraga *et al.*, 1976). In principle, $S(-1)$ is influenced by outer shell correlation lacking in Hartree–Fock calculations. If we use our lower limit for $S(-1)$, we see that it is 18.5% lower than the Hartree–Fock value (5.9638 Ry units), the usual direction of H–F error, but in excellent agreement with a calculation using a more correlated wave function (Müller, 1996). The $S(+1)$ calculated value is sensitive to inner shell correlation. Here, the spectral sum lies below the H–F value, which is counter to expectation, and suggests that the values of Henke *et al.* (1993) (the primary contributor) may be slightly low. An increase of $\sim 5\%$ in the latter would result in a plausible value of $S(+1)$, and also improve $S(6)$. The quantity $S(+2)$, dependent on electron density at the nucleus, is usually given quite well by Hartree–Fock calculations using Slater-type (rather than Gaussian) orbitals. The spectral sum is higher, but only by 1.2%.

2.9 Argon

2.9.1 The data

The ionization potential of argon, forming the $^2P_{3/2}$ state, has been determined recently by a combination of laser experiments (Velcher *et al.*, 1999) to be $127\,109.842(4)\,\mathrm{cm}^{-1}$, an improvement in precision over the classical spectroscopic value of $127\,109.8(1)\,\mathrm{cm}^{-1}$ (Minnhagen, 1973). Thus, IP $(^2P_{3/2}) = 15.759\,610\,3(5)\,\mathrm{eV}$ and IP $(^2P_{1/2}) = 15.937\,103\,9(5)\,\mathrm{eV}$. Minnhagen (1973) provides accurate energies for the five electric dipole-allowed Rydberg series converging to these two limits.

a *The discrete spectrum*

As with neon, three of the dipole-allowed Rydberg series, $ns(3/2)_1^o$, $nd(1/2)_1^o$ and $nd(3/2)_1^o$ (J_cK notation) converge on the $^2P_{3/2}$ ground state. We shall refer to these as ns, nd and n$\underline{\mathrm{d}}$, respectively. The nd series is weak, the n$\underline{\mathrm{d}}$ strong. The other two, $ns'(1/2)_1^o$ and $nd'(3/2)_1^o$ (hereafter ns' and nd'), converge on the $^2P_{1/2}$ state. Their higher members (above IP $^2P_{3/2}$) are broadened by autoionization. Information on the oscillator strengths is still rather limited; the resonance transitions, to 4s and 4s′, have been studied extensively, the 5s, 5s′, 3d, 3$\underline{d}$, 3d′ considerably less so, and the 4d, 4$\underline{d}$, 4d′, 5d, 5$\underline{d}$, 6s, 6s′ and 7s only by electron impact energy loss spectroscopy. Unfortunately, the limited resolution of the electron impact method results in substantial overlap of peaks, e.g. 5s′−3d′, 6s−4$\underline{d}$, 6s′−4d′, 5d−7s−5$\underline{d}$. See Chan *et al.* (1992b). In such cases, deconvolution is necessary, with its attendant hazards. In order to make reasonable estimates of the oscillator strengths to higher series members, we impose some plausible constraints. The absolute photoabsorption cross section at the $^2P_{1/2}$ limit is about 31.5 Mb (Samson, 1966). The ratio of intensities, $^2P_{3/2} : {}^2P_{1/2}$, has been reported to be 1.93, and more-or-less constant with energy (Samson *et al.*, 1975). Therefore, we apportion 20.75 Mb to the continuum cross section at the $^2P_{3/2}$ limit. In the histograms that we construct, and discuss below, this imposes a limit on the sum of the values of df/dE for the three series converging on the $^2P_{3/2}$ threshold. This is a useful constraint on the n$\underline{d}$ series, which is dominant. We can estimate the oscillator strengths of some higher ns′ members converging on $^2P_{1/2}$ by normalizing the high resolution relative photoion yield (Berkowitz, 1971; Radler and Berkowitz, 1979) to the absolute photoabsorption cross section in the open continuum beyond the $^2P_{1/2}$ threshold. For the higher, broad nd′ members, the 'excess' oscillator strength is determined from the expression given in Sect. 2.2.1.b.2. Some parameters applicable to this equation are also provided by Wu *et al.* (1990). The values of $f/\Delta E$ for high ns′ and nd′ members help to establish the corresponding limiting values of df/dE at the $^2P_{1/2}$ threshold.

a.1 The ns series Chan *et al.* (1992b) review the extensive earlier work for the transition to 4s; their value for this oscillator strength is 0.0662 ±0.0033. Subsequently, Ligtenberg *et al.* (1994) obtained 0.0616 ±0.0021, and later Gibson and Risley (1995) reported $f = 0.0580$ ±0.0017. These latter two groups used the method of electron beam excitation and detection of the emitted radiation as a function of gas density, as did Tsurubuchi *et al.* (1990), who obtained $f = 0.057$ ±0.003. We choose $f = 0.0580$, partly because it is the most precise, and also because these authors (Gibson and Risley) found excellent agreement with the theoretically well-known oscillator strength of the helium resonance line. For the transition to 5s, the contending values are 0.0264 ±0.0026 (Chan *et al.*, 1992b), 0.025 ±0.002 (Westerveld *et al.*, 1979) and 0.0268 ±0.002 (Wiese *et al.*, 1969), the latter a reinterpretation of lifetime data. We choose $f = 0.026$. Oscillator strengths for the 6s and 7s transitions are available only from electron impact data. We take 0.0144 ±0.0014 for 6s from (Chan *et al.*, 1992b), but reject 0.0426 ±0.0043 for 7s. (There may be a misprint in Chan *et al.*, since identical values are given for 7s and 5$\underline{d}$.) Instead, we choose 0.0139 from Natali *et al.* (1979). Even with this choice, the histogram (Fig. 2.27) displays an abrupt increase at $n = 7$. It will be recalled that the 5d, 5$\underline{d}$ and 7s transitions are heavily overlapped in the electron energy loss curve. Consequently, little weight can be assigned to

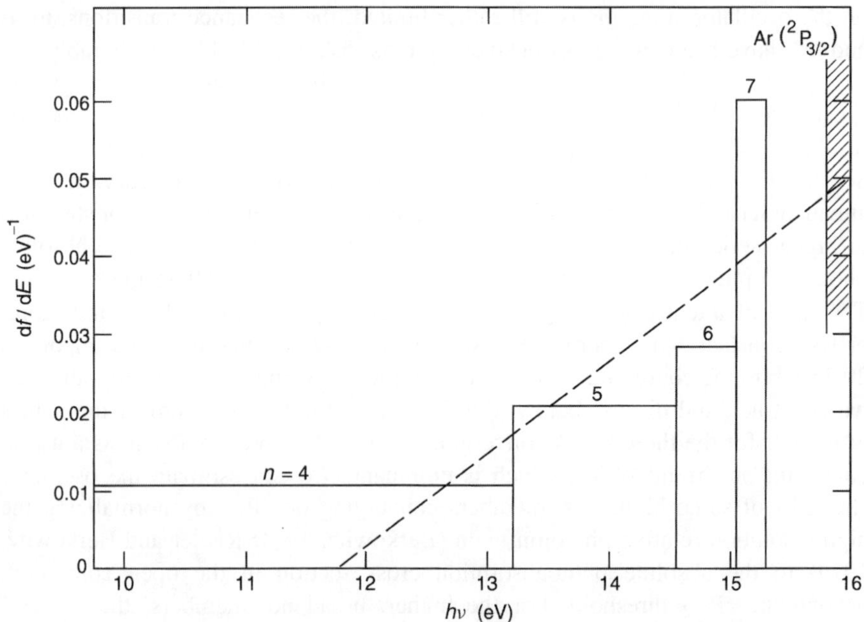

Fig. 2.27 Histogram for the series $\ldots 3p^6$, $^1S_0 \rightarrow \ldots 3p^5$ $(^2P_{3/2})ns$, ^1P in argon

these higher f values when estimating the limiting value of df/dE at the $^2P_{3/2}$ threshold.

a.2 The nd (weak) series The f values to be considered for the transition to 3d are 0.0013 ±0.0001 (Chan *et al.*, 1992b), 0.000 89 ±0.000 07 (Westerveld *et al.*, 1979) and 0.0010 (Natali *et al.*, 1979). We have chosen $f = 0.0011$. For 4d and 5d, where electron impact data and overlapping peaks are involved, we take 0.0019 ±0.0002 (4d) and 0.0041 ±0.0004 (5d) from Chan *et al.* (1992b). The histogram (Fig. 2.28) implicates $f(5d)$ as particularly suspect. However, because these oscillator strengths are small, the effect on the sum rules is not very significant, even with a somewhat arbitrary choice of the limiting value of df/dE.

a.3 The n̲d̲ (strong) series The proffered f values for the 3d̲ transition are 0.0914 ±0.0091 (Chan *et al.*, 1992b), 0.079 ±0.006 (Westerveld *et al.*, 1979), 0.092 (Natali *et al.*, 1979) and 0.093 ±0.006 (Wiese *et al.*, 1969). We have chosen 0.090. For 4d̲, the electron impact values 0.0484 ±0.0048 (Chan *et al.*, 1992b) and 0.048 (Natali *et al.*, 1979) are in substantial agreement. With 5d̲, the value of 0.0426 ±0.0043 (Chan *et al.*, 1992b) appears too large on the histogram (Fig. 2.29). Since this series is the largest contributor of the three converging to $^2P_{3/2}$, and we have constrained the sum of df/dE at this limit for these three,

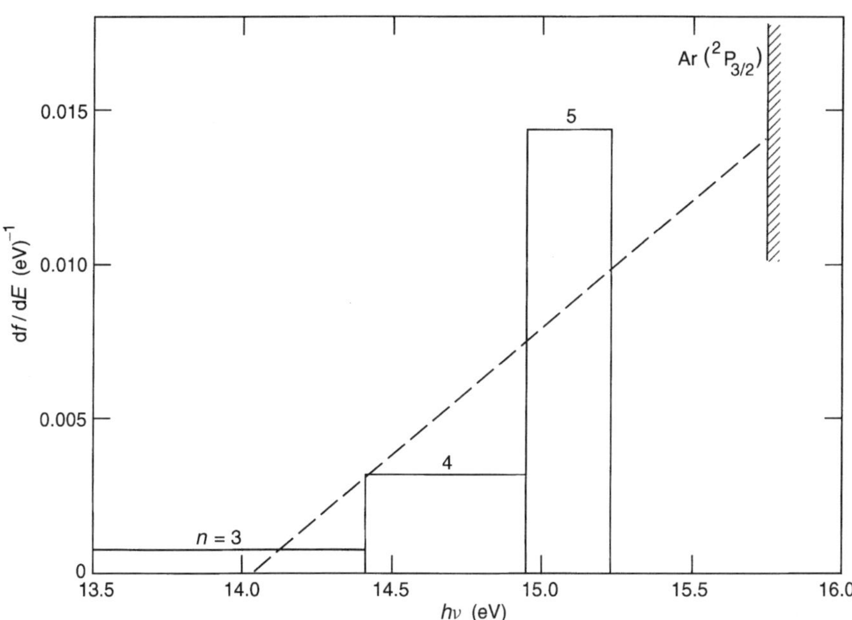

Fig. 2.28   Histogram for the series $\ldots 3p^6$, $^1S_0 \rightarrow \ldots 3p^5$ $(^2P_{3/2})nd$, weak, in argon

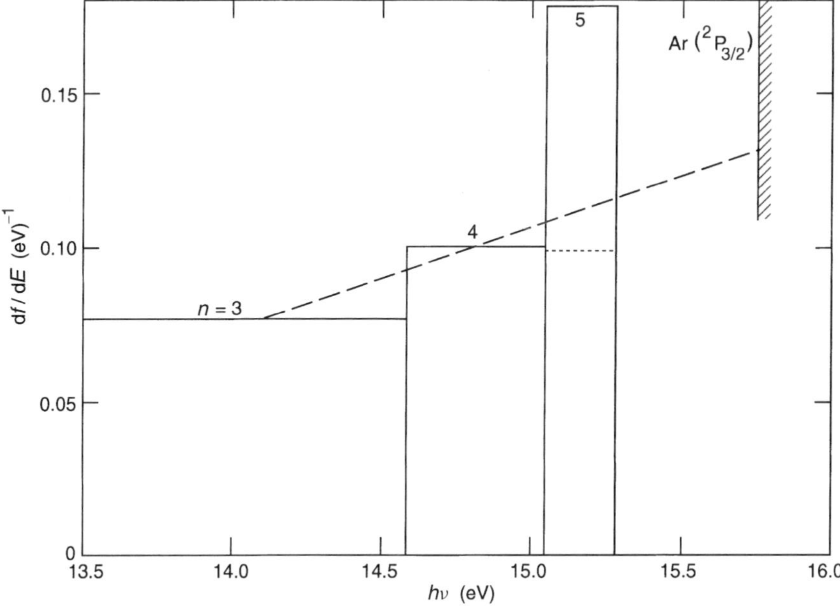

Fig. 2.29   Histogram for the series $\ldots 3p^6$, $^1S_0 \rightarrow \ldots 3p^5$ $(^2P_{3/2})nd$, strong, in argon

it is possible to select a plausible limit for df/dE of the n$\underline{d}$ series, as drawn in Fig. 2.29.

a.4 The ns′ series For the resonance transition to 4s′, the recent values are 0.2214 ±0.0068 (Gibson and Risley, 1995) and 0.2297 ±0.0093 (Ligtenberg *et al.*, 1994). Earlier values using a similar method are 0.213 ±0.011 (Tsurubuchi *et al.*, 1990) and 0.240 ±0.020 (Westerweld *et al.*, 1979). The electron impact value of Chan *et al.* (1992b) is 0.265 ±0.013. We choose the most recent value, 0.2214. There is a precipitous decline in oscillator strength for the 5s′ transition, where the optically based values are 0.0106 ±0.0008 (Westerveld *et al.*, 1979), 0.0119 (Wiese *et al.*, 1969), and the electron energy loss values are 0.0126 ±0.0013 (Chan *et al.*, 1992b) and 0.0124 (Natali *et al.*, 1979). We choose 0.012. The f values for 6s′, again electron-impact based, are 0.0221 ±0.0022 (Chan *et al.*, 1992b) and 0.0224 (Natali *et al.*, 1979). In the histogram (Fig. 2.30), the 6s′ transition looks anomalously high, especially in light of the 11s′ value deduced from ion yield curves. We have chosen to retain the oscillator strength for 6s′, with the rationale that even though the allocation to 6s′ may not be entirely correct, there is likely to be an oscillator strength of this magnitude at this approximate energy, though a larger fraction may be due to 4d′. However, the extrapolation to the $^2P_{3/2}$ limit is heavily weighted toward the inferred value of 11s′ from Berkowitz (1971) and Radziemski and Kaufman (1979).

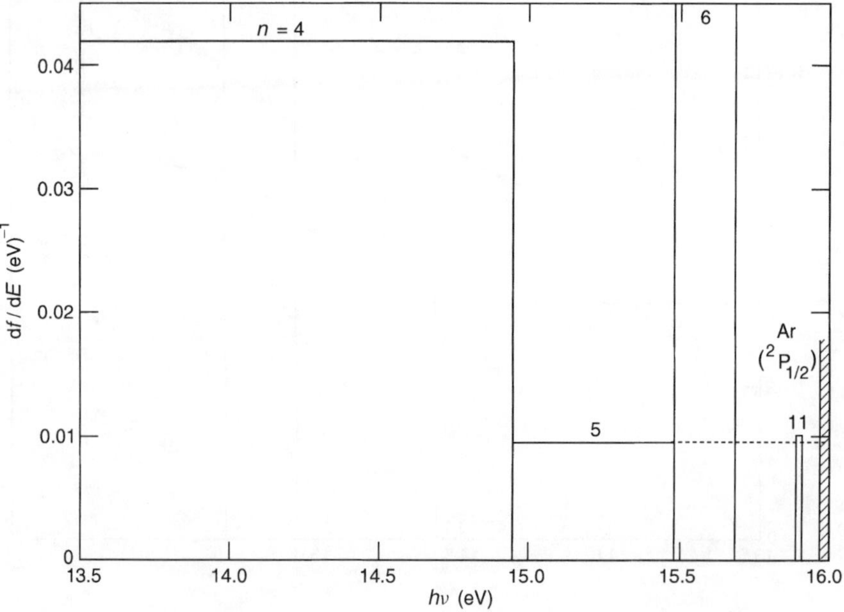

Fig. 2.30 Histogram for the series …3p^6, $^1S_0$ → …3p^5 ($^2P_{1/2}$)ns, ^{1}P in argon

a.5 The nd' series For 3d', the optically based values are 0.086 ±0.007 (Westerveld *et al*., 1979) and 0.106 (Wiese *et al*., 1969). The electron impact based values are 0.106 ±0.011 (Chan *et al*., 1992b) and 0.110 (Natali *et al*., 1979). We choose 0.106. For the 4d' transition, we take 0.0209 ±0.0021 from Chan *et al*. From the ion yield curves (Berkowitz, 1971; Radziemski and Kaufman, 1979), we infer values for 9d' and 10d'. The histogram (Fig. 2.31) clearly shows a reversal, declining between 3d' and 4d', then increasing to 9d' and 10d'. If $f(6s')$ were diminished and 4d' increased in the blended electron impact peak, as intimated in 2.9.1.a.4. above, it would alleviate the apparent discontinuities in Figs. 2.30 and 2.31. The effect on sum rule analysis is to change the slope of the extrapolation to df/dE at the $^2P_{1/2}$ threshold, which influences the estimated contributions of higher nd'. With the available data, we may be underestimating $S(0)$ by ~0.008, and $S(-2)$ by ~0.006 Ry units.

To complete the oscillator strength distribution up to the $^2P_{1/2}$ threshold, we consider the underlying continuum between $^2P_{3/2}-{}^2P_{1/2}$ to have a constant cross section, $\sigma = 20.75$ Mb. This, and each of the linearly extrapolated regions to their respective thresholds, are integrated to determine their contributions to $S(p)$. The results are recorded in Table 2.33.

b   The continuum

b.1   15.937–29.239 eV; 29.239–48.0 eV The photoabsorption cross section increases from the onset of $^2P_{1/2}$ to a maximum at ~22 eV of ~36.5 Mb and then

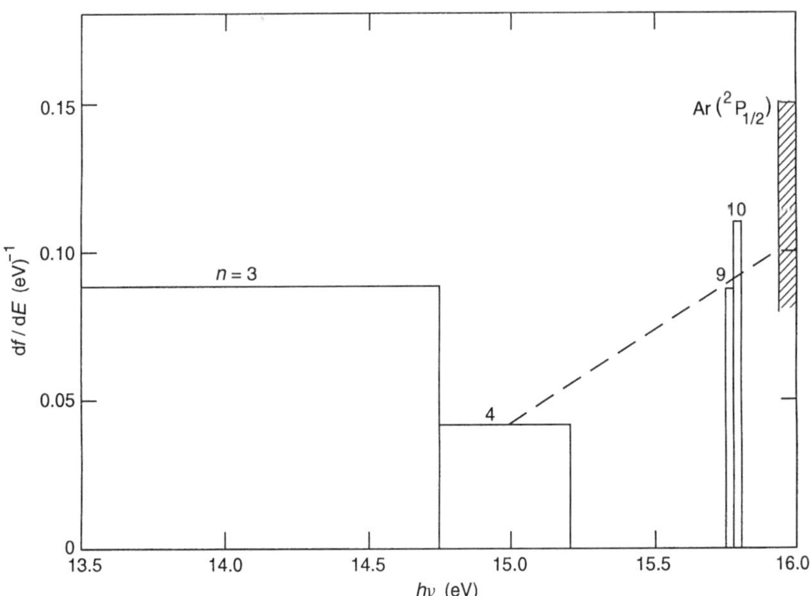

Fig. 2.31 Histogram for the series …3p^6, ^{1}S$_0$ → …3p^5 (^{2}P$_{1/2}$)nd in argon

Table 2.33   Contributions of discrete spectrum to $S(p)$ sums in argon ($S(p)$ in Ry units)

	$S(-2)$	$S(-1)$	$S(0)$	$S(+1)$	$S(+2)$
a. ns series					
4	0.0795	0.0679	0.058	0.0496	0.0423
5	0.0242	0.0251	0.026	0.0269	0.0279
6	0.0118	0.0128	0.014	0.0153	0.0167
7	0.0112	0.0125	0.0139	0.0155	0.0173
8−∞	0.0162	0.0185	0.0211	0.0240	0.0274
b. nd series					
3	0.0011	0.0011	0.0011	0.0011	0.0011
4	0.0016	0.0018	0.0019	0.0021	0.0022
5	0.0033	0.0037	0.0041	0.0046	0.0051
6-∞	0.0048	0.0054	0.0062	0.0071	0.0080
c. n$\underline{d}$ series					
3	0.0832	0.0865	0.090	0.0936	0.0974
4	0.0402	0.0440	0.048	0.0524	0.0573
5	0.0342	0.0382	0.0426	0.0476	0.0531
6-∞	0.0441	0.0503	0.0574	0.0655	0.0748
Sub-total to $^2P_{3/2}$	0.3554	0.3678	0.3843	0.4053	0.4306
Continuum, $^2P_{3/2}$–$^2P_{1/2}$	0.0247	0.0288	0.0335	0.0391	0.0455
d. ns$'$ series					
4	0.2929	0.2547	0.2214	0.1925	0.1673
5	0.0109	0.0115	0.012	0.0126	0.0132
6	0.0181	0.0200	0.0221	0.0244	0.0269
7−∞	0.0053	0.0060	0.0069	0.0079	0.0090
e. nd$'$ series					
3	0.0959	0.1008	0.106	0.1114	0.1172
4	0.0172	0.0190	0.0209	0.0230	0.0254
5−∞	0.0427	0.0490	0.0562	0.0645	0.0740
Total, discrete	0.3554	0.3678	0.3843	0.8807	0.9091

declines to a minimum ($\sim$0.7 Mb) at $\sim$48 eV. This minimum can be associated with ionization from the 3s orbital, which has its onset at $235\,831.33\,\mathrm{cm}^{-1} \equiv 29.239\,37$ eV (Minnhagen, 1971). The 3s $\rightarrow$ np Rydberg series approaching this limit displays window resonances. We shall treat these in the following section. Here, we consider the smooth continuum, and partition it into 15.937–29.239 eV and 29.239–48.0 eV.

A number of experimental groups have studied this region (Samson, 1966; Madden *et al.*, 1969; Carlson *et al.*, 1973; Samson *et al.*, 1991, Chan *et al.*, 1992b). All agree on the general shape, but differ in detail. For example, Samson (1966) and Samson *et al.* (1991) observe lower cross sections approaching and at the minimum. Samson used line sources, whereas Madden *et al.* (1969) and Carlson *et al.* (1973) used synchrotrons. The latter must be corrected for higher-order radiation, which is particularly problematic where the cross section for primary radiation is very low. The electron impact, inelastic scattering results of

Chan *et al.* (1992b) also have higher cross sections approaching the minimum. In addition, they utilize a sum rule in their calibration. These considerations favor selecting the Samson data. However, there is an odd kink in the Samson data in the region of the 3s–np resonances and just beyond the 3s edge, which is not present in the other data. Carlson *et al.* (1973), in particular, comment about this discrepancy. We have utilized the Carlson data to smooth the continuum between 26–30 eV. Otherwise, the data used are from Samson (1966) and Samson *et al.* (1991). The two sections have been separately fitted by regression to a 4-term polynomial.

The $S(p)$ derived from these polynomials are listed in Table 2.34, and the coefficients can be found in Table 2.35.

b.2 Resonances, 26.6–29.2 eV Madden *et al.* (1969) observed a window resonance series corresponding to the transitions $\dots 3s^2 3p^6\ {}^1S_0 \rightarrow \dots 3s3p^6 np$ ${}^1P_1^o$. They record in their Table VI the Fano parameters σ, ρ, Γ, and q for the first three members. From the formula for f_{xs} (see Sect. 2.2.1.b.2) we find $f_{xs} = -0.0328, -0.010, -0.0044$ for $n = 4$–6, i.e. the 'excess' oscillator strength is a

Table 2.34 Spectral sums, and comparison with expectation values for argon ($S(p)$ in Ry units)

Energy, eV	$S(-2)$	$S(-1)$	$S(0)$	$S(+1)$	$S(+2)$
Discrete $\rightarrow {}^2P_{1/2}$					
0–15.937 11[a]	0.8631	0.8576	0.8633	0.8807	0.9091
${}^2P_{1/2} \rightarrow$ 3s edge					
15.937–29.239[b]	1.6462	2.5599	4.0951	6.7379	11.3828
Resonances					
26.6–29.2[c]	−0.0135	−0.0274	−0.0550	−0.1106	−0.2224
3s edge $\rightarrow$ minimum,					
29.239–48.0[d]	0.1951	0.4734	1.1613	2.8833	7.2634
48.0–79.3[e]	0.0152	0.0697	0.3258	1.5499	7.5112
79.3–243.0[e]	0.0144	0.1243	1.1756	12.1848	139.6512
243.0–253.0[f]	0.0006	0.0110	0.1941	3.4474	61.6797
253–264[f]	0.0011	0.0211	0.4004	7.6074	144.5352
264–271[f]	0.0004	0.0087	0.1696	3.3100	64.6725
271–321[f]	0.0030	0.0638	1.3817	29.9907	652.4789
321–336[f]	0.0007	0.0165	0.3993	9.6408	232.7708
336–500[g]	0.0030	0.0892	2.6418	79.1206	2 422.8461
(336–500)[f]	(0.0036)	(0.1067)	(3.0789)	(92.6671)	(2 812.5596)
500–929.7[g]	0.0011	0.0487	2.2991_5	105.5415	5 629.5519
929.7–3203[g]	0.0001	0.0121	1.1995	131.7934	16 350.619
1s $\rightarrow$ 4p,					
3203.3[h]	–	–	0.0022	0.518	122.00
3206–6199.3[i]	–	0.0038	1.1495	361.391	117 710.31
6199.3–10 000[g]	–	0.0005	0.2940	166.003	95 500.6
10^4–10^5[j]	–	0.0002	0.1865	248.223	459 589.1
10^5–10^6[k]	–	–	0.0015	18.567	305 382.0

Table 2.34   (*Continued*)

Energy, eV	$S(-2)$	$S(-1)$	$S(0)$	$S(+1)$	$S(+2)$
$10^6-10^{7\mathrm{k}}$	–	–	–	0.764	121 124.0
$10^7-10^{8\mathrm{k}}$	–	–	–	0.026	41 373.0
$10^8-10^{9\mathrm{k}}$	–	–	–	0.001	13 415.0
$10^9-\infty^{\mathrm{k}}$	–	–	–	–	6 259.0
Total	2.7305	4.3331	17.8853	1190.07	1 186 201.0
Expectation values	2.7729(5)[l]	4.27[m]	18.0		1 157 879.0[n]
	2.770(1)[o]	–	–	–	1 158 048.0[p]
Other values	(2.770)[q]	4.384[q]	(18.0)[q]	1175.8[q]	1 152 400.0[q]
	2.808[f]				
	2.770[r]	4.350[r]			

[a]See Table 2.33 and text.
[b]Mostly from Samson (1966), but smoothing at higher energy from Carlson *et al.* (1973).
[c]Madden (1969); Berrah (1996).
[d]Samson (1966); Samson *et al.* (1991).
[e]Samson *et al.* (1991); Watson (1972).
[f]Chan *et al.* (1992b).
[g]Henke *et al.* (1993).
[h]Deslattes *et al.* (1983).
[i]Wuilleumier (1965).
[j]Chantler (1995).
[k]Using the hydrogenic equation of Bethe and Salpeter (1977).
[l]Coulon *et al.* (1981).
[m]Naon *et al.* (1975).
[n]Hartree–Fock value; Fraga *et al.* (1976).
[o]Orcutt and Cole (1967).
[p]Bunge *et al.* (1993).
[q]Kumar and Meath (1985a).
[r]Olney *et al.* (1997).

Table 2.35   Coefficients of the polynomial $\mathrm{d}f/\mathrm{d}E = ay^2 + by^3 + cy^4 + dy^5$ fitted to data at various energies[a]

Energy range, eV	a	b	c	d
15.9371–29.2395	−25.428 1	170.7881	−247.886	106.5586
29.2395–48.0	76.976 89	−573.622	1 358.922	−976.888
48.0–79.3	14.430 74	−40.8325	−115.985	347.5945
79.3–243	5.617 571	128.2189	−1 203.47	2 660.151
243–336	see text and Fig. 2.32			
336–500[b]	−11.876 8	8 371.694	−109 963	–
500–929.7	35.655 84	4 922.702	8 315 576	−1 757 750
929.7–3203	20.596 92	6 151.107	−2 513.47	−2 337 467
3203–3206	see text			
3206–6199.3	−1004.53	954 912.7	−220 652 027	17 883 565 552
6199.3–10 000	12.263 08	132 886.9	−5 911 229	−624 237 063

[a]$\mathrm{d}f/\mathrm{d}E$ in Ry units, $y = B/E$, $B = $ IP, $^2\mathrm{P}_{1/2} = 15.9371$ eV.
[b]The data of Henke *et al.* (1993) have been used here.

deficit. Recently, Berrah *et al.* (1996) re-examined these resonances with higher resolution, and fitted resonances up to $n = 9$. Their values of ρ, Γ, and q for $n = 3$–6 are very close to those of Madden *et al.* They do not give σ, but using the Madden values of σ, the f_{xs} calculated for $n = 4$–6 are almost identical to those obtained from Madden *et al.* Higher members are considered by extrapolation. The results are included in Table 2.34. The parameters for two observed 2-electron excitation resonances in this region (one at 30.847 eV) are also given by Madden *et al.* The values of f_{xs} are of opposite sign, and almost cancel.

b.3   48.0–79.3 eV   The cross section rises gradually from the minimum at 48 eV to a plateau at $\sim$79.3 eV. We use the data of Samson *et al.* (1991) and Watson (1972), which are in very good agreement and lie about 7% below those of Chan *et al.* (1992b). The values of $S(p)$ from the polynomial fit are given in Table 2.34, while the coefficients are assembled in Table 2.35.

b.4   79.3–243.0 eV   From the plateau at 79.3 eV the photoabsorption cross section declines monotonically until it nears the 2p edge. (The L_{III} ionization potential is 248.6 eV (Jolly *et al.*, 1984).) Beginning at $\sim$243 eV, there is evidence of structure (see Fig. 2.32) heralding the sharp increase at the edge, which we

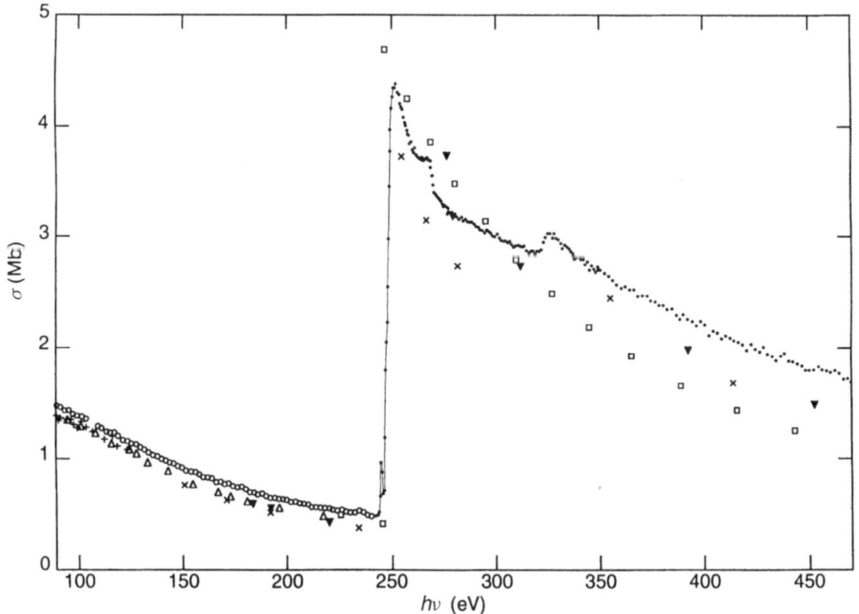

Fig. 2.32 Absolute photoabsorption spectrum of argon, 100–450 eV ($L_{II,III}$ edge). + Samson *et al.* (1991); △ Watson (1972); •○ Chan *et al.* (1992); □ Martin and Wiese (1976); ▽ Henke *et al.* (1993); × Lukirskii *et al.* (1963)

defer to the next section. Between 79.3–243 eV, we again utilize the data of Samson *et al.* (1991) and Watson (1972), which are mutually concordant and 7–10% lower than the cross sections of Chan *et al.* (1992b).

b.5   243–336 eV; 336–500 eV   A glance at Fig. 2.32 reveals a paucity of optical data in the vicinity of the $L_{II,III}$ edge. Clearly, there is structure just below the L_{III} edge and at higher energies, as observed in the electron inelastic scattering data of Chan *et al.* Also, there is a 7% rise at the L_I edge (326.3 eV). However, the sparse optical data confirm an abrupt increase at the edge, and little else. Hence, we adopt the inelastic electron scattering data between 243–336 eV. We have seen that these data are about 7–10% higher than the optical data below 243 eV. Above 336 eV, they are also higher than the few data points of Henke *et al.* (1993) and of Lukirskii and Zimkina (1963). The compilation of Marr and West (1976) lists points above 336 eV which have a much lower cross section than other data, but between 245–295 eV they are higher. The provenance of these points is not clear; their reference set indicates Lukirskii and Zimkina (1963), but they depart significantly from this source.

We have determined the contributions to $S(p)$ between 243–336 eV by graphical integration of the data of Chan *et al.*, recognizing that they may be 7–10% too large. For the smooth continuum between 336–500 eV (limit of the Chan data), which also contain substantial oscillator strength, we have calculated $S(p)$ based on the Chan data, and alternatively by interpolated data of Henke *et al.* (1993), both of which are recorded in Table 2.34. We shall rely on the sum rule analysis as a guide in selecting between these two alternatives.

b.6   500–929.7 eV; 929.7–3203 eV   The data of Henke *et al.* (1993) are fitted in this domain, after partitioning into two sections. The corresponding $S(p)$ are listed in Table 2.34, the coefficients in Table 2.35.

b.7   3203–3206 eV;   3206–6199.3 eV   The K-edge of argon occurs at 3206.3 eV (Breinig *et al.*, 1980). It is preceded by a 1s → 4p excitation, barely resolvable from the K-edge jump (Deslattes *et al.*, 1983). The oscillator strength of this excitation, though small, is recorded in Table 2.34. Higher, two-electron excitations are observed by Deslattes *et al.* (1983), but they are weak modulations of the gradual decline of the $(1s)^{-1}$ ionization continuum. This latter behavior was studied by Wuilleumier (1965). Data are also available from Henke *et al.* (1993). They are in very good agreement (see Fig. 2.33). The more detailed data of Wuilleumier have been fitted by a 4-term polynomial, up to the limit of Wuilleumier's data, 6199.3 eV. The coefficients of this polynomial are given in Table 2.35, the resulting contributions to $S(p)$ in Table 2.34.

b.8   6199.3–10 000 eV   Data from Henke *et al.* (1993) are fitted in this gap.

b.9 10^4–10^5 eV   The calculated cross sections of Chantler (1995) are used.

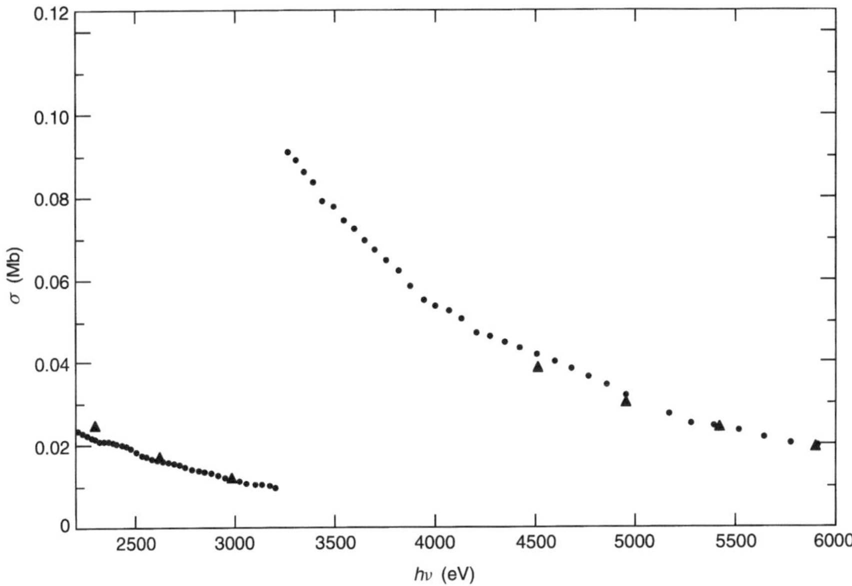

Fig. 2.33 Absolute photoabsorption spectrum of argon, K-edge. ● Wuilleumier (1965);
▲ Henke *et al.* (1993)

2.9.2 The analysis

Our selection of data yields $S(0) = 17.885$, just 0.64% lower than that required (18.0) by the TRK sum rule. Use of the inelastic electron scattering data of Chan *et al.* (1992b), even for just the 336–500 eV region, would have exceeded the TRK value by 1.8%. The value of $S(-2)$ seems well established, by both dielectric constant (2.770 ±0.001, Orcutt and Cole (1967)) and molar refraction (2.7730 ±0.00005, Newell and Baird (1965), 2.7729 ±0.0005, Coulon *et al.* (1981)) measurements, although a subsequent refractivity study (Burns *et al.*, 1986) which may not be as precise yields a slightly lower value. The current spectral sum is about 1.4% too low. Here, the uncertainties in the oscillator strengths in the discrete spectrum and their extrapolations may be suspect. However, $S(0)$ for the discrete region (0.8633) is very close to that obtained by Chan *et al.*, 0.859. Other empirical approaches have found lower values: 0.795 by Kumar and Meath (1985a) and 0.793 by Eggarter (1975). A more likely source of the shortfall in $S(-2)$ is the region between $(3p)^{-1}$ and $(3s)^{-1}$, 15.937–29.239 eV. This region contains the largest contribution to $S(-2)$, and required some adjustment in the data of Samson (1966).

Our spectral sum for $S(-1)$, 4.3331, is ∼25% smaller than the Hartree-Fock value (5.4739) of Fraga *et al.* (1976). The calculated value depends strongly on correlation. Naon *et al.* (1975) used the Bethe–Goldstone method, which implies a high degree of correlation, and obtained $S(-1) = 4.27$, a value 1.5% <u>lower</u> than our spectral sum. The empirical value of Kumar and Meath is higher than ours by $\sim$1.2%, and is probably close to the correct value, since this difference is close to our shortfall in $S(-2)$ and $S(0)$.

Correlation is also important in calculating $S(+1)$. Here, our spectral sum is $\sim$3.6% higher than the Hartree–Fock value (1148.60 Ry units), the direction expected for a more correlated wave function. Our spectral sum is $\sim$1.2% higher than the empirical value of Kumar and Meath.

The Hartree–Fock value for $S(+2)$ should be fairly reliable, since correlation plays a negligible role here. Our spectral sum is $\sim$2.4% too high. The value obtained by Kumar and Meath is actually $\sim$0.5% lower than the Hartree–Fock value.

3

An Aside: The Quantum Yield of Ionization

3.1 Introduction

Rydberg series exist approaching every ionization limit of an atom or molecule. At least some members converging to higher limits have energies above the first limit, the adiabatic ionization potential. These members are degenerate with the ionization continuum accompanying the lowest limit, and consequently are subject to configuration interaction. Some valence states and two-electron excited states, in addition to Rydberg states, may fit this criterion. For atoms, this configuration interaction manifests itself as structure in the photoabsorption spectrum above the IP, and is called autoionization. Fano (1961) has analyzed this interaction, and has defined a parameter q describing the various profiles this structure can assume. In almost all cases involving atoms, the absolute photoabsorption cross section (σ_a) and the absolute photoionization cross section (σ_i) at those resonances are the same. The ratio σ_i/σ_a is defined as η_i, the quantum yield of ionization. Thus, for atoms, $\eta_i \cong 1$, except in a few localized cases, because the autoionization rate is of order 10^{13}/s, whereas the competitive fluorescence rate is approximately 10^8/s. In isolated instances, the absorption process may be electric-dipole allowed, but the autoionization process may be L–S forbidden, but still occur by spin-orbit interaction. In such cases, autoionization occurs at a slower rate, and fluorescence may be more competitive. Examples include atomic oxygen (Dehmer *et al.*, 1973; 1977) and atomic sulfur (Gibson *et al.*, 1986; Chen and Robicheaux, 1994).

Molecules have the freedom to dispose of the absorbed energy by dissociating, at rates comparable with autoionization. Fluorescence must now compete with two rapid processes, and is likely to be observed only in the rare cases where both are forbidden, e.g., certain levels of $^1\Pi_u^-$ symmetry in H_2 (Breton *et al.*, 1980). Weissler and collaborators (Weissler, 1956) made some pioneering measurements of η_i for simple molecules (O_2, CO_2, N_2O) and indeed observed $\eta_i < 1$. Platzman (1960; 1962a; 1962b) recognized the relationship between this behavior and the effect of ionizing radiation (α,β rays) on gaseous matter, and popularized the term 'superexcited states', meaning excited states above the IP. With the limited information available to him, Platzman (1960) conjectured that such states (having $\eta_i < 1$) could exist a few Rydbergs above their ionization potential. He was very careful to define 'dissociation' not necessarily as a true splitting of the molecule, but any internal reorganization of the constituent atoms that disposes of the excitation energy without resulting in ionization (Platzman, 1962a). This caveat may be important in discussing very large molecules.

About 25 years ago, the present author was compiling data on η_i as a function of photon energy for selected molecules, and found in almost all cases that η_i approached unity at $\lesssim 20\,eV$ (Berkowitz, 1979a). The two exceptions at that time (N_2O and SF_6) have since been re-measured, and shown to conform to this behavior (vide infra).

In Figs. 3.1–3.15, examples of quantum yields are displayed for diatomic, triatomic and increasingly larger polyatomic molecules. Each molecule has its unique tale of competition between ionizing and non-ionizing events, but remarkably, they all appear to converge to $\eta_i \approx 1$ at $h\nu \lesssim 20\,eV$ ($\lambda \gtrsim 600\,\text{Å}$). Of course, our selection of molecules is not random. It is dictated by available data, which tend to concentrate on molecules of low Z elements, particularly carbon compounds. Nevertheless, it seems to be an observation which warrants an explanation.

Let us outline a framework for analyzing the events which can take place for individual molecules above their respective IPs.

1. Direct, or prompt ionization. This should be much more rapid than nuclear motion, which can be estimated as the characteristic time of a vibrational period, 10^{-12}–10^{-13} s (faster for hydrides).
2. Direct dissociation (10^{-12}–10^{-13} s).
3. Autoionization. Typically, $(n^*)^3\Gamma$ is roughly constant, where $n^* = n - \delta$, δ = quantum defect, Γ = resonance width. For order of magnitude estimates, $(n^*)^3\Gamma = 11.4 \times 10^{14}\,Hz$ for the broad spin-orbit (s/o) resonances in Xe, $0.085 \times 10^{14}\,Hz$ for sharp s/o resonances in Ne. For $n^* \sim 3$, this corresponds to a range of lifetimes of 2.4×10^{-14}–3.2×10^{-12} s.
4. Predissociation. Can vary from vibrational period to longer times depending upon Landau–Zener coupling strength. Also varies as $(n^*)^{-3}$.
5. Shape resonance. Typically spans few eV. Characteristic time $\sim 10^{-14}$–10^{-15} s.
6. Internal conversion, or radiationless transition. Can become important for large polyatomic molecules, which can dissipate the absorbed energy into many modes, neither ionizing nor dissociating on a laboratory time scale.

In the above outline, we ignore primary fluorescence, but not fluorescence from fragments. We also ignore Auger processes, since the experimental results for η_i direct our attention to lower energy behavior, whereas Auger processes typically involve inner-shell excitation.

One item in the above list, autoionization, requires further elaboration. For molecules, the three modes of energy storage are electronic, vibrational and rotational, and energy exchange from each of these modes to a Rydberg electron can lead to a form of autoionization. The signature of vibrational autoionization is a propensity rule, $\Delta v = -1$, while that for electronic autoionization consists of Franck–Condon factors connecting the resonant, quasi-discrete state with a lower electronic state. This was known more than 20 years ago (Berkowitz 1979a). An early test of this behavior for vibrational (H_2) and electronic (N_2) autoionization was performed by Berkowitz and Chupka (1969). Shortly thereafter, Smith (1970) presented a more detailed analysis of the latter.

In the last two decades, more complex mechanisms have been inferred to rationalize observations. Baer, Guyon and co-workers (Baer *et al.*, 1979; Guyon *et al.*, 1983) observed near-zero energy photoelectrons, as well as the anticipated higher energy ones, in photoionization of N_2O (and later other cases). They called the process producing low energy electrons 'resonant autoionization'. They proposed a model in which the superexcited state was converted via a dissociative state to high vibrational levels of ground state Rydbergs, which vibrationally autoionize. This was later modified by the inclusion of rotational autoionization in the final step (Chupka, 1988). Giusti-Suzor and Jungen (1984) applied multichannel quantum defect theory (MQDT) supplemented by spectroscopic data for NO to deduce that the excited Rydberg state can couple to a valence state which is responsible for predissociation. The dissociation continuum can then couple to the ionization continuum at short range, providing a mechanism for 'vibrational autoionization' with $\Delta v \leq -2$ which can be stronger than the discrete-continuum coupling responsible for conventional vibrational autoionization. This mechanism may offer an explanation for other violations of the propensity rule (Berkowitz and Greene, 1984; Kimman, 1986).

More recently, Kong and Hepburn (1994) studied O_2 photoionization with rotational resolution using a VUV laser and pulsed field ionization (Dixit *et al.*, 1989), zero kinetic energy (PFI-ZEKE) photoelectron spectroscopy. They observed very high vibrational levels of the O_2^+ electronic ground state, well outside the Franck–Condon region. Furthermore, the energies of these high vibrational states were not correlated with any initially excited Rydberg state converging to a higher electronic limit, as invoked by the Baer/Guyon and Giusti-Suzor/Jungen models. To interpret their results, they proposed the existence of a dissociative state (or states) which couples to very high n, very high v Rydberg states of the $(X^2\Pi_g)$ electronic ground state of O_2^+. These very high n states are observed in the usual fashion by PFI-ZEKE. The rotational line intensity for these very high v states was similar to those in the Franck–Condon region, which was offered as support for their model.

Of course, all these delineations could be subsumed by an all-inclusive theory, such as MQDT or R-matrix, but at this time sufficiently accurate applications are limited to diatomics, particularly H_2. Hence, we shall retain the above categories to describe what has been learned in recent years about the photoionization behavior of a number of molecules. At the outset, we shall simplify the discussion by eliminating two categories from consideration – rotational autoionization, because it encompasses much smaller energy ranges than we are concerned with here, and shape resonances, which typically occur outside our spectral range, though their short lifetimes imply rapid ionization.

3.2 Detailed Studies in the Domain of Competitive Processes

3.2.1 Diatomic Molecules

a H_2

H_2 is the prototypical molecule exhibiting vibrational autoionization. Electronic autoionization has been discussed in the context of multiphoton ionization (Dixit *et al.*, 1989), but the state involved is not accessible by single photon ionization. $H_2{}^+$ has a bound ground state ($X^2\Sigma_g^+$, $1s\sigma_g$) and a repulsive state ($2p\sigma_u$) with ground state asymptotes. Franck–Condon factors between $X^1\Sigma_g^+$ and $X^2\Sigma_g^+$ encompass a broad range up to $v' = 17$, with maximum at $v' = 2$. In photoabsorption, Rydberg excitations of $np\sigma$ and $np\pi$ type are observed as resonances which converge to the various vibronic states of the ion. (This designation applies to low n.) Each resonance experiences its own characteristic competition between autoionization and predissociation. These local variations are superposed upon a weak dissociation continuum (Glass-Maujean, 1987), and the inexorable increase of direct ionization with photon energy, rising roughly in step-like fashion as successive vibrational levels of $H_2{}^+$ are accessed. At $\sim$18 eV, η_i appears to reach unity (Backx *et al.*, 1976, Chung *et al.*, 1993).

b N_2

Several PES studies (West *et al.*, 1990; Zubeck *et al.*, 1988; Holland and West, 1987a) seem to agree with the earlier work (Berkowitz and Chupka, 1969) that electronic autoionization, as inferred from Franck–Condon factors, is the prevailing decay mode for Rydberg resonances. The quantum yield rises rapidly at threshold, corresponding to the dominant (0,0) transition to $X^2\Sigma_g^+$ (Fig. 3.1), then oscillates through the Rydberg states approaching $A^2\Pi_u$, $v' = 0$–3, where predissociation and autoionization compete. Fragment fluorescence is not anticipated below 20 eV. The quantum yield becomes 0.975 when the photon energy exceeds most of the Franck–Condon span of the $A^2\Pi_u$ state ($\sim$17.5 eV $\cong$ 710 Å), then increases to unity upon passing through the $B^2\Sigma_u^+$ state (18.8 eV).

c O_2

More recent studies (Cubric *et al.*, 1996; Holland and West, 1987b) verify earlier PES measurements of the I, I′, I″ bands (14.75–15.05 eV) and the H, H′, M, M′ bands (12.4–13.7 eV), and conclude that electronic autoionization is the dominant decay mode. The quantum yield of ionization displays a broad dip between $\sim$700–900 Å (13.8–17.7 eV) with a minimum $\eta_i = 0.25$ at 805 Å (Matsunaga and Watanabe, 1967; Berkowitz, 1979a; Holland *et al.*, 1993). The variation of the quantum yield of the underlying continuum is not very different from that of the superposed Rydberg resonances, implying that direct dissociation and direct ionization are as much in competition as predissociation and autoionization. This suggests the presence of one or more repulsive states in the 700–900 Å region,

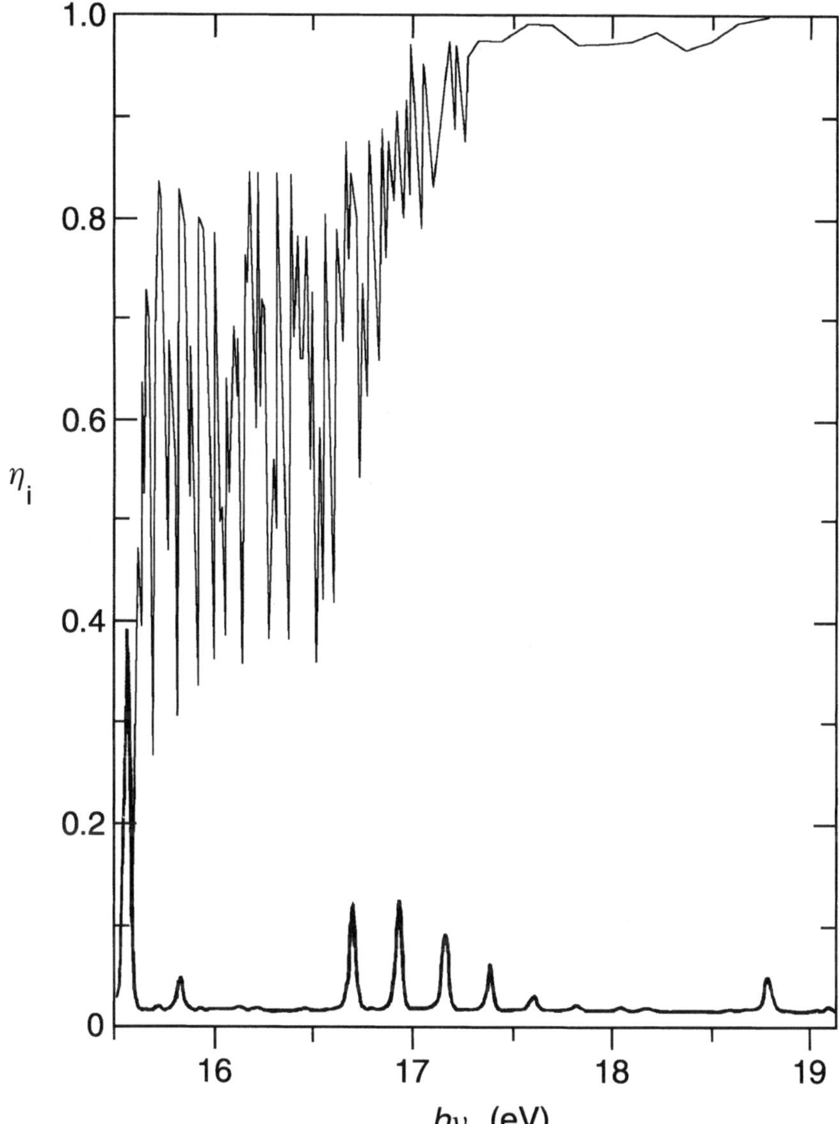

Fig. 3.1 Quantum yield of N_2 (from Shaw *et al*. (1992a)). A He I photoelectron spectrum (from Turner *et al*. (1970)) is shown for guidance

providing some support for the ionization model proposed by Kong and Hepburn (1994). The nature of the dissociating and predissociating states has received considerable attention. From the IP (12.07 eV) to 14.6 eV, predissociation does not lead to fluorescence on energetic grounds. Carlson (1974) has shown that most of the photodissociation between 14.6–16.1 eV is also dark, although Ukai *et al*.

(1992a) do record a structured spectrum for the O 3s ^{3}S $\rightarrow$ ^{3}P emission here. Karawajczyk *et al.* (2000) have measured infrared fluorescence in this region and presented calculations of candidate dissociating states. Between $\sim$670–730 Å (17–18.5 eV), ion-pair formation ($O^+ + O^-$) and fluorescence from some higher energy O atom states, formed by predissociation of Rydbergs converging to $b^4\Sigma_g^-$ is observed. The quantum yield η_i dips at these peaks, but remains $\geq$0.8.

d CO

The consensus of several groups (Leyh *et al.*, 1987; Hardis *et al.*, 1988; Shaw *et al.*, 1997), using Franck–Condon analysis, is that electronic autoionization dominates. Rydbergs converging to the $B^2\Sigma^+$ state autoionize to the $A^2\Pi$ state, and possibly also to the $X^2\Sigma^+$ state, whereas Rydbergs (R_A) converging on $A^2\Pi$ autoionize to the $X^2\Sigma^+$ state. There is a restricted region of R_A states, with $v' = 4$, that appear to autoionize to $A^2\Pi$, $v^+ = 0$, i.e. vibrational autoionization with $\Delta v = -4$, violating the propensity rule (Leyh *et al.*, 1987; Shaw *et al.*, 1997). The quantum yield of ionization resembles that of N_2 (Berkowitz, 1979). The value near threshold, where the (0,0) component of direct ionization to $X^2\Sigma^+$ is dominant, is relatively high ($\sim$0.95 at $\sim$14.76 eV). The Franck–Condon domain for ionization to the $A^2\Pi$ state spans at least 9 vibrational levels, and the Rydbergs approaching this limit are strongly predissociated ($\eta_i \sim 0.5$ at 16.0 eV). When the photon energy surpasses the vibronic range of $A^2\Pi$, at 18 eV, η_i is almost unity. Fluorescence of fragments is not expected below 18.6 eV.

e NO

Nitric oxide, with a valence orbital sequence of $\ldots (1\pi)^4(5\sigma)^2(2\pi)^1$, has a more complex He I photoelectron spectrum than N_2, O_2 and CO, a more intricate variation of η_i with energy (Berkowitz, 1979), and more mechanisms for explaining this variation. Ionization of the singly occupied antibonding 2π orbital results in a low IP (9.264 eV) with a broad Franck–Condon span. Higher ($v' = 1$–4) Rydbergs converging to this limit can autoionize, according to Giusti-Suzor and Jungen (1984), by $\Delta v < -1$ (vide supra). However, autoionization is overwhelmed by predissociation in this domain (Watanabe *et al.*, 1967). Nevertheless, direct ionization, which dominates over direct dissociation (Giusti-Suzor and Jungen, 1984), creeps up incrementally, so that $\eta_i \cong 0.9$ at 10.5 eV, where the Franck–Condon span of $(2\pi)^{-1}$ has just been surpassed. There is a large gap in the photoelectron spectrum between 10.5–15.6 eV, the onset of $(1\pi)^{-1}$, a $^3\Sigma^+$. In this interval, η_i plunges from $\sim$0.9 to $\sim$0.4. Recently, Erman *et al.* (1997) have combined experiment with ab initio theory to interpret the 10.5–13.6 eV region. They suggest photoabsorption to five NO valence states (one attractive, four repulsive) which autoionize by the electronic mechanism. Mitsuke *et al.* (1996) interpret their two-dimensional photoelectron spectrum in the same interval invoking only the single valence bound state, and support their results with Franck–Condon calculations.

Between 13.6–15.6 eV, the dominant photoabsorption and photoionization peaks are Rydbergs converging to $(5\sigma)^{-1}$, $b^3\Pi$, with weaker peaks attributable to $(1\pi)^{-1}$, $w^3\Delta$, $a^3\Sigma^+$ and $W^1\Delta$. Remarkably, the quantum yield does not differ greatly from peaks to valleys. Southworth $et\ al.$ (2000) have recently reported on the photoelectron spectra of $5\sigma \rightarrow 3p\pi$, $4p(\pi,\sigma)$ and $5p(\pi,\sigma)$ all $v' = 0$, converging to $b^3\Pi$. They each display a broad vibrational envelope, characteristic of electronic autoionization to the ground state (Berkowitz, 2000), with some weak contamination at high vibrational energies. This trend continues to higher energy (Erman $et\ al.$, 1995), with $5\sigma \rightarrow$ np, $n = 6$–10 converging to $b^3\Pi$, $v' = 0$, and $5\sigma \rightarrow$ np, $n = 3$–5 converging to $A^1\Pi$, $v' = 0$ being the strongest peaks between 15.6–17.7 eV. In this interval, η_i increases from 0.4 to unity (Watanabe $et\ al.$, 1967), but does not display a marked dependence on autoionization peaks. If anything, the quantum yield at the largest peaks is lower than in the valleys. This suggests that predissociation remains effective, while direct ionization gains, and eventually becomes the dominant player.

f HCl

The valence shell orbital sequence of HCl is $\ldots(3p\sigma)^2 (3p\pi)^4$, or $\ldots (5\sigma)^2 (2\pi)^4$, $^1\Sigma^+$. The quantum yield, depicted in Fig. 3.2, is based on photoabsorption and photoionization data of Frohlich and Glass-Maujean (1990). These authors stress an uncertainty of 30%, and have normalized their photoionization cross sections to photoabsorption for ≤ 700 Å, or ~ 18 eV. Nonetheless, the relative behavior of η_i should have significance, and is roughly supported by (e,2e) measurements (Daviel $et\ al.$, 1984). Common to both is a low value of η_i in the threshold region (12.7–13.6 eV). The He I photoelectron spectrum is dominated by the (0,0) component of $(2\pi)^{-1}$. Since η_i is only ~ 0.3–0.4, direct ionization must have competition, presumably from direct dissociation, although there is some structure indicating autoionization. Spin-orbit and rotational autoionization play a role in the first 0.2–0.3 eV above threshold (Zhu $et\ al.$, 1993; Drescher, 1993), but this is inconsequential in the present context. Similarly, vibrational autoionization near threshold should be insignificant, given the strong (0,0) Franck–Condon factor. The large fractional dissociation could conceivably be attributed to coupling with a $^3\Sigma^+$ $(\sigma \rightarrow \sigma^*)$ repulsive curve calculated by Bettendorff $et\ al.$ (1982). Fragment fluorescence $(H + Cl^*)$ cannot be expected below 13.36 eV $(\sim 928$ Å), and in fact commences there (Frohlich and Glass-Maujean, 1990).

At ~ 13.7 eV (905 Å), the quantum yield ascends rapidly. Terwilliger and Smith (1975) observed the onset of prominent peaks in their absorption spectra at this energy, which they assigned to two electronic states with vibrational progressions. They recognized them to be low Rydbergs converging to $(5\sigma)^{-1}$, $A^2\Sigma^+$, and they were subsequently (Lefebvre-Brion and Keller, 1989) assigned to $(5\sigma)^{-1}$ $4p\pi\,^1\Pi$ and $(5\sigma)^{-1}$ $3d\pi\,^1\Pi$ (and at higher energies, $4d\pi$ and $5d\pi$). By fitting electronic band contours, they obtained the Fano parameters q, Γ and ρ^2 (Fano and Cooper, 1965), which they used to predict the photoelectron spectra of the (presumed) autoionizing bands, assuming electronic autoionization. No direct

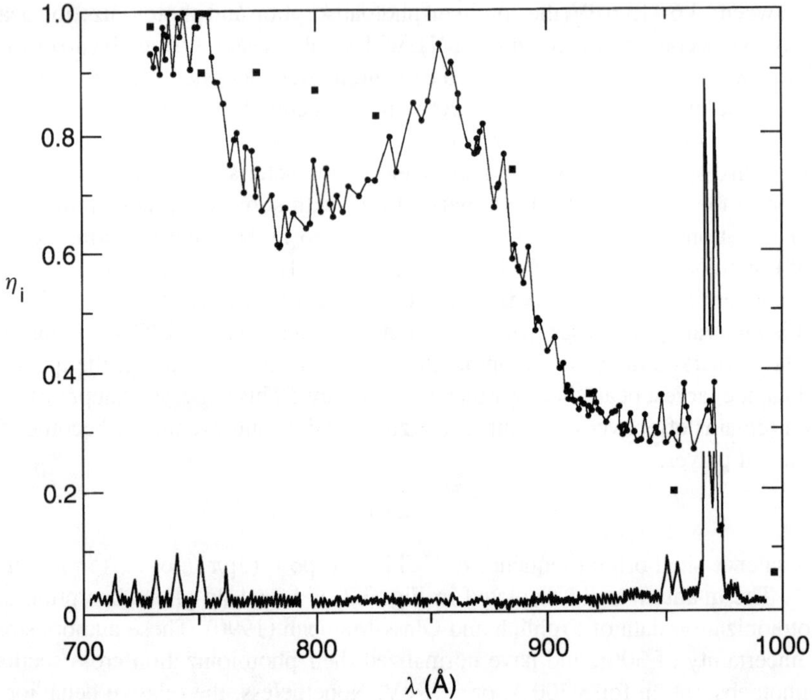

Fig. 3.2   Quantum yield of HCl. ∘ Frohlich and Glass-Maujean (1990); □ Daviel *et al.* (1984). A He I PES, from Turner *et al.* (1970) (approximate wavelength scale) is shown for guidance

test of this prediction is known to this author. Indirect tests include an MQDT treatment incorporating electronic autoionization (Lefebvre-Brion *et al.*, 1988; Lefebvre-Brion and Keller, 1989), which shows fair agreement with experimental photoionization spectra, and an analysis of a threshold photoelectron spectrum (Frohlich *et al.*, 1991) which yields good agreement with the experiment when using the Franck–Condon factors of Terwilliger and Smith. The dip in η_i between ∼750–850 Å (Fig. 3.2) is probably exaggerated (Frohlich 2000), but fragment fluorescence is observed in this region, indicating predissociation. Peaks in photoabsorption and photoionization do not correlate with peaks in fluorescence. Hence, a direct competition between autoionization and predissociation from the same states is not occurring, but rather different (and weaker) excited states are predissociated. Their assignments are in dispute (Lefebvre-Brion and Keller, 1989; Frohlich and Glass-Maujean, 1990). White *et al.* (1987) inferred an additional vibrational autoionization process involving high v', n' Rydberg members converging to $A^2\Sigma^+$, yielding low v^+ members of $A^2\Sigma^+$, but with $\Delta v \ll -1$. However, in overview, electronic autoionization of Rydberg states converging on $A^2\Sigma^+$ predominates, with strong competition from dissociation near threshold, and weak competition from other states than autoionizing ones (750–850 Å). All structure wanes below 750 Å ($>16.5\,\text{eV}$), where direct ionization prevails and η_i approaches unity as the $A^2\Sigma^+$ state is transcended.

3.2.2 Triatomic molecules

a H_2O

The valence orbital sequence of H_2O is $\ldots(1b_2)^2\ (3a_1)^2\ (1b_1)^2,\ \tilde{X}^1A_1$. Our understanding of competing autoionization and predissociation processes in H_2O is still in an early stage. Although experimental photoabsorption and photoionization spectra have been known for almost three decades, theorists began to express interest only when cooled, rotationally resolved spectra became available (Page $et\ al.$, 1988; Dehmer and Holland, 1991). Initial efforts employing MQDT concerned only the region within 0.3 eV of the ionization threshold, where rotational ionization is expected (Child and Jungen, 1990; Child $et\ al.$, 1991). Vrakking $et\ al.$ (1993) have extended this study to somewhat higher energy, but without incorporating a vibrational autoionization mechanism. Since the (0,0,0) transition is dominant for $(b_2)^{-1}$ ionization, vibrational autoionization is unlikely to be significant near threshold. More recently, some of the discrete features in the photoabsorption and photoionization spectra have been assigned as vibrational progressions (bending mode) of 3d, 4d and 5d Rydbergs converging on $(3a_1)^{-1}$, $\tilde{A}^2A_1$ (Child, 1997). Child argues that these linear $(3a_1 \rightarrow ndb_1)^1B_1$ states and the bent $(1b_1 \rightarrow nda_1)^1B_1$ states are completely mixed in linear geometry, but the admixture is progressively quenched as the molecule bends. The rate of autoionization above the bent series limit (i.e., the $\tilde{X}^2B_1$ state) is related to the product of an electronic matrix element and a Franck–Condon factor connecting the Rydberg state with the $\tilde{X}^2B_1$ state. Child describes this as a 'specifically polyatomic vibronic mechanism' involving 'purely electronic interaction', but it has elements similar to electronic autoionization in diatomic molecules.

The quantum yield of ionization (Fig. 3.3) extracted from Katayama $et\ al.$ (1973) and Haddad and Samson (1986) differ in detail, partly because one (Katayama $et\ al.$, 1973) used a continuum source, and the other (Haddad and Samson, 1986) used a multi-line light source. The photoabsorption peaks tend to produce lower values of η_i. There is a local maximum at ~800 Å and a broad minimum at ~720 Å (Katayama $et\ al.$, 1973) which is not apparent in the line source data. The features common to both are an abrupt onset ($\eta_i \sim 0.35$) and a gradual increase to $\eta_i \sim 1.0$ at 620 Å. The relatively low value near threshold implies competition from both direct dissociation and predissociation. The threshold for the OH ($X \rightarrow A$) fluorescence is ~1359 Å, and it is indeed observed (Lee and Suto, 1986; Dutuit $et\ al.$, 1985), but at the ionization threshold it accounts for only 3% of the non-ionizing processes, the remainder being dark (presumably H(1s) + OH(X)). In fact, the quantum yield of fluorescence remains low from IP to 600 Å, having a broad peak at ~700 Å roughly matching the dip in η_i, but accounting for only $\sim2\%$, whereas the yield of neutral processes implied by the data of Katayama $et\ al.$ (1973) is $\sim30\%$.

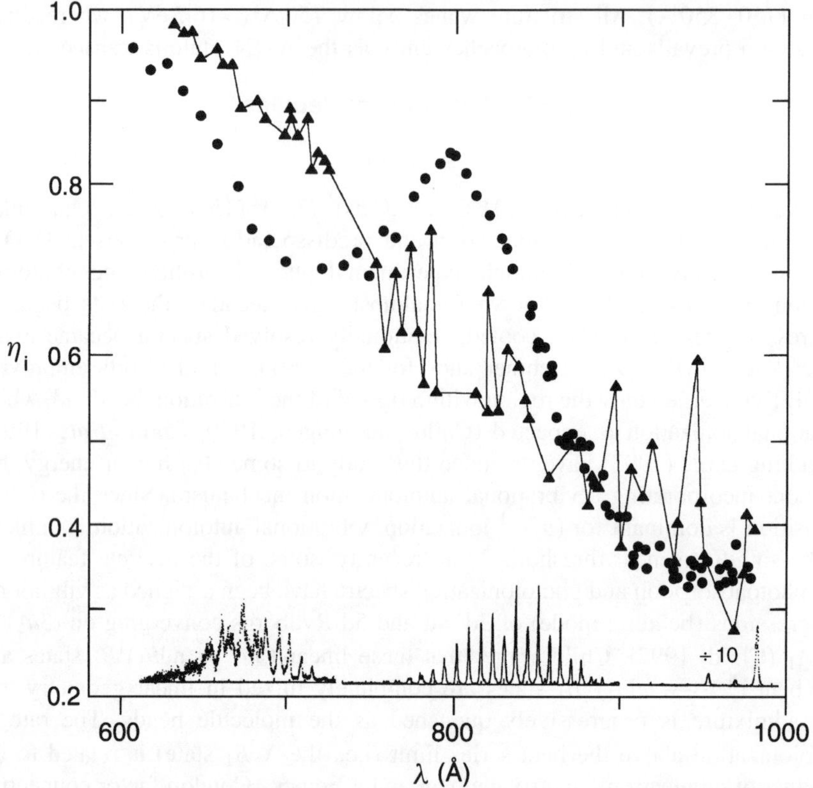

Fig. 3.3 Quantum yield of H_2O. ● Katayama *et al*. (1973); △ Haddad and Samson (1986). A He I PES, from Reutt *et al*. (1986a) (approximate wavelength scale) is also shown

Near the bottom of Fig. 3.3, we plot the He I photoelectron spectrum, taken from Reutt *et al*. (1986a). The energy scales are approximately commensurate. Qualitatively, we can see η_i abruptly rise to ~ 0.35 for the sharp $(1b_1)^{-1}$ threshold, gradually increase to ~ 0.7 as we transcend the broad $(3a_1)^{-1}$ band, and make its final ascent to ~ 1.0 as the photon energy passes across the broad $(1b_2)^{-1}$ ionization. It is roughly the behavior one might expect from direct ionization and direct dissociation, although autoionization and predissociation are super-posed. However, quasi-discrete structure in photoabsorption and photoionization becomes barely detectable below $\sim 670\,\text{Å}$, and the value of η_i tells us that direct ionization is pre-eminent below $\sim 620\,\text{Å}$.

b CO_2

The occupied molecular orbitals of CO_2 in its valence shell are $\ldots(4\sigma_g)^2\,(3\sigma_u)^2\,(1\pi_u)^4\,(1\pi_g)^4$. Only $(1\pi_u)^{-1}$ leads to an extended Franck–Condon region, the other orbitals being antibonding or non-bonding, with dominant (0,0,0) transitions. The quantum yield of ionization rises rapidly from threshold ($\sim$900 Å) to reach $\sim$0.85 at 840 Å (Shaw $et\ al.$, 1995; Nakata $et\ al.$, 1965; Barrus $et\ al.$, 1979). Some, but not all of the non-ionizing moiety appears as fluorescence (Ukai $et\ al.$, 1992b) attributed to $CO(A^1\Pi \to X^1\Sigma^+)$. Between $\sim$830–760 Å, there is a broad dip in η_i (to $\sim$0.7 at 800 Å), with superposed structure corresponding to the Tanaka–Ogawa series (converging to $(1\pi_u)^{-1}$, $A^2\Pi_u$, designated R_A) and the Henning series (converging to $(3\sigma_u)^{-1}$, $\tilde{B}^2\Sigma_u^+$, designated R_B). This behavior is reflected in the fluorescence yield (Ukai $et\ al.$, 1992b), which displays peaks corresponding to minima in η_i, especially in the R_B sharp series ($\lambda \sim$ 690–707 Å). A bent valence state and some R_A states may be implicated as the sources of dissociation/predissociation in the 830–760 Å region (Shaw $et\ al.$, 1995).

Several photoelectron spectroscopic studies have been performed on the auto-ionizing peaks, with inconsistent results. Baer and Guyon (1986) found 'the photoelectron energy distribution obtained upon excitation of the Rydberg states is consistent with that given by the Bardsley–Smith model', while West $et\ al.$ (1996) found 'rather poor agreement between calculations and experiment in the resonance region', indicating that the Born–Oppenheimer approximation is not valid, and that 'the method outlined by Smith does not look promising for CO_2'. Their experiment showed that a considerable fraction of the intensity goes into vibrational modes other than the fundamental symmetric stretch. The Franck–Condon calculation does not include effects of vibronic coupling, which 'is clearly going to be important in any theoretical analysis'. Even West $et\ al.$ find approximate agreement with experiment using the Franck–Condon approach, but evidently complications arise as we go from diatomic molecules (one vibrational mode) to triatomics (3 or 4 vibrational modes). However, in the present context these details need not concern us. More consequential is the observation (Ukai $et\ al.$, 1992b) that there is a significant decrease in fluorescence as the excitation energy exceeds the $\tilde{A}^2\Pi_u$ and $\tilde{B}^2\Sigma_u^+$ thresholds, which coincides with $\eta_i \simeq 1.0$. The implication is that direct ionization takes over after passing these thresholds.

c N_2O

N_2O, isoelectronic with CO_2 but with an asymmetric structure, has the valence orbital sequence $\dots(6\sigma)^2\ (1\pi)^4\ (7\sigma)^2\ (2\pi)^4$. Here, the bonding orbital is (1π), and ejection of an electron from this orbital corresponds to the third IP (second for CO_2). The quantum yield of ionization, based on the data of Shaw $et\ al.$ (1992b) is seen (Fig. 3.4) to approach unity at $\sim$615 Å. This supercedes earlier data (see Berkowitz, 1979), where η_i was only $\sim$0.7 at 600 Å. Similarly to CO_2, η_i increases rapidly to $\sim$0.8 within $\sim$0.3 eV of its IP, followed by a broad dip between $\sim$940 – 810 Å ($\eta_i \sim$0.55 at $\sim$900 Å). Toward shorter wavelengths, η_i gradually increases, but with pronounced fine structure, also visible in fluorescence (Ukai $et\ al.$, 1994).

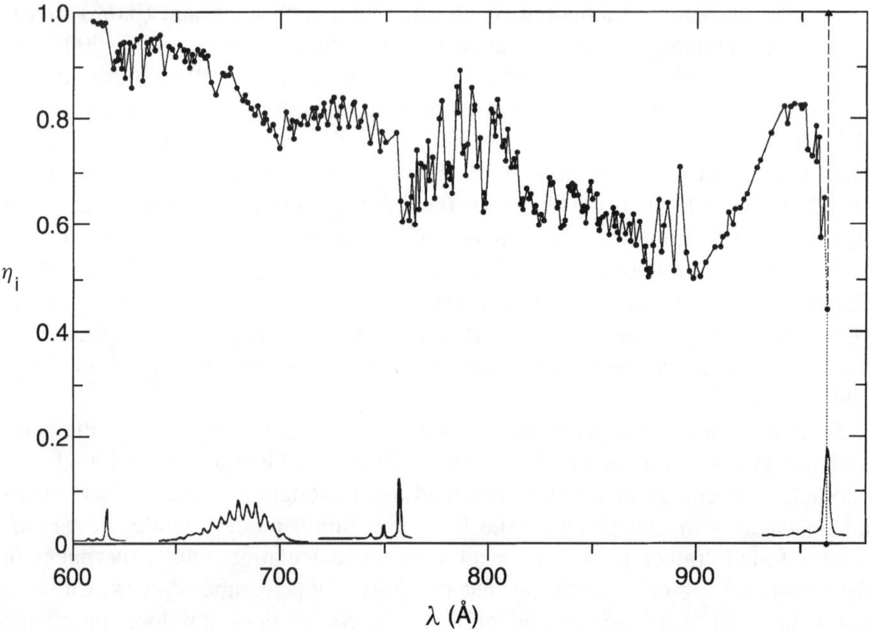

Fig. 3.4 Quantum yield of N_2O (from Shaw *et al.* (1992b)). A He I PES, from Turner *et al.* (1970) (approximate wavelength scale) is also shown

Two-dimensional (electron and photon) photoelectron spectroscopic studies have been performed on Rydberg states converging on $(7\sigma)^{-1}$ $(\tilde{A}^2\Sigma^+)$ states (Sokell *et al.*, 1996) and on $(6\sigma)^{-1}\tilde{C}^2\Sigma^+$ Rydberg states (Sokell, 1997). The former focused not only on the Franck–Condon region of photoexcitation (primarily (0,0,0) and (1,0,0)), but also on the Franck–Condon gap. Even in the Franck–Condon allowed region, vibrational selectivity was observed which was attributed to electronic–vibrational coupling, or interference between Rydberg states belonging to different series. Progressions in the symmetric stretch (ν_1) were prominent in the photoelectron spectrum, the decrease in intensity with quantum number ν_1 being 'qualitatively consistent with the Franck–Condon principle'. Thus, as in CO_2, electronic autoionization appears to be a partial explanation of the data, but a more complete interpretation requires inclusion of vibronic coupling. The Franck–Condon gap was covered more comprehensively than earlier work, and displayed long vibrational progressions, primarily in ν_1. The tentative interpretation (Sokell *et al.*, 1996) was autoionization from dissociative states which predissociate R_A states, rather than vibrational autoionization of high v, high n R_X states coupled to these dissociative states.

Rydberg states converging to the $\tilde{C}^2\Sigma^+$ state were found (Sokell *et al.*, 1997) to exhibit both electronic and vibrational selectivity. The $np\sigma$ series autoionized preferentially to the $\tilde{X}^2\Pi$ ionic state, while the $np\pi$ series strongly favored the

$\tilde{A}^2\Sigma^+$ ionic state, a behavior rationalized by the authors as a symmetry-based (ℓ-conserving) propensity rule. Decay to the $\tilde{B}^2\Pi$ ionic state was most strongly seen from the ndπ/nsσ (incompletely resolved) series.

In the He I photoelectron spectrum, the $\tilde{X}^2\Pi$ and $\tilde{C}^2\Sigma^+$ states have the most dominant (0,0,0) components. Thus, Rydberg series converging to $\tilde{C}^2\Sigma^+$ which autoionize to $\tilde{X}^2\Pi$ would, from Franck–Condon considerations, favor conservation of vibrational quanta. This appears to be the dominant effect, although Sokell *et al.* (1997) note that these transitions exhibit far less vibrational selectivity than the R_A states (Sokell *et al.*, 1996). The R_C Rydbergs decaying to the $\tilde{A}^2\Sigma^+$ state show even less vibrational selectivity, although (0,0,0) is still dominant, and the weaker transitions are almost entirely single quantum excitations. Hence, while there is ample evidence for vibronic coupling, the electronic autoionization mechanism appears to qualitatively explain the decay of R_C states to $\tilde{A}^2\Sigma^+$ and $\tilde{X}^2\Pi$. By contrast, the R_C states autoionizing to $\tilde{B}^2\Pi$ appear to be formed by non-Franck–Condon resonant processes. They occur at $h\nu > 18.5\,\mathrm{eV}$, where η_i, is already quite high ($\sim$0.9). Their presence is seen as weak minima in η_i, and maxima in fluorescence (Ukai *et al.*, 1994), implying predissociation.

In a similar vein, R_B Rydbergs appear as minima in η_i and maxima in fluorescence, as well as peaks in the O^+ fragment from N_2O (Berkowitz and Eland, 1977).

d NO$_2$

Although little is known about the decay mechanisms of superexcited states in NO_2, we include it here because of some unusual properties. The neutral ground state, with an unpaired electron, has a bond angle of $\sim$134°, while the ionic ground state is linear. The connecting Franck–Condon factors near the IP (9.586 eV) are expected to be poor. This manifests itself as an exceedingly low quantum yield of ionization, $\sim$4 $\times$ 10^{-5} at 9.72 eV (Nakayama *et al.*, 1959). Figure 3.5 displays η_i versus energy from two sources: photoabsorption (Nakayama *et al.*, 1959) from 10–11.5 eV, and inelastic electron scattering (Au and Brion, 1997) from 15.0–24.5 eV. Superficially, there appears to be a monotonic increase in η_i, with a plateau between $\sim$16–17.5 eV. There could conceivably be structure in the missing region (11.5–15.0 eV), since here the He I photoelectron spectrum contains four sharp bands. The plateau roughly corresponds to a Franck–Condon gap. Nine bands have been observed in the valence shell photoelectron spectrum (Baltzer *et al.*, 1998), the highest one at 21.3 eV. Somewhat surprisingly, η_i is only $\sim$0.88 upon traversal of this energy, and does not achieve a value of unity until 24.5 eV.

Vibrational autoionization mechanisms have been identified (Matsui and Grant, 1996) in a limited region near threshold (IP–9.88 eV). With so many ionic states and concomitant Rydberg series in the valence region, the relative importance of direct and indirect mechanisms for ionization and dissociation await future research.

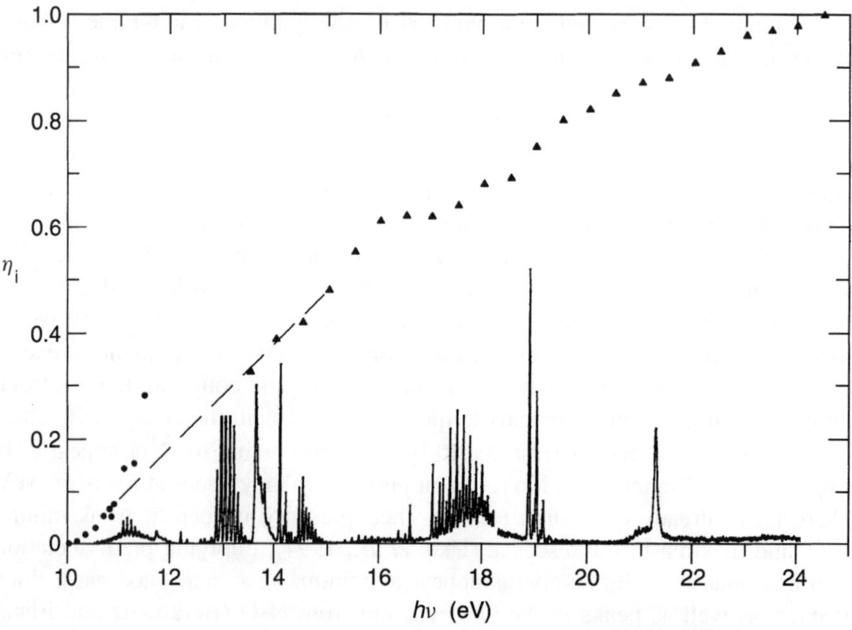

Fig. 3.5　Quantum yield of NO_2. △ Au and Brion (1997); ○ Nakayama *et al*. (1959). Also shown is a composite of the He I and He II photoelectron spectra from Baltzer *et al*. (1989)

3.2.3 **Polyatomic molecules**

a *NH$_3$*

NH_3 is pyramidal in its neutral ground state. The occupied valence orbitals, in C_{3v} symmetry, are $\ldots(1e)^4\,(3a_1)^2$. Excitation from the $3a_1$ orbital leads to planar Rydberg states, and electron ejection forms the planar NH_3^+ ground state. The change of geometry in these electronic transitions results in a vibrational progression in the out-of-plane bending, or umbrella mode. Figure 3.6, culled from several sources (Samson *et al*., 1987b; Watanabe and Sood 1965; Xia *et al*., 1991) depicts the variation of quantum yield with energy. Also shown, on an approximately commensurate energy scale, is the He I photoelectron spectrum corresponding to $(3a_1)^{-1}$ from Edvardsson *et al*. (1999), and $(1e)^{-1}$ from Rabalais *et al*. (1973). The $(3a_1)^{-1}$ photoelectron spectrum clearly shows the umbrella vibrational progression. The integrals of these vibrational peaks would be steps, and such steps are observed in the photoionization spectrum (see, for example, Berkowitz, 1979, pp. 128–130). They imply that the direct ionization mechanism prevails across this band. However, the absolute photoabsorption cross section bears little resemblance to the absolute photoionization cross section. The quantum yield of ionization grows from 0.02 at the IP to ∼0.35 at the band terminus. Weak peak structure is observed superposed on

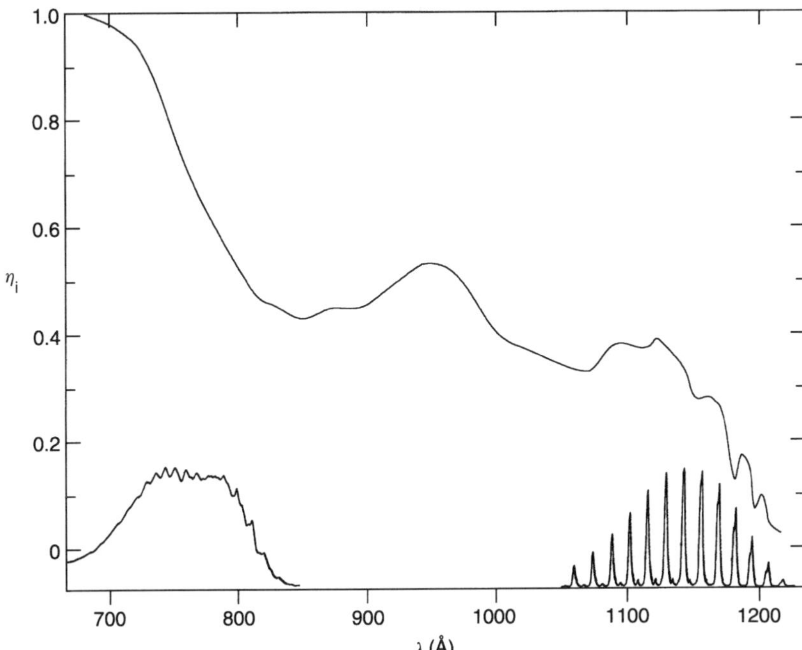

Fig. 3.6 Quantum yield of NH_3. 680–925 Å: Samson *et al.* (1987b); 925–1120 Å: Watanabe and Sood (1965); 1120–1217 Å: Xia *et al.* (1991). He I PES: $(3a)^{-1}$, 1st band, Edvardsson *et al.* (1999); $(1e)^{-1}$, Rabalais *et al.* (1973)

the steps in photoionization, but it is difficult to assign in a room-temperature spectrum. Using supersonic cooling to collapse the rotational population and a VUV laser confined to narrow wavelength regions (1178–1190; 1155–1165 Å), Miller *et al.* (1988) observed Rydberg series converging to $v^+ = 3$ and 5, and concluded that vibrational autoionization was occurring with $\Delta v = -1$. Bacon and Pratt (2000) used two-photon resonant, three photon excitation via the $\tilde{C}'^1 A'_1$ intermediate state to probe the region between $v^+ = 0$–2. In the Rydberg series studied, the $n = 12$–14 Rydberg states were found to vibrationally autoionize 95% with $\Delta v = -1$, but the $n = 11$ states did so with 75–90% probability, the remainder being $\Delta v = -2$. However, the dominant feature of the region corresponding to the first photoelectron band is dissociation and/or predissociation, with direct ionization gaining as the band is traversed.

There is a broad peak in η_i between the photoelectron bands (900–1000 Å), which is attributable to a dip in the photoabsorption spectrum. Then, at about the onset of $(1e)^{-1}$, $\tilde{A}^2 E$, η_i begins an ascent from ~ 0.4 to ~ 1.0 at 680 Å (18.2 eV), near the high-energy end of this band. A plausible inference is that this increase is due to direct ionization resulting from $(1e)^{-1}$.

Locht *et al.* (1991) have performed photoionization studies of several isotopomers of NH_3. They have interpreted their data in terms of vibrational autoionization with Δv up to -9, but their analysis is weakened because they arrived at an incorrect adiabatic IP (10.072 eV), to be compared with 10.1864 eV given in Sect. 6.1, and hence the assigned quantum defects are in error.

b C_2H_2

The valence orbitals of C_2H_2 are $\ldots(2\sigma_u)^2\ (3\sigma_g)^2\ (1\pi_u)^4(1\pi_g)^0\ (3\sigma_u)^0$. The adiabatic IPs from He I photoelectron spectroscopy (Reutt $et\ al.$, 1986b) are 11.40 eV ($\tilde{X}^2\Pi_u$), 16.30 eV ($\tilde{A}^2A_g$, trans-bent) and 18.39 eV ($\tilde{B}^2\Sigma_u^+$). The first band is strong, with the (0,0,0) component being dominant, but there is evidence for vibrational autoionization (Ono $et\ al.$, 1982). After a gap of almost 5 eV, a weaker, broader second band appears, and then a still weaker third band. The quantum yield of ionization rises rapidly at onset (Fig. 3.7) to $\eta_i \cong 0.88$,

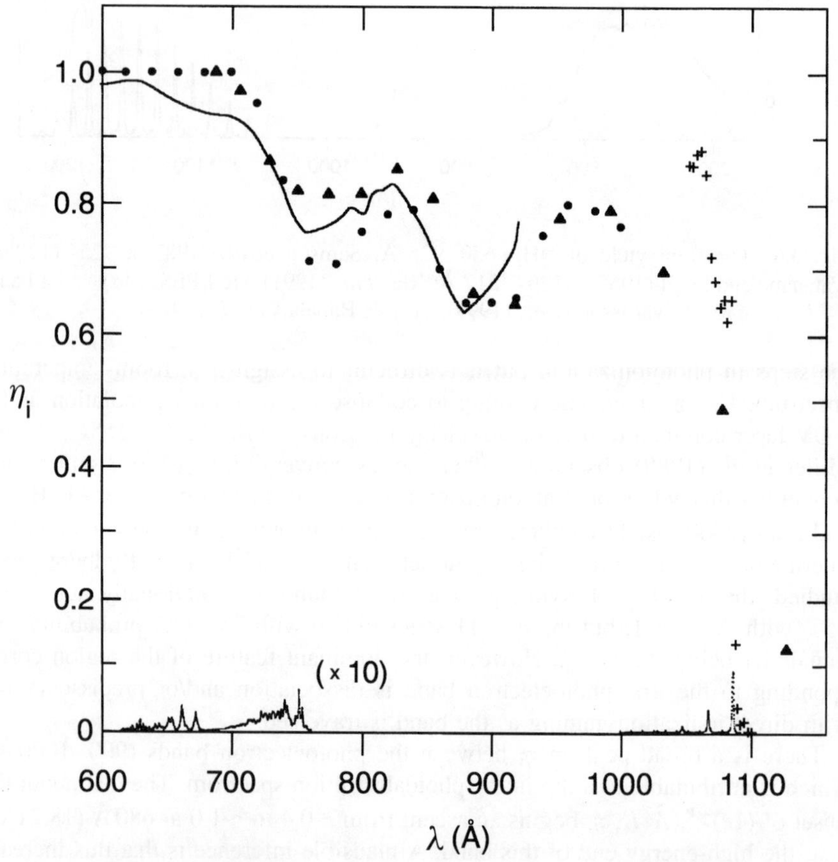

Fig. 3.7   Quantum yield of C_2H_2. • Ukai $et\ al.$ (1991); ○ Metzger and Cook (1964); + Person and Nicole (1970); △ Cooper $et\ al.$ (1988). He I PES, approximate wavelength scale, from Reutt $et\ al.$ (1986b)

suggesting an important role for direct ionization. In the gap between the first two bands in PES, the photoabsorption and photoionization cross sections display two broad peaks at ~ 13.3 and $15.3\,\mathrm{eV}$ (930 and $810\,\text{Å}$), with some fine structure (Metzger and Cook, 1964; Ukai $et\,al.$, 1991). The maxima of these peaks are offset in σ_a and σ_i, and hence their appearance in Fig. 3.7 is closer to that in σ_i. There has been much debate about the origin of the two broad peaks, from both a calculational and experimental standpoint. One school attributes the broad peaks to autoionization from valence states ($3\sigma_g \rightarrow 3\sigma_u$, $2\sigma_u \rightarrow 1\pi_g$) calculated by Hayaishi $et\,al.$ (1982). Recent adherents of this view are Hattori $et\,al.$ (1997) and Avaldi $et\,al.$ (1995). The breadth of the peaks is said to be caused by large geometry changes between lower and upper states.

The other school attributes these broad peaks, directly or indirectly, to shape resonances. The calculational underpinning is from Lynch $et\,al.$ (1984). Experimental supporters are Ukai $et\,al.$ (1991), who find evidence in η_i for three shape resonances (σ^*, π^*, σ^*) and Holland $et\,al.$ (1999), who seem to favor autoionizing resonances interacting with a shape resonantly enhanced background continuum. Recent theoretical work, incorporating interchannel coupling (Wells and Lucchese, 1999) or using RPA (Yasuike and Yabushita, 2000) agrees with the earlier consensus that the higher-energy feature ($\sim 15.3\,\mathrm{eV}$) is mainly due to a $2\sigma_u \rightarrow 1\pi_g$ valence transition with superimposed Rydberg structure, but the lower-energy hump is attributed to a low Rydberg $3\sigma_g \rightarrow 3p\sigma_u$ transition which, according to Yasuike and Yabushita (2000), is perturbed by the $3\sigma_g \rightarrow 3\sigma_u$ valence transition. This latter interpretation agrees with the 2D photoelectron spectrum of Hattori $et\,al.$ Both decays imply electronic autoionization.

The quantum yield η_i increases rather rapidly from ~ 0.8 to 1.0 as the photon energy crosses the $\tilde{A}^2A_g$ ionic state, which occurs just above the second broad peak. This final ascent appears to conform to a direct ionization mechanism. The quantum yield of fluorescence peaks at $\sim 16.5\,\mathrm{eV}$ ($750\,\text{Å}$) with $\eta_d \sim 0.05$–0.07 (Ibuki $et\,al.$, 1995; Han $et\,al.$, 1989), implying that dark dissociation channels make a substantial contribution.

The confusion and controversy encountered in arriving at a detailed understanding of the behavior of superexcited states (and hence η_i) for a relatively simple polyatomic molecule (C_2H_2) is perhaps a harbinger of the complexities to be expected for larger polyatomic molecules.

c C_2H_4

The most recent determination of η_i is by Holland $et\,al.$ (1997). In agreement with earlier work, η_i exhibits a broad hump near threshold, roughly reflecting the first photoelectron band, but extending beyond it. This behavior suggests some contribution from autoionization, in addition to direct ionization. The quantum yield rises rapidly through the second photoelectron band, with a weak minimum between the second and third photoelectron bands. The competing process of photodissociation is particularly strong between the first and second photoelectron bands. Detailed information regarding decay mechanisms is limited. Baudais and

Taylor (1980) found autoionizing resonances in the 14–15 eV region (near the third electronic state) decaying to the ionic ground state, verified by Grimm *et al*. (1991). This implies electronic autoionization, but occurs when η_i is already ~0.9.

d C_6H_6

The η_i of C_6H_6 (Fig. 3.8) has a pronounced hump near threshold, and subsequent plateaus. However, the σ_i curve (Rennie *et al*., 1998) may be more revealing. Roughly it displays monotonic increases across photoelectron bands, and flat regions between them. The Franck–Condon gap between the first and second photoelectron bands, a valley in η_i, corresponds to a maximum in σ_d. This behavior suggests the prevalence of direct ionization. Staib and Domcke (1991) have considered the influence of two Rydberg series converging to the Jahn–Teller split ionic ground state, and conclude that autoionization is strongly quenched by radiationless-decay channels. This latter study encompasses only the first 0.3 eV above threshold.

e SF_6

The quantum yield of ionization (Fig. 3.9), from Holland *et al*. (1992), is punctuated by step-like features in its monotonic ascent from ~0–1. The He I photoelectron spectrum distorts the relative intensity (presumably due to autoionization), but this need not concern us, since $\eta_i \approx 1$ at 21.2 eV. Comparison of

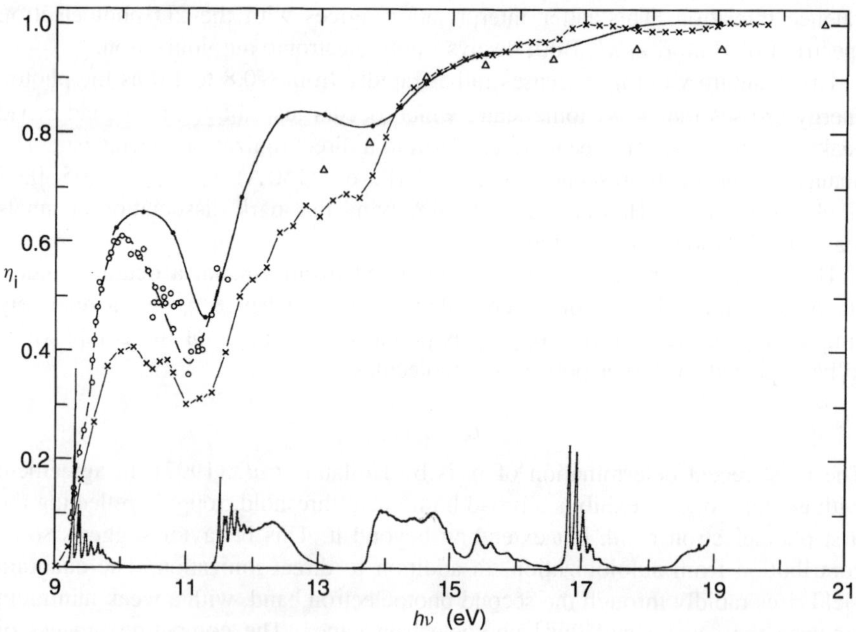

Fig. 3.8   Quantum yield of C_6H_6. ● Rennie *et al*. (1998); ○ Person (1965); △ Yoshino *et al*. (1973); × Jochims *et al*. (1996). He I PES from Baltzer *et al*. (1997)

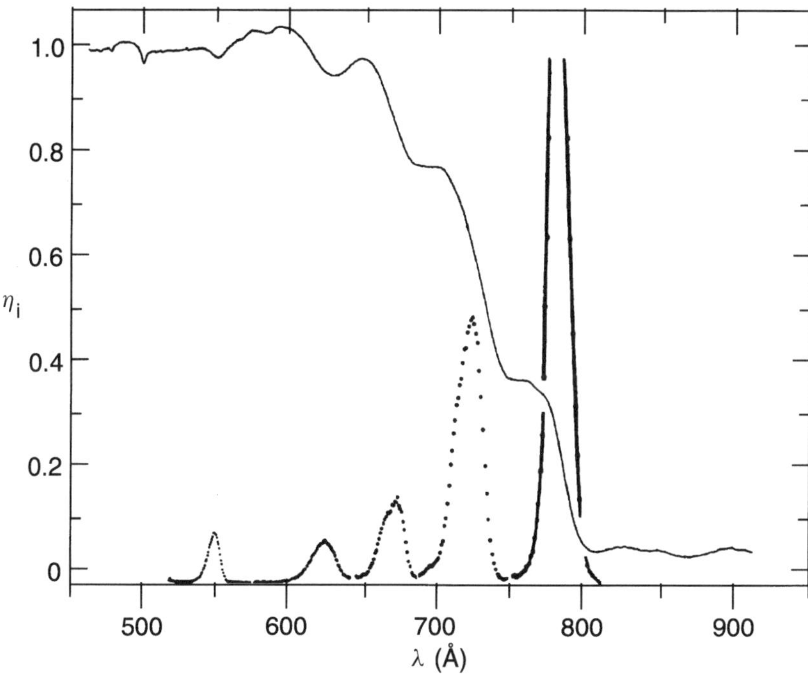

Fig. 3.9 Quantum yield of SF_6 (from Holland *et al.* (1992)). He II PES from Karlsson *et al.* (1976). (First band appears anomalously strong in He II PES, but is weak in He I PES)

the He II photoelectron spectrum (Karlsson, 1976) with η_i (Fig. 3.9) reveals that the ascents of the steps correspond to the peaks in PES, and the plateaus to the gaps between the peaks. This is just the behavior expected for direct ionization.

Let us briefly summarize our observations to this point. For diatomic molecules, electronic autoionization is usually the prevalent ionization mechanism at photoabsorption peaks. Exceptions are H_2, where Rydberg states converging to higher IPs are non-existent, some R_A states of CO and HCl, where $\Delta v \ll -1$ is inferred, and near the lowest IP (e.g., NO), where $\Delta v = -1$ and $\Delta v < -1$ are observed. At $h\nu \simeq 20\,\mathrm{eV}$, photoabsorption structure wanes or vanishes, $\eta_i \approx 1$, and direct ionization reigns supreme. For triatomic molecules, although electronic autoionization appears to be the most common decay mode, Franck–Condon analysis gives only approximate agreement. The appearance of additional vibrations implies vibronic coupling. Some vibrational ionization is observed near threshold (NO_2), as is autoionization from predissociated R_A states. The polyatomic molecules we have examined thus far (which contain noteworthy structure in their η_i curves) display weak vibrational autoionization (NH_3), some electronic autoionization (C_2H_2, perhaps C_2H_4), but increasingly direct ionization appears to become the controlling ionization mechanism.

3.2.4 General observations of other polyatomic molecules

We now turn to those molecules which have little or no observable structure in their η_i curves. The simplest are methane and silane. The valence shell occupied orbitals of CH_4 are $\ldots(2a_1)^2\,(1t_2)^6$, and of SiH_4 are $\ldots(3a_1)^2\,(2t_2)^6$. Excitation or ionization from the outer t_2 orbitals results in large Jahn–Teller splittings, which span the range $\sim$12.9–16 eV for CH_4 and $\sim$11.6–14 eV for SiH_4 (Potts and Price, 1972). The quantum yield η_i increases monotonically, nearly linearly from 0–1 over just these respective ranges for CH_4 (Samson *et al.*, 1989), Fig. 3.10 and SiH_4, Fig. 3.11. Nishikawa and Watanabe (1973) have had some success in predicting the ascent of the photoionization curve for CH_4 as a Franck–Condon overlap between the neutral ground state and the Jahn–Teller-split ion ground state, i.e. essentially direct ionization. There appears to be little effect on η_i from $(2a_1)^{-1}$ excitation. Presumably, a similar analysis could be applied to SiH_4.

The He I photoelectron spectra of ethane and propane display several bands in the valence region (Kimura *et al.*, 1981), unlike methane. Kameta *et al.* (1996) have obtained more modern values of η_i for these molecules (Figs. 3.12, 3.13). For ethane, η_i increases rapidly from threshold ($\sim$11.6 eV), followed by a plateau or slight dip ($\sim$13.8–14.6 eV). Between 14.6–16.5 eV, there is a final ascent to $\eta_i \sim$1.0. The photoelectron spectrum is interpreted as three unresolved bands (11.6–14 eV), a valley ($\sim$14–14.5 eV), followed by another broad band

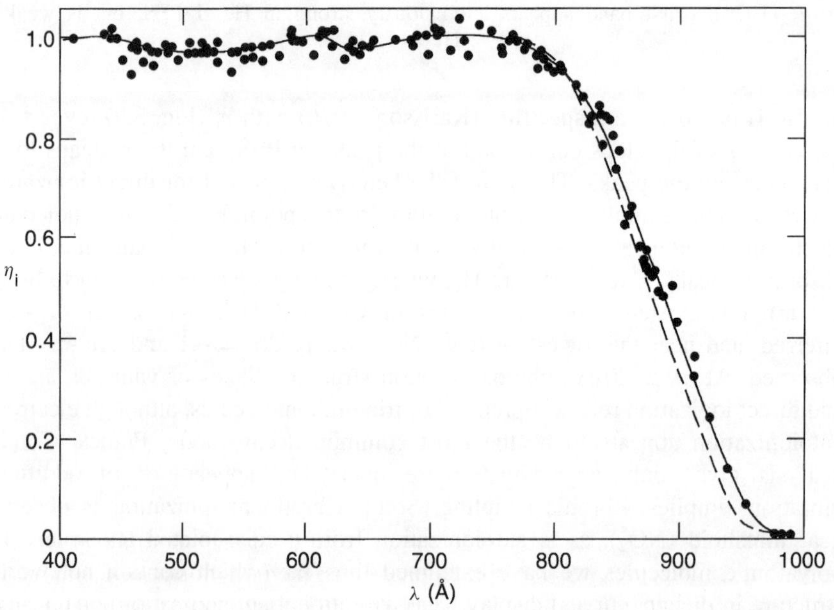

Fig. 3.10 Quantum yield of CH_4 (from Samson *et al.* (1989)). ∘ Backx *et al.* (1975); • Samson *et al.* (1989); —— Nishikawa and Watanabe (1973)

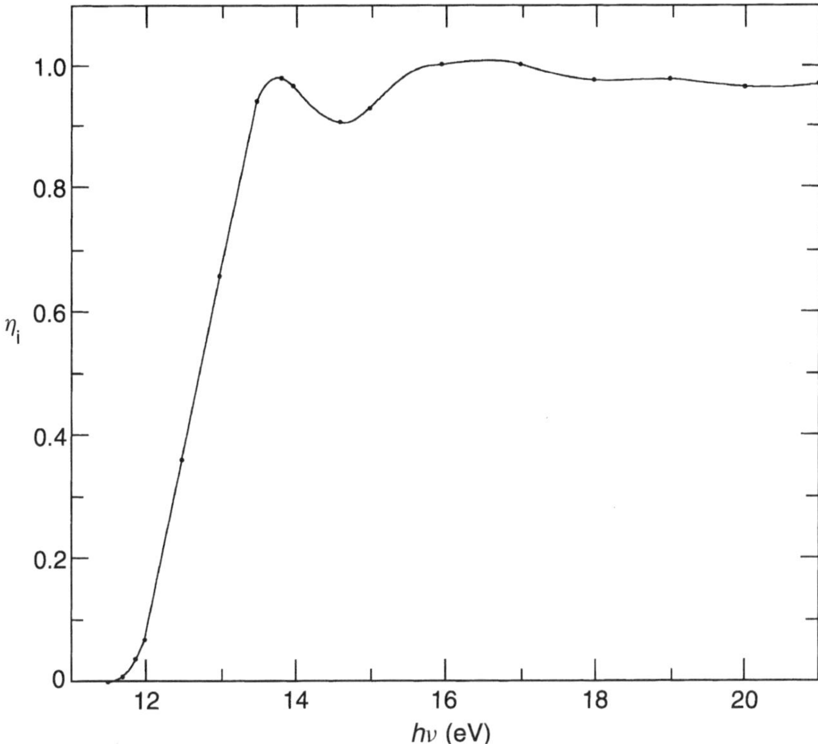

Fig. 3.11 Quantum yield of SiH_4. 11.5–13.8 eV: Derived from Hayaishi *et al.* (1987); 13.8–21.0 eV: Kameta *et al.* (1991)

($\sim$14.5–16.5 eV). This near-perfect correlation provides strong support for the direct ionization mechanism. The η_i curve for propane consists of a monotonic increase from 11.2–16.8 eV, with a short plateau at $\sim$14.6 eV. The photoelectron spectrum, spanning the range $\sim$11–16.8 eV, contains as many as seven bands bundled into three groups, with the deepest valley at $\sim$14.6 eV, again conforming to a direct ionization prescription. Koizumi (1991) has generalized this behavior to higher alkanes (cyclopropane, cyclohexane, n-butane), alkenes, alkynes, alcohols, ethers and water, but only for the first 2 eV above threshold. He focuses on the gap between the first and second IP. When this gap is small (alkanes) there is a monotonic increase, with alkenes one finds a plateau and with alkynes (large gap), he infers a dip. This much is plausible, but he is perhaps too inclusive in his choice of H_2O, C_2H_2 and C_2H_4 as exhibiting only direct ionization, as we have seen. He also seems to distinguish those molecules whose superexcited states autoionize and those that only predissociate, based on the magnitude of the gap between the dissociative ionization and primary ionization threshold. The dissociative states need not ionize, and in almost all instances, the dissociative threshold is lower than the onset of ionization.

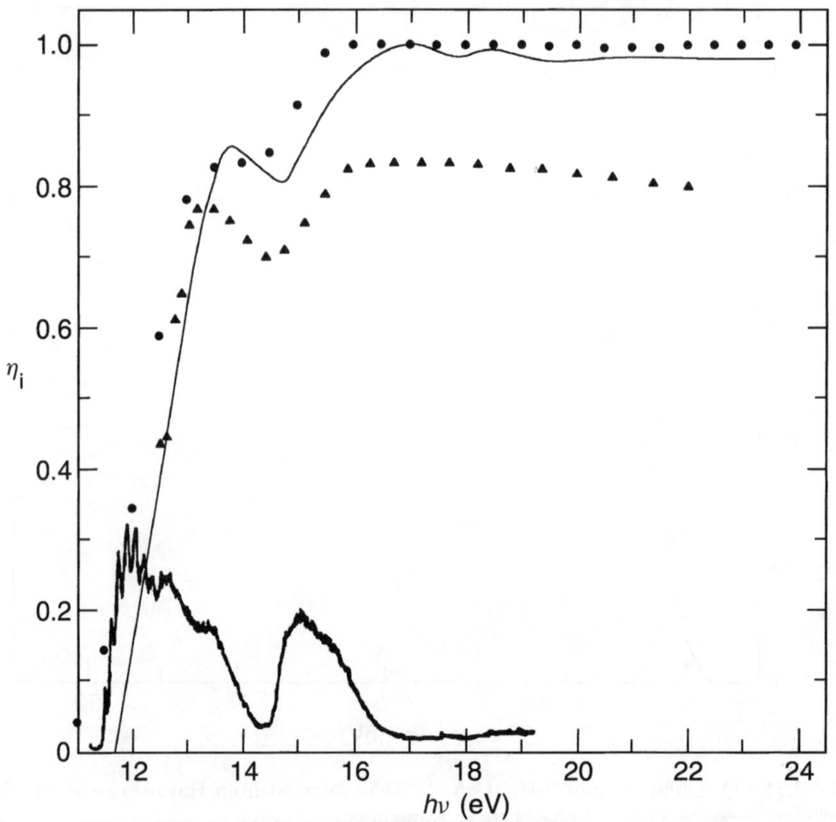

Fig. 3.12 Quantum yield of C_2H_6. • Au *et al.* (1993a; 1993b); ○ Metzger and Cook (1964); —— Kameta *et al.* (1996). He I PES from Kimura *et al.* (1981)

More recently, Jochims *et al.* (1996) have extended our knowledge of quantum yields to the successively larger polyaromatic hydrocarbons (PAHs) naphthalene ($C_{10}H_8$), azulene ($C_{10}H_8$), anthracene ($C_{14}H_{10}$), phenanthrene ($C_{14}H_{10}$) and perdeuterated benz(a)anthracene ($C_{18}D_{12}$), of which we reproduce naphthalene (Fig. 3.14) and phenanthrene (Fig. 3.15). Coronene ($C_{24}H_{12}$) has been studied by others (Verstraete *et al.*, 1990). In these experiments, there is one caveat: the relative quantum yield was measured, and normalized to unity at higher energies, where the relative yield of ionization to absorption reached constancy. For naphthalene, we can compare η_i with a He I photoelectron spectrum (Turner *et al.*, 1970, p. 321). The span of the valence shell PES (8–17 eV) is also the range of η_i from 0–1. The expected structure of η_i for direct ionization (ascent, then plateau from peak to valley) corresponds well, except between ∼9–10 eV, where a gradual increase is seen in η_i, rather than a plateau. This interpretation is contrary to that of Rühl *et al.* (1989), who infer from their threshold photoelectron

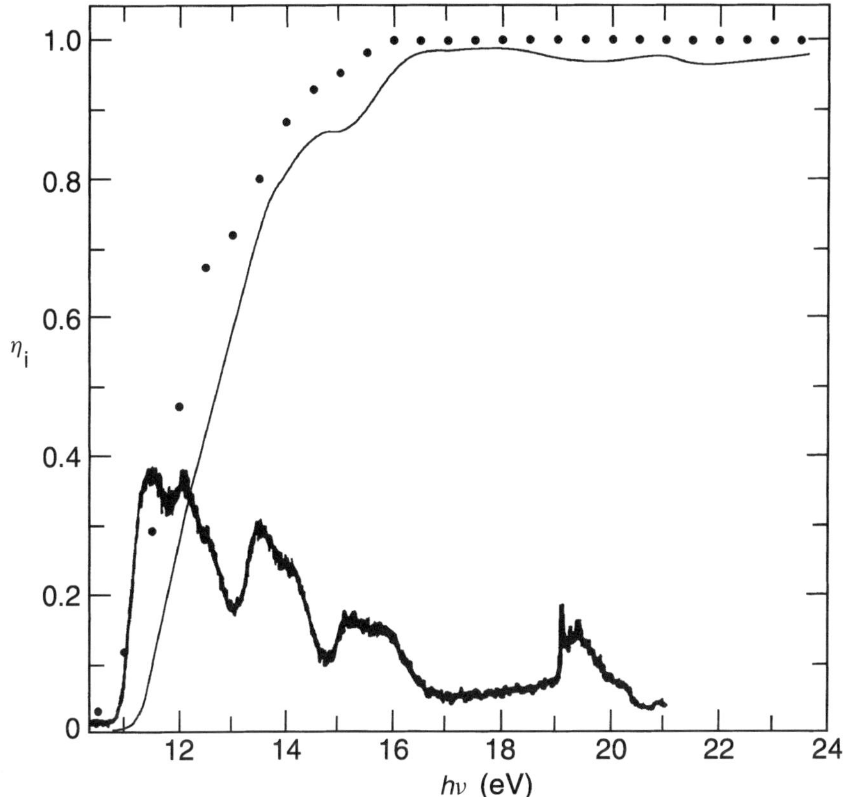

Fig. 3.13   Quantum yield of C_3H_8. ● Au *et al.* (1993a; 1993b); —— Kameta *et al.* (1996). He I PES from Kimura *et al.* (1981)

spectrum (TPES) that autoionization is the overwhelming ionization mechanism in naphthalene.

For phenanthrene, there is almost a smooth linear increase ($\eta_i = 0 \rightarrow 1$) between ~7.5–16.5 eV, which has prompted Jochims *et al.* to suggest that η_i may be modeled for PAHs by a linear function over a span of 9.2 eV from the IP. Since η_i is almost devoid of structure, comparison with a He I spectrum would not be very revealing. Instead, we present an integrated He I PES (from Boschi *et al.*, 1972) to demonstrate how closely it approximates the η_i curve. (Such integral curves were directly obtained in the early years of photoelectron spectroscopy, using retarding-field analyzers.) Examples of polyatomic molecules whose integral spectra can be compared with η_i curves include NH_3 (Frost *et al.*, 1967a), C_6H_6 (Clark and Frost, 1967) and SF_6 (Frost *et al.*, 1967b). Of course, there are many reasons why such curves should bear little relationship to one another. If only direct ionization were contributing, then comparing σ_i with the integral PES might be reasonable. But if σ_a is approximately constant, then $\eta_i \propto \sigma_i$.

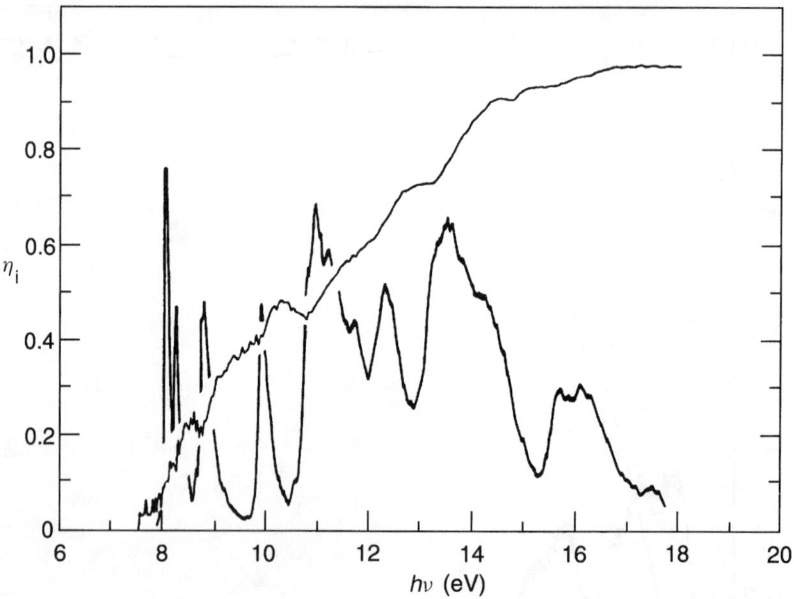

Fig. 3.14   Quantum yield of $C_{10}H_8$ (naphthalene) from Jochims *et al.* (1996); He I PES from Turner *et al.* (1970)

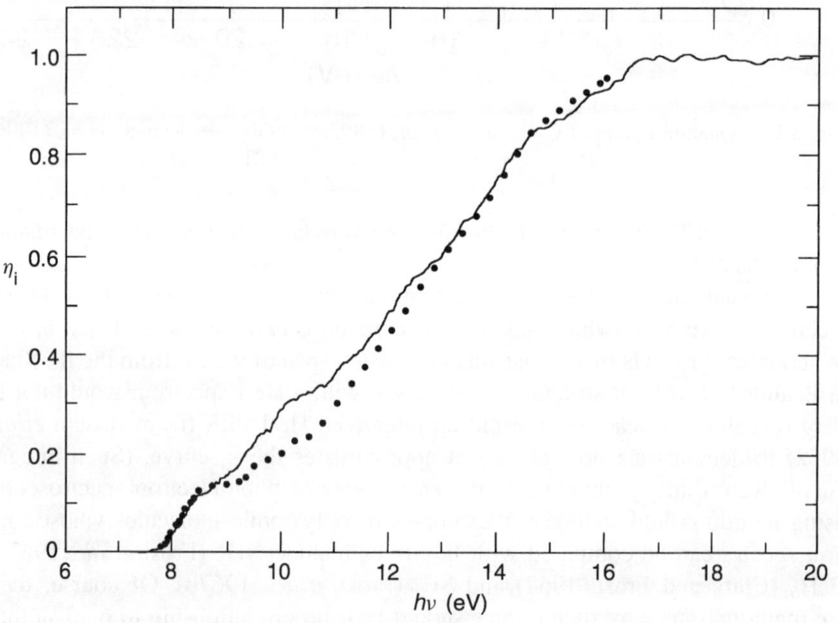

Fig. 3.15   Quantum yield of $C_{14}H_{10}$ (phenanthrene) from Jochims *et al.* (1996); • Integrated He I PES from Boschi *et al.* (1972)

Step-function threshold behavior is implicitly assumed to hold from each IP to 21.2 eV. The collection efficiency of the photoelectron spectrum is assumed to be independent of electron energy, and the photoelectron angular dependence is neglected. Despite these criticisms (arguments can be presented to minimize their effect) the similarity in shape, especially toward the upper end of the valence region, is suggestive of the dominance of direct ionization. Of relevance to the present monograph is our assertion (see Sect. 6.8) that for C_{60}, η_i increases from $0 \rightarrow 1$ between 8–21 eV, with weak structure. Since it has been demonstrated (Verstraete et al., 1990) that for $C_{24}H_{12}$, η_i varies from $0 \rightarrow 1$ between $\sim$7.5–17.6 eV, the C_{60} result is a plausible extrapolation.

Our tentative conclusion is that most superexcited states for large molecules do not ionize. That there is some autoionization is undeniable, since TPES, which is sensitive to near-zero energy electrons, invariably detects not only the states seen in PES (direct ionization), but also others, attributed to resonant ionization, which involves a form of autoionization. However, TPES detects only that component of direct ionization which corresponds to near-zero-energy electrons. Our survey suggests that direct ionization is dominant, since otherwise it is not clear why $\eta_i \rightarrow 1$. What happens to those superexcited states which do not ionize? For small molecules, we have seen that predissociation and dissociation are competitive with ionization. However, this simple picture becomes clouded as the size of the molecule increases. Even with triatomic molecules, there is evidence of vibronic coupling. With still larger molecules, the number of vibrational modes increases linearly with the number of atoms, but the density of vibronic states increases far more rapidly. This manifests itself in at least two ways. First, if the initially prepared vibronic state, which has barely enough energy to ionize, evolves into the dense vibrational manifold, it is most unlikely to restore itself to a condition where most (or all) of the energy is available for the electron to detach. This is often referred to as a very long Poincaŕe recurrence time (Thoss and Domcke, 1997). Secondly, even though the available energy is sufficient for dissociation into neutral products, there may be only a small probability that enough energy is assembled into a dissociation mode. Given enough time, it will dissociate, unless another form of competition arises. Here we encounter two times, the experimental duration (typically how long the molecule is viewed by the detector, of order μs) and the competitive vibrational fluorescence lifetime, of order ms. (We recall here Platzman's careful definition of 'dissociation' as not necessarily a true splitting of the molecule.)

Recently, Thoss and Domcke (1997) have presented a mathematical model for near-threshold photoionization of large molecules, which includes the possibility of intramolecular vibrational relaxation and predissociation. Their results indicate 'that vibrational relaxation may be an important mechanism suppressing the autoionization of high Rydberg states of large molecules'. They also conclude that with increasing photon energy 'the quantum yield tends to unity because an increasing part of the ionization signal comes from direct ionization well above the corresponding threshold'.

Schlag *et al*. (1992) and Schlag and Levine (1992) examined the 'ionization efficiency' of species in the 200–2000 amu mass range experimentally and with a model. The experimental portion described three methods of ionization: electron impact, 2-photon ionization ($\lambda = 236$ nm) and single photon ionization ($\lambda = 118$ nm). Only the last mode concerns us here, but all three were shown to have a monotonic decrease of parent ion intensity with molecular weight. The exact meaning of the terms 'ionization efficiency' and 'intensity' is not made clear. The molecules studied (tryptophan and tryptophan-containing peptides) were apparently desorbed from a substrate by a CO_2 laser, then ionized by the pulsed VUV laser (Köster and Grotemeyer, 1992). There is no mention of relative vapor density or light intensity, let alone photoionization cross sections. Only a single VUV wavelength was used. Nevertheless, they set out to explain why the ionization efficiency decreases with molecular weight.

Their model involves coupling of the excited electron's motion to the nuclear rovibrational degrees of freedom, damping (but not thermalizing) the electron's motion. (From an RRK estimate, they conclude that a completely equilibrated, or thermalized electron leads to a rate of ionization several orders of magnitude lower than the rate of dissociation, and hence too low.) The escape probability of this tethered electron is reduced as the size of the molecule (and hence its density of states) increases. In other words, this is a variant of the superexcited states description for large molecules, with no account taken of direct ionization. They have interpreted some experiments in which delayed ionization has been observed in C_{60} and some metal clusters to mean that prompt ionization is no longer observed with larger molecules, and that the delay increases with molecular size. Hence, their provocative question 'Do large molecules ionize?', and their conclusion that they can, but with increasing difficulty as the system becomes larger. If the question were 'Do large molecules autoionize?', their conclusion would be in line with the other inferences presented above.

Throughout the following section on molecular oscillator strengths, we employ atomic additivity to estimate molecular cross sections for inner shell ionization. Where comparison with direct molecular measurements is available, this estimate has been found to be very satisfactory. Consequently, the inner shell cross sections will increase with molecular size. For outer shells, the molecular orbitals are more diffuse, and atomic additivity is no longer appropriate. Bobeldijk *et al*. (1994) have had some success in assigning partial cross sections σ_{X-Y} to atom pairs $X-Y$ so that the molecular cross section $\sigma_M(E)$ is given by

$$\sigma_M(E) = \sum_{\text{all } X-Y} \sigma_{X-Y}(E)$$

Although this expression is at best semi-empirical, it does indicate that even the outer shell cross sections should increase with molecular size. Here, we are concerned primarily with this outer, or valence shell, since we have seen that $\eta_i \approx 1$ when the photon energy traverses this range. As the molecular size increases, the number of occupied valence molecular orbitals increases, but they squeeze together in ionization energy between the IP and approximately 20–21 eV. If the photoionization cross section per orbital (with due account for degeneracies) is very roughly constant, as Koizumi (1991) shows for some alkanes, alkenes, alcohols and ethers, then the photoionization cross section at $\sim$20–21 eV should also increase with molecular size.

When the earliest He I photoelectron spectra appeared, it was clear that they corresponded to direct ionization, and the vast majority of subsequent measurements confirmed this view. The observation that $\eta_i \rightarrow 1$ at $\sim$20 eV is another manifestation of the same underlying fact. At 20–21 eV, the photon energy exceeds the IP of most valence orbitals, and there is little possibility for formation of such superexcited states.

3.3 Involvement of Quantum Yield (η_i) with M_i^2

The quantum yield of ionization also enters into the determination of the square of the dipole moment for ionization of molecules by charged particles at high incident energy. Bethe (1930) derived the equation describing the variation of the ionization cross section with projectile energy, in the limit of the first Born approximation. His derivation, though general, was limited in its immediate application to atomic hydrogen. Fano (1954) and later Miller and Platzman (1957) showed how experimental data could be analyzed to extract the squared dipole moment, which the latter authors called M_i^2. Asymptotically, the Bethe formula can be written (Miller and Platzman, 1957)

$$\sigma_i \longrightarrow \left(\frac{4\pi a_0^2}{T/R}\right) M_i^2 \ln(4T c_i / R)$$

where σ_i is the total ionization cross section, T is the incident (electron) energy, c_i is a constant depending on non-dipole properties of the target, and R and a_0 have their usual meanings. Thus, a plot of $(T/R)(\sigma_i/4\pi a_0^2)$ versus $\ln(T/R)$, often called a Fano plot (Inokuti, 1971), should have M_i^2 as a slope and $M_i^2 \ln 4c_i$ as intercept. Empirically, asymptotic conditions appear to be valid for $T > 1\,\text{keV}$ for electrons. For $T > 10\,\text{keV}$, relativistic effects become increasingly important. This is particularly significant when beta particles are the projectiles. Bethe (1933) derived the corresponding equation, which can be written (Inokuti, 1971)

$$\sigma_i = \left(\frac{8\pi a_0^2}{mv^2/R}\right)\left\{M_i^2\left[\ln\left(\frac{\beta^2}{1-\beta^2}\right) - \beta^2\right] + C_i\right\}$$

Here, T has been replaced by $mv^2/2$, where m is the electron mass, and β has its usual meaning, (v/c) and C_i is a constant related to, but differing from, c_i. Thus, for proton impact, asymptotic behavior would require an incident energy almost 2000 times higher than for electrons. For incident structureless particles of charge Z, there is an additional factor Z^2.

At such high energies, large impact parameters and very low momentum transfers dominate, validating the optical approximation. Thus (Fano 1954; Miller and Platzman, 1957) M_i^2 corresponds to that portion of $S(-1)$ leading to ionization, which we (Berkowitz, 1979) have dubbed $S_i(-1)$. The same interpretation for M_i^2 is valid for both non-relativistic and relativistic equations. For atoms, this involves all of the transitions (with rare exceptions) above the IP, i.e.

$$S_i(-1) = \int_{IP}^{\infty} (1/E)(df/dE)\,dE$$

where E is in Ry units. For molecules, we have seen that some excitations above the IP do not result in ionization. The modified equation, involving the quantum yield of ionization, becomes

$$S_i(-1) = \int_{IP}^{\infty} (\eta_i(E)/E)(df/dE)dE$$

In principle, this relation provides another test of oscillator strength distributions, just as independent measurements of electric dipole polarizabilities can be compared with $S(-2)$, and the TRK rule with $S(0)$. In practice, the relevant polarizabilities are often known to $\sim 1\%$ accuracy, and $S(0)$ is rigorous, within the non-relativistic dipole approximation. However, $M_i^2 \equiv S_i(-1)$ is usually not known to such accuracy. Exceptions are atomic hydrogen and helium, where precise calculations are available, and lithium, which is almost as well known. However, for heavier atoms and for molecules, where analysis of experimental data is required, the prevailing evidence (Table 3.1) indicates that accuracies better than $\sim 10\%$ cannot be expected.

The identification of $S_i(-1)$ with M_i^2 stems from the relation between σ_i and M_i^2. Usually, σ_i refers to the cross section per event. If multiple ionization occurs, and the experiment measures total current, then σ_i takes on a new meaning, σ_i', i.e.

$$\sigma_i' = \sigma_i^+ + 2\sigma_i^{++} + 3\sigma_i^{+++} + \cdots$$

The quantity σ_i' is called the 'gross' ionization cross section, and in general is larger than the simple sum of ionizing events σ_i, which is referred to as the 'counting' cross section. In principle, it is possible to infer σ_i from σ_i', if a separate measurement determines the relative cross sections for multiple ionizations. Some experiments (Rieke and Prepejchal, 1972; Sorokin, 1998; 2000) are designed for counting cross sections, others (Schram et al., 1965; 1966; Rudd et al., 1983) for gross cross sections.

Cross sections decline as $(1/T)\ln T$, and can become quite small above 1 keV, contributing to experimental uncertainty. Then there are the usual problems of collection efficiency, and where appropriate, pressure and path length determinations. In experiments conducted at higher pressures, e.g. Rieke and Prepejchal (1972), Penning ionization may distort the meaning of M_i^2. Thus, when measuring helium in the necessary presence of low concentrations of additives, ionization is registered by discrete excited states of He upon collision with the impurity species, whose IPs are lower than even the first excited state of He. Consequently, the value of M_i^2 obtained (Table 3.1) actually corresponds to M^2(total), or $S(-1)$.

The optical values listed in Table 3.1 are culled from the present monograph, except where noted. Those labelled Rudd *et al.* (1983; 1985 a and b) derive from proton impact studies, the others from electron impact. The data of Rieke and Prepejchal (1972) and Sorokin (1998; 2000) were obtained from counting cross sections, the others from gross ionization measurements. Rudd *et al.* (1983) and Sorokin *et al.* (1998; 2000) do not give M_i^2 directly, but report cross sections as a function of energy. The values shown were obtained from the high-energy slopes of Fano plots. The review article by Rudd *et al.* (1985a) incorporated several sources in their selection of M_i^2, including earlier optical data from Berkowitz (1979).

Table 3.1 Comparison of M_i^2 from photoionization and charged particle ionization

	Optical[a]	Schram, et al. (1965; 1966)	Rieke and Prepejchal (1972)	Nishimura and Tawara (1994)	Sorokin et al. (1998, 2000)	Rudd et al. (1983)[b]	Rudd et al. (1985a and b)
H	(0.2834)[c]						0.28
He	0.4896	0.489 ±0.003[d]	(0.75)[e]			0.47	0.49
Ne	1.72	1.87 ±0.01[d]	2.02 ±0.05		1.92[f]	2.15	1.63
Ar	3.48	4.50 ±0.04[d]	3.69 ±0.12		4.30[g]	3.02	3.85
H_2	0.642	0.721 ±0.003[d]	0.695 ±0.015			0.57	0.71
N_2	3.36(3.47)[h]	3.85 ±0.02[d]	3.74 ±0.14			3.46	3.82
CO	3.26		3.70 ±0.15			3.32	3.67
O_2	3.98(3.97)[i]	4.75 ±0.03[d]	4.20 ±0.18			4.47	4.77
CO_2	5.42(5.28)[j]		5.75 ±0.10			6.17	6.55
CH_4	3.76	4.28[k]	4.23 ±0.13	4.8		4.02	4.55
C_2H_4	6.1	7.32[k]	6.75 ±0.10	9.3			
C_2H_6	7.3	8.63[k]	6.80 ±0.36	12			
NH_3	3.40(3.62)[l]		3.58 ±0.35				4.01
H_2O	2.90	3.14 ±0.047[m]	3.24 ±0.15				2.98[n]

[a]From current monograph, unless indicated separately.
[b]From Fano plots of cross sections given in Rudd *et al.* (1983).
[c]Obtained from theory.
[d]Schram *et al.* (1965).
[e]Interpreted as M_{total}^2; see text.
[f]From Fano plot of cross sections given in Sorokin *et al.* (1998).
[g]From Fano plot of cross sections given in Sorokin *et al.* (2000).
[h]Shaw *et al.* (1992a).
[i]Holland *et al.* (1993).
[j]Shaw *et al.* (1995).
[k]Schram *et al.* (1966).
[l] Edvardsson *et al.* (1999).
[m]Schutten *et al.* (1966b).
[n]Rudd *et al.* (1985b).

A comparison of M_i^2 from charged particle ionization with $S_i(-1)$ from optical data reveals that usually, $M_i^2 > S_i(-1)$. If we assume the optical data as a secondary standard, we find that the most recent results are not the best. The largest deviations occur with the data of Nishimura and Tawara (1994). The clever strategy of Sorokin $et\ al.$ (1998; 2000) still leads to apparently high values of M_i^2 for neon and argon. In the older work, the values based on gross ionization (Schram 1965; 1966) have about twice the average deviation from the optical data than the M_i^2 from counting cross sections given by Rieke and Prepejchal (1972). Of the proton impact results, those extracted from Fano plots are, on average, closer to the optical values.

Before the advent of extensive photoabsorption (and quantum yield) measurements in the vacuum ultraviolet, measurements of M_i^2 provided useful additional information for estimating the oscillator strength distribution. At the present time, given the scatter of the values in Table 3.1, it is the oscillator strength distribution, characterized by $S_i(-1)$, which provides a measure of the accuracy of M_i^2. In the detailed molecular studies that follow, we shall largely confine ourselves to comparison of the optical results with the values of Rieke and Prepejchal (1972). Though not claiming the highest precision, they appear to be the most accurate, and extensive.

4

Diatomic Molecules

4.1 Molecular Hydrogen (H$_2$)

4.1.1 Preamble

Molecular hydrogen and atomic helium, both 2-electron systems, provide excellent tests of sum rules using modern data. In both cases, high-quality, highly correlated ground state wave functions have been calculated (Pekeris (1959) for He, Kolos and Wolniewicz for H$_2$), enabling one to extract accurate values for $S(-1)$, $S(+1)$ and $S(+2)$. In addition, accurate values of the static electric dipole polarizability (α) have been computed, allowing one to infer $S(-2)$. These quantities, together with $S(0)$ given by the Thomas–Reiche–Kuhn sum rule, provide five well-known sums to compare with available information on the corresponding oscillator strength distribution. The best available data involve a combination of theoretical calculations and experimental values for both He and H$_2$.

Molecular hydrogen introduces more complications than He, since each np Rydberg is split into npσ and npπ by the molecular field, and each Rydberg member has vibrational structure. Nonetheless, accurate wavefunctions are available from Kolos and Wolniewicz for 2pσ, 2pπ, 3pσ and 3pπ (usually referred to as B, C, B$'$ and D) which enable one to calculate the oscillator strength of transitions from the ground state to each of these upper states, subdivided into their vibrational components. We report below on the sources for these calculations and supporting experiments. Unlike the case of He, higher Rydberg states require recourse to experiment. Such experiments exist for $n = 4$–6, though (as we shall see) they require calibration. Also, unlike He and other atoms, photoabsorption peaks above the IP for H$_2$ can not only autoionize, but also predissociate and occasionally fluoresce. For the five sum rules cited above, only the total photoabsorption is relevant, but with molecules we shall have occasion to utilize an additional quantity, referred to as $S_i(-1)$, which is the ionized component of $S(-1)$. In assessing this quantity, we must take into account not only predissociation, but also direct photodissociation.

In H$_2$, for the first $\sim$1.2 eV above the IP there is competition between autoionization, predissociation, direct ionization and direct photodissociation. The recent photoabsorption cross sections reported by Samson and Haddad (1994) ignore this region ($\sim$15.42–16.67 eV), focusing their attention on the 18–300 eV domain. Thus, we are left with an interval (15.426–18 eV) where we must estimate the contribution of four channels to the oscillator strength. The manner in which these have been treated is given in detail below.

Typically, at energies of 5 keV or more above an absorption edge, the photo-effect is basically an atomic process for most molecules, but H_2 is an exception because the core is also the valence shell, and is intimately involved in bonding. Both experiment and theory have demonstrated the consequences in the unique case of H_2, which is taken into account in the data selection that follows.

4.1.2 The data

The most accurate (current) adiabatic ionization potential of H_2 is based on the sum of frequencies for the transitions X $\to$ EF and EF $\to$ high Rydberg states. The former has been calibrated accurately against several I_2 lines (Shiner *et al.*, 1993). The latter now dominates the residual error in the IP. Shiner *et al.* (1993) give 124 417.507 (12) and 124 417.484 (17) cm^{-1} as experimental IP values, and 124 417.471 cm^{-1} as a recent calculated value by W. Kolos (private communication). We take the average of the experimental values, 124 417.496 (17) cm^{-1}, or 15.425 801 9 (21) eV.

a   The discrete spectrum and transitions below the IP

a.1 $X^1\Sigma_g^+ \to 2p\sigma_u, B^1\Sigma_u^+$ and $2p\pi_u, C^1\Pi_u$; Lyman (11.183–14.672 eV) and Werner (12.292–14.672 eV) bands   Chan *et al.* (1992c) have recently measured the high energy (3 keV) electron impact energy loss spectrum of H_2 at a resolution of 0.048 eV FWHM in the energy loss range 11–20 eV. They normalized their results in the discrete spectrum to values established in the continuum which had, in turn, been calibrated by use of the Thomas–Reiche–Kuhn sum rule. They report oscillator strengths for the first 23 vibrational bands of the Lyman system. Their results are in very good agreement with calculated values of Allison and Dalgarno (1970) and in good agreement with most earlier experimental data. They also obtain oscillator strengths for the first seven vibrational bands of the Werner system, again in very good agreement with Allison and Dalgarno.

Allison and Dalgarno utilized the electronic transition moments of the Lyman and Werner systems calculated by Wolniewicz (1969) as a function of inter-nuclear distance, together with available high quality potential energy curves of $X^1\Sigma_g^+, B^1\Sigma_u^+$ and $C^1\Pi_u$, to compute accurately the oscillator strengths of 37 vibrational transitions in Lyman bands, and 14 vibrational transitions in the Werner bands. (See also Dressler and Wolniewicz (1985)). We display the contributions of these adopted oscillator strengths to $S(-2)$, $S(-1)$ and $S(0)$ in Table 4.1a, b. (The contributions to $S(+1)$ and $S(+2)$ are not given here, to save space, but are taken into account in the subsequent analysis.)

a.2 $X^1\Sigma_g^+ \to 3p\sigma_u, B'^1\Sigma_u^+$ and $3p\pi_u, D^1\Pi_u$; 13.698–14.672 eV and 13.994–16.516 eV   The transition probabilities between the ground state $X^1\Sigma_g^+$ and the individual vibrational levels of the $B'^1\Sigma_u^+$ and $D^1\Pi_u$ states have been

Table 4.1 Contributions to $S(p)$ of the Lyman and Werner bands of H_2

v'	E eV[a]	$S(-2)$[b]	$S(-1)$[b]	$S(0)$[b,c]	v'	E, eV[a]	$S(-2)$[b]	$S(-1)$[b]	$S(0)$[b,c]
a. Lyman bands ($2p\sigma_u$, $B^1\Sigma_u^+$)									
0	11.184	2.500	2.055	1.689	19	13.610	3.937	3.938	3.939
1	11.347	8.326	6.943	5.790	20	13.702	3.174	3.196	3.219
2	11.506	16.164	13.670	11.56	21	13.791	2.562	2.597	2.632
3	11.661	23.893	20.477	17.55	22	13.877	2.071	2.112	2.154
4	11.811	29.857	25.919	22.50	23	13.960	1.678	1.721	1.766
5	11.957	33.287	29.254	25.71	24	14.040	1.362	1.405	1.450
6	12.100	34.190	30.405	27.04	25	14.117	1.108	1.150	1.193
7	12.238	33.04	29.72	26.73	26	14.190	0.902	0.941	0.981_5
8	12.372	30.51	27.75	25.23	27	14.261	0.733	0.769	0.805_7
9	12.503	27.21	25.01	22.98	28	14.328	0.595	0.627	0.660_3
10	12.630	23.62	21.92	20.35	29	14.391	0.486	0.514	0.543_2
11	12.753	20.08	18.82	17.64	30	14.451	0.400	0.424	0.450_8
12	12.872	16.80	15.90	15.04	31	14.506	0.326	0.347	0.370_2
13	12.988	13.89	13.26	12.66	32	14.557	0.259	0.277	0.296_1
14	13.100	11.38	10.96	10.55	33	14.601	0.197	0.212	0.227_3
15	13.209	9.263	8.993	8.730	34	14.637	0.140	0.150	0.161_6
16	13.314	7.504	7.343	7.185	35	14.662	0.074	0.080	0.0865
17	13.416	6.059	5.974	5.891	36	14.672	0.011	0.011	0.0124
18	13.514	4.885	4.853	4.820	Total		372.473	339.697	310.594
b. Werner bands ($2p\pi_u$, $C^1\Pi_u$)									
0	12.285	58.39	52.72	47.60	8	14.119	6.59	6.84	7.098
1	12.571	85.33	78.84	72.84	9	14.273	4.17	4.38	2.592
2	12.840	78.39	73.98	69.82	10	14.408	2.65	2.81	2.976
3	13.094	59.08	56.86	54.72	11	14.522	1.68	1.79	1.909
4	13.332	40.35	39.54	38.74	12	14.611	1.02	1.09	1.171
5	13.553	26.18	26.08	25.98	13	14.672	0.48	0.52	0.559
6	13.758	16.63	16.81	17.00					
7	13.947	10.46	10.72	10.99	Total		391.40	372.98	355.995

[a]Energies from Monfils (1968).
[b]Values listed should be divided by 10^3.
[c]The oscillator strengths are from Allison and Dalgarno (1970).

calculated by Glass-Maujean (1984) using dipole moment functions for the $X \rightarrow B'$ transitions from Wolniewicz (1975) and for the $X \rightarrow D$ transitions from Rothenberg and Davidson (1975) and high-quality potential energy curves for $X^1\Sigma_g^+$, $B'^1\Sigma_u^+$ and $D^1\Pi_u$. Experimentally, Glass-Maujean *et al.* (1985) found good agreement with these calculations for $X \rightarrow B'$, $v' = 5$–8. Lewis (1974) obtained lower oscillator strengths for $v' = 1$ and 3. On the other hand, Arrighini *et al.* (1980), employing several calculations, obtained about twice the total band oscillator strength that we compute using Glass-Maujean's transition probabilities. A similar situation exists for the $X \rightarrow D$ transitions. Glass-Maujean *et al.* (1984) obtained experimental transition probabilities in good agreement with the calculated values for 10 of the vibrations, Lewis had lower oscillator strengths for

$v' = 0, 2$, while Arrighini *et al.* calculated a total band oscillator strength about 14% higher than we obtain using the transition probabilities of Glass-Maujean (1984). We conclude that there is reasonable support for these calculations, and adopt them. In Table 4.2a,b, we list the corresponding oscillator strengths to the vibrational levels of the B$'$ and D states, and their contributions to $S(p)$.

a.3 $X^1\Sigma_g^+ \to$ higher Rydberg states, up to IP   Glass-Maujean *et al.* (1987) present photoabsorption spectra for H_2 between 850–803 Å in their Figs. 3–5.

Table 4.2   Contributions to $S(p)$ of $X^1\Sigma_g^+ \to B'^1\Sigma_u^+$ and $D^1\Pi_u$ of H_2

v'	E, eV[a]	$S(-2)$[b]	$S(-1)$[b]	$S(0)$[b,c]	$S(+1)$[b]	$S(+2)$[b]
a. $X^1\Sigma_g^+ \to 3p\sigma_u, B'^1\Sigma_u^+$						
0	13.698	2.88	2.90	2.92	2.94	2.96
1	13.931	5.71	5.85	5.99	6.13	6.27
2	14.144	6.79	7.06	7.34	7.63	7.93
3	14.333	6.31	6.65	7.00	7.37	7.77
4	14.494	4.89	5.21	5.55	5.91	6.30
5	14.613	2.69	2.89	3.11	3.34	3.59
6	14.651	0.29	0.31	0.34	0.36	0.39
7	14.664	0.28	0.30	0.33	0.35	0.38
8	14.672	0.19	0.20	0.22	0.23	0.25
Total		30.03	31.37	32.80	34.26	35.84
b. $X^1\Sigma_g^+ \to 3p\pi_u, D^1\Pi_u$						
0	13.994	8.52	8.76	9.01	9.27	9.53
1	14.270	13.85	14.52	15.23	15.97	16.75
2	14.530	13.94	14.89	15.90	16.98	18.13
3	14.775	11.40	12.38	13.44	14.59	15.85
4	15.003	8.35	9.20	10.15	11.19	12.34
5	15.218	5.76	6.44	7.20	8.05	9.01
6	15.418	3.88	4.39	4.98	5.64	6.39
7	15.602	2.58	2.96	3.39	3.89	4.46
8	15.772	1.70	1.97	2.28	2.64	3.06
9	15.928	1.13	1.32	1.55	1.81	2.12
10	16.068	0.76	0.90	1.06	1.25	1.48
11	16.191	0.52	0.61	0.73	0.87	1.03
12	16.299	0.35	0.42	0.50	0.60	0.72
13	16.390	0.23	0.28	0.34	0.41	0.49
14	16.462	0.16	0.19	0.23	0.28	0.34
15	16.516	0.10	0.12	0.14	0.17	0.21
Total		73.23	79.35	86.13	93.61	101.91

[a]Energies from Namioka (1964). The later paper by Monfils (1968) has slightly different numbers, but terminates at lower v'.
[b]Values listed should be divided by 10^3.
[c]The oscillator strengths are from transition probabilities calculated by Glass-Maujean (1984).
[d]Energies from Monfils (1968). The later paper by Takezawa (1970) has slightly different numbers, but terminates at $v' = 13$.

This covers the region of higher Rydberg transitions up to the IP, 803.745 Å. Glass-Maujean *et al*. do not give absolute cross sections, but within each figure there exist calibrants, i.e. transitions to B′ and D states whose oscillator strengths we have already adopted. With this cross calibration, we deduce the following (state/oscillator strength): B″/0.0223; D′/0.0145; 5pσ/0.0119; D″/0.0026; 6pσ/0.0061; 6pπ/0.004; 7pσ/0.005. The total, $f = 0.0664$, is given as an $S(0)$ contribution in Table 4.3, which lists the cumulative $S(p)$ of various bands and energy regions.

a.4 Dissociation continuum below IP Glass-Maujean *et al*. (1987) also present absolute cross sections for a photodissociation continuum between 850–770 Å in their Fig. 11. We partition this curve into two components – below the IP (the bulk) and above the IP. By graphical integration, we obtain $S(0) = 0.0210$, and corresponding values of $S(p)$ given in Table 4.3. This completes the oscillator strength contributions below the IP.

b The ionization continuum and transitions above the IP

For practical purposes, we take the free, smooth ionization continuum to begin at 18 eV. Recent absolute photoabsorption cross sections (Samson *et al*., 1994) are available from 18–113 eV, with a stated accuracy of ±2 to ±3%. Furthermore, apart from some weak departures between 33–41 eV, the quantum yield of ionization is essentially unity (see Chung *et al*., 1993). Between the adiabatic IP (15.4258 eV) and 17.0 eV there exist the remnants of a photodissociation continuum, peak structures that contribute to predissociation and autoionization, and an underlying photoionization continuum. The gap between 17.0–18.0 eV is essentially ionization continuum not encompassed by the data of Samson and Haddad (1994). We consider these in turn.

b.1 Dissociation continuum above IP The data of Glass-Maujean *et al*. (1987) utilized in 4.1.2.a.4. is accessed once again, this time between IP and 770 Å. By graphical integration, we obtain $S(0) = 0.0058$, and corresponding values of $S(p)$ given in Table 4.3.

b.2 Predissociation above IP In 4.1.2.a.3, we described how absolute photoabsorption cross sections below the IP were extracted from Glass-Maujean *et al*. (1987), Figs. 3–5. For photoabsorption cross sections above the IP, we continue this procedure for Figs. 6–9. Here, we wish to determine only the predissociation cross sections. In their Table 4, Glass-Maujean *et al*. (1987) list their measured predissociation yields. (These are essentially the complement of the autoionization yields given in Table VI of Dehmer and Chupka (1976) if we ignore the fluorescence yields.) In the present Table 4.4, we present the predissociation oscillator strengths, obtained by combining the absorption cross sections and predissociation yields. The corresponding contributions to $S(p)$ are given in Table 4.3.

Table 4.3 Spectral sums, and comparison with expectation values for H_2[a]

Energy, eV	$S(-2)$	$S(-1)$	$S(0)$	$S(+1)$	$S(+2)$
Lyman bands, 11.18–14.67[b]	0.3725	0.3397	0.3106	0.2842	0.2537
Werner bands, 12.29–14.67[b]	0.3911	0.3728	0.3560	0.3405	0.3262
B′, 13.70–14.67[c]	0.0301	0.0313	0.0328	0.0342	0.0360
D, 13.99–16.52[c]	0.0732	0.0794	0.0861	0.0938	0.1017
Higher Rydbergs, up to IP[d]	0.0539	0.0598	0.0664	0.0740	0.0821
Dissociation continuum, below IP[d]	0.0168	0.0190	0.0210	0.0230	0.0250
Dissociation continuum, above IP[d]	0.0045	0.0049	0.0058	0.0066	0.0077
Predissociation above IP[e]	0.0125	0.0147	0.0170	0.0196	0.0231
Autoionization[f]	0.0252	0.0295	0.0347	0.0406	0.0481
Ionization continuum, IP–17.0[d]	0.0719	0.0868	0.1048	0.1268	0.1533
17.0–18.0[d]	0.0580	0.0746	0.0959	0.1234	0.1588
18–40[g]	0.2448	0.4137	0.7282	1.3424	2.6016
40–80[g]	0.0092	0.0335	0.1255	0.4865	1.9595
80–160[g]	0.0005	0.0034	0.0254	0.1979	1.6041
160–300[g]	–	0.0004	0.0055	0.0843	1.3304
300–1800[h]	–	0.0001	0.0019	0.0688	3.0687
1800–10 000[h]	–	–	–	0.0065	1.6269
10^4–10^5	–	–	–	0.0006	0.9188
10^5–∞	–	–	–	–	0.4464
Total	1.3642	1.5636	2.0176	3.3537	14.7721
Expectation values	1.3573[i]	1.5487[j]	(2.00)	3.4008[j] 3.4024[j]	15.4034–15.4197[j]

[a] $S(p)$ in Ry units.
[b] From Allison and Dalgarno (1970). See Table 4.1.
[c] From Glass-Maujean (1984). See Table 4.2.
[d] See text.
[e] See Table 4.4 and text.
[f] See Table 4.5 and text.
[g] From Samson *et al.* (1994). See text.
[h] The absorption cross sections $\sigma(H_2)$ between 300 eV (where $\sigma(H_2) = 1.26 \times 2\sigma(H)$) and 5.41 keV (where $\sigma(H_2) = 1.444 \times 2\sigma(H)$) have been interpolated from a plot of $E^{3.5} \times \sigma(H_2)$ versus $E^{-1/2}$. At higher energies, $\sigma(H_2) = 1.444 \times 2\sigma(H)$, where $\sigma(H)$ is from the analytical formula. See text.
[i] From Kolos and Wolniewicz (1967).
[j] See Appendix.

b.3 Autoionization Relative photoionization cross sections for H_2 in the autoionization region (IP to $\sim$740 Å) have been reported by Chupka and Berkowitz (1969) at a resolution of 0.04 Å FWHM, and later by Dehmer and Chupka (1976) at a resolution of $\sim$0.02 Å FWHM. Either, or both, require an

Table 4.4 Oscillator strengths attributed to predissociation above the IP

λ, Å	Assignment	$f(\times 10^3)$	λ, Å	Assignment	$f(\times 10^3)$
796	$6p\pi$ (2,0)	1.0	783.3	D′ (5,0)	0.2
795.5	$5p\pi$ (3,0)	1.0	780	D″ (4,0)	0.6
794.7	B″ (5,0)	2.4	779.5	B″ (7,0)	1.3
794.7	D′ (4,0)	1.5	773.5	D′ (6,0)	1.5
791.5	$7p\sigma$ (2,0)	0.7	773	B″ (8,0)	1.5
791	D″ (3,0)	1.6	769.5	D″ (5,0)	0.4
787	B″ (6,0)	1.4	766.8	B″ (9,0)	0.4
786	$6p\sigma$ (3,0)	0.2	765	D′ (7,0)	0.7
784	$5p\sigma$ (4,0)	0.6			
				Total	17.0

absolute calibration for the present purposes. This has been provided by Raoult and Jungen (1981) from their multichannel quantum defect (MQDT) calculation and fitting to the data of Dehmer and Chupka (1976). Toward this end, they used an absolute calibration near threshold from Backx *et al*. (1976), to fit the $8p\sigma$, $v = 2$ peak at 787.9 Å. In a contemporary work, Jungen and Raoult (1981) give the effective peak heights (see their Table 1) of several transitions, in absolute units, after convolution with the experimental resolution. We have chosen $5p\pi$, $v = 3$ (790.9 Å), $7p\sigma$, $v = 2$ (791.4 Å), 21 po, $v = 1$ (791.4 Å), 16 p2, $v = 1$ (791.5 Å) and 22 po, $v = 1$ (791.3 Å). Comparing these absolute peak heights with measured peak heights (mm) from Fig. 1 of Dehmer and Chupka (1976), we infer a calibration of 10.1 ±0.1 Mb/mm from threshold to 750 Å. Below 750 Å, there are additional weak features which require different calibrations. These have been deduced from known values of the underlying ionization continuum (see below). From the calibrated peak heights and areas, we have computed the autoionization oscillator strengths and record them in Table 4.5. Their sum, $f = 0.0347$, is about double that of the predissociation oscillator strengths, $f = 0.0170$. (We note parenthetically that, although some of the autoionization peaks are as intense as 400 Mb, the autoionization oscillator strength represents about 1.7% of $S(0)$.)

b.4 **Underlying ionization continuum, IP–17.0 eV; 17–18 eV** Jungen and Raoult (1981) give the differential oscillator strength of the unperturbed H_2^+ continuum between $v = 0$ and $v = 1$ as 0.0130/eV in their Table 1, and that between $v = 4$ and $v = 5$ as 0.092/eV in their Table 3. Interpolated values for $v = 1$–2 (0.037/eV), $v = 2$–3 (0.060/eV), $v = 3$–4 (0.076/eV) and extrapolated values of $v = 5$–6 (0.1056/eV) and $v = 6$–7 (0.1167/eV) have been deduced by utilizing the Franck–Condon factors given by Berkowitz and Spohr (1973). The thresholds for $v = 1$–7 are 15.698, 15.954, 16.195, 16.420, 16.632, 16.829 and 17.0. The summed oscillator strength for the underlying continuum, IP–17.0 eV, is 0.1048.

Table 4.5 Autoionization oscillator strengths in H_2

λ, Å	Assignment	$f(\times 10^3)$	λ, Å	Assignment	$f(\times 10^3)$
803.26	$5p\pi$, $v = 2$	1.20	791.58	$20p_0$, $v = 1$	0.16
803.16	$7p\pi$, $v = 1$	0.17	791.50	$16p_2$, $v = 1$	0.31
800.85	$8p\sigma$, $v = 1$	0.94	791.40	$7p\sigma$, $v = 2$	0.73
799.81	$8p\pi$, $v = 1$	0.05	791.37	$21p_0$, $v = 1$	0.38
798.45	$9p\sigma$, $v = 1$	0.53	791.24	$22p_0$, $v = 1$	0.32
797.73	$9p\pi$, $v = 1$	0.26	791.18	$17p_2$, $v = 1$	0.16
797.31	$6p\sigma$, $v = 2$	0.27	791.12	$23p_0$, $v = 1$	0.08
796.72	$10p\sigma$, $v = 1$	0.57	791.01	$24p_2$, $v = 1$	0.16
795.92	$10p\pi$, $v = 1$	0.04	790.95	$18p_2$, $v = 1$	0.30
794.46	$11p\sigma$, $v = 1$	0.45	790.86	$5p\pi$, $v = 3$	0.61
795.0	$6p\pi$, $v = 2$	0.05	790.77	$27p_0$, $v = 1$	0.03
794.75	$12p_0$, $v = 1$	0.01	790.64	$29p_0$, $v = 1$	0.03
794.48	$11p_2$, $v = 1$	0.31	790.60	$30p_0$, $v = 1$	0.03
794.35	$5p\sigma$, $v = 3$	0.02	790.55	$31p_0$, $v = 1$	0.03
793.92	$13p_0$, $v = 1$	0.30	790.14	$7p\pi$, $v = 2$	0.27
793.82	$4p\pi$, $v = 4$	0.08	787.90	$8p\sigma$, $v = 2$	0.81
793.30	$14p_0$, $v = 1$	0.16	786.97	$8p\pi$, $v = 2$	0.23
792.98	$13p_2$, $v = 1$	0.03	786.55	$4p\sigma$, $v = 6$	0.04
792.82	$15p_0$, $v = 1$	0.08	785.84	$6p\sigma$, $v = 3$	0.34
792.51	$16p_0$, $v = 1$	0.08	785.43	$9p\sigma$, $v = 2$	0.60
792.18	$17p_0$, $v = 1$	0.09	784.70	$9p_2$, $v = 2$	0.09
791.97	$18p_0$, $v = 1$	0.02	783.85	$10p_0$, $v = 2$	0.03
791.86	$15p_2$, $v = 1$	0.09	783.68	$5p\sigma$, $v = 4$	1.89
791.72	$18p_0$, $v = 1$	0.14	783.26	$4p\pi$, $v = 5$	0.66
783.11	$10p_2$, $v = 2$	0.22	773.69	$4p\pi$, $v = 6$	0.05
782.89	$6p\pi$, $v = 3$	0.05	773.46	$5p\sigma$, $v = 5$	1.35
782.60	$11p_0$, $v = 2$	0.28	772.95	$9p_2$, $v = 3$	0.06
782.00	$11p_2$, $v = 2$	0.12	772.58	$4p\sigma$, $v = 8$	0.10
781.69	$12p_0$, $v = 2$	0.16	772.15	$10p_0$, $v = 3$	0.84
781.18	$13p_0$, $v = 2$	0.17	771.95	$6p\pi$, $v = 4$	0.07
780.93	$12p_2$, $v = 2$	0.05	770.93	$11p_0$, $v = 3$	0.27
780.56	$14p_0$, $v = 2$	0.15	770.00	$12p_0$, $v = 3$	0.15
780.08	$15p_0$, $v = 2$	0.16	769.43	$5p\pi$, $v = 5$	0.69
779.83	$16p_0$, $v = 2$	0.13	768.86	$14p_0$, $v = 3$	0.39
779.71	$5p\pi$, $v = 4$	0.72	768.60	$13p_2$, $v = 3$	0.21
779.64	$14p_2$, $v = 2$	0.60	768.46	$15p_0$, $v = 3$	0.13
779.43	$7p\sigma$, $v = 3$	0.14	768.18	$16p_0$, $v = 3$	0.18
779.23	$15p_2$, $v = 2$	0.03	767.85	$17p_0$, $v = 3$	0.15
779.16	$4p\sigma$, $v = 7$	0.05	767.60	$18p_0$, $v = 3$	0.07
778.63	$21p_0$, $v = 2$	0.09	767.46	$7p\pi$, $v = 4$	0.30
778.53	$17p_2$, $v = 2$	0.05	764.88	$4p\pi$, $v = 7$	0.29
778.46	$22p_0$, $v = 2$	0.02	764.53	$8p\pi$, $v = 4$	0.43
778.38	$23p_0$, $v = 2$	0.08	763.81	$6p\sigma$, $v = 5$	0.62
778.30	$24p_0$, $v = 2$	0.10	763.00	$9p_0$, $v = 4$	0.93
778.22	$7p\pi$, $v = 3$	0.17	761.97	$6p\pi$, $v = 5$	0.41
776.19	$8p\sigma$, $v = 3$	0.30	761.40	$10p_0$, $v = 4$	0.22
775.56	$6p\sigma$, $v = 4$	0.06	760.27	$11p_0$, $v = 4$	0.24
775.10	$8p\pi$, $v = 3$	0.14	760.16	$5p\pi$, $v = 6$	0.47

Table 4.5 (*Continued*)

λ, Å	Assignment	$f(\times 10^3)$	λ, Å	Assignment	$f(\times 10^3)$
759.40	$12p_0$, $v = 4$	0.13	740.25	$13p_0$, $v = 6$	0.02
758.86	$13p_0$, $v = 4$	0.33	739.89	$14p_0$, $v = 6$	0.03
758.65	$12p_2$, $v = 4$	0.05	739.74	$13p_2$, $v = 6$	0.03
758.31	$14p_0$, $v = 4$	0.20	739.36	$14p_2$, $v = 6$	0.02
757.86	$15p_0$, $v = 4$	0.15	739.10	u^b	0.01
757.65	$14p_2$, $v = 4$	0.18	739.02	u^b	0.01
755 (complex)[a]	$6p\sigma$, $v = 6$ plus others	2.43	738.82	$17p_0$, $v = 6$	0.01
753.33	$9p_0$, $v = 5$	0.57	738.46	u^b	0.08
752.88	$6p\pi$, $v = 6$	0.23	738.27	u^b	0.12
751.90	u^b	0.37	738.19	u^b	0.04
751.74	$10p_0$, $v = 5$	0.22	738.11	$21p_0$, $v = 6$	0.02
750.53	$11p_0$, $v = 5$	0.23	738.02	$22p_0$, $v = 6$	0.08
749.88	u^b	0.20	737.90	$23p_0$, $v = 6$	0.01
749.72	$12p_0$, $v = 5$	0.10	737.83	$24p_0$, $v = 6$	0.04
749.05	$13p_0$, $v = 5$	0.06	737.22	(complex)	0.14
748.73	$14p_0$, $v = 5$	0.21	733.63	u^b	0.08
748.25	$15p_0$, $v = 5$	0.04	732.26	u^b	0.03
747.86	$16p_0$, $v = 5$	0.04	731.78	u^b	0.03
744.70	u^b	0.49	731.69	u^b	0.02
744.0	u^b	0.09	731.43	u^b	0.04
742.75	$10p_0$, $v = 6$	0.16	731.24	u^b	0.03
741.71	$11p_0$, $v = 6$	0.10	731.12	u^b	0.04
740.88	$12p_0$, $v = 6$	0.08	730.90	u^b	0.05
740.54	$12p_2$, $v = 6$	0.10	730.67	u^b	0.05
			730.45	u^b	0.02
			730.39	u^b	0.01
				Total	34.71

[a]See Jungen and Raoult (1981) for a discussion of this complex resonance.
[b]Unidentified peak.

Between 17–18 eV, the photoabsorption cross section is declining. We take a mean value of $10.53\,\mathrm{Mb} \equiv 0.096/\mathrm{eV}$ from Backx *et al.* (1976) and Samson and Haddad (1994), which contributes 0.0959 to the oscillator strength. These quantities, and the corresponding $S(p)$, are given in Table 4.3.

b.5 **Ionization continuum, 18–300 eV** The photoabsorption cross section continues to decline monotonically, in this range and on to infinity. Samson and Haddad (1994) have presented their own data between 18–113 eV, stated to be accurate to 2–3%. They are in excellent agreement with rather accurate calculations of Flannery *et al.* (1977) in the calculational range, to 28 eV. The latter use a two-parameter Weinbaum function for H_2 and an exact electronic ground state function for H_2^+. Points at 43.6 and 70.8 eV from Richards and Larkins (1986) are also in very good agreement with the experimental data. Richards and Larkins used a numerical two-dimensional continuum wavefunction at the Hartree–Fock level with exchange. Samson and Haddad also examine higher energy data, to 300 eV. They select the experimental points of Denne (1970) and Alaverdov and Podolyak (1982), which are in good agreement with one another, and with calculated points by Richards and Larkins at 125.2 and 179.7 eV. Samson and Haddad estimate an uncertainty of 3–4% for the data above 113 eV.

We partition the tabulated cross sections of Samson and Haddad into four regions, 18–40, 40–80, 80–160 and 160–300 eV. Each segment is fitted to a four-term polynomial, $df/dE = \sum_{n=2}^{5} a_n y^n$, where $y = B/E$, and $B = $ IP. The fitted functions are used to evaluate $S(p)$ in each region, and are recorded in Table 4.3. The coefficients of the polynomials are given in Table 4.6. Samson and Haddad find that the expression

$$\sigma(\text{Mb}) = 96\,000 E^{-3.15}$$

where E is in eV, fit their selection of data between 80–300 eV within 2%. From their expression, we calculate $\Delta S(0) = 0.0310$, whereas our fitted polynomials give 0.0309.

b.6   300–10 000 eV   Samson and Haddad (1994) note that no experimental photoabsorption cross sections have been reported between 300 eV and 3 keV, and earlier measurements at higher energies are questionable. In such circumstances, it is tempting to approximate the high-energy cross sections by additivity, which is commonly done well above an absorption edge. However, Crasemann $et\,al.$ (1974) measured the photoelectric cross section of H_2 at two x-ray energies, and obtained the 'rather remarkable result' that the cross section exceeded that of two hydrogen atoms by 44% at 5.4 keV and essentially the same amount at 8.4 keV. Such an 'excess' cross section was predicted theoretically by Kaplan and Markin (1973) using a relativistic generalization of the Heitler–London wavefunction for H_2 and a relativistic plane wave for the ejected electron wavefunction. They found an excess cross section of about 70% in the 5–10 keV region, later reduced to 57–58% using an improved (Weinbaum) function for H_2 (Kaplan and Markin, 1975).

Upon learning the results obtained by Crasemann $et\,al.$, Cooper (1974) was able to deduce a simple and accurate explanation. He noted the high-energy

Table 4.6   Coefficients of the polynomial $df/dE = ay^2 + by^3 + cy^4 + dy^5$ fitted to data at various energies[a]

Energy range, eV	a	b	c	d
18–40	−2.260 82	12.840 03	−15.044	6.312 276
40–80	−0.408 43	5.334 694	−10.216	12.117 78
80–160	−0.399 36	9.862 324	−57.1393	132.2687
160–300	0.018 496	0.314 785	18.527 21	−86.1512
300–1740	−0.019	2.966 907	−48.3045	478.3061
1740–9886	−0.000 79	0.822 739	5.121 865	1641.146

[a] df/dE in Ry units, $y = B/E$, $E = $ IP $= 15.4258$ eV.

asymptotic form of the cross-section for a two-electron system (Kabir and Salpeter, 1957; Salpeter and Zaidi, 1962),

$$\sigma = C Z^2 \delta^3(\vec{r}_1)\varepsilon^{-7/2},$$

where $\delta^3(\vec{r}_1)$ is a measure of the initial-state charge density of a single electron at the nucleus and ε is the energy (photon or photoelectron). (This expression has been used by us for He. The quantity $\delta^3(\vec{r}_1)$ enters into the determination of $S(+2)$ and will be used subsequently.) The same asymptotic form applies to atomic hydrogen. Therefore, the high-energy cross section ratio $\sigma(H_2)/2\sigma(H)$ is given by the ratio of charge density at the nuclei for H_2 divided by twice the charge density of H atom. At this point Cooper is a little imprecise in his notation. He gives this ratio as $\delta^3_{H_2}(\vec{r}_{1a})/\delta^3_H(r_1)$, where $\vec{r}_{1a}$ refers to one hydrogen nucleus. He obtains $\delta^3_{H_2}(\vec{r}_{1a}) = 0.4598$ from Kolos and Wolniewicz (1964). However, in the definition of $\delta^3_{H_2}(\vec{r}_{1a})$ used by Kolos and Wolniewicz, we obtain half this number, as will be shown (see Appendix) in our determination of $S(2)$ for H_2. Therefore, using the Kolos–Wolniewicz definition of $\delta^3_{H_2}(\vec{r}_{1a})$, $\sigma(H_2)/2\sigma(H) = 2 \cdot \delta^3_{H_2}(\vec{r}_{1a})/\delta^3_H(\vec{r}_1)$. With $\delta^3_H(\vec{r}_1) = 1/\pi$, the ratio is 1.444, in excellent agreement with the experimental results of Crasemann et al. (The author acknowledges the assistance of Dr. Mitio Inokuti in clarifying this point.) Thus, in the unique case of H_2, the core is the valence shell, and molecular bond formation distorts the charge density around the nuclei, enhancing the high energy cross section. Cooper (1974) considers the cases of Li_2 and LiH, and concludes that departure from additivity for the core should be negligible.

From Crasemann et al., we know that the asymptotic form is appropriate at 5.4 keV, but we must bridge the gap between 300 eV and 5.4 keV. At 300 eV, $\sigma(H_2)/2\sigma(H) = 1.26$ (Samson and Haddad, 1994), while at 5.4 keV this ratio is 1.444. A plot of $E^{3.5} \cdot \sigma(H_2)$ vs. $E^{-0.5}$ provides a basis for a linear interpolation between the data of Samson and Haddad and the points of Crasemann et al. At higher energies, we compute $\sigma(H_2)$ from the asymptotic ratio. Between 300–10 000 eV, these cross sections have been fitted by regression to two 4-term polynomials (300–1800 eV; 1800–10 000 eV), whose coefficients are given in Table 4.6. Each fitted function is used to calculate $S(p)$, which is listed in Table 4.3.

b.7   10 keV → infinity   The photoabsorption cross section of atomic hydrogen is given by

$$\frac{df}{d\varepsilon} = \left(\frac{2^7}{3}\right)(1+k^2)^{-4} \cdot e^{-\frac{4}{k}\arctan k}\left(1 - e^{-\frac{2\pi}{k}}\right)^{-1}$$

where $\varepsilon = k^2$ is the kinetic energy of the electron, in Ry units. (See Sect. 2.1.)

We have used this expression to calculate $\sigma(H)$ from 10 keV to infinity, and thereupon $\sigma(H_2) = 2.888 \times \sigma(H)$. The contribution of this high-energy domain is insignificant for all $S(p)$ except $S(+2)$, where it adds $\sim$10%, as seen in Table 4.3.

4.1.3 The analysis

We must now evaluate theoretical $S(p)$ to compare with the spectral sums given in Table 4.3. We use the well-known relation $\alpha = a_0^3 \, S(-2)$, together with $\alpha = 0.8045 \times 10^{-24}\,\mathrm{cm}^3$ calculated by Kolos and Wolniewicz (1967) to obtain $S(-2) = 1.3573$. Here we have used the rotational average at 293 K given by Kolos and Wolniewicz. Experimental values available for comparison are 1.3596 ± 0.0007 by Orcutt and Cole (1967) and 1.3569 ± 0.0005 by Newell and Baird (1965). The theoretical value is 0.5% smaller than the spectral sum. The theoretical $S(-1)$, $S(+1)$ and $S(+2)$ require more detailed consideration of the papers by Kolos and Wolniewicz, and are derived in the Appendix.

For $S(-1)$, the theoretical value (1.5487) is about 1% lower than the spectral sum, while for $S(0)$ it is 0.88% lower. Thus, all three spectral sums are higher than the expectation values by 0.5–1%. The agreement is quite good and within the recorded experimental error bars. For this light molecule, the contributions from $\sim$11–40 eV to $S(-2)$, $S(-1)$ and $S(0)$ are rather evenly distributed (see Table 4.3), so that no one spectral region in this domain can be identified as the major source of the residual discrepancy.

The deviations between spectral sums and expectation values for $S(+1)$ and $S(+2)$ are in the opposite direction – the theoretical value is larger. This is somewhat surprising for $S(+1)$, since only $\sim$25% of its contribution comes from $h\nu > 40$ eV, and the implication from the previous sum rules is that the spectral sum is too large for $11\,\mathrm{eV} < h\nu < 40\,\mathrm{eV}$, yet the shortfall for $S(+1)$ is about 1.4%.

The quantity $S(+2)$ receives $\sim$61% of its magnitude from $h\nu > 80$ eV. Hence, it is strongly dependent on the 44% enhancement from the correction factor found by Crasemann $et\ al.$ (1974). Even with this augmentation, the spectral sum is lower than the theoretical sum by 4.1–4.2%. An expectation value of $S(+2) = 14.77\,\mathrm{Ry}$ units has been obtained by other authors, using the same source material (see Appendix). Such a value would be in quite good agreement with the spectral sum. However, our rather exhaustive discussion of this point in the Appendix appears to show that the higher value, 15.40–15.42, is correct. The implication of the $S(+1)$ and $S(+2)$ comparisons is that the cross sections for H_2 above 80 eV need to be increased by several per cent.

The value of $S_i(-1)$ is 0.642 from Table 4.3. Rieke and Prepejchal (1972) obtained $M_i^2 = 0.695 \pm 0.015$ from the high-energy dependence of electron impact ionization. Their value is about 8.3% higher than the spectral sum, whereas the full $S(-1)$ spectral sum is about 1% higher than the theoretical value. With other molecules, the magnitudes of M_i^2 found by Rieke and Prepejchal are usually too high. In this particular case, their deviation is well documented by the precision of the cross sections, and the excellent agreement with the sum rules.

Recently, Yan $et\ al.$ (1998a) have reported a sum rule analysis for H_2. They made the approximation of assuming that the discrete oscillator strengths of H_2 are 1.7-times those for atomic H. Our detailed analysis is likely to be more accurate. For high energies, they used 2.8-times the asymptotic $\sigma(H)$, as did we.

After trying the Samson and Haddad (1994) data between 18–300 eV and finding $S(+2)$ too small, they arbitrarily but smoothly increased the cross sections from 113 eV to 'the asymptotic limit at high energies' to achieve better agreement for $S(+2)$. The measurements of Samson and Haddad extended only up to 113 eV, and had a stated accuracy of 2–3%, but their selected data from 113–300 eV were estimated to be accurate to 3–4%. Yan *et al.* have increased these cross sections between $\sim$150–300 eV by about 13–14%. From $\sim$525–10 000 eV, the cross sections of Yan *et al.* are within $\sim$3% of ours. Thus, the enhanced cross sections of Yan *et al.* are primarily in a region where experimental data exist, and the increase is about four times larger than the estimated error. Although this adjustment improves $S(+2)$, it is unlikely to be the source of the discrepancy. Even with this adjustment, their value of $S(+1)$ remains farther from the theoretical value than the present spectral sum.

4.1.4 Appendix

1. $S(-1)$

   By standard derivation,

   $$S(-1) = \tfrac{1}{3}\langle \vec{r}_1 + \vec{r}_2 \rangle^2 \text{ in Ry units}$$

   $$= \tfrac{1}{3}[\langle (x_1 + x_2)^2 \rangle + \langle (y_1 + y_2)^2 \rangle + \langle (z_1 + z_2)^2 \rangle]$$

   $$= \tfrac{2}{3}\langle (x_1 + x_2)^2 \rangle + \tfrac{1}{3}\langle (z_1 + z_2)^2 \rangle (x, y \text{ equivalent})$$

   $$\langle (x_1 + x_2)^2 \rangle = \langle x_1^2 + 2x_1 x_2 + x_2^2 \rangle = 2\langle x_1^2 \rangle + 2\langle x_1 x_2 \rangle$$

   $$\langle (z_1 + z_2)^2 \rangle = 2\langle z_1^2 \rangle + 2\langle z_1 z_2 \rangle$$

   $$S(-1) = \tfrac{4}{3}\left[\langle x_1^2 \rangle + \langle x_1 x_2 \rangle \right] + \tfrac{2}{3}\left[\langle z_1^2 \rangle + \langle z_1 z_2 \rangle \right]$$

There are several references by Kolos and/or Wolniewicz where these quantities are tabulated with slightly different values, including Kolos and Wolniewicz (1965) and Wolniewicz (1966). We choose the latter. In his Table VI, for $v = 0$:

$$\langle x_1 x_2 \rangle = -0.056\,06$$

$$\langle x_1^2 \rangle = 0.7743$$

$$\langle z_1 z_2 \rangle = -0.1774$$

$$\langle z_1^2 \rangle = 1.064$$

$$S(-1) = 1.5487$$

2. $S(+1)$

   By standard derivation,

   $$S(+1) = \tfrac{4}{3}\left(0 \,\middle|\, \vec{p}_1^2 + \vec{p}_1 \cdot \vec{p}_2 \,\middle|\, 0 \right) \text{ in a.u.}$$

   $$= -\tfrac{4}{3}[(0|\Delta_1|0) + (0|\vec{\nabla}_1 \cdot \vec{\nabla}_1|0)]$$

Kolos and Wolniewicz (1964) give in their Table III at $R = 1.4011$ a.u.

$$-\frac{\Delta_1}{2M} = 70.193\,\mathrm{cm}^{-1}, \quad -\frac{\vec{\nabla}_1 \cdot \vec{\nabla}_2}{2M} = 6.025\,\mathrm{cm}^{-1},$$

where $M = m_\mathrm{p}/m_\mathrm{e} = 1836.152\,701$

Thus, $-\Delta_1 = 1.174\,487$ a.u., $-\vec{\nabla}_1 \cdot \vec{\nabla}_2 = 0.100\,811\,8$ a.u.

The first term, which is the kinetic energy, should be equal in magnitude, and opposite in sign to the potential energy, according to the virial theorem. Kolos and Wolniewicz give $-1.174\,470\,1$ for the latter, which is close. Hence,

$$S(+1) = \tfrac{4}{3}(1.174\,487 + 0.100\,811\,8) = 1.700\,399 \text{ a.u., or } 3.400\,797 \text{ Ry units.}$$

In a later publication which treats adiabatic corrections, Wolniewicz (1993) gives

$$E_2' = -\frac{1}{4\mu}(\Delta_1 + \vec{\nabla}_1 \cdot \vec{\nabla}_2) = 76.253\,\mathrm{cm}^{-1} \text{ at } R = 1.40 \text{ a.u.,}$$

where $\mu = \dfrac{M}{2} = 918.076\,35.$

Thus, $S(+1) = -\dfrac{4}{3}\langle\Delta_1 + \vec{\nabla}_1 \cdot \vec{\nabla}_2\rangle$

$$= \frac{16}{3}\,\frac{\mu \cdot (76.253)}{2.194\,746 \times 10^5\,\mathrm{cm}^{-1}/\mathrm{a.u.}}$$

$$= 1.701\,180 \text{ a.u., or } 3.402\,359 \text{ Ry units.}$$

3. $S(+2)$

By standard derivation,

$S(+2) = \tfrac{4}{3} \cdot 4\pi\,\delta(r_{1\mathrm{a}})$ in a.u., where $\delta(r_{1\mathrm{a}})$ is the charge density at the nucleus.

In particular, this is the expression used by Wolniewicz (1993), eq. (47). We proceed by extracting $\delta(r_{1\mathrm{a}})$, and consequently $S(+2)$, from several papers of Kolos and Wolniewicz.

a. From Kolos and Wolniewicz (1964)

$$\varepsilon_4 = \pi\alpha^2 \int \Phi\left[2\delta^{(3)}(r_{1\mathrm{a}}) - \delta^{(3)}r_{12}\right]\Phi\,d\tau_1\,d\tau_2$$

$$\varepsilon_5 = 2\pi\alpha^2 \int \Phi\delta^{(3)}(r_{12})\Phi\,d\tau_1\,d\tau_2$$

$$\therefore\ 2\varepsilon_4 + \varepsilon_5 = 4\pi\alpha^2 \int \Phi\delta^{(3)}(r_{1\mathrm{a}})\Phi\,d\tau_1\,d\tau_2 = 4\pi\alpha^2\delta(r_{1\mathrm{a}}),$$

where α is the fine-structure constant.

In their Table IV, they give ε_4 and ε_5 in units of $\alpha^2 \times$ a.u. $= 11.687\,15\,\mathrm{cm}^{-1}$. Our interpretation is: $\varepsilon_i \times 11.687\,15\,\mathrm{cm}^{-1} \times$

$$\frac{(137.035\,989\,5)^2}{2.194\,746\,3 \times 10^5\,\mathrm{cm}^{-1}/\mathrm{a.u.}} = \varepsilon_i(1.0)\text{ a.u.}$$ Therefore, the values of ε_4 and ε_5 given in the Table are effectively in a.u., and the fine-structure constant can be ignored.

At $R = 1.4011$, $\varepsilon_4 = 1.391\,38$, $\varepsilon_5 = 0.106\,25$

$$2\varepsilon_4 + \varepsilon_5 = 4\pi\,\delta(r_{1a}) = 2.889\,01$$

$$S(+2) = 3.852\,01 \text{ a.u.} = 15.408\,05 \text{ Ry units}$$

$$\delta(r_{1a}) = 0.229\,90$$

b. From Kolos and Wolniewicz (1964), Table V.

This time, α is a variational parameter and is 1.0, the values of ε_4 and ε_5 are in a.u. at $R = 1.4$ a.u. and we choose the 54-term wavefunction.

$$\varepsilon_4 = 1.392\,194, \quad \varepsilon_5 = 0.106\,425$$

$$2\varepsilon_4 + \varepsilon_5 = 4\pi\,\delta(r_{1a}) = 2.890\,813$$

$$S(+2) = 3.854\,417 \text{ a.u.} = 15.417\,67 \text{ Ry units}$$

$$\delta(r_{1a}) = 0.230\,044$$

c. Wolniewicz (1966), p. 521 (top) gives

$$16\pi\,\langle\delta(r_{1a})\rangle = 11.5648 \text{ a.u.}$$

There is no ambiguity in units here.

$$S(+2) = 3.854\,933 \text{ a.u.} = 15.4197 \text{ Ry units}$$

$$\delta(r_{1a}) = 0.230\,074$$

d. From Wolniewicz (1993).

Table III. Lowest-order relativistic corrections

$$\varepsilon_4 = 1.391\,376, \quad \varepsilon_5 = 0.105\,394$$

$$2\varepsilon_4 + \varepsilon_5 = 4\pi\,\delta(r_{1a}) = 2.888\,146$$

$$S(+2) = 3.850\,86 \text{ a.u.} = 15.4034 \text{ Ry units}$$

$$\delta(r_{1a}) = 0.229\,831$$

We have gone to this excruciating detail because of two misunderstandings in the literature. One is the value given by Cooper (1974) for $\delta(r_{1a})$, 0.4598. We agree with his final result, $\sigma(\mathrm{H}_2)/2\sigma(\mathrm{H}) = 1.444$ asymptotically, but with the Kolos/Wolniewicz definition of $\delta(r_{1a})$, Cooper must have meant $2\delta(r_{1a})$. The other is a frequently quoted value of $S(+2)$ which is slightly, but significantly lower than the one we deduce. Garcia (1966), citing Kolos and Wolniewicz (1964) as the source, gives $S(+2) = 3.692\,86$ a.u. Victor and Dalgarno (1969), also citing Kolos and Wolniewicz (1964), give $S(+2) = 3.693$ a.u. Finally, Meath and Kumar (1990), again citing Kolos and Wolniewicz (1964), give $S(+2) = 3.693$ a.u. Our values from Kolos and Wolniewicz (1964) are $S(+2) = 3.8544$ and 3.8549. We do not understand the discrepancy. The recent sum rule analysis by Yan *et al.* (1998a) lists $S(+2) = 3.851$ a.u., citing Wolniewicz (1993) as the source, in agreement with Appendix 3.d., above.

4.2 Molecular Nitrogen (N$_2$)

4.2.1 Preamble

Nitrogen has a slightly higher ionization potential (15.58 eV) than H$_2$ (15.43 eV). In both molecules, the onset of photoabsorption occurs at relatively high energies, and gives rise to npσ and npπ Rydberg series, each with its vibrational complement. Unlike H$_2$, where high-quality calculations were available (as well as experimental data) to determine the oscillator strengths of the resulting bands, the calculations for N$_2$ are not of sufficiently high quality. Hence, we must resort to experimental information. Fortunately, two experimental studies have appeared recently which reduce the uncertainties that existed with prior data. Shaw *et al.* (1992a) have measured the absolute photoabsorption cross section from the ionization threshold to 485 Å (25.56 eV). At roughly the same time, Chan *et al.* (1993a) reported on oscillator strengths in the valence region (12–22 eV) using dipole (e,e) spectroscopy with 0.048 eV FWHM resolution, and out to 200 eV using low resolution, 1 eV FWHM. Each of the groups was apparently unaware of the other's work. As a consequence, Shaw *et al.* used older oscillator strength data for the sub-ionization region (Gürtler *et al.*, 1977a) in their sum rule analysis, which yielded a contribution of 0.61 to $S(0)$, whereas Chan *et al.* obtained 1.173. The flawed data of Gürtler *et al.* was recognized by Shaw *et al.* in a 'note added in proof', and $S(0)$ was increased, but not to the extent implied by Chan *et al.* The latter authors emphatically make the point that photoabsorption in the sharp line discrete region is subject to saturation error, whereas the (e,e) experiment is not. On the other hand, normalization is required in the (e,e) experiment, typically by some form of the TRK sum rule, whereas photoabsorption in regions of broad structure or smooth continuum can rely on the Beer–Lambert law, obviating the need for an auxiliary normalization. Hence, in principle the photoabsorption data of Shaw *et al.* are preferred above the ionization potential.

Chan *et al.* have compared their oscillator strengths for the first 5 bands (transitions to b$^1\Pi_u$, b$'^1\Sigma_u^+$, c$^1\Pi_u$, c$'^1\Sigma_u^+$, o$^1\Pi_u$) with calculations and prior experimental data. They find that the calculations are not sufficiently accurate, as indicated earlier. However, they find that the <u>relative</u> oscillator strengths obtained in earlier electron impact experiments and their data are 'reasonably consistent'. One of these earlier experiments, by Zipf and McLaughlin (1978) finds oscillator strengths about 10–15% higher than those of the Vancouver group (Chan *et al.*, 1993a). In a later review, the Vancouver group (Olney *et al.*, 1997) have slightly increased their oscillator strengths, after finding that their initial data present a shortfall for $S(-2)$. Using the data of Chan *et al.* (1993a) we obtain spectral sums for $S(-2)$ and also $S(0)$ less than anticipated. We believe that the values most consistent with these sum rules lie somewhere between the oscillator strengths of Chan *et al.* (1993a) and those of Zipf and McLaughlin (1978), and in the final analysis we shall try to specify where they lie.

The autoionization region encompasses the range from the ionization potential to ~ 18.79 eV. Here we turn to the data of Shaw *et al.* (1992a) which have considerably higher resolution than the (e,e) experiment, have an absolute calibration based on the Beer–Lambert law, and utilize the smooth continuum of synchrotron radiation. Above 18.79 eV, in the unstructured continuum, we switch to the data of Samson *et al.* (1987a) and Samson and Haddad (1984).[1] The Samson data are given digitally, whereas Shaw *et al.* present graphical information in compressed form. Also, the Samson data extend to 107 eV, whereas the experiments of Shaw *et al.* terminate at 25.6 eV.

Between 107–200 eV, we find that the values of Chan *et al.* (1993a) differ only slightly (within experimental uncertainty) from atomic additivity, using atomic nitrogen cross sections from Henke *et al.* (1993). Apart from some pre-K-edge resonances, we remain with atomic additivity and the atomic data of Henke *et al.* to 10 keV.

4.2.2 The data

The most accurate adiabatic ionization potential of N_2, $125\,667.032(65)$ cm^{-1} $\equiv$ $15.580\,725\,(8)$ eV, stems from the spectroscopic analysis of Huber and Jungen (1990). Merkt and Softley (1992) used ZEKE to get $125\,668 \pm 0.25$ cm^{-1}, not as accurate but in substantial agreement. Later studies by Kong *et al.* (1993) and Hepburn (1997) say they agree with Merkt and Softley, but do not present new numbers.

a   *The discrete spectrum and transitions below the IP*

The data of Chan *et al.* (1993a) represent a complete set of oscillator strengths, from the onset of absorption to the IP, arrived at by a single technique. Hence the relative oscillator strengths should be fairly accurate, a conclusion supported by their observation that previously reported relative data using the inelastic electron scattering method are 'reasonably consistent' with theirs. There are minor distinctions arising from deconvolution of overlapping peaks and their assignments, but these effects cancel when assessing the total oscillator strength. Hence, in Table 4.7 we list their oscillator strengths for the vibronic transitions to two valence states, $b^1\Pi_u$ and $b'^1\Sigma_u^+$, and three Rydberg states, $c^1\Pi_u$, $c'^1\Sigma_u^+$ and $o^1\Pi_u$, and in Table 4.8 these summed band intensities are tabulated, as well as higher

---

[1] These data sets are not identical, but close. Regression fitting to both data sets yield contributions to $S(0)$, $S(-1)$ and $S(-2)$ which differ by 0.0034, 0.0001 and -0.0002, respectively. We use the more current 1987 data.

Table 4.7 Contributions to $S(p)$ of transitions to the valence states $(b^1\Pi_u, b'^1\Sigma_u^+)$ and the lowest Rydberg states $(c^1\Pi_u, c'^1\Sigma_u^+, o^1\Pi_u)$ of N_2[a]

v'	E, eV	$S(-2)$	$S(-1)$	$S(0)$	v'	E, eV	$S(-2)$	$S(-1)$	$S(0)$
a. $X^1\Sigma_g^+ \to b^1\Pi_u$ (Birge–Hopfield bands)									
0	12.500[b]	3.01	2.77	2.54	8	13.258	–	–	–
1	12.578	13.22	12.22	11.30	9	13.461	4.84	4.75	4.66
2	12.665	31.39	29.22	27.20	10	13.437	15.07	14.89	14.70
3	12.754	59.86	56.12	52.60	11	13.529	4.90	4.87	4.84
4	12.838	96.70	91.25	86.10	12	13.617	1.81	1.81	1.81
5	12.981	6.73	6.43	6.13	13	13.704	–	–	–
6	13.061	5.43	5.21	5.00	14	13.788	2.82	2.86	2.90
7	13.156	25.35	24.51	23.70		Total	271.13	256.88	243.48
b. $X^1\Sigma_g^+ \to b'^1\Sigma_u^+$									
6	13.390[c]	2.23	2.20	2.16	16	14.228	57.25	59.86	62.6
7	13.508	–	–	–	17	14.304	28.77	30.25	31.8
8	13.582	–	–	–	18	14.408	2.91	3.08	3.26
9	13.663	12.69	12.75	12.8	19	14.467	14.68	15.61	16.6
10	13.755	2.15	2.18	2.2	20	14.532	15.17	16.20	17.3
11	13.834	6.33	6.43	6.54	21	14.591	–	–	–
12	13.916	28.97	29.63	30.3	22	14.690	3.90	4.21	4.55
13	13.999	–	–	–	23	14.750	7.63	8.27	8.97
14	14.077	31.85	32.96	34.1	24	14.808	3.07	3.34	3.63
15	14.155	37.79	39.31	40.9	–	Total	255.38	266.27	277.71
c. $X^1\Sigma_g^+ \to c^1\Pi_u$									
0	12.912[d]	70.51	66.91	63.5	3	13.737	–	–	–
1	13.208	67.91	65.93	64.0	4	13.992	1.99	2.04	2.10
2	13.476	15.80	15.65	–	–	Total	156.21	150.53	145.10
d. $X^1\Sigma_g^+ \to c'^1\Sigma_u^+$									
0	12.934[c]	215.8	205.1	195.0	4	13.982	46.97	48.27	49.6
1	13.188	1.57	1.52	1.47	5	14.237	0.55	0.57	0.6
2	13.458	–	–	–	6	14.482	11.92	12.68	13.5
3	13.720	18.69	18.84	19.0	–	Total	295.45	287.00	279.17
e. $X^1\Sigma_g^+ \to o^1\Pi_u$									
1	13.345[e]	21.93	21.51	21.1	4	14.048	5.82	6.01	6.2
2	13.584	27.79	27.74	27.7	5	14.275	1.41	1.48	1.55
3	13.818	22.88	23.24	23.6	–	Total	79.83	79.98	80.15

[a] $S(p)$ in Ry units. The numbers given should be divided by 10^3. The oscillator strengths are from Chan *et al.* (1993a). Analysis (Sect. 4.2.3) indicates that multiplying these oscillator strengths by 1.034 will provide more accurate results.

[b] Energies from Carroll and Collins (1969).

[c] Energies from Carroll *et al.* (1970).

[d] Energies from Yoshino (1983).

[e] Energies from Yoshino *et al.* (1975).

Table 4.8 Spectral sums, and comparison with expectation values for N_2[a]

Energy, eV	$S(-2)$	$S(-1)$	$S(0)$	$S(+1)$	$S(+2)$
$\rightarrow b^1\Pi_u,$					
12.500–13.788[b]	0.2711	0.2569	0.2435	0.2309	0.2190
$\rightarrow b'^1\Sigma_u^+,$					
12.854–14.808[b]	0.2554	0.2663	0.2777	0.2897	0.3024
$\rightarrow c^1\Pi_u,$					
12.912–13.992[b]	0.1562	0.1505	0.1451	0.1399	0.1349
$\rightarrow c'^1\Sigma_u^+,$					
12.934–14.482[b]	0.2954	0.2870	0.2792	0.2719	0.2653
$\rightarrow o^1\Pi_u,$					
13.103–14.275[b]	0.0798	0.0800	0.0802	0.0803	0.0806
$\rightarrow e^1\Pi_u,$					
14.330–14.860[b]	0.0204	0.0216	0.0229	0.0243	0.0257
$\rightarrow e'^1\Sigma_u^+,$					
14.364[b]	0.0093	0.0099	0.0104	0.0110	0.0116
$\rightarrow n=5,\ ^1\Pi_u,$					
14.839[b]	0.0095	0.0104	0.0113	0.0123	0.0134
$\rightarrow$ IP,					
14.92–15.58[b]	0.0882	0.0991	0.1114	0.1253	0.1407
IP–18.786[c]	0.5449	0.6792	0.8493	1.0653	1.3403
18.786–107.07[d]	1.1555	2.6092	6.9920	22.5273	91.6872
107.07–200[b]	0.0071	0.0699	0.7120	7.4868	81.3411
107.07–200[e]	0.0074	0.0728	0.7395	7.7477	83.8099
200–407.4[e]	0.0009	0.0163	0.3118	6.2050	128.5748
401.0[f]	0.0002	0.0071	0.21	6.1894	182.4209
405.6[f]	–	0.0001	0.0028	0.0835	2.4884
406.8[f]	–	0.0002	0.0067	0.2003	5.9896
407.4–430[f]	0.0003	0.0104	0.3242	9.9030	305.79
430–1253.6[e]	0.0014	0.0605	2.7289	133.5771	7163.52
1253.6–3691.7[e]	0.0000_3	0.0035	0.4434	61.5084	9347.24
3691.7–10 000[e]	–	0.0001	0.0505	19.8607	8431.11
10^4–10^{5}[g]	–	–	0.0064	8.0116	13 618.06
10^5–10^{6}[h]	–	–	–	0.3930	6 249.14
10^6–10^{7}[h]	–	–	–	0.0134	2 154.36
10^7–10^{8}[h]	–	–	–	0.0004	700.68
10^8–10^{9}[h]	–	–	–	–	223.57
10^9–∞[h]	–	–	–	–	103.72
Total	2.8956	4.6382	13.8197	278.21	48 792.2
Revised total[i]	(2.9360)	(4.6813)	(13.8874)	(278.51)	(48 794.7)
Expectation value	2.9360[j]	–	14.0	–	48 225.9[k]
Other values	(2.935)[l]	4.742[l]	(14.0)[l]	276.0[l]	49 200.0[l]
	2.35[c]	4.20[c]	13.65[c]	–	
	2.885[b]			–	
	2.938[m]	4.629[m]	–		

[a]In Ry units.
[b]Chan *et al.* (1993a).
[c]Shaw *et al.* (1992a).
[d]Samson *et al.* (1987a).
[e]Henke *et al.* (1993).
[f]Zhadenov *et al.* (1987).
[g]Chantler (1995).
[h]Using the hydrogenic equation of Bethe and Salpeter (1977).
[i]Sub-ionization region multiplied by 1.034. See text.
[j]From $\alpha(N_2) = 1.7403(9) \times 10^{-24}\,cm^3$. See text.
[k]From $\delta(N_2) = 205.591$ a.u., obtained by Bader *et al.* (1967).
[l]Zeiss *et al.* (1977b).
[m]Olney *et al.* (1997).

energy bands up to the IP. This will enable us to change the normalizing factor (if called for) in the final analysis, as they have done in their subsequent review (Olney *et al.*, 1997). From their data, we obtain 1.182 for the integrated oscillator strength below the IP, whereas they give 1.173. The slight difference probably arises from our interpolation of their data, which is partitioned at 15.54 eV, whereas the IP is 15.58 eV.

b   The autoionization region, IP–18.786 eV (660 Å)

This region contains autoionization structure attributed to three series converging on the excited $A^2\Pi_u$ state of N_2^+ at 16.698 eV and three series converging on the $B^2\Sigma_u^+$ state of N_2^+ at 18.751 eV, according to Shaw *et al.* (1992a). Photoabsorption cross section curves in this domain have been given by these authors in their Figs. 1–3. We have scanned, digitized and integrated these curves, using trapezoidal integration with a fine mesh. The resulting values of $S(p)$ are listed in Table 4.8.

c   The continuum, 18.786–107.07 eV

The tabulated data of Samson *et al.* (1987a) have been fitted to a four-term polynomial by regression. The fitted function has been appropriately integrated to yield the contributions $S(p)$, which are given in Table 4.8. The coefficients of the polynomial are listed in Table 4.9.

d   The continuum, 107.07–200 eV

The photoabsorption data of Cole and Dexter (1978) are about 8% lower than those of Samson *et al.* (1987a) in their region of overlap below 107 eV. At higher energy, they continue to be lower than the (e,e) data of Chan *et al.* (1993a) and the doubled atomic cross sections of Henke *et al.* (1993), the latter two being rather close. Hence, we have ignored the values of Cole and Dexter, and fitted the cross sections of Chan *et al.* and Henke *et al.* separately to the same form of polynomial as in 4.2.2.c. above. Both sets of $S(p)$ values are given in Table 4.8. The Henke values are higher, but only by 3.7%. For reasons described in Sect. 4.2.3, we give the polynomial coefficients based on the Henke data in Table 4.9.

Table 4.9   Coefficients of the polynomial $\mathrm{d}f/\mathrm{d}E = ay^2 + by^3 + cy^4 + dy^5$ fitted to data at various energies[a]

Energy range, eV	a	b	c	d
18.786–107.07	6.745 049	37.682 68	−86.7847	45.236 78
200–407.4	−9.444 86	601.6067	−7409.83	33 538.62
430–1253.6	4.070 194	6126.04	−91 338.9	300 335.4
1253.6–3691.7	−3.870 22	7477.125	−166 966	1818 713
3691.7–10 000	−1.5637	6085.626	122 665.1	−18 521 139

[a]$\mathrm{d}f/\mathrm{d}E$ in Ry units, $y = B/E$, $B = \mathrm{IP} = 15.5807$ eV.

e   The continuum, 200–407.4 eV

The photoabsorption cross sections of Chan *et al.* terminate at 200 eV. The data of
Cole and Dexter remain lower (up to their limit, 248 eV) than the doubled atomic
cross sections of Henke *et al.* while the inelastic scattering results are higher than
the Henke values by about 20%. Consequently, we traverse the 200–407.4 eV
region with twice the atomic cross sections of Henke *et al.* The data are fitted to
a four-term polynomial, as before, and the calculated $S(p)$ are given in Table 4.8,
while the coefficients of the polynomial are recorded in Table 4.9. The K-edge
of N_2 occurs at 409.9 eV (Jolly *et al.*, 1984). Preceding the K-edge are some
resonances (treated below) and some unresolved structure, beginning at 407.4 eV,
which we also consider separately. Hence, for practical reasons we extend the
above integration to 407.4 eV, rather than the K-edge at 409.9 eV.

f   Resonances preceeding the K-edge

Zhadenov *et al.* (1987) have measured the oscillator strengths of a strong π_g
resonance[2] at 401.0 eV, and two Rydberg transitions, σ_g 3s at 405.6 eV and π_g
3p at 406.8 eV. These are recorded, with their corresponding $S(p)$, in Table 4.8.

g   K-edge structure, 407.4–430 eV

Zhadenov *et al.* (1987) present the spectral dependence of the photoabsorption
cross section up to 430 eV. We have graphically integrated the curve, and present
the $S(p)$ in Table 4.8. The more recent data of Kempgens *et al.* (1996) are
presented on a coarser scale in their Fig. 1.

h   Post K-edge continuum, 430–10 000 eV

We fit twice the atomic cross section data of Henke *et al.* in three segments
(430–1253.6, 1253.6–3691.7, 3691.7–10 000 eV) to four-term polynomials. The
derived $S(p)$ are included in Table 4.8, the polynomial coefficients in Table 4.9.
At 430 eV, the calculated cross section is 9% lower than the value extracted from
Zhadenov *et al.*

i 10^4–10^5 eV

The calculated atomic nitrogen cross sections of Chantler (1995) are doubled,
and the $S(p)$ are evaluated. The results are recorded in Table 4.8. This region
contributes $\sim$38% to $S(+2)$, $\sim$3% to $S(+1)$ and insignificantly to the other $S(p)$.

4.2.3   The analysis

The refractive index of N_2 has been measured in the visible and infrared (Peck and
Khanna, 1966) and the microwave (Newell and Baird, 1965) regions. The dielectric constant has also been determined (Orcutt and Cole, 1967). When converted

---

[2] Spectra exist in this region which display beautiful vibrational resolution, but without an
absolute scale. See e.g. Vondrácek *et al.* (1999). These authors estimate 50 Mb for their
highest peak, from which we infer $f \cong 0.217$ for the entire band, in excellent agreement
with the value from Zhadenov *et al.* (1987).

to static electric dipole polarizability, these measurements are remarkably consistent—1.7402 ±0.0006 × 10^{-24} cm^3 (Peck and Khanna, 1966), 1.7404 ±0.0005 × 10^{-24} cm^3 (Newell and Baird, 1965) and 1.7403 ±0.0008 × 10^{-24} cm^3 (Orcutt and Cole, 1967). The corresponding value of $S(-2)$, 2.9360 ±0.0009, is 1.4% larger than our spectral sum, 2.8956. The spectral sum for $S(0)$, 13.8197, is 1.3% smaller than required by the TRK sum rule.

About 42% of the total $S(-2)$ is attributable to the sub-ionization region. If we attribute all of the shortfall to this region, an enhancement of 3.4% would bring agreement with polarizability measurements. Chan *et al*. (1993a) estimate an uncertainty of 5–10% for strong and well-separated peaks, 10–20% for others. Olney *et al*. (1997), the same group as Chan *et al*. have increased their oscillator strengths in their review article to achieve agreement with experimental polarizability. If we apply the same procedure (i.e. multiply all oscillator strengths in Table 4.7 by 1.034, and the sub-ionization region in Table 4.8 by the same factor), then $S(0)$ is increased to 13.8535; in the same spirit, choosing the doubled atomic cross sections of Henke *et al*. (1993) rather than the data of Chan *et al*. for the 107–200 eV region boosts $S(0)$ to 13.8874, just 0.8% below the TRK requirement. We offer these corrected values as the best current estimate for the oscillator strength distribution in N$_2$.

For most molecules, ab initio values of $S(+2)$ are unavailable. However, for N$_2$, Bader *et al*. (1967) calculated the charge density at the nuclei (δ) and obtained 205.591 a.u. This calculation was performed to Hartree–Fock accuracy. They point out that the Hartree–Fock one-electron density distributions are correct to second order. From the Reference Table,

$$S(+2) = \frac{16\pi}{3} \cdot Z \cdot \delta(r_{N_2})$$

where Z (nuclear charge) is 14, and $S(+2)$ is in Ry units. This yields $S(+2) =$ 48 225.9. The value of $S(+2)$ for atomic nitrogen, calculated at the Hartree–Fock level, is 24 156.3 (Fraga *et al*., 1976) or 24 156.8 (Bunge *et al*., 1993). Thus, atomic additivity yields a value just 0.2% larger than the molecular value. This level of agreement is found for all the homonuclear diatomic molecules treated by Bader *et al*. (1967), i.e. Li$_2$, B$_2$, C$_2$, N$_2$, O$_2$ and F$_2$. This behavior has previously been pointed out by Meath and co-workers (Zeiss *et al*., 1980; Mulder and Meath, 1981). On this basis, we shall utilize atomic additivity to estimate $S(+2)$ for other molecules. In the present case, the value of $S(+2)$ obtained from the spectral sum (48 794) is 1.2% higher than that based on theory.

Zeiss *et al*. (1980) also demonstrate that atomic additivity works fairly well for $S(+1)$. For nitrogen, the atomic value of Fraga *et al*. (1976) yields $S(+1)$N$_2 =$ 273.2, about 1.9% lower than our spectral sum. In principle, applying additivity to the determination of $S(+1)$ for molecules is less well established, since the definition of $S(+1)$ involves inner shell correlation. Nevertheless, in the absence of other data, it represents a useful first approximation.

Zeiss *et al*. (1977b) obtained a dipole oscillator strength distribution (DOSD) for N$_2$ by utilizing experimental data available at that time, which were entirely different from those used here. However, they adjusted the input data by applying constraints, such that established values for $S(-2)$ and $S(0)$ were satisfied. As they point out, such DOSDs are not totally reliable in local detail. 'The constraint procedures... cannot completely offset the errors that are inherent in the input information used to construct the DOSD'. Their adjusted sums are shown in Table 4.8. Not surprisingly, they are in good agreement with the current results, but the latter should be more reliable in 'local detail'. Also shown are the more recent results of Shaw *et al.* (1992a). As pointed out in Sect. 4.2.1, they are severely affected for $S(-2)$ and $S(-1)$ by flawed data used for the discrete region. This should not affect their $S_i(-1)$, the ionization component of $S(-1)$. They obtain 3.47, whereas the current selections (based partly on their quantum yields of ionization) yield 3.36. The measured value is 3.74 ± 0.14 (Rieke and Prepejchal, 1972).

4.3 Molecular Oxygen (O$_2$)

4.3.1 Preamble

The oxygen molecule has two electrons in the valence anti-bonding π_g orbital. This results in a reduction of the adiabatic ionization potential relative to N_2 (12.07 versus 15.58 eV) and also a diminution in the dissociation energy (5.12 versus 9.76 eV). As a consequence, some excited states are broadened by predissociation, others (Schumann–Runge bands) are dissociative. Hence, photoabsorption methods employing the Beer–Lambert law are not as seriously influenced by band width effects. This can be noted in the generally good agreement between photoabsorption measurements and inelastic electron scattering, the latter inherently unaffected by band width.

For example, it will be shown that the oscillator strength of the low-energy Schumann–Runge bands from photoabsorption is 0.162, whereas inelastic electron scattering has provided 0.169 (Chan *et al.*, 1993b) and 0.161 (Huebner *et al.*, 1975a). For the next higher feature, often referred to as the 'longest' band, the photoabsorption data of Ogawa and Ogawa (1975) yield $f = 0.0083$, whereas the electron scattering data of Chan *et al.* arrive at $f = 0.0084$. Such agreement between photoabsorption and electron scattering data continues (where data are available) up to the ionization potential. Thus, we are enabled to implement photoabsorption data, which use the more rigorous Beer–Lambert law for normalization rather than indirectly invoking a Thomas–Reiche–Kuhn sum rule, for the sub-ionization region.

Prominent autoionization structure is manifest from the IP to $\sim$24.5 eV. This encompasses the ionic states $X^2\Pi_g$, $a^4\Pi_u$, $A^2\Pi_u$, $b^4\Sigma_g^-$, $B^2\Sigma_g^-$, $2^2\Pi_u$, $3^2\Pi_u$, and $c^4\Sigma_u^-$, the latter at 24.56 eV. The absolute photoabsorption cross sections in this region have been reported recently by Holland *et al.* (1993). They are stated to be accurate to 2.1% and in good agreement throughout the region of overlap (IP–21.4 eV) with earlier measurements by Matsunaga and Watanabe (1967) performed at comparable resolution ($\sim$3 meV). We find that the contribution to oscillator strength in this region differs by $\leq$3% in the two studies. Here, we utilize the more recent data (Holland *et al.*, 1993) which extend to higher energies, although the tabular presentation of Matsunaga and Watanabe will be useful to some investigators.

Between 24.53–107.08 eV, where structure is largely absent, we turn to the data of Samson and Haddad (1984), which were obtained with line sources. They differ only slightly from Samson *et al.* (1982). The overlap with Holland *et al.* (1993) is fairly good at the low end, and with Mehlman *et al.* (1978b) at the high end. Here, good agreement is already seen with the doubled atomic oxygen cross sections of Henke *et al.* (1993). The data of Cole and Dexter (1978) are distinctly lower between 36–248 eV, and are not used.

Around the K-edge (506–600 eV), the data of Barrus *et al.* (1979) are graphically integrated. Above and below this region, atomic additivity is employed, and overlaps reasonably well with the Barrus data.

4.3.2 The data

The adiabatic ionization potential of O_2 is 12.070 14 (15) eV, according to Merkt *et al.* (1998).

a   The sub-ionization region

a.1 $X^3\Sigma_g^- \rightarrow B^3\Sigma_u^-$ (Schumann–Runge bands), 7.1–9.8 eV   Figure 4.1 displays the results of three photoabsorption studies, Gibson *et al.* (1983), Ogawa and Ogawa (1975) and Watanabe *et al.* (1953a). The mutual concordance is rather good except near the maximum, where the Ogawa cross sections are about 7% higher than the others. Our consensus is generally an average of the Gibson and the Watanabe data. Graphical integration of the selected data in Fig. 4.1 leads directly to $\Delta S(-2) = 0.4190$. Corresponding integrations yield $\Delta S(-1) = 0.2600$ and $\Delta S(0) = 0.1620$. As mentioned earlier, Chan *et al.* (1993b) give 0.169 and Huebner *et al.* (1975a), 0.161 for $\Delta S(0)$.

a.2 $X^3\Sigma_g^- \rightarrow$ valence + Rydberg $E^3\Sigma_u^-$ (mixed), $v' = 0$–2, 9.75–10.62 eV   The three bands in this region, described in the spectroscopic literature as the 'longest', 'second' and 'third' bands, have been identified with transitions to $v' = 0$, 1 and 2 of a mixed valence-Rydberg ($E^3\Sigma_u^-$) state by ab initio calculations (Buenker *et al.*, 1976; Yoshimine, 1976). The 'longest' and 'second' band regions, shown in Figs. 4.2 and 4.3 taken from the data of Ogawa and Ogawa (1975), each contribute approximately 10-times more to the oscillator strength than the 'third' band. Lewis *et al.* (1988a) present a more detailed graph of the doublet in the 'longest' band, for which they give $f = 0.006\,25$. The corresponding region from Fig. 4.2 yields $f = 0.0064$. The 9.75–10.17 eV domain contributes 0.008 33 to $\Delta S(0)$, whereas Chan *et al.* (1993b) report 0.008 44.

For the 'second' band, our integration of Fig. 4.3 yields $f = 0.007\,07$. Lewis *et al.* (1988b) give $f = 0.0071$ for the $^{16}O\,^{18}O$ isotope, while Chan *et al.* (1993b)

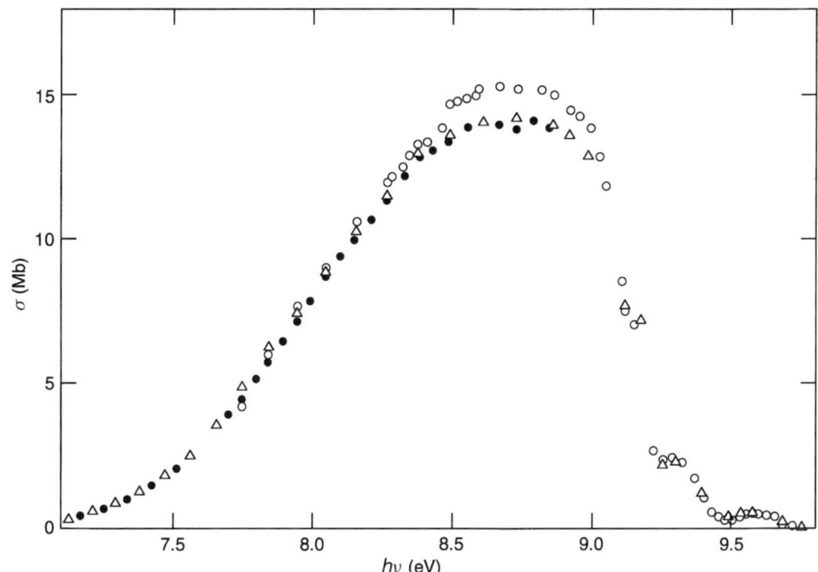

Fig. 4.1 Absolute photoabsorption spectrum of O_2 – the Schumann–Runge system. • Gibson *et al.* (1983); ○ Ogawa and Ogawa (1975); △ Watanabe *et al.* (1953)

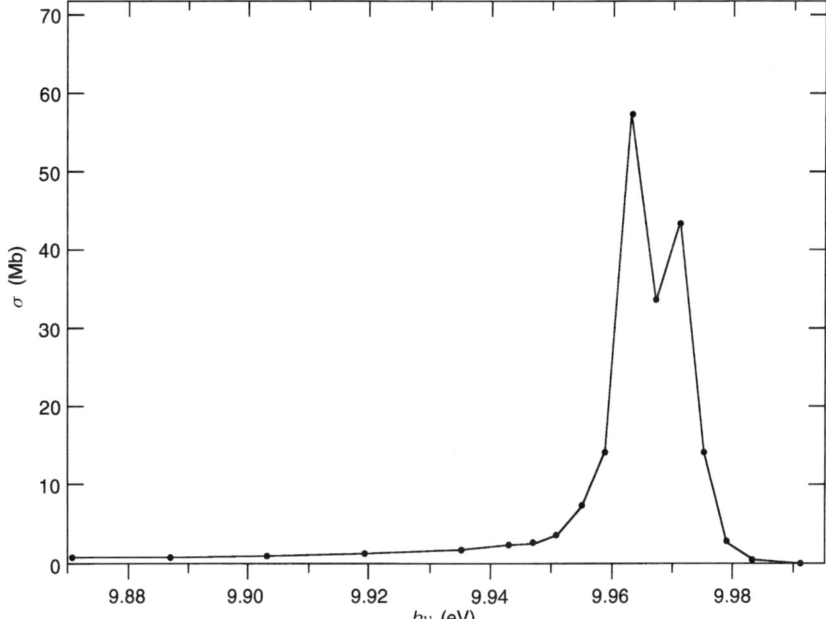

Fig. 4.2 Absolute photoabsorption spectrum of O_2 – the longest band. • Ogawa and Ogawa (1975)

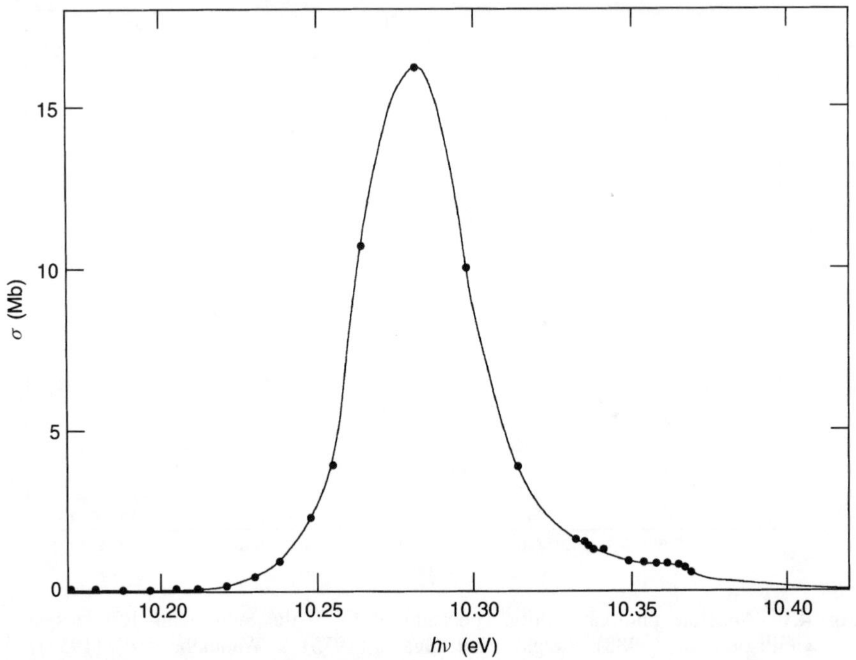

Fig. 4.3 Absolute photoabsorption spectrum of O_2 – the second band. • Ogawa and
Ogawa (1975)

obtain $f = 0.007\,59$ from electron energy loss experiments. The weak 'third'
band contributes $f = 0.000\,77$ from our integration of the corresponding region
of Ogawa and Ogawa (1975). Lewis *et al.* (1988b) obtain $f = 0.000\,78$, while
Chan *et al.* give $f = 0.000\,827$.

a.3 Transitions to excited states, 10.62–12.07 eV (IP) These transitions,
some of which are to the $2\,^3\Pi_u$ state, all have relatively small oscillator strengths.
We use the data of Ogawa and Ogawa from 10.62–11.33 eV (its limit), then
utilize the (e,e) results of Chan *et al.* from 11.33–11.59 eV, but return to photoabsorption values (Matsunaga and Watanabe, 1967) from 11.59 eV to the IP. The
contributions to $S(p)$ from individual regions, including the lower energy states,
are listed in Table 4.10. We note that the sub-ionization region is dominated by
the Schumann–Runge bands, which contribute $\sim 90\%$ to the magnitude of $S(-2)$
below the IP.

b The autoionization region, IP–24.53 eV (505 Å)

The excited states of $O_2{}^+$ in this region are $a\,^4\Pi_u$ (16.101), $A\,^2\Pi_u$ (17.045), $b\,^4\Sigma_g^-$
(18.171), $B\,^2\Sigma_g^-$ (20.296), $c\,^4\Sigma_u^-$ (24.564) and the recently characterized $3\,^2\Pi_u$
(21.32), where the numbers in parentheses are adiabatic IPs, and they are taken
from Baltzer *et al.* (1992). Rydberg·members have been identified converging

Table 4.10 Contributions to $S(p)$ of sub-ionization transitions in O_2[a]

Energy range, eV	$S(-2)$	$S(-1)$	$S(0)$	$S(+1)$	$S(+2)$
7.125–9.75[b]	419.2	260.2	162.0	101.2	63.4
9.75–10.17[c,d]	15.55	11.38	8.33	6.10	4.47
10.17–10.44[c,e]	12.36	9.35	7.07	5.34	4.04
10.44–10.62[c,f]	1.28	0.99	0.77	0.60	0.46
10.62–10.71[c]	1.08	0.84	0.66	0.52	0.41
10.71–10.84[c]	2.23	1.77	1.40	1.11	0.87
10.84–10.98[c]	1.26	1.01	0.81	0.65	0.52
10.98–11.17[c]	1.15	0.94	0.76	0.62	0.50
11.17–11.33[c]	0.74	0.61	0.50	0.41	0.34
11.33–11.52[g]	2.08	1.75	1.47	1.23	1.04
11.52–11.59[g]	0.58	0.49	0.419	0.36	0.30
11.59–12.07[h]	7.287	6.328	5.496	4.774	4.147
$\Sigma \to$ IP	464.8	295.7	189.7	122.9	80.5

[a] $S(p)$ in Ry units. The numbers given should be divided by 10^3.
[b] From consensus of data shown in Fig. 4.1. See text.
[c] Ogawa and Ogawa (1975).
[d] 'Longest' band.
[e] 'Second' band.
[f] 'Third' band.
[g] Chan *et al.* (1993b)
[h] Matsunaga and Watanabe (1967).

to all of these limits, sometimes with complex vibrational progressions. Holland *et al.* (1993) give a brief review, together with references to earlier work.

For the present purposes, we have scanned, digitized and integrated the absolute photoabsorption curves in Figs. 1–5 of Holland *et al.* As noted above, these data (to 21.4 eV) are in good agreement with the earlier work of Matsunaga and Watanabe (1967). Previously (Berkowitz, 1979) the latter data were manually extracted and integration yielded $\Delta S(0) = 0.724$ between 1027–750 Å. The present scan of the data of Holland *et al.* gives $\Delta S(0) = 0.6175$ for this interval, compared to 0.587 given by Chan *et al.* (1993b). For the range selected here, IP–24.53 eV, our scan of the data of Holland *et al.* results in $\Delta S(0) = 2.4900$. To check the accuracy of the scanning, digitizing and integration, we have evaluated $\Delta S(0)$ for the interval IP–25.30 eV (490 Å) and obtain $\Delta S(0) = 2.66$; Holland *et al.* give $\Delta S(0) = 2.684$. The $S(p)$ in this range are recorded in Table 4.11.

c   The continuum, 24.53–107.07 eV

For reasons discussed in 4.3.1 (above), we utilize the tabulated data of Samson and Haddad (1984) in this energy range. They have been fitted by regression to a four-term polynomial. The fitted function is used to calculate contributions to $S(p)$, which are listed in Table 4.11. The coefficients of the polynomial are given in Table 4.12. This is a very sensitive region for $S(-1)$ and $S(0)$, accounting for nearly half their respective totals.

Table 4.11   Spectral sums, and comparison with expectation values for O_2[a]

Energy, eV	$S(-2)$	$S(-1)$	$S(0)$	$S(+1)$	$S(+2)$
7.125–12.07 (IP)[b]	0.4648	0.2957	0.1897	0.1229	0.0805
12.07–24.53[c]	1.3926	1.8307	2.4900	3.4967	5.0546
24.53–107.07[d]	0.7869	2.2668	7.4841	28.4261	127.4863
107.07–206.6	0.0126[e]	0.1251[e]	1.2816[e]	13.5964[e]	149.4800[e]
	(0.0135)[f]	(0.1347)[f]	(1.3873)[f]	(14.7901)[f]	(163.6077)[f]
206.6–506[f]	0.0015	0.0298	0.6220	13.8138	327.5557
530.9[g]	0.00004	0.0014	0.0549	2.1419	83.5787
506–600[g]	0.0004	0.0171	0.7045	29.1541	1155.5279
600–2622.4[f]	0.0007	0.0424	2.7998	211.7083	18913.17
2622.4–10000[f]		0.0007	0.1947	59.2970	20612.72
10^4-10^5[h]	–	–	0.0116	14.7586	25283.26
10^5–10^6[i]	–	–	–	0.7652	12202.02
10^6–10^7[i]	–	–	–	0.0272	4261.46
10^7–10^8[i]	–	–	–	0.0008	1392.03
10^8–10^9[i]	–	–	–	–	444.79
10^9–∞[i]	–	–	–	–	206.46
Total	2.6595	4.6097	15.8329	377.31	85164.6
Expectation value	2.6478[j]		16.0		83849.5[k]
	2.6678[l]				
	2.6416[m]				
Other values	(2.648)[n]	4.652[n]	(16.0)[n]	374.4[n]	84840.0[n]
	2.702[c]	4.598[c]	16.066[c]		
	2.635[o]				
	2.648[p]	4.591[p]			

[a]In Ry units.
[b]See Table 4.10.
[c]Holland *et al*. (1993).
[d]Samson and Haddad (1984).
[e]Mehlman *et al*. (1978b).
[f]Twice atomic cross section from Henke *et al*. (1993).
[g]Barrus *et al*. (1979).
[h]Twice atomic cross section from Chantler (1995)
[i]Using the hydrogenic equation of Bethe and Salpeter (1977).
[j]Consensus of refractive index and dielectric constant measurements. See text.
[k]Bader *et al*. (1967).
[l]Newell and Baird (1965)
[m]Holm and Kerl (1990).
[n]Zeiss *et al*. (1977); Kumar *et al*. (1996).
[o]Chan *et al*. (1993b).
[p]Olney *et al*. (1997).

d   *The continuum, 107.07–206.6 eV*

Here we compare the directly measured molecular cross sections of Mehlman *et al*. (1978b) with doubled atomic oxygen cross sections of Henke *et al*. (1993), to provide a basis for using the latter at still higher energies. Each has been fitted to the four-term polynomial, and the corresponding $S(p)$ have been evaluated. The doubled atomic cross sections are 7–8% higher.

Table 4.12 Coefficients of the polynomial $\mathrm{d}f/\mathrm{d}E = ay^2 + by^3 + cy^4 + dy^5$ fitted to data at various energies[a]

Energy range, eV	a	b	c	d
24.53–107.07	22.528 01	146.4725	−717.439	768.943
107.07–206.6	−31.4765	1795.228	−18 844.4	68 269.45
206.6–506	−2.311 83	1151.734	−21 573.6	169 397.3
600–2622.4	2.545 171	24 744.01	−743 485	8 954 206
2622.4–10 000	−2.545 72	22 944.27	520 329.6	−132 525 383

[a] $\mathrm{d}f/\mathrm{d}E$ in Ry units, $y = B/E$, $B = \mathrm{IP} = 12.0701\,\mathrm{eV}$.

e The continuum, 206.6–506 eV

We follow the decline of the O_2 cross section until the approach of the K-edge, using the doubled atomic cross sections of Henke *et al.* (1993). The K-edge occurs at $\sim$543.7 eV (Jolly *et al.*, 1984) but there are resonances preceding the edge. Since directly measured cross sections across the K-edge are available (Barrus *et al.*, 1979), which begin at 506 eV, we temporarily terminate the Henke data at this energy. The $S(p)$ obtained from the standard four-term polynomial, shown in Table 4.11, contribute $\sim$4% to $S(0)$ and $S(+1)$, much less to the other $S(p)$. The coefficients of the polynomial are given in Table 4.12.

f The K-edge region, 506–600 eV

The data of Barrus *et al.* (1979) have been re-plotted in more convenient units in Fig. 4.4. Some isolated points from Henke *et al.* are also shown. At 572.8 eV, the doubled Henke cross section is $\sim$6% lower than the Barrus cross section, whereas it was found to be 7–8% higher than the values of Mehlman *et al.* at $\sim$200 eV. The cross sections depicted in Fig. 4.4 have been graphically integrated to determine the $S(p)$ contributions given in Table 4.11. The sharp π_g resonance at 530.9 eV is listed separately.

g Post K-edge continuum, 600–10 000 eV

The doubled atomic oxygen cross sections of Henke *et al.* are fitted, in two segments (600–2622.4, 2262.4–10 000 eV) to four-term polynomials. The derived $S(p)$ are included in Table 4.11, the polynomial coefficients in Table 4.12. This region contributes about 3 electrons to $S(0)$, $\sim$75% of $S(+1)$ and $\sim$45% of $S(+2)$.

h 10^4–10^5 eV

The calculated atomic oxygen cross sections of Chantler (1995) are doubled and the $S(p)$ are evaluated.

4.3.3 The analysis

Numerous measurements have been performed which can be analyzed to provide a static electric dipole polarizability, or equivalently $S(-2)$, for O_2. Langhoff and Karplus (1969) used the method of Padé approximants to the Cauchy equation,

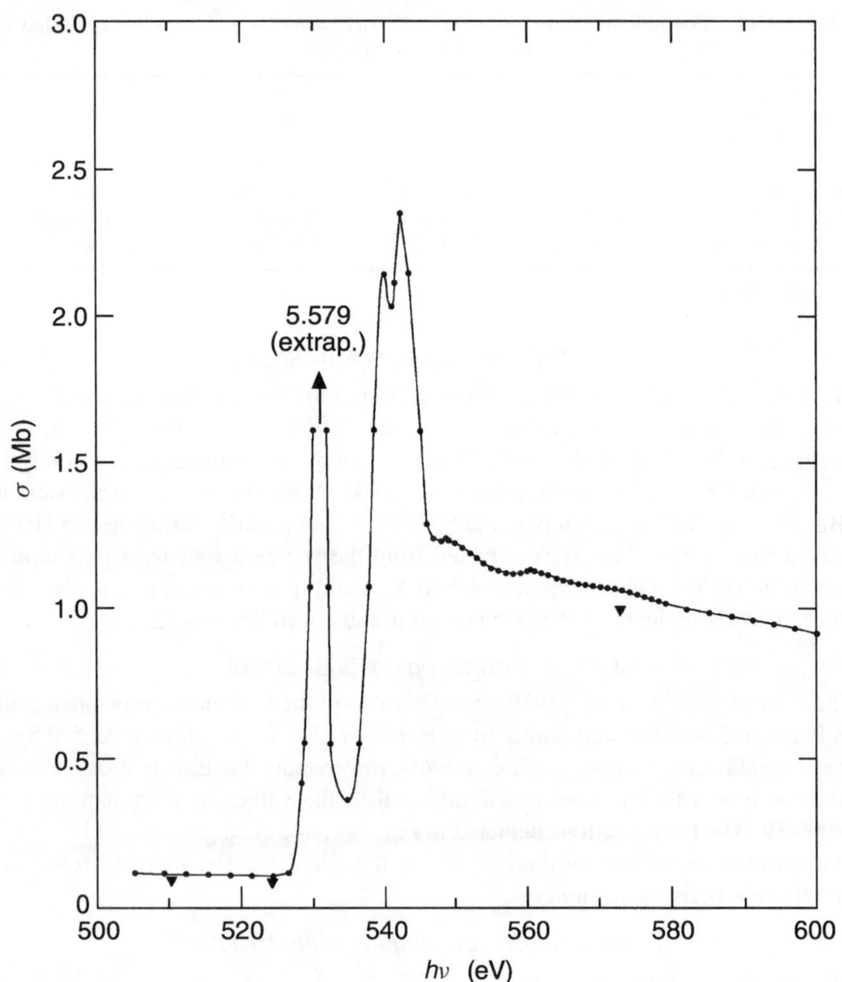

Fig. 4.4 Absolute photoabsorption spectrum of O_2–K-edge region. ● Barrus *et al.* (1979); ▼ Henke *et al.* (1993)

and the refractive index data of Stoll (1922) in the visible and near infrared, to deduce $S(-2) = 2.650$ Ry units. Ladenburg and Wolfsohn (1932) fitted their refractive index measurements in the visible and ultraviolet to an equation which readily yields the refractive index at infinite wavelength and one atmosphere ($(n-1)10^6 = 265.27$), the molar refractivity ($R = 3.9639$), the polarizability ($\alpha = 1.571\,43 \times 10^{-24}\,\mathrm{cm}^3$) and $S(-2) = 2.6510$ Ry units. Much more recently, Hohm and Kerl (1990) measured the dispersion at four wavelengths in the visible. Their extrapolated static polarizability ($1.5658(14) \times 10^{-24}\,\mathrm{cm}^3$) corresponds to $S(-2) = 2.6416$ Ry units.

Dunn (1964) performed dielectric constant measurements and reviewed earlier work. He obtained $\varepsilon = 530.61 \pm 0.2 \times 10^{-6}$ at one atmosphere, corresponding to $R = 3.9644$, $\alpha = 1.5716 \times 10^{-24}$ cm^3 and $S(-2) = 2.6514$ Ry units. Younglove (1972) studied the dielectric constant of O_2 at various pressures, and proffered a low density Clausius–Mosotti function of 0.1236 cm^3/g, equivalent to $R = 3.9651$, $\alpha = 1.5679 \times 10^{-24}$ cm^3, $S(-2) = 2.6452$ Ry units.

The above five results span the range $S(-2) = 2.6416\text{--}2.6510$, and average 2.6478. Thus, it is surprising to find that Newell and Baird (1965) obtained a refractive index $(n-1)\ 10^6 = 266.95 \pm 0.05$ at 47.736 GHz, equivalent to $S(-2) = 2.6678(5)$. Their concomitant measurements of N_2 were much closer to other values. In their sum rule analysis, Zeiss and Meath (1977a) and later Kumar $et\ al.$ (1996) chose $S(-2) = 2.6475$, essentially the average value found above. Holland $et\ al.$ (1993) base their expectation value of $S(-2)$ on 'a direct experimental value' of α, 1.59×10^{-24} cm^3, citing Bridge and Buckingham (1966). These latter authors do, in fact, give $\alpha = 1.59_8 \times 10^{-24}$ cm^3, but mention that it came from earlier sources, and refers to $\lambda = 6328$ Å. This is about 2% larger than the static value.

Our spectral sum, $S(-2) = 2.6595$, is $\sim$0.4% above our selected average of expectation values. Holland $et\ al.$ used somewhat different data sources in the sub-ionization region, and obtained $S(-2) = 2.702$. These authors misinterpreted the numbers for $S(-2)$ and $S(-1)$ of Zeiss $et\ al.$ (1977b), which are given in a.u., and hence concluded that they are substantially higher than the experimental results. Actually, they are very close to the spectral sum, which is not surprising, since their fitting program uses their chosen polarizability and the TRK sum rule for $S(0)$ as constraints. For $S(-1)$, the current spectral sum (4.6097) is 25% higher than that of Holland $et\ al.$ but 0.9% lower than that of Zeiss $et\ al.$ (1977b) and Kumar $et\ al.$ (1996). The latter investigators have also presented ab initio calculations incorporating correlation, but the result, $S(-1)$ $\sim$4.8, is still too high.

Bader $et\ al.$ (1967) have calculated the total charge density at the nuclei for O_2 at the Hartree–Fock level, from which we obtain $S(+2) = 83\,459.5$. This is very nearly twice the atomic value (41 776.9, Fraga $et\ al.$ (1976) or 41 775.4, Bunge $et\ al.$ (1993)). This observation provides further support for the use of additivity when alternative sources of $S(+2)$ are unavailable. Our spectral sum for $S(+2)$ is $\sim$1.6% higher than expectation, whereas the value obtained by Zeiss $et\ al.$ (1977b) is 1.2% higher.

For $S(+1)$, the doubled atomic Hartree–Fock value is 367.76 (Fraga $et\ al.$, 1976). Correlation is expected to increase this quantity. The ab initio calculations of Kumar $et\ al.$ (1996) do indeed show an increase, to $\sim$372.0. The spectral sum of Zeiss $et\ al.$ (1977b) is 374.4, and the current spectral sum is still higher, 377.3 Ry units. The major contributions to $S(+1)$ in the present study are the doubled atomic cross sections of Henke $et\ al.$ They are experimentally based, and are probably more accurate than the calculated values (Chantler, 1995) at still higher energies.

To evaluate $S_i(-1)$, the ionized component of $S(-1)$, we have utilized the ionization cross sections of Matsunaga and Watanabe (1967) in the autoionization region, rather than extracting both the quantum yield of ionization (η) and the absorption cross sections from Holland *et al.* (1993) and determining their product. Here we assume that η is unity below 650 Å ($>19.07\,\mathrm{eV}$). Our deduced value of $S_i(-1)$, 3.980, is almost identical to that of Holland *et al.*, 3.97. In this case, the experimental value of $M_i^2 \equiv S_i(-1)$ of Rieke and Prepejchal (1972), 4.20 ± 0.18, is characteristically larger, but within their stated uncertainty almost agrees with the spectral sum.

The overall agreement with all the sum rules, with the caveats mentioned en passant, provides strong support for the cross sections utilized in this study.

4.4   Carbon Monoxide (CO)

4.4.1   Preamble

The CO molecule is isoelectronic with N_2. This implies similarities in orbital *aufbau* and excitation, but there are noteworthy differences. The dissociation energy (D_0) is higher in CO than in N_2 (11.11 versus 9.76 eV), but the IP is lower (14.01 versus 15.58 eV). This latter observation suggests lower excitation energies in CO than in N_2, which is indeed observed. The total discrete oscillator strength is about one unit in both cases, but in CO the lower excitation energies make $S(-2)$ in this domain relatively larger, which accounts in part for the polarizability (α) being larger in CO than in N_2.

The discrete region consists of transitions to the $A^1\Pi$, $B^1\Sigma^+$, $C^1\Sigma^+$ and $E^1\Pi$ states (8.0–11.8 eV and higher excitations from 12.4–14.01 eV. There are at least three recent determinations of discrete oscillator strengths by inelastic electron scattering, but only one extensive photoabsorption study. The electron scattering data are in fairly good agreement with one another, and could be improved by a consistent absolute calibration. The photoabsorption studies yield f values about 7% larger for the $A^1\Pi$ state, but are much lower for the $C^1\Sigma^+$ and $E^1\Pi$ states, the latter probably due to pressure saturation. Hence, we select the electron scattering data for the sub-ionization region, with a caveat. The Vancouver group (Olney *et al.*, 1997) have recently revised their original oscillator strength calibration (Chan *et al.*, 1993c) downward by 2.5%, noting an error in their energy calibration. In addition, their 1997 paper suggests an absolute calibration based on polarizability data. Here we use their original (1993) data, since it is well-documented, with the proviso that the normalization can be adjusted to optimize agreement with the sum rules, especially $S(-2)$.

The photoabsorption region from 14.01–20 eV is structured, due to autoionization and predissociation. The structure is attributable to Rydberg series converging to the $X^2\Sigma^+$, $A^2\Pi$ (AIP = 16.544 eV) and $B^2\Sigma^+$ (AIP = 19.672 eV) states of CO^+, with their vibrational complements. Recent absolute photoabsorption measurements are unavailable. Samson and Haddad (1984)

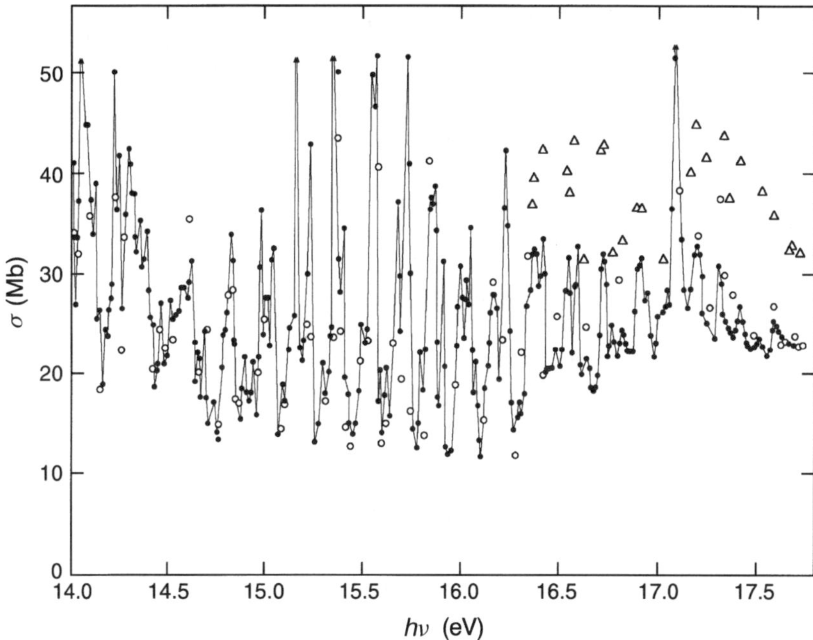

Fig. 4.5 Absolute photoabsorption spectrum of CO – autoionization region. • Cook *et al.*
(1965); ○ Samson and Haddad (1984); △ Huffman *et al.* (1964)

cover this region using line sources, which are a haphazard match for the
autoionization structure. Two studies in the mid-1960s, Huffman *et al.* (1964)
and Cook *et al.* (1965) used a helium continuum light source, with comparable
resolution, but differed in their photoabsorption cross sections by more than 30%
at some energies. Our modus operandi here is to compare both of the latter
measurements with those of Samson and Haddad, as shown in Fig. 4.5. From
previous experience, the error in the Samson/Haddad data is not expected to
exceed 3–5%. Taking into account the wavelength resolution in the continuum
source, and slight possible mis-matches in wavelength, the data of Cook *et al.*
appear to be in fairly good agreement with those of Samson and Haddad, while
those of Huffman *et al.* are distinctly higher (only the higher energy points are
shown, for clarity). This pattern persists at the Ne I lines, 16.671 and 16.848 eV,
where the cross section was measured with a claimed accuracy of ±0.8% by
Samson and Yin (1989). However, beyond 19.42 eV in the weakening tail of the
helium continuum, the data of Cook *et al.* fall distinctly below those of Samson
and Haddad. In this region, the autoionization structure is waning, and we can
safely switch to the Samson/Haddad data. We remain with Samson and Haddad
(1984) in the relatively flat region (20.68–26.84 eV) and the subsequent decline
to 124.37 eV. Atomic additivity, using the data of Henke *et al.* (1993) provides
a smooth juncture with the Samson/Haddad data and is continued to the vicinity

of the carbon K-edge. Auxiliary data sources are utilized for $\sim$60 eV domains around both the carbon and oxygen K-edges. Spot checks are provided by atomic additivity, which is also used in the inter-edge region and beyond to 10 000 eV.

4.4.2 The data

The best currently available adiabatic ionization potential of CO is $113\,027.5(3)\,\text{cm}^{-1} \equiv 14.013\,62(4)\,\text{eV}$, from Mellinger *et al.* (1996). Kong *et al.* (1993) used ZEKE to obtain $113\,025.6 \pm 1.5\,\text{cm}^{-1} \equiv 14.0134(2)\,\text{eV}$, not as precise but in substantial agreement.

a   *The discrete spectrum and transitions below the IP*

Following the early work of Lassettre and Skerbele (1971), there have been three recent determinations of the oscillator strengths in the discrete region of CO, using inelastic electron scattering: Zhong *et al.* (1997), Wu *et al.* (1997), and Chan *et al.* (1993c). We shall tentatively employ the data of Chan *et al.* They represent a complete set of oscillator strengths, from the onset of absorption to the IP, arrived at by a single technique. We implicitly assume that their relative oscillator strengths in this domain are accurate, but we have the option of adjusting the entire set by a scale factor. In Table 4.13, we list their oscillator strengths for the vibronic transitions to $A^1\Pi$, $B^1\Sigma^+$, $C^1\Sigma^+$, and $E^1\Pi$. In Table 4.14 the summed band intensities are tabulated as well as higher energy transitions up to the IP. Note that these higher energy transitions account for $\sim$2/3 of $S(0)$ and $\sim$1/2 of $S(-2)$ below the IP.

The most extensive photoabsorption studies of CO in the sub-ionization region have been performed by Eidelsberg *et al.* (1992) on the $X^1\Sigma^+ \rightarrow A^1\Pi$ band, and by Letzelter *et al.* (1987) and Stark *et al.* (1999) on the $X^1\Sigma^+ \rightarrow B^1\Sigma^+$ and higher bands. There are also some less extensive data obtained by Fock *et al.* (1980), Jolly *et al.* (1997), Stark *et al.* (1998; 1999) and others (see Sect. 4.4.3). We shall compare with the photoabsorption data after we have arrived at the best normalization of the recorded data of Chan *et al.* (1993c).

b   *The autoionization region, IP–20.68 eV ($\sim$600 Å)*

As discussed in Sect. 4.4.1 (see Fig. 4.5), we utilize the data of Cook *et al.* (1965) given in their Table II, supplemented by points from their Fig. 3, to evaluate the $S(p)$ between the IP and 17.71 eV. Between 17.71–19.42 eV, we find that the data points of Cook *et al.* (Tables II and III) and those of Samson and Haddad (1984) are commensurate, and can be combined. From 19.42–20.68 eV, we turn to the Samson and Haddad data, as discussed earlier. The data have been processed by trapezoidal integration. The $S(p)$ are recorded in Table 4.14.

c   *The continuum, 20.68–124.37 eV*

In this region, the data of Samson and Haddad are used exclusively. For the 20.68–26.84 eV interval, where the cross section decreases slightly from 22.8 to

Table 4.13 Contributions to $S(p)$ of transitions to the $A^1\Pi$, $B^1\Sigma^+$, $C^1\Sigma^+$ and $E^1\Pi$ states of CO[a]

v'	E, eV[b]	$S(-2)$	$S(-1)$	$S(0)$	v'	E, eV	$S(-2)$	$S(-1)$	$S(0)$
a. $X^1\Sigma^+ \to A^1\Pi$									
0	8.0278	46.53	27.46	16.2	7	9.2234	9.01	6.11	4.14
1	8.2115	96.36	58.16	35.1	8	9.3771	4.25	2.93	2.02
2	8.3907	105.70	65.19	40.2	9	9.5266	1.94	1.36	0.95
3	8.5659	87.54	55.12	34.7	10	9.6718	0.81	0.58	0.41
4	8.7367	58.69	37.69	24.2	11	9.8130	0.35	0.25	0.18
5	8.9032	33.86	22.16	14.5	12	9.9498	0.17	0.12	0.09
6	9.0654	18.13	12.08	8.05	Total		463.34	289.21	180.74
b. $X^1\Sigma^+ \to B^1\Sigma^+$									
0	10.7762	12.80	10.14	8.03					
1	11.0344	2.01	1.63	1.32					
Total		14.81	11.77	9.35					
c. $X^1\Sigma^+ \to C^1\Sigma^+$									
0	11.3965	167.75	140.52	117.7					
1	11.6626	4.85	4.15	3.56					
Total		172.60	144.67	121.26					
d. $X^1\Sigma^+ \to E^1\Pi$									
0	11.5219	98.45	83.37	70.6					
1	11.7887	4.70	4.07	3.53					
Total		103.15	87.44	74.13					

[a] $S(p)$ in Ry units. The numbers given should be divided by 10^3. The oscillator strengths are from Chan *et al.* (1993c). Analysis (Sect. 4.4.3) indicates that multiplying these oscillator strengths by 0.9546 will provide more accurate results.
[b] Energies from Tilford and Simmons (1972) except $E^1\Pi$, $v' = 1$ which is from Letzelter *et al.* (1987).

20.5 Mb, the $S(p)$ are evaluated by trapezoidal integration. For the monotonic decline between 26.84–124.37 eV, the data have been fitted by regression to a 4-term polynomial. The contributions to $S(p)$ are listed in Table 4.14, and the coefficients of the polynomial in Table 4.15.

d   The continuum, 124.37–292.5 eV

The carbon K-edge of CO occurs at 296.2 eV (Jolly *et al.*, 1984). Some structure appears before the K-edge, as we shall discuss shortly. Atomic additivity, using the atomic cross sections of Henke *et al.* (1993) smoothly joins with the Samson/Haddad data at 124.37 eV, and is continued to 292.5 eV (just before the aforementioned structure). The sparse Henke data are fitted to a 4-term polynomial, whose coefficients appear in Table 4.15. The corresponding $S(p)$ are listed in Table 4.14.

Table 4.14 Spectral sums, and comparison with expectation values for CO[a]

Energy, eV	$S(-2)$	$S(-1)$	$S(0)$	$S(+1)$	$S(+2)$
$\rightarrow A^1\Pi$,					
8.028–9.950[b]	0.4633	0.2892	0.1807	0.1131	0.0709
$\rightarrow B^1\Sigma^+$,					
10.776–11.034[b]	0.0148	0.0118	0.0094	0.0074	0.0059
$\rightarrow C^1\Sigma^+$,					
11.397–11.662[b]	0.1726	0.1447	0.1213	0.1016	0.0852
$\rightarrow E^1\Pi$,					
11.522–11.789[b]	0.1032	0.0874	0.0741	0.0629	0.0533
$\rightarrow$ IP,					
12.130–14.013[b]	0.6727	0.6577	0.6437	0.6306	0.6185
IP–17.712[c]	0.6677	0.7557	0.8902	1.0168	1.1880
17.712–20.68[c,d]	0.3175	0.4456	0.6267	0.8832	1.2471
20.68–26.84[d]	0.4012	0.6915	1.1984	2.0886	3.6606
26.84–124.37[d]	0.5156	1.5586	5.4062	22.0207	107.1671
124.37–292.5[e]	0.0056	0.0066	0.8343	11.0399	154.8934
287.4[f]	0.0004	0.0081	0.17	3.5910	75.8544
292.5–350[g]	0.0009	0.0203	0.4741	11.1059	260.8341
350–539.5[e]	0.0008	0.0247	0.7635	24.0176	767.2078
534.1[f]	0.0001	0.0019	0.076	2.9835	117.1206
539.5–598[h]	0.0002	0.0101	0.4202	17.5393	732.7273
598–1486.7[e]	0.0005	0.0273	1.6491	105.8993	7271.48
1486.7–3691.7[e]		0.0023	0.3414	53.4133	8964.76
3691.7–10000[e]		0.0002	0.0591	23.2706	9890.05
10^4–10^{5}[i]			0.0074	9.3340	15942.02
10^5–10^{6}[j]				0.4726	7527.23
10^6–10^{7}[j]				0.0167	2616.17
10^7–10^{8}[j]				0.0005	853.22
10^8–10^{9}[j]					272.49
10^9–∞[j]					126.46
Total	3.3371	4.7437	13.9458	289.61	55686.6
Revised total[k]	(3.2723)	(4.6896)	(13.8991)	(289.57)	(55686.6)
Expectation value	3.2723[l]		14.0		54587.7[m]
					54588.2[n]
Other values[o]	(3.270)[o]	4.8635[o]	(14.0)[o]	281[o]	55160[o]
	3.265[p]	4.543[p]			

[a] $S(p)$ in Ry units.
[b] Chan *et al.* (1993c).
[c] Cook *et al.* (1965).
[d] Samson and Haddad (1984).
[e] Sum of atomic carbon and oxygen cross sections given by Henke *et al.* (1993).
[f] Hitchcock *et al.* (1990).
[g] McLaren *et al.* (1987).
[h] Barrus *et al.* (1979). See Fig. 4.7.
[i] From Chantler (1995), summing atomic carbon and oxygen cross sections.
[j] Using the hydrogenic equation of Bethe and Salpeter (1977) for both C and O K-shells.
[k] Sub-ionization region multiplied by 0.9546.
[l] Parker and Pack (1976).
[m] Atomic additivity, using $S(+2)$ from Fraga *et al.* (1976).
[n] Atomic additivity, using $S(+2)$ from Bunge *et al.* (1993).
[o] Jhanwar and Meath (1982).
[p] Olney *et al.* (1997).

Table 4.15 Coefficients of the polynomial $df/dE = ay^2 + by^3 + cy^4 + dy^5$ fitted to data at various energies[a]

Energy range, eV	a	b	c	d
26.84–124.37	6.187 55	112.8605	−410.21	394.6003
124.37–292.5	−15.2523	815.3072	−7 770.61	25 104.52
350–539.5	25.616 92	286.9811	34 112.31	−534 748
598–1486.7	−1.201 99	10 620.67	−244 932	2 122 862
1486.7–3691.7	−5.064 26	12 462.84	−500 609	13 909 834
3691.7–10 000	−0.645 46	7 945.036	963 143.4	−140 547 087

[a] df/dE in Ry units, $y = B/E$, $B = IP = 14.0136\,eV$.

e   The region around the carbon K-edge, to 350 eV

A number of photoabsorption studies have been performed in this region. See, for example, Sham *et al.* (1989), Domke *et al.* (1990), Ma *et al.* (1991), Schmidbauer *et al.* (1992), Shigemasa *et al.* (1993), Kempgens *et al.* (1997a) and Carravetta *et al.* (1997). Some of these experiments display very good resolution, but typically they lack a useful absolute intensity calibration. As a consequence, we must resort to inelastic electron scattering measurements. In Fig. 4.6, the (e,e) data of McLaren *et al.* (1987) are plotted, together with isolated points from Henke *et al.* (1993) based on atomic additivity. The agreement is fair. Not shown is the very strong $1s \rightarrow \pi^*$ resonance at 287.4 eV, whose f-value is given as 0.17 by Hitchcock *et al.* (1990). The 292.5–350 eV region in Fig. 4.6 has been graphically

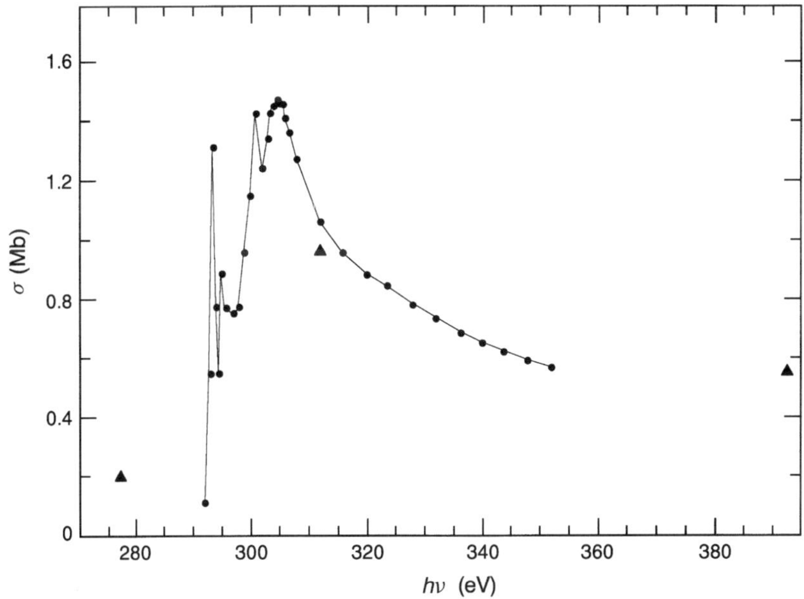

Fig. 4.6   Absolute photoabsorption spectrum of CO – carbon K-edge. • McLaren *et al.* (1987); ▲ Henke *et al.* (1993) + additivity

integrated, and yields $\Delta S(0) = 0.474$. These values, and the other $S(p)$, are documented in Table 4.14.

f Inter-edge continuum, 350–539.5 eV

The oxygen K-edge of CO occurs at 542.5 eV (Jolly *et al*., 1984). The pre-edge $1s \rightarrow \pi^*$ resonance, at 534.1 eV, is treated separately. We choose 539.5 eV as a convenient terminus for the inter-edge region, as will become evident below. Using the atomic cross sections for carbon and oxygen from Henke *et al*. and additivity, we calculate a cross section function by regression, by fitting a 4-term polynomial.

g The region around the oxygen K-edge, 539.5–598 eV

Figure 4.7 compares the photoabsorption cross section in this region (Barrus *et al*., 1979) with inelastic scattering data (McLaren *et al*., 1987). Also shown are isolated points from Henke *et al*, using atomic additivity. Using the latter as a

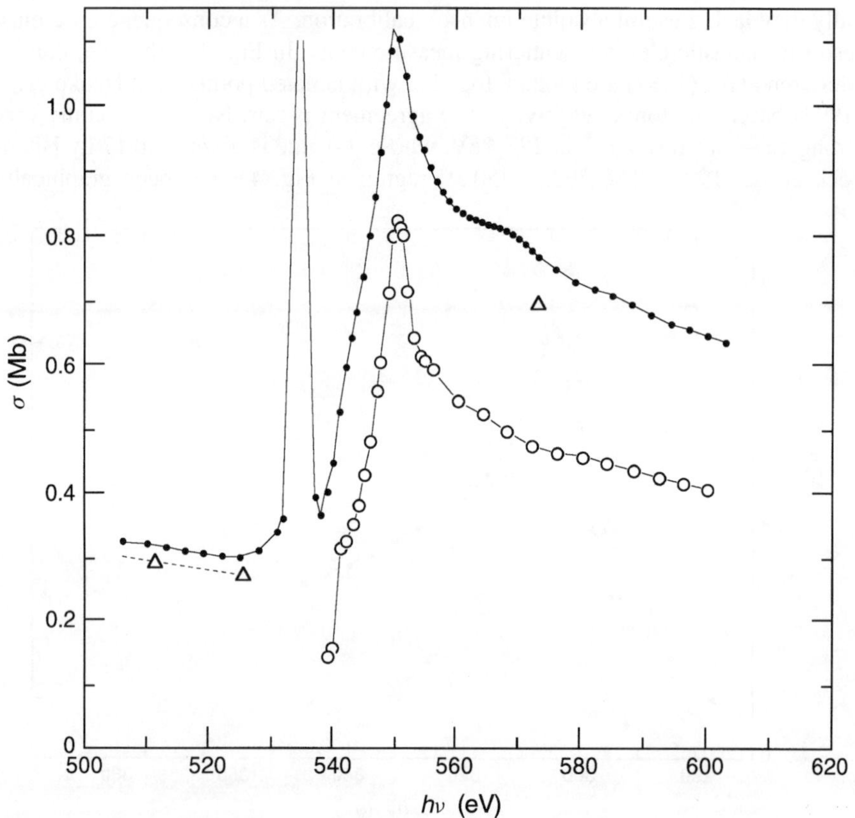

Fig. 4.7 Absolute photoabsorption spectrum of CO–oxygen K-edge. • Barrus *et al*. (1979); ○ McLaren *et al*. (1987); △ Henke *et al*. (1993)

guide, we conclude that the McLaren data may be too low, the Barrus data perhaps too high, but closer to Henke. (At 572.8 eV, the cross section based on additivity is 10% lower than the value of Barrus *et al.*) We choose the Barrus data, and obtain $\Delta S(0) = 0.420$ by graphical integration, excising the $1s \rightarrow \pi^*$ resonance at 534.1 eV, for which Hitchcock *et al.* (1990) explicitly provide $f = 0.076$. The McLaren data would lower $S(0)$ by ~ 0.015 (approximately 1%) and $S(+1)$ by ~ 6.4 Ry units (approximately 2.2%). The contributions to $S(p)$ from the data of Barrus *et al.* (1979) are listed in Table 4.14.

h Post K-edge continuum, 598–10 000 eV

We fit the summed atomic cross sections of Henke *et al.* to four-term polynomials in three segments (598–1486.7, 1486.7–3691.7 and 3691.7–10 000 eV). The derived $S(p)$ are included in Table 4.14, the polynomial coefficients in Table 4.15.

i 10^4–10^5 eV

We sum the calculated atomic cross sections of Chantler (1995). As with N_2, this region contributes $\sim 39\%$ to $S(+2)$, $\sim 3\%$ to $S(+1)$ and insignificantly to the other $S(p)$.

4.4.3 The analysis

Parker and Pack (1976) used refractive index measurements at 94 wavelengths, together with a Cauchy expansion to arrive at $S(-2) = 3.272\,27\,(5)$ Ry units for the ultraviolet portion. (There is also an infrared component ($\sim 1\%$), which need not concern us here.) Jhanwar and Meath (1982) performed a similar analysis and obtained $S(-2) = 3.270(3)$ Ry units. Our spectral sum, $S(-2) = 3.3371$, is distinctly larger, although the spectral sum for $S(0) = 13.9459$, is slightly less than the TRK expectation value. This dichotomy suggests two corrections: 1) a diminution in the low energy oscillator strengths, as anticipated in Sect. 4.4.1, and 4.4.2) an increase in some higher-energy oscillator strengths. First, we re-normalize the sub-ionization oscillator strengths, all from Chan *et al.* (1993c). A reduction factor of 0.9546 will make the spectral sum match the value of Parker and Pack. This reduces $S(0)$ to 13.8991, still within 0.7% of the required value. Given the overall uncertainty of the data, further fine tuning is unwarranted. The correction factor reduces $S(-1)$ by $\sim 1.2\%$, but has a trivial impact on $S(+1)$ and $S(+2)$.

It is instructive to compare the adjusted discrete oscillator strengths with photoabsorption measurements and ab initio calculations. For the $X^1\Sigma^+ \rightarrow A^1\Pi$ transition, we had noted (Sect. 4.4.2.a) that the photoabsorption measurements of Eidelsberg *et al.* were 7% larger than the original data of Chan *et al.*, and hence the discrepancy is now increased to 12%. More recent measurements appear to be confined to higher values of v' ($A^1\Pi$), where f is small. Thus, the oscillator strengths of Smith *et al.* (1994) are in good agreement with our corrected data for $v' = 11$, 12, Jolly *et al.* (1997) are about 6% larger for $v' = 9$–11, but Federman *et al.* (1997) are about 12% smaller for $v' = 7$–12, although they are in good agreement for $v' = 5$, which has a much larger oscillator strength. Wu *et al.* (1999) measured an f value for X-A$^1\Pi$ ($v' = 1$) which is $\sim$5% lower than Chan *et al.*, consistent with the current finding. Kirby and Cooper (1989) have presented two sets of calculations of oscillator strength, one based on an experimental (but abbreviated) dipole moment function obtained by Field *et al.* (1983), the other from ab initio calculations. The f values from the experimental dipole moment function are approximately 8.7% larger than our corrected set, but those derived from the ab initio calculations are quite close, only 3% lower. In summary, both photoabsorption measurements and calculations display deviations above and below our selected values, but the deviations are not large, providing some justification for our strategy.

For the $X^1\Sigma^+ \rightarrow B^1\Sigma^+$ transitions, essentially limited to $v' = 0, 1$, the oscillator strengths are small. The measured values of Letzelter *et al.* are only 58% of our selected values, the calculations of Kirby and Cooper (1989) still lower, only 27%. However, the recent high resolution measurements of Stark *et al.* (1999) are much closer, almost within the combined error limits. For the $X^1\Sigma^+ \rightarrow C^1\Sigma^+$ transitions, which are 13-times more intense than to $B^1\Sigma^+$, the experimental values of Letzelter *et al.* are again low, 56% of our corrected values, but the calculated strengths (Kirby and Cooper, 1989) are actually 3.5% higher. A similar pattern persists for the $X^1\Sigma^+ \rightarrow E^1\Pi$ excitation, the f values of Letzelter *et al.* remaining 55% of Chan *et al.* (1993c), corrected, but the calculations of Kirby and Cooper remain low, 76% of our selection. It seems plausible to conclude that the experimental values of Letzelter *et al.* for the B, C and E excitations may be too low because of saturation, since the Chan values receive support from the other recent electron scattering results (Zhong *et al.*, 1997). Morton and Noreau (1994) examined existing photoabsorption and electron scattering data below 12.4 eV. They adopted the oscillator strengths of Chan *et al.* for the $X \rightarrow A$ and $X \rightarrow C$ transitions, but preferred the photoabsorption data of Letzelter *et al.* for the $X \rightarrow B$ and $X \rightarrow E$ transitions. With their choices, we would realize better agreement for $S(-2)$, without adjustment, but the electron scattering data of Chan *et al.* exhibit good signal-to-noise for the $X \rightarrow B$ and $X \rightarrow E$ transitions, and are not masked by overlapping features. For the weak $X \rightarrow B$ transition, the recent results (Stark *et al.*, 1999; Zhong *et al.*, 1997) support an oscillator strength mid-way between Letzelter *et al.* and Chan *et al.*, but for the stronger $X \rightarrow E$ transition, the data of Zhong *et al.* (1997) clearly favor Chan *et al.* Hence, although the adjusted oscillator strength of Chan *et al.* for $X \rightarrow B$ may be $\sim$15% too high, it has an insignificant effect on the sum rules, but the $X \rightarrow E$ oscillator strength has an influence, leading us to retain the Chan data here, in contrast to Morton and Noreau.

We noted earlier (Sect. 4.4.2.a) that a substantial fraction of the sub-ionization oscillator strength resides in the higher-energy transitions (12.13–14.01 eV). Therefore, some test of this region is desirable. We find that Letzelter *et al.* obtain only $\sim$80% of our corrected oscillator strengths in the region of overlap, and Stark *et al.* (1991) only $\sim$75%. Letzelter *et al.* note that the figures of

Fock *et al.* (1980) imply values as much as an order of magnitude larger. The cross section measurements of Cook *et al.* (1965), which we had adopted for the autoionization region, extend below the IP to 12.697 eV. In the region of overlap, the oscillator strengths are lower than Chan (corrected) by $\sim 10\%$, but they are wavelength sensitive, larger in some regions and smaller in others. If saturation is responsible for this variation, it should be less problematical above the IP, where an additional mechanism of line broadening exists. We conclude that our selection is plausible, though the supporting evidence from photoabsorption is not overwhelming. The above discussion underscores the difficulty we would have encountered if we had chosen to deduce the sub-ionization oscillator strengths entirely from photoabsorption measurements.

Our spectral sum for $S(+2)$, 55 686.6 Ry units, is 2% larger than the value based on the sum of the electron charge densities for atomic carbon and oxygen, at the Hartree–Fock limit, 54 587.7 (Fraga *et al.*, 1976) or 54 588.2 (Bunge *et al.*, 1993). Jhanwar and Meath (1982) obtained $S(+2) = 55\,160$ Ry units, about 1% larger than the sum of Hartree-Fock atomic charge densities. Their value of $S(+1)$, 281.0 Ry units, is 3% lower than the present spectral sum, 289.6 Ry units, but barely above the Hartree–Fock sum of 280.67 Ry units (Fraga *et al.*, 1976). Correlation effects usually increase $S(+1)$ above atomic additivity by a larger percentage than indicated by Jhanwar and Meath.

Our corrected value of $S(-1)$, 4.690 Ry units, is 3.7% lower than that obtained by Jhanwar and Meath (1982) (4.864) using older data with their fitting and constraint procedure. Carravetta *et al.* (1993) have calculated $S(-2) \cong 3.18$, $S(-1) \cong 4.65$ and $S(0) \cong 13.74$, using the random phase approximation and/or linear response theory. Their calculated values for $S(-2)$ and $S(0)$ are low. If we apply a rough correction based on the shortfall in $S(-2)$ and $S(0)$, $S(-1)$ will lie between the present value and that of Jhanwar and Meath, only 1.5% higher than our modified spectral sum.

Cook *et al.* (1965) have also measured photoionization cross sections in the autoionization region, from which we compute S_i (-1). In the primary autoionization region depicted in Fig. 4.5, IP–17.71 eV, approximately 59% of the photoabsorption results in photoionization. We obtain $S_i(-1) = 3.263$, which may be compared to a directly measured value of Rieke and Prepejchal (1972), 3.70 ± 0.15. Their values are typically higher than those obtained from spectral sums, this time by 13%. However, the spectral sum is probably an underestimate, since Samson and Gardner (1976) have measured photoionization cross sections and quantum yields with line sources between 16.6–20 eV which are typically larger than those of Cook *et al.*

Jhanwar and Meath have criticized our earlier treatment of CO (Berkowitz, 1979), with some justification. Their approach was to incorporate the best available data at that time, and then adjust these data by least squares, subject to constraints imposed by molar refraction data and the TRK sum rules. They point out that their procedure should yield reliable results for various dipole properties ('global distributions'), but does not guarantee reliable cross sections at specific energies ('local detail'). Upon examining their recommended values, we find that the sub-ionization oscillator strengths used (Berkowitz 1979) were too large, partially due to the use of older electron scattering data. The current results, based entirely on the modified data of Chan *et al.* (1993c) but compared with other sources, are closer to the recommended values of Jhanwar and Meath. We differ locally; our integrated oscillator strength between 7.7–12.5 eV is lower, but that between 12.5–13.5 eV and 13.5–15 eV is higher. The high value of $S(-2)$ obtained previously (Berkowitz 1979) and recognized as such is now corrected, as we have already noted.

The other criticism involved integration of the structured data of Huffman *et al.* and Cook *et al.* between 12–20.7 eV. The former was in error, but was not used in the sum rules. As for the data of Cook *et al.*, we initially reported $\Delta S(0) = 2.30$, and their integration yielded 2.25. In the present analysis, the result is 2.12, but using the revised data of Chan *et al.* between 12–14 eV.

For the smooth continuum between 20.7–115 eV, Jhanwar and Meath selected older data of Cairns and Samson (1966) and Lee *et al.* (1973) together with additivity at the higher energies, whereas here we had available the more recent results of Samson and Haddad (1984). Their recommended value of $\Delta S(0)$ for this range is 6.98, ours is 6.48. We agree that the earlier data of de Reilhac and Damany (1977) are too low.

The present analysis explicitly uses molecular data in the vicinity of the carbon and oxygen K-edges; the approach of Jhanwar and Meath used mixture rules, i.e. $\sigma(CO) = \sigma(C) + 1/2\sigma(O_2)$.

In summary, there were ample reasons for revising the analysis of CO reported previously (Berkowitz, 1979), including new and better data. Our current sum rule results do not differ greatly from those of Jhanwar and Meath, but our only adjustment was a scaling factor for the sub-ionization data of Chan *et al.* (1993c), whereas theirs was a global adjustment. Therefore, it is likely that the local distribution of oscillator strengths reported here is closer to reality. This furthers our goal, which is to provide the best selection of cross sections at specific energies.

4.5 Nitric Oxide (NO)

4.5.1 Preamble

In the independent particle approximation, the sequence of occupied molecular orbitals in NO is: $(1\sigma)^2 (2\sigma)^2 (3\sigma)^2 (4\sigma)^2 (1\pi)^4 (5\sigma)^2 2\pi$. The 1σ and 2σ orbitals are essentially the K-shells of oxygen and nitrogen, while the 3σ and 4σ should have large contributions from O_{2s} and N_{2s}, respectively. The 1π and 5σ are nominally the bonding valence orbitals.

With a single valence electron in the outermost antibonding π_g-like orbital, NO has a significantly lower ionization potential than either O_2 or N_2. However, its polarizability is comparable to that of N_2, and only slightly larger than O_2.

Its oscillator strength in the sub-ionization region is only approximately 0.07. Although some photoabsorption data in this region existed from earlier work, a more detailed understanding (particularly in the region approaching the IP) has become available from the inelastic electron scattering results of Chan *et al.* (1993d). The lower energy transitions from the $X^2\Pi_{1/2}$ ground state are to $A^2\Sigma^+$ (γ bands), $B^2\Pi$ (β bands), $C^2\Pi$ (δ bands) and $D^2\Sigma^+$ (ε bands). The photoabsorption oscillator strengths of Bethke (1959) and the (e,e) data of Chan *et al.*, both assigned uncertainties of $\sim$10%, are in fairly good agreement and have been averaged. (The Bethke measurements were performed with pressure broadened lines.) For the higher-energy region (7.5 eV $\rightarrow$ IP), the photoabsorption data of Marmo (1953) are subject to saturation, and hence we opt for the results of Chan *et al.*

Autoionization is observed from the IP (9.264 eV) to $\sim$20.93 eV. Here we prefer the tabulated data of Watanabe *et al.* (1967) to those of Metzger *et al.* (1967), both for their more extended range and because they are in significantly better agreement with the results of Gardner *et al.* (1973) in the region of overlap. The cross sections of Watanabe *et al.* are on average about 6.5% larger than those of Metzger *et al.* However, even the data of Watanabe *et al.* begin to suffer from weak incident light at their highest energies, attributable to the waning helium continuum, whereas Gardner *et al.* used synchrotron radiation. Resonance structure observed between 18.9–20.8 eV by Gardner *et al.* is missing in the helium continuum-based data. Consequently, we switch to the cross sections of Gardner *et al.* at 18.44 eV, and continue to 32 eV. The 21–32 eV region is essentially devoid of structure. The low resolution (e,e) data of Iida *et al.* (1986) agree rather well with the cross sections of Gardner *et al.* in this region, whereas the values of Lee *et al.* (1973) are at least 10% lower.

Between 80–140 eV, the differential oscillator strengths of Iida *et al.* merge fairly well with summed atomic cross sections taken from Henke *et al.* (1993), but severe fluctuations appear toward the limit of their data, 140–180 eV. Therefore, we turn to atomic additivity between 150 eV and the vicinity of the nitrogen K-edge, 405 eV. Kosugi *et al.* (1992) and others present relative photoion yield spectra encompassing both the nitrogen and oxygen K-edge regions of NO, which we normalize in the post K-edge regions to atomic sums from Henke *et al.*

4.5.2 The data

Miescher (1976) obtained $74\,721.5 \pm 0.5$ cm^{-1} for the adiabatic IP of NO, using classical absorption spectroscopy, focusing on extrapolation of a Rydberg f series. This identical value was confirmed by Fredin *et al.* (1987) using laser optical double resonance through the C state, and extrapolating s and d Rydberg series. Early ZEKE studies were a few cm^{-1} lower, but later Reiser *et al.* (1988) obtained $74\,721.7 \pm 0.4$ cm^{-1} by a variant of ZEKE, Biernacki *et al.* (1988) reported $74\,721.67 \pm 0.10$ cm^{-1} by optical double-resonance and Rydberg f-series extrapolation, and Strobel *et al.* (1992) found $74\,721.7$ cm^{-1} (no error given), in 'perfect agreement' with Biernacki *et al.*, using non-resonant two-photon ZEKE. All these values agree, within their respective error limits. We choose the one offering the highest precision, $74\,721.67 \pm 0.10\,\mathrm{cm}^{-1} \equiv 9.264\,305 \pm 0.000\,013\,\mathrm{eV}$.

a   *The discrete spectrum and transitions below the IP*

Table 4.16 lists oscillator strengths for vibronic transitions to the $A^2\Sigma^+$, $B^2\Pi$, $C^2\Pi$ and $D^2\Sigma^+$ states. They are primarily averages of the photoabsorption f-values of Bethke (1959) and the (e,e) data of Chan *et al.* (1993d). Other sources are summarized by Chan *et al.*, but they are less complete. For $v' = 11$ and 14 of the β bands ($X^2\Pi \rightarrow B^2\Pi$), the high values of Chan *et al.* have been rejected, since the authors surmise that they may arise from deconvolution errors. In some

Table 4.16   Contributions to $S(p)$ of transitions to the valence states ($A^2\Sigma^+$, $B^2\Pi$) and the lowest Rydberg states ($C^2\Pi$, $D^2\Sigma^+$) of NO[a]

v'	E, eV	$S(-2)$	$S(-1)$	$S(0)$
a. $X^2\Pi \rightarrow A^2\Sigma^+$ (γ bands)				
0	5.4800[b]	2.527	1.018	0.410[c,d]
1	5.7703[b]	4.481	1.900	0.806[c,d]
2	6.0566[b]	3.543	1.577	0.702[c,d]
3	6.3389[b]	1.649	0.768	0.358[c,d]
	Total	12.200	5.263	2.276
b. $X^2\Pi \rightarrow B^2\Pi$ (β bands)				
0	5.641[e]	0.000 143	0.000 059 3	0.000 024 6[e]
1	5.764[e]	0.001 254	0.000 531	0.000 225[e]
2	5.893[e]	0.008 26	0.003 58	0.001 55[c]
3	6.010[e]	0.023 63	0.010 44	0.004 61[c]
4	6.135[e]	0.0679	0.0306	0.0138[c]
5	6.256[f]	0.1310	0.0602	0.0277[c,d]
6	6.375[f]	0.1895	0.0888	0.0416[c,d]
7	6.495[f]	1.580	0.754	0.36[g]
8	6.605[f]	0.509	0.247	0.12[g]
9	6.720[f]	1.377	0.680	0.336[c,d]
10	6.783[f]	0.121	0.060	0.03[g]
11	6.938[f]	1.392	0.710	0.362[c]
12	7.037[f]	8.22	4.25	2.20[c,d]
13	(7.168)[f]	0.036	0.019	0.01[g]
14	7.259[f]	0.706	0.377	0.201[c]
15	7.398[f]	2.672	1.453	0.790[d,g]
	Total	17.035	8.744	4.499
c. $X^2\Pi \rightarrow C^2\Pi$ (δ bands)				
0	6.493[b]	9.726	4.641	2.215[c,d]
1	6.782[b]	23.725	11.826	5.895[c,d]
2	7.062[b]	10.801	5.606	2.910[c,d]
3	7.367[b]	3.336	1.806	0.978[d,g]
	Total	47.588	23.879	11.998

Table 4.16 (*Continued*)

v'	E, eV	$S(-2)$	$S(-1)$	$S(0)$
d. $X^2\Pi \to D^2\Sigma^+$ (ε bands)				
0	6.6072[b]	10.707	5.200	2.525[c,d]
1	6.8899[b]	17.957	9.094	4.605[c,d]
2	7.1669[b]	12.596	6.635	3.495[c,d]
3	7.4374[b]	5.99	3.27	1.79[d]
	Total	47.250	24.199	12.415

[a] $S(p)$ in Ry units. The numbers given should be divided by 10^3.
[b] Energies from Miescher and Huber (1976).
[c] Bethke (1959).
[d] Chan *et al.* (1993d).
[c,d] An average from refs. c and d.
[e] Hasson and Nicholls (1971).
[f] Lagerqvist and Miescher (1958).
[g] Gallusser and Dressler (1982).
[d,g] An average from refs. d and g.

cases, neither Bethke nor Chan *et al.* provide values, due to overlapping or very weak transitions. For the very weak $v' = 0$, 1 of the β bands, we utilize the measurements of Hasson and Nicholls (1971). For $v' = 7$, 8 and 10, we avail ourselves of the oscillator strengths calculated by Gallusser and Dressler (1982), who adjusted the electronic transition moments to the data of Bethke. We note that the total oscillator strengths to the A and B states, which are characterized as valence states, are significantly weaker than those to C and D, which are low Rydberg states.

In the congested energy region $7.5\,\text{eV} \to \text{IP}$, we resort to the incremental oscillator strengths listed by Chan *et al.* The contributions to $S(p)$ for this domain, and for the lower energy region, are summarized in Table 4.17.

b The autoionization region, IP–20.93 eV

The complexity of the autoionization region can be rationalized from a knowledge of the vacuum ultraviolet photoelectron spectrum (see, for example, Edqvist *et al.* (1971)). There is a large gap between the first band (identified with the ionic ground state $X^1\Sigma^+$, vertical IP $= 9.5\,\text{eV}$) and the second (AIP $= 15.65\,\text{eV}$). However, between 15.65–20 eV the spectrum is highly congested, with at least seven states having been identified. Many Rydberg series must exist which converge to these ionic limits. The two most prominent peaks, at 16.56 and 18.33 eV, are assigned to $b^3\Pi$, $v' = 0$ and $A^1\Pi$, $v' = 0$, respectively. Both can be associated with electron emission from the 5σ orbital. Huber (1961) found that the limits of two Rydberg series, identified as β and γ series, occur at 16.56 and 18.33 eV, which we recognize as the $b^3\Pi$ and $A^1\Pi$ ionic states. (The β and γ series, named by Tanaka (1942), should not be confused with the β and γ bands discussed in Sects. 4.5.1 and 4.5.2.a). At somewhat higher energy (21.72 eV)

Table 4.17   Spectral sums, and comparison with expectation values for NO[a]

Energy, eV	$S(-2)$	$S(-1)$	$S(0)$	$S(+1)$	$S(+2)$
X → A,B,C,D,					
5.480–7.437[b]	0.1241	0.0621	0.0312	0.0157	0.0080
7.517–9.264(IP)[c]	0.0988	0.0614	0.0383	0.0239	0.0150
IP–18.44[d]	1.3065	1.3672	1.4637	1.6018	1.7897
18.44–20.93[e]	0.2610	0.3775	0.5467	0.7929	1.1513
20.93–32.0[e]	0.6891	1.2883	2.4433	4.7020	9.1778
32.0–80[f]	0.3754	1.2264	4.2569	15.7732	62.4064
80–150[f]	0.0242	0.1790	1.3628	10.7077	86.9837
150–405[g]	0.0037	0.0538	0.8311	13.7845	247.2514
399.7[h]	0.0001	0.0018	0.053	1.5570	45.7407
405–435[h]	0.0002	0.0069	0.2123	6.5589	202.7096
435–539[g]	0.0004	0.0139	0.4928	16.1456	622.7639
532.7[i]	–	0.0006	0.0247	0.9686	37.9248
539.565[i]	0.0001	0.0049	0.1973	8.0097	325.1998
565–2042.4[g]	0.0007	0.0399	2.4303	165.0061	12 746.91
2042.4–10 000[g]	–	0.0012	0.2558	63.8410	19 040.47
10^4–10^{5}[j]	–	–	0.0090	11.3851	19 450.66
10^5–10^{6}[k]	–	–	–	0.5791	9 225.58
10^6–10^{7}[k]	–	–	–	0.0203	3 207.91
10^7–10^{8}[k]	–	–	–	0.0006	1 046.35
10^8–10^{9}[k]	–	–	–	–	334.18
10^9–∞[k]	–	–	–	–	155.09
Total	2.8843	4.6849	14.6492	321.47	66 850.3
Expectation value	2.880(3)[l]		15.0		65 933.2[m]
					65 932.2[n]
Other values	(2.880)[o]	4.752[o]	(15.0)[o]	324.6[o]	67 040[o]
	2.865[c]				
	2.882_5^p	4.581_5^p			

[a]In Ry units.
[b]See Table 4.16.
[c]Chan *et al.* (1993d).
[d]Watanabe *et al.* (1967).
[e]Gardner *et al.* (1973).
[f]Iida *et al.* (1986).
[g]Summed atomic cross sections from Henke *et al.* (1993).
[h]See text and Fig. 4.8.
[i]See text and Fig. 4.9.
[j]Summed atomic cross sections, Chantler (1995).
[k]Using the hydrogenic equation of Bethe and Salpeter (1977) for both N and O K-shells.
[l]Nielson *et al.* (1976).
[m]Sum of atomic Hartree-Fock values, Fraga *et al.* (1976).
[n]Sum of atomic Hartree-Fock values, Bunge *et al.* (1993).
[o]Zeiss *et al.* (1977).
[p]Olney *et al.* (1997).

there is another prominent peak in the photoelectron spectrum, identified as $c^3\Pi$ and associated with $(4\sigma)^{-1}$. At least two Rydberg series have been found which converge to this limit (Narayana and Price 1972a, Sasanuma *et al.*, 1974).

For reasons discussed in Sect. 4.5.1, we utilize the data of Watanabe *et al.* from IP–18.44 eV, then transfer to the values in Fig. 1 of Gardner *et al.* between 18.44–20.93 eV. The integrations are performed trapezoidally, and summarized in Table 4.17.

c   The continuum, 20.93–32.0 eV

In this region, the cross section rises to a broad maximum at ~23.5 eV, and then begins its descent as it approaches the nitrogen K-edge. We have fitted a 4-term polynomial to the data of Gardner *et al.* in this interval. The coefficients are assembled in Table 4.18, the corresponding values of $S(p)$ in Table 4.17. Trapezoidal integration using the extracted data points yields essentially the same result.

d   The continuum, 32.0–405 eV

We utilize the (e,e) data of Iida *et al.* between 32–150 eV for reasons given in Sect. 4.5.1. The data are fitted in two segments, 32–80 eV and 80–150 eV, to 4-term polynomials by regression. The nitrogen K-edge has a $^3\Pi$ component at 410.3 eV and a weaker $^1\Pi$ component at 411.7 eV (Jolly *et al.*, 1984). Resonances precede the edge, so that a convenient terminus is 405 eV. Atomic additivity, using the cross sections of Henke *et al.* (1993) provides points between 150–405 eV, which are also fitted to a 4-term polynomial by regression. The contributions to $S(p)$ are given in Table 4.17, the various polynomial coefficients in Table 4.18.

e   Nitrogen K-edge structure, 400–435 eV

Recently, several groups have performed high resolution photoabsorption or photoion yield measurements in the vicinity of the nitrogen K-edge of NO, e.g. Ma *et al.* (1991), Kosugi *et al.* (1992b), Remmers *et al.* (1993), Erman *et al.* (1996). Unfortunately, these are all relative cross sections. A plausible scaling method is to normalize to summed atomic cross sections at 25 eV or more above the K-edge IP, as suggested by Hitchcock and Mancini (1994). However, none

Table 4.18   Coefficients of the polynomial $\mathrm{d}f/\mathrm{d}E = ay^2 + by^3 + cy^4 + dy^5$ fitted to data at various energies[a]

Energy range, eV	a	b	c	d
20.93–32.0	41.013 33	−45.5039	91.921 66	−262.136
32.0–80.0	−58.6547	1609.223	−8137.48	12 502.78
80.0–150.0	−151.158	6225.524	−68 135.4	247 337.6
150.0–409	30.100 96	−1150.0	44 757.05	−390 507.0
565–2042.4	−9.406 33	47 105.27	−1 928 662	33 795 762
2042.4–10 000	−18.7319	57 578.19	−5 358 701	390 453 871

[a] $\mathrm{d}f/\mathrm{d}E$ in Rydberg units, $y = B/E$, $B = \text{IP} = 9.2643$ eV.

of the high-resolution spectra published extends that far. Kosugi *et al.* (1992b) display an ion-yield spectrum already resolved into 1s → π and 1s → σ components, which can be summed, and extends to 435 eV. Alternatively, one can resort to an older electron energy loss spectrum (Wight and Brion, 1974a) which has the desired energy range, but its background seems ill-defined. None of the alternatives is very satisfactory. In Fig. 4.8, we have attempted a composite absolute spectrum, based loosely on the extended spectra of Kosugi *et al.* (1992) and Wight and Brion (1974a), which roughly matches the sum of atomic cross sections (estimated from Henke *et al.*, 1993) at 435 eV. The sharp, pre-threshold structure has been incorporated from the high-resolution experiments. A plausible consequence of the normalization chosen in Fig. 4.8 is that the integrated oscillator strength is about half that in the corresponding region of N_2. The $S(p)$ derived from Fig. 4.8 are given in Table 4.17. The intensity of the sharp 1s → π^* resonances centered at 399.7 eV can then be related to the higher-energy features from the high-resolution absorption measurements. In this fashion, we obtain $f = 0.053$ for this sharp peak. Kosugi *et al.* have calculated $f = 0.1345$ for the several transitions within this peak, but the calculation cannot be expected to be very accurate. Their calculated f-values for the Rydberg transitions are about a factor 4 lower than those implied by Fig. 4.8.

f Inter-edge region, 435–539 eV

The $^3\Pi$ component of the oxygen K-edge in NO occurs at 543.3 eV (Jolly *et al.*, 1984). Apart from the π^* resonance at 532.7 eV, which we treat separately, some

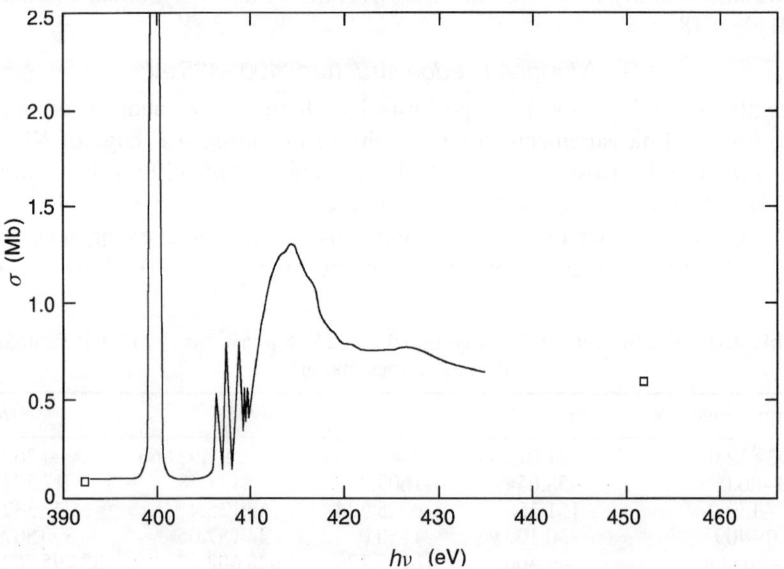

Fig. 4.8 Absolute photoabsorption spectrum of NO – nitrogen K-edge. —— Kosugi *et al.* (1992); □ Henke *et al.* (1993) + additivity

pre-edge structure attributable to Rydberg excitations commences at $\sim$539 eV. The inter-edge region, 435–539 eV, is essentially a linearly declining function, based on summed atomic cross sections from Henke *et al.* (1993) and Chantler (1995). The function

$$\sigma(\mathrm{NO}) = -0.002\,455\,E + 1.7155$$

with σ in Mb and E in eV, is a suitable approximation, and has been used to compute the $S(p)$ given in Table 4.17.

g Oxygen K-edge structure, 539–565 eV

The problems encountered in establishing absolute photoabsorption cross sections here are similar to those discussed for the nitrogen K-edge. To construct a composite spectrum, we require that the structured region merges with summed atomic cross sections at its lower and upper limits. The electron energy loss spectrum of Wight and Brion does not meet these requirements. The ion-yield spectrum of Kosugi *et al.* (1992), resolved into 1s $\rightarrow$ π and 1s $\rightarrow$ σ components, must be summed and normalized to atomic sums at the upper limit. At the lower limit, there is a continuum background ($\sim$0.4 Mb) which must be added. The composite spectrum appears in Fig. 4.9. The 1s $\rightarrow$ π^* resonance at 532.7 eV has been scaled to the higher-energy structure, and Rydberg resonances have been incorporated from high-resolution data. The 'experimental' f-value of the

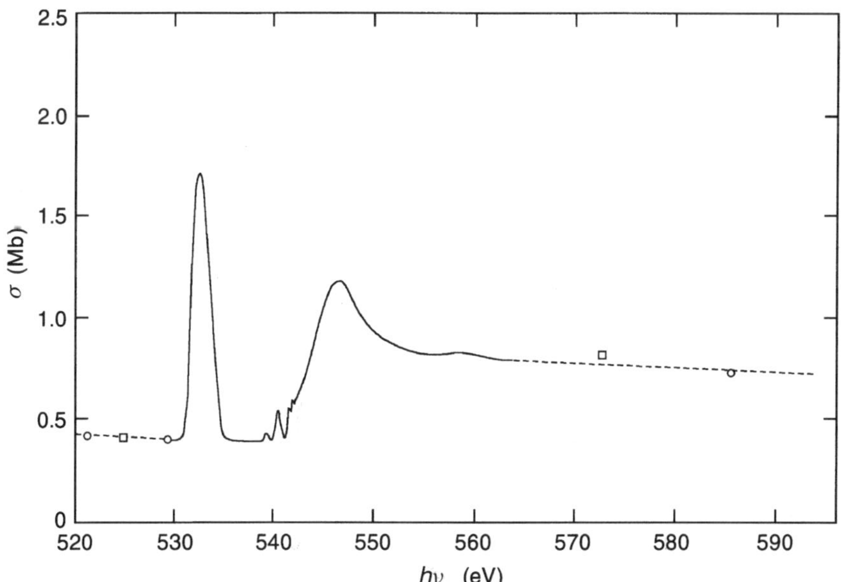

Fig. 4.9 Absolute photoabsorption spectrum of NO – oxygen K-edge. —— Kosugi *et al.* (1992), Remmers *et al.* (1993), Wight and Brion (1974a); o Henke *et al.* (1993) + additivity; □ Chantler (1995) + additivity

532.7 eV resonance is 0.0247, this time rather close to the sum of $^2\Sigma^-$, $^2\Delta$ and $^2\Sigma^+$ components calculated by Kosugi *et al.*, 0.0277. The integrated oscillator strength between 539–565 eV is 0.1973, comparable to a similar region around the nitrogen K-edge, and roughly half that of the oxygen K-edge in O_2.

h Post oxygen K-edge, 565–10 000 eV

The summed atomic cross sections of Henke *et al.* have been fitted by regression to two 4-term polynomials, partitioned between 565–2042.4 eV and 2042.4–10 000 eV. The coefficients of these polynomials are given in Table 4.18, the deduced values of $S(p)$ in Table 4.17.

i 10^4–10^5 eV

The calculated atomic cross sections of Chantler are summed, and the $S(p)$ are recorded in Table 4.17. This region contributes $\sim$39% to $S(+2)$, $\sim$4% to $S(+1)$, and insignificantly to the other $S(p)$.

4.5.3 The analysis

Probably the best current estimate of $S(-2)$ comes from Nielson *et al.* (1976), who fitted refractive index values at 22 wavelengths in a Cauchy series and obtained $S(-2) = 11.518(13)$ a.u. $\equiv 2.880(3)$ Ry units. Essentially the same value was found by Zeiss and Meath (1977). Our spectral sum, $S(-2) = 2.8843$, is in almost perfect agreement with the expectation value. The bulk of this quantity comes from the region IP–50 eV. Major contributions to $S(-2)$ derive from the data of Watanabe *et al.*, Gardner *et al.* and to a lesser extent, Iida *et al.*. The good agreement provides some justification for these choices.

The spectral sum for $S(0)$, 14.6492, is 2.3% lower than that required by the TRK sum rule. Large contributions to this sum come from Iida *et al.* between 32–150 eV (where sensitivity to $S(0)$ is greater than to $S(-2)$), from the summed atomic cross sections of Henke *et al.*, and to a lesser extent, the K-edge regions of nitrogen and oxygen, which required manipulation of existing data.

Zeiss *et al.* (1977) adapted existing data together with a fitting procedure which constrained $S(-2)$ and $S(0)$ to known values to infer $S(-1)$, and with mixing rules, e.g. $\sigma(NO) = 1/2\sigma(N_2) + 1/2\sigma(O_2)$, to calculate $S(+1)$ and $S(+2)$. They had available the important data of Watanabe *et al.* and Gardner *et al.*, but not those of Iida *et al.* and Chan *et al.*. In addition, most of the nitrogen and oxygen K-edge structural information was not available. Their value of $S(-1)$, which is 1.4% higher than our spectral sum, is probably more accurate, since their $S(0)$ and $S(-2)$ are constrained to expectation values.

As shown in other examples (N_2, O_2), the summed Hartree–Fock values for electron density at the nucleus offer an accurate prediction for $S(+2)$. Thus, we obtain $S(+2) = 65\,933.2$ from Fraga *et al.* (1976) and $65\,932.2$ from Bunge *et al.* (1993). Our spectral sum is 1.4% larger; the value of $S(+2)$ obtained by Zeiss *et al.* (1977) is 1.7% larger.

Applying atomic additivity to $S(+1)$ from Fraga *et al.* (1976) yields 320.5, rather close to our spectral sum of 321.47. Zeiss *et al.* (1977) obtain $S(+1) = 324.6$, which may be closer to the true value, although insufficient information is currently available.

The value of $S_i(-1)$ can be computed from photoionization cross sections given by Watanabe *et al.* from IP-20 eV, and the assumption that at higher energies the quantum yield of ionization is unity. These data yield $S_i(-1) = 3.984$. The measured value of $M_i^2 \equiv S_i(-1)$ given by Rieke and Prepejchal (1972) is 4.31 ± 0.48, characteristically larger than the spectral sum, but in agreement within the substantial error limits.

In the autoionization region IP-18.44 eV, approximately 58.5% of photoabsorption leads to ionization. The current analysis is a substantial improvement over that reported earlier (Berkowitz, 1979), aided in large measure by the subsequent (e,e) data of Iida *et al.* and Chan *et al.*

4.6   Hydrogen Chloride (HCl)

4.6.1   Preamble

In the independent particle approximation, the electronic ground state of HCl has the configuration

$$(1\sigma)^2(2\sigma)^2(3\sigma)^2(1\pi)^4(4\sigma)^2(5\sigma)^2(2\pi)^4, \, X^1\Sigma^+$$

The lowest unoccupied molecular orbital, 6σ, is an antibonding combination of H(1s) + Cl(3p). The bonding combination is 5σ, while 2π is the degenerate set of Cl(3p) lone pairs. The 4σ, dominantly Cl(3s), manifests the typical breakdown of the independent particle approximation for inner valence orbitals in its photoelectron spectrum. The 1π and 3σ orbitals are the Cl(2p) spin-orbit components, while 2σ and 1σ are essentially Cl(2s) and Cl(1s), respectively.

Although optical and photoelectron spectroscopies have proceeded apace, the situation regarding absolute photoabsorption cross sections is less satisfactory. Kumar and Meath (1985a) performed a sum rule analysis which depended heavily on inelastic electron scattering data of Daviel *et al.* (1984) obtained at low resolution, 1 eV FWHM. Higher-resolution data from this group have been in preparation (Olney *et al.*, 1997), but as of this writing these are unavailable. Absolute photoabsorption cross sections from absorption onset ($\sim$6 eV) to $\sim$11.7 eV were initially presented by Myer and Samson (1970) and later by Nee *et al.* (1986). These measurements were performed at much higher resolution, 0.002–0.004 eV FWHM, and displayed sharp structure obscured in the (e,e) data of Daviel *et al.* The cross sections of Myer and Samson and Nee *et al.* agree within $\sim$10% for broad bands, but at sharp peaks Nee *et al.* obtain maximum cross sections as much as a factor 3 larger than those of Myer and Samson, although the latter resolution was higher. Nee *et al.* suspect saturation effects, which Myer and

Samson appear to acknowledge by noting that their values are lower limits. For this reason, and also because Nee *et al.* present their data in expanded form, we shall utilize their spectra here, although they estimate an uncertainty of 15%.

There is a gap in absolute photoabsorption measurements between $\sim$11.7 eV and the IP (12.746 eV). Between the IP and $\sim$17 eV, Frohlich and Glass-Maujean (1990) have obtained both absolute photoabsorption and absolute photoionization cross sections, with a resolution of 0.003 eV FWHM. Their estimated uncertainties in cross section range from 10% on structures to 30% in continuum regions. Above 17 eV, direct photoabsorption measurements are sparse, although summed atomic cross sections (primarily Cl) are probably adequate between $\sim$90 eV and the pre-Cl(2p) structure at $\sim$198 eV. Daviel *et al.* tabulate cross sections up to 40 eV, which we shall use tentatively, pending sum rule analysis. Carlson *et al.* (1983) found a Cooper minimum in the partial cross sections of both $(2\pi)^{-1}$ and $(5\sigma)^{-1}$ at $\sim$45 eV. However, ionization from 4σ generates a number of ionic states between $\sim$22–35 eV (Adam, 1986) and hence will contribute to the total cross section. In principle, it is possible to combine the relative intensities of Adam (1986) with the absolute measurement of $(2\pi)^{-1}$ at 21.2 eV given by Carlson *et al.* to estimate absolute cross sections at several energies between 30–90 eV, but apart from the neighborhood of the Cooper minimum (40–50 eV) these estimated values are significantly higher than those of Daviel *et al.*, and also summed atomic cross sections. There is no obvious indication of a Cooper minimum in the total cross section at $\sim$45 eV in Fig. 1 of Daviel *et al.*, and their tabulated values terminate at 40 eV.

The L_{III} edge in HCl occurs at 207.39 eV (Jolly *et al.*, 1984). Absolute cross sections for the structured region between $\sim$200–210 eV are given by Hayes and Brown (1972) and Ninomiya *et al.* (1981). Both acknowledge an uncertainty of 20%. The cross sections of Hayes and Brown are approximately 25% higher, except for the 208–210 eV region. Ninomiya *et al.* extended their measurements to 280 eV, and hence can be compared with summed atomic cross sections (essentially atomic Cl). At 25 eV above the L_{III} edge, the fitted atomic Cl cross section (Henke *et al.*, 1993) is about a factor 1.5 larger than the value extracted from Ninomiya *et al.* This observation favors the larger cross sections of Hayes and Brown, which we utilize, between 198–208 eV. Above 208 eV, augmented cross sections from Ninomiya *et al.* are utilized, such that they match the atomic cross section (Henke *et al.*, 1993) at 232 eV.

The absolute photoabsorption cross section of HCl in the Cl K-edge region has been reported by Bodeur *et al.* (1990). It covers the range 2820–2852 eV, with the K-edge given as 2829.8 eV. Thus, it ends just shy of the recommended 25 eV above the IP recommended for comparison with the summed atomic cross sections. The measured cross section is about 75% of atomic additivity, but a concomitant study of Cl_2 comes much closer. Hence, we modify the spectrum of Bodeur *et al.* slightly, changing the slope between 2830–2852 eV so that it agrees with the fitted atomic value (Henke *et al.*, 1993) at 2855 eV.

4.6.2 The data

The adiabatic ionization potential (AIP) of HCl was determined by Tonkyn *et al.* (1992) to be 102 802.8 $\pm 2\,\mathrm{cm}^{-1} \equiv 12.7459 \pm 0.0002\,\mathrm{eV}$ by pulsed-field ZEKE. Subsequently, Drescher *et al.* (1993) obtained 102 801.5 $\pm 1\,\mathrm{cm}^{-1} \equiv 12.7458 \pm 0.0001\,\mathrm{eV}$ from high-resolution photoionization using a VUV laser. Edvardsson *et al.* (1995) reported 12.7447 eV from a high resolution He I photoelectron spectrum. We choose 12.7458 eV.

a   The discrete spectrum, and transitions below the IP

The data of Nee *et al.* have been utilized from absorption onset ($\sim$6.7 eV) to their limit, 11.627 eV. Most of their cross sections are scanned from figures, but between 9.3–10.0 eV explicit oscillator strengths are listed in their Table I. There is a small region between 12.527 eV and the AIP contained in a figure from Frohlich and Glass-Maujean. The gap between 11.627–12.527 eV is interpolated from the tabulated (e,e) data of Daviel *et al.* We note parenthetically that the points from Daviel *et al.* are at least 10% higher than the mean of the much higher resolution, structured cross sections of both Nee *et al.* and Frohlich and Glass-Maujean.

b   The continuum

b.1   IP–16.97 eV   Figures from Frohlich and Glass-Maujean have been scanned, digitized and trapezoidally integrated to generate the $S(p)$ gathered in Table 4.19.

b.2   16.97–38.0 eV   Tabulated (e,e) data of Daviel *et al.* are tentatively utilized, by default.

b.3   38.0–91.5 eV   This is the controversial region where little, if any, reliable data exist. The cross section is dropping rapidly at 38 eV, it may reach a Cooper minimum at $\sim$45 eV, and should be fairly well described by atomic additivity at 91.5 eV (a point taken from Henke *et al.* (1993)). We have opted to interpolate between 38–91.5 eV with a smooth curve. The saving feature in a sum rule analysis is that this region has a modest influence on all the $S(p)$, as seen in Table 4.19. If a Cooper minimum does exist, we estimate that the $S(0)$ contribution will be reduced by $\sim$0.1, and $S(-2)$ by $\sim$0.01 Ry units.

b.4   91.5–198 eV   The atomic H and Cl cross sections of Henke *et al.* (1993) are summed, and fitted to a 4-term polynomial. The coefficients of the polynomial are given Table 4.20. The function is analytically integrated to yield the $S(p)$ found in Table 4.19.

b.5   198–208 eV   Figure 7 of Hayes and Brown (1972) is scanned and trapezoidally integrated. Data above 208 eV, which disagree with Ninomiya *et al.*, are ignored.

Table 4.19 Spectral sums and comparison with expectation values for HCl[a]

Energy, eV	$S(-2)$	$S(-1)$	$S(0)$	$S(+1)$	$S(+2)$
6.71–9.25[b]	0.1379	0.0814	0.0483	0.0288	0.0173
9.301–9.336[b]	0.0055	0.0038	0.0026	0.0018	0.0012
9.581–9.634[b]	0.2569	0.1813	0.1280	0.0904	0.0638
9.649–9.671[b]	0.0097	0.0069	0.0049	0.0035	0.0025
9.918–9.999[b]	0.0323	0.0236	0.0173	0.0127	0.0093
10.021–10.925[b]	0.0131	0.0100	0.0077	0.0059	0.0045
10.925–11.627[b]	0.2680	0.2217	0.1834	0.1518	0.1257
11.627–12.527[c]	0.5896	0.5235	0.4651	0.4133	0.3674
11.627–12.527[d]	(0.5012)	(0.4450)	(0.3953)	(0.3513)	(0.3123)
12.527–12.7458[e] (IP)	0.1194	0.1109	0.1030	0.0957	0.0889
IP–14.415[e]	0.6213	0.6173	0.6143	0.6121	0.6108
14.415–14.423[e]	0.0025	0.0027	0.0029	0.0030	0.0032
14.423–16.97[e]	0.9823	1.1255	1.2924	1.4871	1.7148
16.97–38.0[c]	1.6728	2.6623	4.3855	7.5066	13.3928
16.97–38.0[d]	(1.4219)	(2.2630)	(3.7278)	(6.3806)	(11.3839)
38.0–91.5[f]	0.0481	0.1898	0.7977	3.5771	17.0724
91.5–198[g]	0.0079	0.0713	0.6744	6.7073	69.9044
198–208[h]	0.0005	0.0080	0.1207	1.8124	27.2229
208–232[i]	0.0028	0.0459	0.7431	12.0355	195.086
232–2,820[g]	0.0102	0.2527	7.4893	297.462	18 173.6
2820–2855[j]	–	0.0002	0.0446	9.290	1936.7
2855–10 000[g]	–	0.0045	1.3780	464.583	177 030.6
10^4–10^{5}[k]	–	0.0001	0.1479	195.940	360 475.0
10^5–∞[l]	–	–	0.0012	14.850	371 143.2
Total	4.7808	6.1434	18.6523	1016.670	929 084.8
Revised[d]	(4.4415)	(5.6656)	(17.9248)	(1015.482)	(929 082.7)
Expectation values	4.352[m]		18.		916 586.1[n]
					916 950.2[o]
Other values	(4.3475)[p]	5.635[p]	(18.)[p]	1024[p]	911 200[p]
	4.355[q]	5.33[q]			

[a]In Ry units.
[b]Nee *et al.* (1986), Figs. 2–5 and Table 1.
[c]Daviel *et al.* (1984).
[d]15% reduction of values from (c).
[e]Frohlich and Glass-Maujean (1990).
[f]Interpolation between Daviel *et al.* (c) and Henke *et al.* (g).
[g]Henke *et al.* (1993).
[h]Hayes and Brown (1972).
[i]Ninomiya *et al.* (1981). The cross sections in Figs. 2 and 4 of this paper have been increased to match Hayes and Brown (1972) at the low end, and Henke *et al.* (1993) at the high end.
[j]Bodeur *et al.* (1990). Figure 2 of their data is extrapolated from 2852 to 2855 eV, and the cross section at that energy increased by 35% to match the fitted data of Henke *et al.* (1993).
[k]Chantler (1995).
[l]Using the hydrogenic equation of Bethe and Salpeter (1977) for K-shell of Cl.
[m]From both experiment and calculations. See text.
[n]Fraga *et al.* (1976).
[o]Bunge *et al.* (1993).
[p]Kumar and Meath (1985a).
[q]Olney *et al.* (1997).

Table 4.20 Coefficients of the polynomial $\mathrm{d}f/\mathrm{d}E = ay^2 + by^3 + cy^4 + dy^5$ fitted to data at various energies[a]

Energy range, eV	a	b	c	d
91.5–198	13.965 07	−11.402	−438.705	1300.734
232–851.5	3.685 704	12 175.54	−237 409	1 144 902
851.5–2820	10.959 25	13 396.75	−423 175	6 602 132
2855–10 000	3.640 223	215 417.8	−18 355 948	196 009 512

[a] $\mathrm{d}f/\mathrm{d}E$ in Ry units, $y = B/E$, $B = 12.7458\,\mathrm{eV}$.

b.6   208–232 eV The shape of the spectrum given by Ninomiya *et al.*, Figs. 2 and 4, is essentially maintained, but the absolute cross sections are increased by a factor 1.5, to match atomic sums at 232 eV, which is 25 eV above the L_{III} edge.

b.7   232–2820 eV The atomic chlorine cross sections (Henke *et al.*, 1993) are fitted by regression to two 4-term polynomials, spanning 232–851.5 eV and 851.5–2820 eV. The contribution of atomic hydrogen is negligible here.

b.8   2820–2855 eV Bodeur *et al.* tabulate the oscillator strengths of the peaks at 2823.9, 2827.0 and 2827.8 eV. The underlying and ensuing continuum is displayed to 2852 eV, but as noted in Sect. 4.6.1, it is only $\sim$75% of the fitted atomic chlorine cross section. The decline in cross section between 2830–2852 eV is very nearly linear. We maintain this linearity to 2855 eV (25 eV above the K-edge), but require the cross section at 2855 eV to equal that of atomic chlorine. This region contributes $\sim$1% to $S(+1)$, and much less to the other $S(p)$.

b.9   2855–10 000 eV The cross sections of atomic chlorine are fitted to another 4-term polynomial to traverse this region, which is a major contributor to $S(+1)$, and contains the bulk of the Cl(1s) oscillator strength.

4.6.3   The analysis

Modern evaluations of the static electric dipole polarizability of HCl are based on three older measurements of the wavelength-dependent refractive index, by Cuthbertson and Cuthbertson (1913a), Frivold *et al.* (1937) and Larsén (1938). Kumar and Meath (1985a) note that the measurements agree within 0.2%, but choose the Frivold data to deduce $S(-2) = 17.39\,\mathrm{a.u.} \equiv 4.348\,\mathrm{Ry}$ units. Russell and Spackman (1997) analyze the Cuthbertson data, and infer $S(-2) = 17.43(2)\,\mathrm{a.u.} \equiv 4.358(5)\,\mathrm{Ry}$ units. Larsén has fitted his data to an equation, which yields $S(-2) = 17.41\,\mathrm{a.u.} \equiv 4.352\,\mathrm{Ry}$ units.

Maroulis (1998a) has performed a systematic study of the dependence of the polarizability of HCl on basis set and electron correlations. He recommends $S(-2) = 17.41 \pm 0.02\,\mathrm{a.u.}$ This is essentially the average of the experimental results, and equal to that of Larsén. Other contemporary calculations (Russell and Spackman, 1997; Hammond and Rice, 1992) are also close to this value.

Our first pass spectral sum gives $S(-2) = 4.781$ Ry units, 9.9% higher, while $S(0)$ is 3.6% higher than the Thomas–Reiche–Kuhn expectation value. A glance at Table 4.19 reveals that the (e,e) data of Daviel *et al.* (1984), which we had noted to have higher cross sections than overlapping photoabsorption measurements, make substantial contributions to $S(-2)$. After brief numerical trials, a reduction of 15% in the cross sections of Daviel *et al*, results in the $S(p)$ shown in parentheses in Table 4.19. (In the more recent work from the Vancouver group (Dyck *et al.*, 1995) they presented a poster on HCl made available to the author, which is much better resolved, and appears to indicate a 15% reduction in cross sections, compared to Daviel *et al.*) This correction brings the spectral sum $S(-2)$ within 2% of the expectation value, and $S(0)$ within <0.5%. This tentative conclusion must be treated with caution, since the UV-VUV photoabsorption cross sections were estimated to be no more accurate than 15%. However, their combined contribution to $S(-2)$ is approximately the same as that attributed to Daviel *et al.* If we decreased these photoabsorption cross sections by 15%, the discontinuity between their values and the (e,e) measurements would be greater. If, alternatively, they were increased by 15%, the discrepancy with the expectation value of $S(-2)$ would be greater, and we would be forced to select lower values in the other energy ranges.

Comparison with Kumar and Meath is not meaningful for $S(-2)$ and $S(0)$, since these are fixed quantities in their optimization procedure. Olney *et al.* (1997) give $S(-2) = 4.355$ Ry units, but this is also probably adjusted to their choice of polarizability in their newer work, awaiting publication. Our adjusted value of $S(-1)$ is 0.5% higher than that arrived at by Kumar and Meath. The value of $S(-1)$ given by Olney *et al.* is understandably lower, since they typically do not include higher energy contributions.

The expectation value for $S(+2)$ can be well approximated as the sum of the charge distributions at the nuclei (predominantly Cl), in appropriate units. The alternative expectation values from Hartree–Fock calculations are 916 586.1 (Fraga *et al.*, 1976) and 916 950.2 (Bunge *et al.*, 1993). Our spectral sum is $\sim$1.3% larger, while that of Kumar and Meath is $\sim$0.6% smaller.

A crude approximation to $S(+1)$, based on atomic additivity and the atomic chlorine contribution from Fraga *et al.*, 1976 yields $S(+1) = 1010.7$ Ry units. Correlation effects typically increase the molecular value. The current spectral sum is $\sim$0.5% larger, while that of Kumar and Meath is $\sim$1.3% larger.

Frohlich and Glass-Maujean (1990) provide the necessary additional information to calculate $S_i(-1)$, i.e. the absolute photoionization cross section from the IP to 16.97 eV. Their data indicate that η_i, the quantum yield of ionization, reaches unity at their upper terminus, and the compressed figure given by Daviel *et al.* seems to approximately agree. Using our adjusted value of $S(-1)$, and substituting $\Delta S_i(-1) = 1.1751$ for $\Delta S(-1) = 1.5889$ between IP and 16.97 eV, we obtain $S_i(-1) = 4.064$ Ry. We are currently unaware of direct electron impact measurements of this quantity.

Between IP–16.97 eV, approximately 70% of absorption leads to ionization.

5

Triatomic Molecules

5.1 Water (H₂O)

5.1.1 Preamble

The electronic ground state of the H_2O molecule has the orbital configuration

$$(1a_1)^2(2a_1)^2(1b_2)^2(3a_1)^2(1b_1)^2, \tilde{X}^1A_1$$

The absolute photoabsorption cross section of H_2O in the UV-VUV region, taken from the inelastic electron scattering data of Chan $et\ al.$ (1993e) is presented in Fig. 5.1. The stated resolution is 48 meV. A similar spectrum obtained by photoabsorption with a resolution <1 meV can be found in Fig. 3 of Gürtler $et\ al.$ (1977b). The lowest-energy band (7.4 eV) is broad, with no apparent structure. The second band (9.7 eV) is also broad, with some evident superstructure, but dominated on its high-energy side by strong, sharp peaks. Various studies (Wang $et\ al.$, 1977; Diercksen $et\ al.$, 1982) have concluded that the lowest-energy band results from excitation of a $1b_1$ electron to a mixed Rydberg-valence state, designated $\tilde{A}(^1B_1)$. It is broadened by predissociation to yield $OH(X^2\Pi) + H(^2S)$. The second band is attributed to excitation of a $3a_1$ electron to a mixed Rydberg-valence state, designated $\tilde{B}(^1A_1)$. It, too, is broadened by predissociation, forming $OH(A^2\Sigma^+) + H(^2S)$. The sharp features at higher energy are associated with excitations to various s, p and d-like Rydberg states.

The 7.4 eV band has been investigated by numerous groups, and reviewed by Chan $et\ al.$ Because it is broad, it is not subject to saturation in photoabsorption measurements, thereby negating the advantage of (e,e) scattering for determining oscillator strengths in the sub-ionization region. Recently, Yoshino (1996a) have re-examined the photoabsorption cross section of the first and second bands, with <1 meV resolution and $<2\%$ uncertainty in the measured cross section. Their results for the first band (tabulated on the internet) are 8% lower than Chan $et\ al.$ (1993e). Consequently, we utilize the data of Yoshino $et\ al.$ (1996a) for the first band, although it has little effect on the total $S(0)$, 0.004/10.0, and on $S(-2)$, 0.01/2.46.

For the second band and beyond to the IP, we are faced with a work in progress. Yoshino $et\ al.$ find agreement with Chan $et\ al.$ for the broad structure near 9.7 eV, but it is clear that the extrapolation of this band into the region

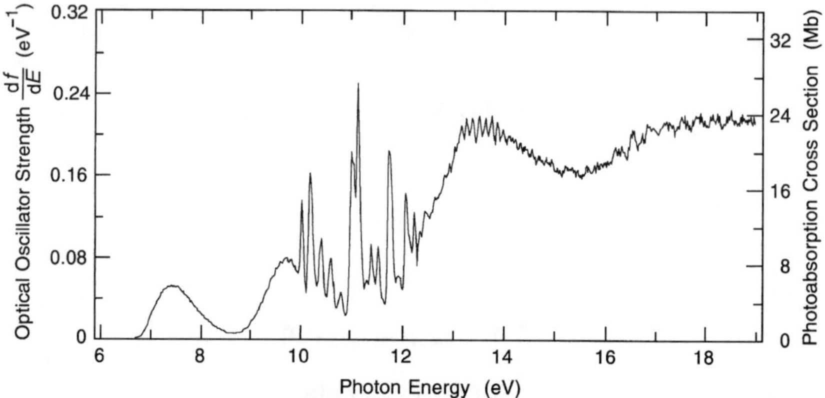

Fig. 5.1   Electron energy-loss spectrum of H_2O, 6–19 eV. Reproduced from Chan *et al.* (1993e), a composite of their Figs. 5 and 7, with permission

of sharp Rydberg structure will depend on the resolution. Yoshino (1998) have begun to explore this region at still higher resolution. At the first sharp Rydberg peak, designated as the $\tilde{C}$–$\tilde{X}$ band, at 10.0 eV, their partitioning of the oscillator strength attributes a significantly larger fraction to the peak, and less to the trailing continuum of the second band, whereas these contributions are nearly equal in the data of Chan *et al.* Nevertheless, the total oscillator strength in this interval does not differ greatly ($f = 0.0102$, Yoshino; $f = 0.00975$, Chan). Thus, pending further detailed studies, we shall utilize the data of Chan *et al.* from the second band to the IP. Gürtler *et al.* present graphical data in this region at rather high resolution (0.3 meV), but it is too compressed to scan accurately.

Haddad and Samson (1986) have obtained absolute cross sections for H_2O from the IP to 124 eV, with a stated accuracy of ±3%. However, their radiation source was a discrete many-line spectrum, which, though well suited for the smooth continuum, may miss the peaks and valleys in the undulating region between the IP and ∼20 eV. Data using a continuum source would be preferred, if sufficiently accurate. In Fig. 5.2, we compare the cross sections of Haddad and Samson with those of Katayama *et al.* (1973) between the IP and 700 Å, (17.71 eV). Within the scatter in the data, they appear to be commensurable between 700–850 Å, but between 850 Å–IP (the more highly structured region) the values of Haddad and Samson are mostly lower. However, between 600–700 Å (vide infra) they are higher than those of Katayama *et al.* by about 5%. Taking the results of Haddad and Samson as a secondary standard, we conclude that the Katayama data are skewed (higher at longer wavelength, lower at shorter wavelength), but they are still preferred because they track the autoionization structure. The graphical data of Gürtler *et al.* appear to be about 8–10% lower than those of Katayama *et al.*, while the (e,e) data of Chan *et al.* (at much coarser resolution) are commensurate with the values of Katayama *et al.*

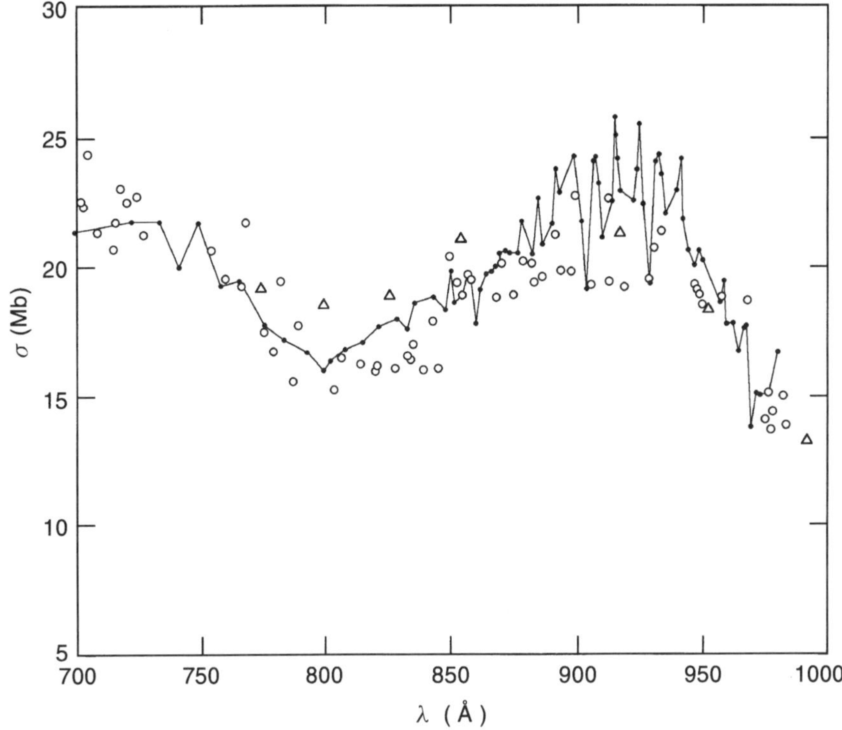

Fig. 5.2 Absolute photoabsorption spectrum of H_2O, 12.4–17.7 eV. ● Katayama *et al.* (1973); ○ Haddad and Samson (1986); △ Chan *et al.* (1993e)

In Fig. 5.3, we compare the measured values of Haddad and Samson with those of various other investigators in the region $100\,\text{Å} < \lambda < 760\,\text{Å}$. Good agreement with Chan *et al.* exists throughout this region, which supports the normalization used by the latter authors. The data of Phillips *et al.* (1977) and of de Reilhac and Damany (1977) appear to be somewhat skewed, i.e. higher cross sections at shorter wavelengths and conversely. The Haddad/Samson data, which we choose in the range 600–$100\,\text{Å}$ (20.66–124 eV), also merge smoothly with the summed atomic cross sections of Henke *et al.* (1993).

Ishii *et al.* (1987) have obtained electron energy loss data for H_2O in the vicinity of the oxygen K-edge, which they have normalized to atomic additivity 25 eV above the K-edge. Since atomic hydrogen has a negligible cross section, the higher energy region beyond the K-edge is essentially that of atomic oxygen.

5.1.2 The data

The adiabatic ionization potential of H_2O, forming the $\tilde{X}^2B_1$ state of H_2O^+, is $101\,766(2)\,\text{cm}^{-1} \equiv 12.6174(2)\,\text{eV}$, according to Tonkyn *et al.* (1991), who performed single photon ZEKE experiments. A slightly refined value of

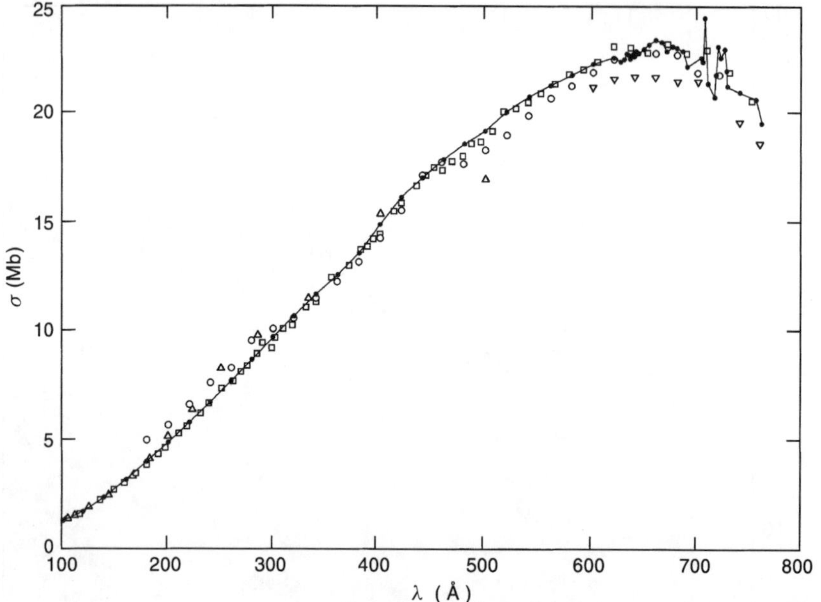

Fig. 5.3 Absolute photoabsorption spectrum of H_2O, 15.5–124 eV. ● Haddad and Samson (1986); ○ Phillips *et al.* (1977); △ Reilhac and Damany (1977); □ Chan *et al.* (1993e); ▽ Katayama *et al.* (1973)

101 766.8 (1.2) cm^{-1} ≡ 12.617 47 (15) eV has recently been reported by Merkt *et al.* (1998).

a The discrete spectrum and transitions below the IP

For the reasons discussed in Sect. 5.1.1, we take the cross sections for the 7.4 eV band from Yoshino *et al.* (1996a). The integrated oscillator strength is 0.0460, which is indeed 8% lower than that of Chan *et al.* For the region 8.631–11.122 eV, we turn to the tabulated data of Chan *et al.* Their resolution, and consequent deconvolution, will tend to increase the oscillator strength of the broad 9.7 eV band and concomitantly decrease f for the sharper Rydberg structure on the high-energy side, but the net effect should be compensatory. Between 11.122 eV and the IP, the graphical data of Chan *et al.* are digitized and integrated. Chan *et al.* obtained 0.3362 for their oscillator strength in the sub-ionization region, while our treatment of their data arrives at 0.3370. This is close enough to assure fairly accurate representations for $S(-2)$ and $S(-1)$. The partitioned data are summarized in Table 5.1.

b The autoionization region, IP–20.664 eV (600 Å)

This region contains autoionizing Rydberg series converging to the $(3a_1)^{-1}$, $^2A_1$ (adiabatic IP = 13.838 eV) and $(1b_2)^{-1}$, $^2B_2$ (AIP = 17.189 eV) states of H_2O^+.

Table 5.1 Contributions to $S(p)$ of transitions below the IP in H_2O^a

Energy, eV	$S(-2)$	$S(-1)$	$S(0)$	$S(+1)$	$S(+2)$
$\tilde{X}^1A_1 \rightarrow \tilde{A}^1B_1$,					
7.4 band	0.1498	0.0829	0.0460	0.0256	0.0143
$\tilde{X}^1A_1 \rightarrow \tilde{B}^1A_1$,					
9.7 band	0.1440	0.1027	0.0732	0.0522	0.0372
9.994	0.0097	0.0071	0.0052	0.0038	0.0028
10.168–10.171	0.0251	0.0187	0.0140	0.0105	0.0078
10.332–10.338	0.0185	0.0141	0.0107	0.0081	0.0062
10.559–10.575	0.0153	0.0118	0.0092	0.0071	0.0055
10.765–10.780	0.0110	0.0087	0.0069	0.0055	0.0043
10.990–11.057	0.0332	0.0269	0.0218	0.0177	0.0143
11.122	0.0334	0.0273	0.0223	0.0182	0.0149
11.2–12.617	0.1601	0.1408	0.1240	0.1093	0.0965
Total	0.6001	0.4410	0.3333	0.2580	0.2038

$^a S(p)$ in Ry units. Data for 7.4 eV band from Yoshino *et al.* (1996a), tabulated data on internet (http://cfa-www.harvard.edu/amdata/ampdata/amdata.html). All other data from Chan *et al.* (1993a).

An analysis has been given by Ishiguro *et al.* (1978). The graphical data of Katayama *et al.* have been electronically scanned, digitized and trapezoidally integrated. The $S(p)$ values are recorded in Table 5.2.

c The continuum, 20.664–123.98 eV (600–100 Å)

Excitation and ionization from the $2a_1$ orbital occurs in this region, at $\sim$32 eV ($\sim$390 Å). There is only a slight change of slope in Fig. 5.3 at $\sim$400 Å marking this channel. The tabulated data of Haddad and Samson (1986) have been fitted to a 4-term polynomial by regression. The coefficients of the polynomial are given in Table 5.3. The expression is analytically integrated to yield the $S(p)$, listed in Table 5.2.

d The continuum, 124–533 eV

The oxygen K-edge of H_2O occurs at 539.9 eV (Jolly *et al.*, 1984). Structure preceeding the K-edge (vide infra) begins to appear at 533 eV. We use the summed atomic cross sections of Henke *et al.* to traverse the region from 124–533 eV. The sparse values are fitted to the usual 4-term polynomial, whose coefficients are given in Table 5.3, while the contributions to $S(p)$ appear in Table 5.2.

e The oxygen K-edge region, 533–566 eV

Ishii *et al.* present the oscillator strength distribution in the vicinity of the oxygen K-edge, normalized to atomic additivity 25 eV above threshold. Schirmer *et al.* (1993) have subsequently obtained a higher-resolution photoabsorption spectrum, but only relative intensities are given. The two spectra agree quite well, although the valleys are deeper between peaks in the higher resolution spectrum, but no vibrational structure is evident. The first two peaks (534.0 and 535.9 eV) are

Table 5.2 Spectral sums, and comparison with expectation values for H_2O[a]

Energy, eV	$S(-2)$	$S(-1)$	$S(0)$	$S(+1)$	$S(+2)$
6.60–IP[b]	0.6001	0.4410	0.3333	0.2580	0.2038
IP–20.66[c]	1.0179	1.2060	1.4587	1.8003	2.2649
20.66–124.0[d]	0.8225	1.9630	5.5812	19.3876	88.6298
124–533[e]	0.0047	0.0592	0.8210	13.0728	246.8782
534.0[f]	–	0.0004	0.014	0.549	21.566
535.9[f]	–	0.0004	0.016	0.630	24.823
537.0[g]	–	0.0001	0.0047	0.186	7.322
537.8[g]	–	–	0.0014	0.057	2.250
538.4[g]	–	–	0.0012	0.048	1.910
533–566[h]	0.0001	0.0047	0.1891	7.6247	307.567
566–2622.4[e]	0.0004	0.0246	1.5489	112.107	9727.9
2622.4–10 000[e]	–	0.0004	0.0976	29.7072	10 324.1
10^4–10^5[i]	–	–	0.0058	7.3793	12 641.63
10^5–10^6[j]	–	–	–	0.3826	6 101.01
10^6–10^7[j]	–	–	–	0.0136	2 130.73
10^7–10^8[j]	–	–	–	0.0004	696.01
10^8–10^9[j]	–	–	–	–	222.39
10^9–∞[j]	–	–	–	–	103.11
Total	2.4457	3.6998	10.0729	193.204	42 650.3
Expectation value	2.406[k]		10.0		41 787.6[l]
	2.45 $\pm$0.02$_5$[m]				41 786.1[n]
Other values[o]	(2.410$_5$)	3.658	(10)	191.2	42 440

[a]In Ry units.
[b]See Table 5.1.
[c]Katayama *et al.* (1973).
[d]Haddad and Samson (1986).
[e]Henke *et al.* (1993) + additivity.
[f]Ishii *et al.* (1987).
[g]Schirmer *et al.* (1993), normalized to f.
[h]Integration of underlying continuum in Fig. 5.2.
[i]Chantler (1995).
[j]Assuming hydrogen-like behavior, bare oxygen atom with screening, K-shell only, from Bethe and Salpeter (1977).
[k]From experiment. See discussion in Sect. 5.1.3.
[l]$S(+2)$ for atomic oxygen from Fraga *et al.* (1976).
[m]From ab initio calculations. See discussion in Sect. 5.1.3.
[n]$S(+2)$ for atomic oxygen from Bunge *et al.* (1993).
[o]Zeiss *et al.* (1977).

Table 5.3 Coefficients of the polynomial $df/dE = ay^2 + by^3 + cy^4 + dy^5$ fitted to data at various energies[a]

Energy range, eV	a	b	c	d
20.66–124.0	15.835 88	−1.533 21	−33.0621	21.187 47
124.0–533.0	−2.164 77	418.8783	−3 813.74	12 178.65
566.0–2622.4	6.850 869	9 461.838	−209 448	1 211 065
2622.4–10 000	−3.070 88	11 634.44	−150 276	−28 450 425

[a]df/dE in Rydberg units, $y = B/E$, $B = $ IP $= 12.6174$ eV

broad and relatively intense. Ishii *et al.* give their oscillator strengths as 0.014 and 0.016, respectively. Schirmer *et al.* ascribe the transitions to upper states which have mixed valence and Rydberg character, essentially the same upper states responsible for the 7.4 and 9.7 eV bands in the discrete spectrum. Three higher states, which are mixtures of Rydberg states, have lower oscillator strengths which we estimate by normalizing the spectrum of Schirmer *et al.* to that of Ishii *et al.* The integrated oscillator strength of the underlying continuum between 533–540 eV, and the post K-edge continuum to 566 eV, has been extracted from Fig. 2 of Ishii *et al.* Table 5.2 lists the detailed contributions to $S(p)$ from the K-edge region.

f Post K-edge, 566–10 000 eV

The atomic oxygen cross section is essentially the only contributor here. The cross sections of Henke *et al.* are partitioned into 566–2622.4 eV and 2622.4–10 000 eV. Each region is fitted by a 4-term polynomial, which is then integrated to yield $S(p)$.

5.1.3 The analysis

Although dielectric constant measurements have been performed on H_2O (Birnbaum and Chatterjee, 1952), the molar polarization is dominated by the dipole moment term in the Debye equation, resulting in poor sensitivity to the polarizability. Several groups have undertaken refractivity measurements, but uncertainties persist at the 2–3% level.

There are several reasons for this discrepancy. One is the departure of water vapor from ideal gas behavior, and how one corrects for it. Another is the tendency of water to form adsorbed layers on surfaces in the light path. A third factor is the conventional reporting of the final results for the hypothetical standard conditions, 0°C and 1 atm. Thus, Russell and Spackman (1995) arrived at $S(-2) = 9.83(2)$ a.u. by fitting the refractive index data of Cuthbertson and Cuthbertson (1913a), while Zeiss and Meath (1975), considering and normalizing various sources, obtained $S(-2) = 9.630$ a.u., and later (Zeiss *et al.*, 1977), 9.642 a.u.

The experiments fall into two categories: the study of steam at relatively high temperatures and pressures (Cuthbertson and Cuthbertson, 1913a; Wüst and Reindel, 1934; Hölemann and Goldschmidt, 1934) and water vapor at ambient conditions (Barrell and Sears, 1940; Newbound, 1949). Cuthbertson and Cuthbertson express their results for the hypothetical 0°C and 1 atm, without describing their correction for non-ideality. Fitting their data does indeed yield $S(-2) = 9.83$ a.u., as found by Russell and Spackman. Wüst and Reindel (1934) obtained refractivities approximately 2% higher. Upon extrapolating to $\lambda = \infty$, they obtained $(n-1) = 250.7 \times 10^{-6}$, or $S(-2) = 10.02$ a.u. However, they suspected some systematic error, because a contemporary measurement by Hölemann and Goldschmidt (1934) had been shown to be in better agreement with the Cuthbertsons at one wavelength, 5461 Å (the refractivities were 3.762

(Cuthbertson and Cuthbertson, 1913a), 3.836 (Wüst and Reindel, 1934) and 3.766 ±0.004 (Hölemann and Goldschmidt, 1934)). From a 'most probable' wavelength dependence based on Cuthbertson and Cuthbertson (1913a) and Wüst and Reindel (1934), Hölemann and Goldschmidt (1934) deduced a refractivity at infinite wavelength of 3.665 cm^3/mol, or $S(-2) = 9.80$ a.u.

These higher pressure experiments retain unresolved questions regarding non-ideality. For the present purposes, the refractive index for a dilute gas is preferable. Barrell and Sears (1940) performed such an experiment by obtaining the refractive index of humid and dry air, and subtracting. They present results appropriate for 10 torr and 20 °C. Assuming ideal gas behavior at these conditions, we obtain $S(-2) = 9.60$ a.u. Owens (1967) and Zeiss and Meath have discussed departure from ideality, but we find that incorporation of these corrections is almost imperceptible at these conditions. However, Erickson (1962) has made more precise relative refractivity measurements which, when normalized to Barrell and Sears at 4679.46 Å, yield $S(-2) = 9.62_5$ a.u. Barrell and Sears also show, by incorporating non-ideality, that their results at 0 °C and 1 atm. are compatible with those of the Cuthbertsons. The latter, in a brief report (Cuthbertson and Cuthbertson, 1936), repeated their earlier work, with no substantial change. The refractivity data of Newbound (1949) correspond to a still lower value of $S(-2)$ than that of Barrell and Sears, but also suffer from scatter.

There have been numerous ab initio calculations, the more recent ones incorporating various forms of correlation. Christiansen *et al.* (1999) used coupled-cluster theory and obtained a vibrationally averaged static polarizability of 9.85 a.u. Kobayashi *et al.* (1999) used Møller-Plesset (MP) perturbation theory to arrive at 9.79 a.u. Maroulis (1991; 1998b) presented the results of several approaches, with values ranging from 9.94 a.u. (MP4) to 9.68 a.u. (coupled electron pair approximation) to still lower values involving less correlation. Spelsberg and Meyer (1998) obtain $S(-2) = 9.77$ a.u. as their best calculation, but acknowledged that '$(i)t$ seems to be very difficult to ascertain an error of less than 2%'. Dougherty and Spackman (1994) used a modest basis set at the MP2 level to calculate $S(-2) = 9.46$ a.u., which increased to 9.792 a.u. with a more elaborate basis set, and 10.084 a.u. (exceeding experiment) after including zero-point vibrational corrections (Russell and Sprackman, 1995).

In summary, the experimental value $S(-2) = 9.62_5$ a.u. $= 2.406$ Ry, obtained from low-pressure data, seems preferable to the other experimental results, but it is primarily based on one source (Barrell and Sears, 1940). The modern ab initio calculations appear to average $\sim 9.8 \pm 0.1$ a.u. $\equiv 2.45 \pm 0.02_5$ Ry units. Both are listed as expectation values in Table 5.2. Our spectral sum lies between these values, but distinctly closer to the higher one. The spectral sum for $S(0)$ is only 0.7% above the TRK sum rule requirement. This remarkable agreement is very likely due to the fortuitous cancellation of errors. The major contributors to $S(-2)$ are the first three entries in Table 5.2, and apart from the oxygen K-shell, this is also largely true for $S(0)$. The sources of data in this region that are stated to be most accurate are the 7.4 eV band ($\sim 2\%$) from Yoshino *et al.* and the line source data, 20.66–124.0 eV ($\sim$3%) from Haddad and Samson. For the autoionization region, IP–20.66 eV, Katayama *et al*. estimate $\pm$7% uncertainty. Hence, agreement to better than 1% for $S(-2)$ and $S(0)$ is unexpected. Nevertheless, since we have selected but not adjusted the input sources, this probably is the best current oscillator strength distribution for H_2O.

The expectation values for $S(+2)$ shown in Table 5.2 are predominantly due to atomic oxygen. The Hartree-Fock values for this atom are from Fraga *et al*. (1976) and from Bunge *et al*. (1993). Together with the small contribution from hydrogen, they should be a fairly accurate representation of $S(+2)$ for H_2O. Our spectral sum, 42 650.3, is higher by 2%. The value of $S(+2)$ arrived at by Zeiss *et al*. is slightly lower, and is based on a mixture rule using their high-energy cross sections for H_2 and O_2.

Our spectral sum for $S(+1)$, 193.2, is 1% higher than that of Zeiss *et al*. Both are substantially larger than 186.6, the value based on Hartree–Fock calculations for atomic oxygen (Fraga *et al*., 1976), the exact value for atomic hydrogen, and additivity.

Our spectral sum for $S(-1)$ is 1.1% higher than that of Zeiss *et al*. A choice between these values depends upon decisions regarding $S(-2)$, since a larger $S(-2)$ will dictate a larger $S(-1)$. The quantity $S_i(-1)$ was determined by digitizing the figures of Katayama *et al*. displaying photoionization cross sections from IP–20.66 eV. The ionization yield was assumed to be unity at higher energies. The resulting $S_i(-1) = 2.90$. Haddad and Samson, using their line source data, obtained $S_i(-1) = 2.93$, in excellent agreement, but to some extent fortuitous since their quantum yield of ionization and that of Katayama *et al*. differ (lower at some wavelengths, higher in others). Rudd *et al*. (1985b) studied the ionization cross section of water with proton impact in the 7–4000 keV range, and obtained $M_i^2 = S_i(-1) = 2.98$, within 3% of the photoionization result. By contrast, Rieke and Prepejchal (1972) deduced $M_i^2 = 3.24 \pm 0.15$ by high-energy electron impact. Between IP–20.66 eV, about 73% of photoabsorption leads to photoionization. This is somewhat larger than that found for first row diatomic molecules, which averaged around 59%.

5.2 Carbon Dioxide (CO_2)

5.2.1 Preamble

This re-visiting of the photoabsorption cross sections of CO_2 offers a long-delayed opportunity to correct the earlier discussion in Berkowitz (1979). There, it was not recognized that the polarizability (α) of CO_2, and (hence $S(-2)$), given in some compilations included a significant contribution from infrared bands (nuclear vibrations). These should be subtracted when comparing with electronic absorption spectra. This separation had already been performed by Pack (1974), and should have been used. At that time, comparative data suggested that the photoabsorption cross sections of Nakata *et al*. (1965) could be increased by as much as 50% between 836–600 Å. Since the true deficit in $S(-2)$ was $\sim$4.5%, rather than 10–15%, this correction for the 'missing' oscillator strength could have been relaxed. The recent synchrotron-based photoabsorption measurements of Shaw *et al.* (1995) largely remove the residual deficit in $S(-2)$ and $S(0)$, and provide support for the earlier conclusion that the cross sections of Nakata *et al.* (1965) were too low at their shorter wavelengths, but by a lesser amount.

Shaw *et al.* also find 'reasonable agreement' with two other contemporary studies in their regions of overlap (13.776–35.9 eV): the electron impact excitation measurements of Chan *et al.* (1993f) and the discrete line photoabsorption cross sections of Samson and Haddad (1984). We shall utilize the data of Chan *et al.* below the IP (13.775 eV), and the cross sections of Samson and Haddad between 35.9–107.08 eV, their upper limit. The new results alter the $S(p)$ below the ionization potential imperceptibly from those in Berkowitz (1979), but there is a noticeable increase above the IP. Absolute photoabsorption cross sections in the vicinity of the carbon and oxygen K-edges have become available from Sivkov *et al.* (1984). Atomic additivity, using the atomic cross sections of Henke *et al.* (1993) merges smoothly with the Samson/Haddad data at 107 eV and is utilized to near the carbon K-edge, between the carbon and oxygen K-edges, and above the oxygen K-edge to 10 keV.

5.2.2 The data

Cossart-Magos *et al.* (1987) obtained 111 121 $\pm 2\,\mathrm{cm}^{-1}$ as the adiabatic ionization potential of CO_2 (to the lower $\tilde{X}^2\Pi_{g,3/2}$ component) from the convergence limit of high members of an nf Rydberg series. Merkt *et al.* (1993) used pulsed field ionization ZEKE, and with an estimate for the field effect, obtained 111 110.0 $\pm 3\,\mathrm{cm}^{-1}$. Wiedmann *et al.* (1995) performed a similar experiment, with smaller pulsed fields and jet-cooled gas. They appear to be cautious in explicitly stating their IP, but from their tables we infer 111 113.2 cm^{-1}, with $\pm 2\,\mathrm{cm}^{-1}$ uncertainty in wavelength calibration. Merkt *et al.* (1993) mention other cases where ZEKE yields lower IPs than Rydberg series extrapolation. They note that their result implies a non-constant quantum defect for the Rydberg series examined by Cossart-Magos *et al.* (1987). Recognizing the slight discrepancy, we select the intermediate value, 111 113 $\mathrm{cm}^{-1} \equiv 13.7763$ eV, with an uncertainty of 3 $\mathrm{cm}^{-1} \cong 0.4$ meV.

a   *The discrete spectrum and transitions below the IP*

The molecular orbital structure in CO_2 may be written

$$(1\sigma_g)^2(1\sigma_u)^2(2\sigma_g)^2(3\sigma_g)^2(2\sigma_u)^2(4\sigma_g)^2(3\sigma_u)^2(1\pi_u)^4(1\pi_g)^4$$

A concise view of the photoabsorption spectrum in the sub-ionization region is depicted in Fig. 1 of Chan *et al.* Two weak bands with maxima at approximately 8.5 and 9.3 eV, are seen to merge with one another. At higher resolution, vibrational fine structure appears, superposed on a continuum, in both regions. Slanger and Black (1978) showed that the quantum yield of dissociation (CO + O) is unity in both bands. Although the expected products are $CO(^1\Sigma^+) + O(^1D)$ from spin-conservation rules, Zhu and Gordon (1990) found evidence for $\sim6\%$ $O(^3P)$, which was substantiated by Stolow and Lee (1993). Knowles *et al.* (1998) performed electronic structure calculations of high accuracy ($\sim0.05\,eV$). They show that in the absorption region of interest, the $^1\Sigma_u^-$, $^1\Delta_u$ and $^1\Pi_g$ states of CO_2 (all dipole forbidden) are nearly degenerate and manifest a conical intersection. The lower energy absorption may be the bent B_2 component of the $^1\Delta_u$ state, and the higher energy absorption involves the Renner–Teller splitting of the $^1\Pi_g$ state, perhaps also to a bent B_2 component. According to Knowles *et al.* this CO_2 absorption region involves 'a very complex situation...which cannot be properly described within the Born–Oppenheimer approximation'. The more intense bands at higher energy are to Rydberg states, whereas these lower bands are to valence, or valence-Rydberg mixture states.

We take the photoabsorption cross sections for the 8.5 and 9.3 eV bands from data of Lewis and Carver (1983). Yoshino *et al.* (1996b) have re-examined this region with comparable resolution, and obtained 'very good agreement' with Lewis and Carver. The sum of the oscillator strengths for these two bands, given in Table 5.4, is about 15% lower than the corresponding quantity given by Chan *et al.*. For the stronger bands between 10.33 eV–IP, we have digitized Fig. 5 of Chan *et al.*, and obtained an integrated oscillator strength about 4% larger than can be inferred from the incremental sums in Tables 3 and 4 of Chan *et al.*. However, the Vancouver group (Olney *et al.*, 1997) have subsequently increased their cross sections by 3.4% to match $S(-2)$ values from refractivity. Thus, our integrated sums appearing in Table 5.4 should be nearly equal to the 'refined' values of Olney *et al.* (1997).

b *The autoionization region and beyond: IP–35.93 eV*

In the region IP–19.6 eV, autoionizing Rydberg series are found that converge to $\tilde{A}^2\Pi_u$ (AIP = 17.286 eV), $\tilde{B}^2\Sigma_u^+$ (AIP = 18.076 eV) and $C^2\Sigma_g^+$ (AIP = 19.394 eV). We have digitized Figs. 1–6 from Shaw *et al.* which include this region and the smooth section beyond, to 35.93 eV. We compare this higher energy region with the cross sections of Samson and Haddad using discrete line sources, and the synchrotron-based data of Watson *et al.* (1975) in Fig. 5.4. We note that, despite the suppressed zero, the cross sections of Shaw *et al.* and Samson and Haddad are in excellent agreement until $\sim25\,eV$. The cross sections of Watson *et al.* are somewhat lower between 20–27 eV. For the region IP–25 eV, trapezoidal integration yields $\Delta S(0) = 3.4087$ (Shaw *et al.*, 1995) and 3.3795 (Samson and Haddad, 1984); from 25–39.5 eV, $\Delta S(0) = 2.7681$ (Shaw *et al.*, 1995) and 2.7751 (Samson and Haddad, 1984). The two results are very close, with no significant evidence of different energy dependencies. For the total domain, IP–39.5 eV, the two results differ by $<0.4\%$. However, Shaw *et al.* explicitly give $\Delta S(0) = 6.034$ from their data, while our digitizing and integration yield 6.177, which is 2.3% higher. Their direct determination is

Table 5.4   Spectral sums, and comparison with expectation values for CO_2[a]

Energy, eV	$S(-2)$	$S(-1)$	$S(0)$	$S(+1)$	$S(+2)$
7.0–10.33[b]	0.0268	0.0174	0.0113	0.0074	0.0049
10.66–IP[c]	1.4970	1.3627	1.2473	1.1478	1.0616
IP–35.93[d]	2.3982	3.7123	6.1768	11.0290	20.9959
	(2.330)[e]	(3.616)[e]	(6.034)[e]	(10.795)	(20.576)
35.93–107.08[f]	0.4359	1.6358	6.6563	29.6766	15.296
107.08–290.0[g]	0.0166	0.1749	1.9612	23.6488	308.472
291.0[h]	0.0003	0.0058	0.125	2.674	57.18
290–341.3[h]	0.0011	0.0247	0.5695	13.1722	305.186
341.3–524.9[g]	0.0010	0.0292	0.8821	27.0327	841.215
534.6[h]	–	0.0012	0.046	1.8074	71.019
524.9–572.8[h]	0.0003	0.0140	0.5692	23.1037	938.203
572.8–2293.2[g]	0.0010	0.0563	3.5152	248.2581	20 344.48
2293.2–10 000[g]	–	0.0012	0.2972	81.3806	26 063.67
10^4–10^{5}[i]	–	–	0.0132	16.7163	28 578.30
10^5–10^{6}[j]	–	–	–	0.8552	13 628.34
10^6–10^{7}[j]	–	–	–	0.0303	4 746.90
10^7–10^{8}[j]	–	–	–	0.0009	1 549.24
10^8–10^{9}[j]	–	–	–	–	494.88
10^9–∞[j]	–	–	–	–	229.69
Total	4.3782	7.0355	22.0703	480.541	98 324.1
	(4.310)	(6.939)	(21.928)	(480.307)	(98 323.7)
Expectation values	4.389[k]		22.0		96 366.7[l]
	4.378[m]				96 363.6[n]
Other values	(4.378)[m]	7.070[m]	(22)[m]	468.8[m]	97 320[m]
	4.205[o]	6.839[o]	21.797[o]		
	4.378[p]	6.870[p]			

[a]$S(p)$ in Ry units.
[b]Lewis and Carver (1983).
[c]Digitized data of Chan *et al.* (1993f) see text.
[d]Digitized data of Shaw *et al.* (1995).
[e]Integrated data directly given in d.
[f]Samson and Haddad (1984).
[g]Summed atomic cross sections from Henke *et al.* (1993)
[h]Digitized data of Sivkov *et al.* (1984) see also Sivkov *et al.* (1987)
[i]Summed atomic cross sections calculated by Chantler (1995).
[j]Using the hydrogenic equation of Bethe and Salpeter (1977) for the high-energy cross sections of the carbon and oxygen K-shells.
[k]Pack (1974).
[l]Summed Hartree–Fock atomic values of $S(+2)$ from Fraga *et al.* (1976).
[m]Jhanwar and Meath (1982).
[n]Summed Hartree–Fock atomic values of $S(+2)$ from Bunge *et al.* (1993).
[o]Values of $S(p)$ given in d.
[p]Adjusted values based on c, above, given by Olney *et al.* (1997).

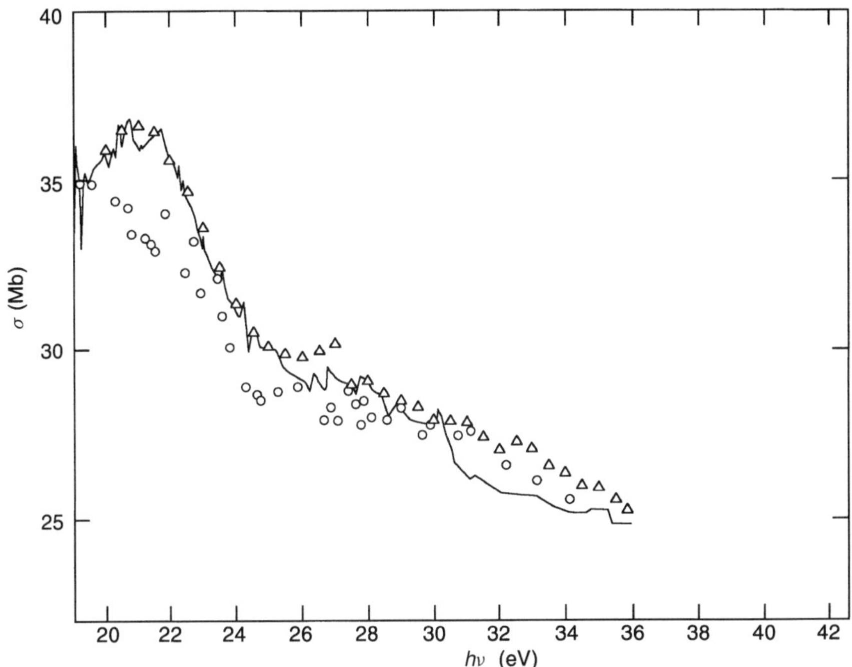

Fig. 5.4 Absolute photoabsorption spectrum of CO_2, 21–36 eV. ○ Watson *et al.* (1975);
△ Samson and Haddad (1984); —— Shaw *et al.* (1995)

certainly more reliable, but considering the excellent agreement of our digitized sum of the data of Shaw *et al.* with the tabulated data of Samson and Haddad, we shall retain both sets of values in Table 5.4, and ultimately examine their effects on $S(p)$.

c The continuum, 35.93–107.08 eV

In Fig. 5.5, we compare the photoabsorption cross sections of Samson and Haddad, the (e,e) measurements of Chan *et al.* and the summed atomic cross sections of Henke *et al.* between 35–200 eV. There is fairly good agreement between Samson and Haddad and Chan *et al.*, although the latter is consistently lower, and does not have as pronounced a bulge at ~50 eV. This bulge is evidently a consequence of excitation from the inner valence orbitals, i.e. $(2\sigma_u)^{-1}$ and $(3\sigma_g)^{-1}$ which result in an array of ionized states ranging from 22.8–40.6 eV, but concentrated between 32–38 eV (see, for example, Freund *et al.* (1986)). Their intensity above 45 eV increases, relative to the outer valence region, accounting for the protrusion. At 91 eV and beyond, the summed atomic cross sections are seen to merge smoothly with the molecular data of Samson and Haddad. The latter have been integrated trapezoidally, and the corresponding $S(p)$ in this region are recorded in Table 5.4.

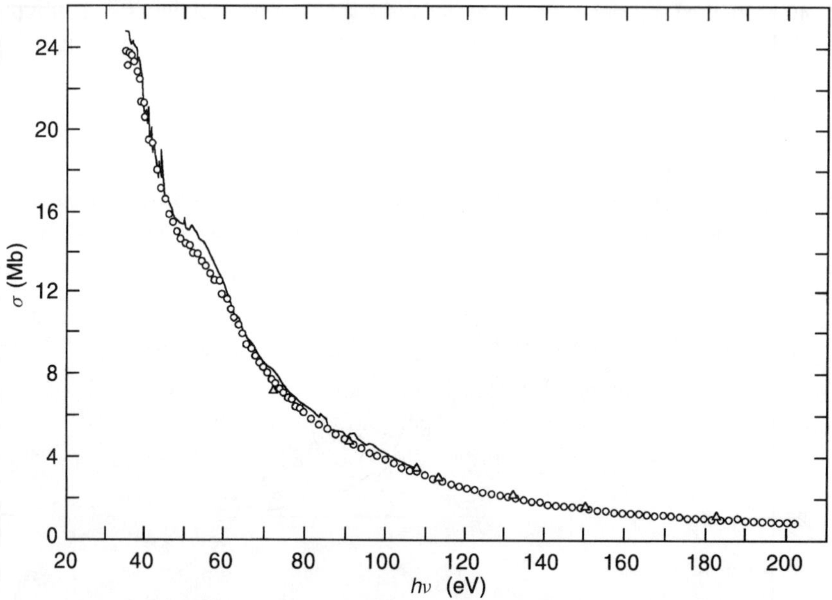

Fig. 5.5   Absolute photoabsorption spectrum of CO_2, 35–200 eV. —— Samson and Haddad (1984); ○ Chan *et al*. (1993f), Table 1; △ Henke *et al*. (1993) + additivity

d   The continuum, 107.08–290.0 eV

The carbon K-edge in CO_2, $(2\sigma_g)^{-1}$, occurs at 297.65 eV (Jolly *et al*., 1984), but prominent structure begins at 290 eV. Accordingly, we fashion a transition region from the summed atomic cross sections of Henke *et al*. between 107–290 eV. The sparse data are fitted by regression to a 4-term polynomial. The $S(p)$ within these limits are evaluated analytically from the polynomial, and shown in Table 5.4; the coefficients of the polynomial are given in Table 5.5.

e   Near carbon K-edge, 290–341.3 eV

The X-ray photoabsorption spectrum of CO_2 in the vicinity of the carbon K-edge is given by Sivkov *et al*. (1984). It consists of a sharp, intense peak at 291.0 eV,

Table 5.5   Coefficients of the polynomial $df/dE = ay^2 + by^3 + cy^4 + dy^5$ fitted to data at various energies[a]

Energy range, eV	a	b	c	d
107–290	−20.1699	1244.259	−11 489	36 074.58
341.3–524.9	37.505 09	−429.43	65 138.77	−902 421
572.8–2293.2	0.731 716	19 329.84	−472 016	4 458 877
2293.2–10 000	−4.331 01	19 022.22	161 688.6	−71 331 572

[a]dF/dE in Ry units, $y = B/E$, $B = $ IP $= 13.7763$ eV.

attributed to excitation to $2\pi_u$, the antibonding first vacant orbital, and weaker excitations to $4\sigma_u$ and Rydberg levels, with a maximum at $\sim$312 eV associated with a shape resonance. A similar spectrum obtained by electron energy loss spectroscopy and normalized to additivity at higher energies is shown by McLaren *et al.* (1987). In a subsequent paper, Sivkov *et al.* (1987) present the integrated oscillator strength as a function of energy. We have digitized the spectrum of Sivkov *et al.* (1984) and compared with the running integrated oscillator strength of Sivkov *et al.* (1987). The good agreement obtained enables us to assign $f = 0.125$ to the 191 eV peak, whereas McLaren *et al.* give $f = 0.14$ from their integration of the Sivkov *et al.* (1984) data and $f = 0.16$ from their own work. The underlying continuum and additional structure from Sivkov *et al.* (1984) has been trapezoidally integrated between 290–341.3 eV, and the corresponding $S(p)$ appear in Table 5.4.

f   The inter-edge continuum, 341.3–524.9 eV

The oxygen K-edge in CO_2 occurs at 541.1 eV (Jolly *et al.*, 1984). A prominent peak at 534.6 eV, and additional structure precedes this edge. It is convenient to consider the near K-edge region as beginning at 524.9 eV, since Sivkov *et al.* (1984) provide a datum here. It is compatible with the summed atomic cross sections of Henke *et al.*, which we utilize in the inter-edge region. The Henke sums are fitted to the usual 4-term polynomial and integrated to yield the corresponding $S(p)$.

g   Near oxygen K-edge, 524.9–572.8 eV

Sivkov *et al.* (1984) also present the absolute photoabsorption cross sections of CO_2 in this region. The large peak at 534.6 eV is attributed to the $1\sigma_g \rightarrow 2\pi_u$ excitation. Our digitizing and integration yields $f = 0.046$, while McLaren *et al.* give $f = 0.062$ from their integration, and $f = 0.14$ from their own data. The underlying continuum and additional structure are trapezoidally integrated; the relevant sums can be found in Table 5.4

h   Post K-edge continuum, 572.8–10 000 eV

We use the summed atomic cross sections of Henke *et al.* in this interval. At the juncture (572.8 eV) the molecular cross section given by Sivkov *et al.* (1984) is about 11% higher than the summed atomic cross sections, which may imply that the K-edge cross sections have not yet subsided to their atomic sums, or it may simply be experimental uncertainty. We partition the Henke data into two regions, 572.8–2293.2 eV and 2293.2–10 000 eV and fit each region by regression to the usual 4-term polynomial. Almost 69% of $S(+1)$ occurs in this interval.

i 10^4–10^5 eV

The calculated atomic cross sections of Chantler (1995) are stoichiometrically summed, and the $S(p)$ are given in Table 5.4. As with earlier cases, this region contains $\sim$39% of $S(+2)$, 3.7% of $S(+1)$ and insignificant contributions to the other $S(p)$.

5.2.3   The analysis

We noted in Sect. 5.2.1 that an inconsistent static polarizability was used earlier (Berkowitz, 1979) to infer $S(-2)$. The value used, $2.91 \times 10^{-24}\,\mathrm{cm}^3$ is essentially that cited by Pack, $2.9126(5) \times 10^{-24}\,\mathrm{cm}^3$ as the total static polarizability. However, he also shows that a significant portion, $0.311 \times 10^{-24}\,\mathrm{cm}^3$, arises from infrared contributions, i.e. nuclear vibrations. The difference, $2.6016(5) \times 10^{-24}\,\mathrm{cm}^3$, is the electronic component appropriate for comparison in this work, and is equivalent to $S(-2) = 4.3890\,\mathrm{Ry}$ units. Jhanwar and Meath (1982) have subsequently made their own analysis of visible and ultraviolet refractivity data, and deduce $S(-2) = 4.377_5\,\mathrm{Ry}$ units. More recent refractivity measurements by Hohm (1994) at 5 wavelengths are compatible with the latter value. The current spectral sum yields $S(-2) = 4.3782\,\mathrm{Ry}$ units, which is fortuitously almost identical with the expectation value deduced by Jhanwar and Meath. The uncertainty in the individual measurements between 7.0–107 eV would suggest at least 3% error. Had we used the $S(-2)$ value directly given by Shaw *et al.* for their data, our result would be 1.5% lower than the expectation value. The overall $S(-2)$ inferred by Shaw *et al.* is 4% lower than the expectation value.

Our spectral sum for $S(0)$ is also remarkably good, deviating from the TRK sum rule by $\pm$0.3%, depending upon our digitizing of the Shaw data or the value directly presented by these authors for their data. Their total $S(0)$ is also quite good, about 1% below the TRK sum. Jhanwar and Meath constrain their distribution of oscillator strengths to match their selection of S(-2) and the TRK sum, so comparison is only for the other sum rules. Their value for $S(-1)$, 7.070 Ry units, lies between our spectral sum based on our digitizing and the value directly given by Shaw *et al.* A value $S(-1) = 6.987\,\mathrm{Ry}$ units, an average of the values based on Shaw (direct) and Shaw (digitized), is suggested because the corresponding average for $S(0)$ gives almost identically the TRK sum.

Our spectral sum for $S(+2)$ is about 2% larger than the summed atomic Hartree–Fock values (Fraga *et al.*, 1976; Bunge *et al.*, 1993). Jhanwar and Meath, who apparently used an earlier compilation by Hubbell (1977) obtain $S(+2) = 97\,200$, less than 1% above atomic additivity.

However, we conjecture that their value for $S(+1)$, 468.8, may be too low. It is only $\sim$4 Ry units above atomic additivity (Fraga *et al.*, 1976) whereas our spectral sum is 11.6 Ry units higher. A more correlated wave function than atomic Hartree–Fock tends to increase $S(+1)$ above atomic additivity, since the total electronic energy will increase. For the case of O_2, Zeiss *et al.* (1977) found a value of $S(+1)$ about 6.6 Ry units larger than additivity. Almost 80% of our spectral sum contributing to $S(+1)$ is based on the experimentally based compilation of Henke *et al.* The true value of $S(+1)$ for CO_2 may lie between our spectral sum and that of Jhanwar and Meath.

The quantum yield of ionization varies from IP–20 eV, but appears to be unity above 20 eV ($\lambda < 617$ Å), according to Shaw *et al.* We wish to compute S_i (−1). Rather than using a compressed figure displaying the spectral dependence of the absolute photoionization cross section, we have digitized the spectral variation of the quantum yield shown by Shaw *et al.*, matched it with the digitized photoabsorption cross section, and determined their product between IP–20 eV. The resulting value of S_i (−1) is 5.421. Shaw *et al.* have obtained 5.284. The difference stems mainly from our digitizing of their data, and hence theirs must be considered more reliable. From the slope of absolute ionization cross sections versus kinetic energy for high energy electron impact, Rieke and Prepejchal (1972) reported $M_i^2 \equiv S_i(-1) = 5.75 \pm 0.10$, about 8% higher. In the more highly structured region between IP–18.0 eV, about 80% of photoabsorption leads to ionization, a higher fraction than has been found for some first row diatomic molecules.

The present analysis agrees with that of Shaw *et al.* in most respects. Our somewhat better agreement for $S(-2)$ is due in part to our enhancement of the cross sections of Chan *et al.* below the IP, which we justify by the subsequent increase of their own data (Olney *et al.*, 1997), but also by our treatment of the data of Shaw *et al.*, which is limited by scanning inaccuracy. Although our values of $S(-2)$, $S(-1)$ and $S(0)$ are in generally good agreement with Jhanwar and Meath, $S(-2)$ and $S(0)$ being constraints in their optimization procedure, our distribution of oscillator strengths is somewhat different from theirs. The oscillator strengths of Jhanwar and Meath are higher than presently selected ones by ~ 0.5 between 28–70 eV, but are lower below 20 eV and above 80 eV by a corresponding amount. Their high values result from the data of Lee *et al.* (1973) and of de Reilhac and Damany (1977), which were available in 1981. Those cross sections are indeed higher than the more recent values of Shaw *et al.* and Samson and Haddad used currently. Their low values below 20 eV are a consequence of their use of the data of Nakata *et al.*, which was discussed in Sect. 5.2.1; the ones above 80 eV were evaluated by a mixture rule, $\sigma(CO_2) = \sigma(C) + \sigma(O_2)$. The data base at higher energy presented here includes measurements extending to 200 eV (see Fig. 5.4), directly measured cross sections in the vicinity of the K-edge, and atomic additivity which smoothly joins these regions.

5.3 Nitrous Oxide (N$_2$O)

5.3.1 Preamble

Nitrous oxide, which is isoelectronic with carbon dioxide, displays some analogous properties. Its adiabatic ionization potential, 12.89 eV, is about 0.9 eV lower than that of CO_2. With its lower excitation energies, its polarizability (α) is correspondingly higher, 2.919×10^{-24} cm^3 versus 2.595×10^{-24} cm^3 for CO_2. This is just the electronic component; both N_2O and CO_2 have non-negligible contributions in the infrared, from nuclear vibrations. Both N_2O and CO_2 are linear in their electronic ground states, and have similar orbital sequences, allowing for the asymmetry of N_2O. The valence orbital sequence in N_2O is... $(\pi)^4 (\sigma)^2 (\pi)^4$, whereas that of CO_2 is ... $(\sigma)^2 (\pi)^4 (\pi)^4$. The first significant absorptions in the ultraviolet region for each molecule are nominally to $^1\Delta$ and $^1\Pi$ excited states, but both are believed to be bent, and the oscillator strengths are weak.

The oscillator strength distribution of N_2O in the UV and VUV has been refined by recent inelastic electron scattering measurements of Chan $et\,al.$ (1994) and photoabsorption cross sections determinations by Shaw $et\,al.$ (1992b). We shall have recourse to the latter between IP and $\sim$25 eV, a region containing autoionization structure. The data of Chan $et\,al.$ (1994) will be compared with earlier photoabsorption data below the IP, and from 25–200 eV.

There do not appear to be any accurate absolute photoabsorption cross section measurements in the vicinity of the nitrogen and oxygen K-edges of N_2O. However, relative cross section data exist, which we normalize to atomic additivity $\sim$30 eV beyond the respective edges. Summed atomic cross sections are also utilized in the approach to the nitrogen K-edge, the inter-edge region, and beyond the oxygen K-edge to $\sim$10 keV. Calculated atomic cross sections are summed to still higher energy.

5.3.2 The data

The adiabatic ionization potential of N_2O is 103 963 $\pm$5 cm^{-1} $\equiv$ 12.8898 $\pm$0.0006 eV, from single-photon ZEKE measurements by Wiedmann $et\,al.$ (1991).

a The discrete spectrum and transitions below the IP

The first significant absorption in N_2O is a broad band from 5.7–8.0 eV, with very weak vibrational structure superposed. It has been assigned as a $^1\Sigma^+ \rightarrow {}^1\Delta$ transition, although the upper state is probably bent. It dissociates (or predissociates) into N_2 ($X^1\Sigma_g^+$) + O(^{1}D) (see Preston and Barr, 1971). Photoabsorption measurements should be reliable here, since there is no danger of saturation. Zelikoff $et\,al.$ (1953) find $f = 0.0015$ from the integral of their photoabsorption spectrum, while Rabalais $et\,al.$ (1971) give $f = 0.0014$. Chan $et\,al.$ get $f = 0.001\,31$, but their inelastic electron scattering curve suffers from statistical scatter. We select $f = 0.0015$. The second band, from 8.0–9.0 eV, is more intense, and is surmounted by more obvious vibrational structure, which is less apparent in the (e,e) data of Chan $et\,al.$ than in photoabsorption measurements. Early photoabsorption measurements yielded disparate oscillator strengths, Zelikoff $et\,al.$ reporting $f = 0.0211$ while Rabalais $et\,al.$ obtained $f = 0.0072$. Subsequently, Lee and Suto (1984) obtained a photoabsorption curve estimated to have $\pm$5% uncertainty. Our scanning and digitizing of their data yields $f = 0.0255$, while Chan et al. extract 0.0253 from the Lee/Suto spectrum and 0.0245 from their own data. Here we retain our digitizing, and determine the other $S(p)$ accordingly. Although nominally a $^1\Sigma^+ \rightarrow {}^1\Pi$ transition, the rather low oscillator strength is attributed to the fact that it derives from the parity forbidden

$^1\Sigma^+ \to {}^1\Pi_g$ transition of a $D_{\infty h}$ molecule such as CO_2 (see Rabalais *et al.*). Prevailing evidence (Black *et al.*, 1975) indicates that the upper state dissociates (or predissociates) along several pathways, yielding $O(^1S)$, $N(^2D)$ and N_2 $(A^3\Sigma_u^+)$.

The first strong (and hence fully allowed) transition occurs between 9–10.2 eV, and is assigned to $\tilde{X}^1\Sigma^+ \to \tilde{D}^1\Sigma^+$. It is analogous to the first strong transition in CO_2 ($\tilde{X}^1\Sigma_g^+ \to {}^1\Sigma_u^+$), whose upper state has $3p\pi_u$ character. The spectrum of Lee and Suto reveals a broad peak, with no superstructure, negating saturation problems in photoabsorption. Various oscillator strength measurements are in good agreement. We choose $f = 0.378$ from the spectrum of Lee and Suto, very close to that obtained by Chan *et al.*, $f = 0.376$. Shapiro (1977) has shown by calculation that the upper state dissociates predominantly to N_2 $(^1\Sigma_g^+) + O(^1S)$, in agreement with earlier experiments.

Between 10.2 eV and 11.8 eV, Lee and Suto identify seven progressions associated with Rydberg excitations. Chan *et al.* note 'excellent agreement' with Lee and Suto except for two sharp peaks where line saturation effects show up in their low resolution data. However, Lee and Suto also present a much higher resolution spectrum. We have digitized the latter, and find generally good agreement with the segmented integrated oscillator strengths given in a table by Chan *et al.* In the interval 10.23–11.80 eV, Chan *et al.* obtain $f = 0.339$, our integration of the Lee and Suto data yields $f = 0.35$, and earlier electron inelastic scattering data (Huebner *et al.*, 1975b) provides $f = 0.380$. In this instance we have chosen the photoabsorption cross sections of Lee and Suto. However, between 11.80 eV and the IP, where comparative photoabsorption data are not available, we utilize the electron scattering results of Chan *et al.* The various $S(p)$ for these intervals, and their provenance, are summarized in Table 5.6. We note that the total oscillator strength below the IP in N_2O is about 1 unit, slightly less than that in CO_2.

b The autoionization region, IP–20.19 eV

Here we turn to the photoabsorption measurements of Shaw *et al.*, which were performed over the structured region with a resolution of ~ 5 meV. Their

Table 5.6 Contributions to $S(p)$ of transitions below the IP in N_2O[a]

Energy, eV	$S(-2)$	$S(-1)$	$S(0)$	$S(+1)$	$S(+2)$
5.70–8.0[b]	0.0062	0.0030	0.0015	0.0007	0.0004
8.0–9.0[c]	0.0650	0.0407	0.0255	0.0160	0.0100
9.0–10.23[c]	0.7558	0.5344	0.3780	0.2673	0.1893
10.23–11.80[c]	0.5341	0.4350	0.3547	0.2895	0.2366
11.80–IP[d]	0.2830	0.2564	0.2324	0.2107	0.1911
Total	1.6441	1.2695	0.9921	0.7842	0.6274

[a] $S(p)$ in Ry units.
[b] Zelikoff *et al.* (1953).
[c] Lee and Suto (1984).
[d] Chan *et al.* (1994).

estimated error in cross sections is 2–3%. There is a slight discontinuity at the IP, where the data of Chan *et al.* appear to give $\sigma \sim 12\,\mathrm{Mb}$, whereas the data of Shaw *et al.* show $\sigma \sim 10\,\mathrm{Mb}$. However, at higher energies (20–25 eV) the Shaw data appear to be higher that the Chan cross sections by $\sim 4\%$. In the interval 13.9–20.1 eV, complex structure is observed by Shaw *et al.*, attributed primarily to autoionizing Rydberg series converging to $(7\sigma)^{-1}\tilde{A}^2\Sigma^+$ (adiabatic IP = 16.38 eV) and $(6\sigma)^{-1}\tilde{C}^2\Sigma^+$ (AIP = 20.11 eV) (see Holland *et al.*, 1990). Weaker structure related to $(2\pi)^{-1}\tilde{B}^2\Pi$ (AIP = 17.65 eV) has also been identified (Berkowitz and Eland, 1977). Shaw *et al.* present their data graphically; the only sums given are for the entire electromagnetic range, which include other data. Consequently, it was necessary to scan and digitize their figures. Trapezoidal integration was performed on the digitized data. The resulting $S(p)$ are listed in Table 5.7.

c   The continuum, 20.19–25.56 eV

This is a smooth region where the data of Shaw *et al.* overlap the values given by Chan *et al.*. The latter give their 'low resolution' cross sections in a table, but also present graphical information at 'high resolution', 48 meV. From our data extraction, the contributions to $S(0)$ of the respective data sets are: Shaw *et al.*, 1.7159; Chan, high resolution, 1.6396; Chan, low resolution, 1.6733. We adopt the last set, which is an approximate mean, is based on tabulated data, and merges with the higher energy data considered below. The integration has been performed trapezoidally.

d   The continuum, 25.56–108.5 eV

In Fig. 5.6, we compare various sources of cross section measurements between 35–207 eV. There is a bulge between ~ 45–65 eV, which is presumably attributable to excitation and ionization from the inner valence orbitals 1π, 5σ and 4σ (see Holland *et al.*, 1990). The photoabsorption measurements of Cole and Dexter (1978) are uniformly lower than the (e,e) data of Chan *et al.*. The cross sections of Lee *et al.* (1973), also obtained by photoabsorption using a synchrotron light source, tend to be higher than the values of Chan *et al.*, though in some regions they are close. They terminate at ~ 69 eV, so it is advantageous to proceed with the cross sections of Chan *et al.*, which appear to merge smoothly with the summed atomic cross sections of Henke *et al.* (1993) above 108.5 eV. Accordingly, the tabulated data of Chan *et al.* in this interval have been trapezoidally integrated and the results are recorded in Table 5.7. This is a critical region for application of the sum rules, since $\sim 9/22$ of $S(0)$ occurs here.

e   The continuum, 108.5–400.0 eV

The nitrogen K-edges in N_2O occur at ~ 408.7 eV (N terminal) and ~ 412.6 eV (N central) (Jolly *et al.*, 1984). However, prominent pre-edge structure has been observed, beginning at ~ 400 eV (Wight and Brion 1974b; Bianconi *et al.*, 1978).

Table 5.7 Spectral sums, and comparison with expectation values for N_2O[a]

Energy, eV	$S(-2)$	$S(-1)$	$S(0)$	$S(+1)$	$S(+2)$
5.70–IP[b]	1.6441	1.2695	0.9921	0.7842	0.6274
IP–20.19[c]	1.6150	1.9513	2.3898	2.9658	3.7282
20.19–25.56[d]	0.6056	1.0044	1.6733	2.8005	4.7084
25.56–108.5[d]	0.9539	2.7518	9.0183	34.3653	152.5994
108.5–400.0[e]	0.0150	0.1645	1.9649	26.3074	402.1237
401.16[f]	0.0003	0.0077	0.226	6.6635	196.5
404.92[f]	0.0003	0.0080	0.239	7.1129	211.7
405–455[f]	0.0007	0.0229	0.7216	22.7585	718.4
455–506[e]	0.0004	0.0139	0.4879	17.1937	606.5
506–533[g]	0.0001	0.0057	0.2168	8.2745	315.9
533–545[g]	0.0002	0.0072	0.2843	11.2353	444.0
545–603[g]	0.0004	0.0153	0.6441	27.0989	1 141.0
603–2042.4[e]	0.0008	0.0466	2.9926	212.3497	16 934.1
2042–10 000[e]		0.0016	0.3491	86.8971	25 867.3
10^4–10^{5}[h]	–	–	0.0122	15.3909	26 259.7
10^5–10^{6}[i]	–	–	–	0.7756	12 350.2
10^6–10^{7}[i]	–	–	–	0.0270	4 285.1
10^7–10^{8}[i]	–	–	–	0.0008	1 396.7
10^8–10^{9}[i]	–	–	–	–	445.9
10^9–∞[i]	–	–	–	–	206.7
Total	4.8368	7.2704	22.212	483.00	91 943.5
Expectation values	4.925[j]		22.0		90 089.5[l]
	4.93(1)[k]				90 089.1[m]
Other values	(4.925)[j]	7.405[j]	(22.0)[j]	462.6[j]	91 640.0[j]
	4.76[n]	7.16[n]	21.98[n]		
	5.058[o]				
	4.925[p]	7.10[p]			

[a] $S(p)$ in Ry units.
[b] See Table 5.6 and text.
[c] Digitized data of Shaw *et al.* (1992b).
[d] Tabulated data of Chan *et al.* (1994).
[e] Using summed atomic cross sections from Henke (1993).
[f] See Fig. 5.7 and text.
[g] See Fig. 5.8 and text.
[h] Summed atomic cross sections from Chantler (1995).
[i] Summed atomic cross sections using hydrogenic calculation, K-shells only.
[j] Zeiss *et al.* (1977).
[k] See discussion in Sect. 5.3.3.
[l] Fraga *et al.* (1976).
[m] Bunge *et al.* (1993).
[n] Values given in c.
[o] Chan *et al.* (1994).
[p] Adjusted data of o, above, given by Olney *et al.* (1997).

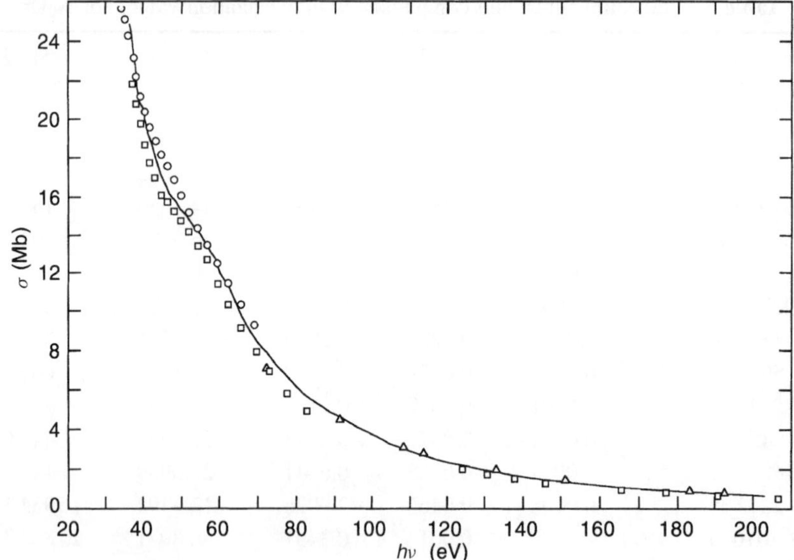

Fig. 5.6   Absolute photoabsorption spectrum of N_2O, 35–200 eV. —— Chan *et al*. (1994); o Lee *et al*. (1973); □ Cole and Dexter (1978); △ Henke *et al*. (1993) + additivity

Figure 5.6 describes a smooth, monotonic decline in the cross section between 108.5–203 eV, which we assume will persist to 400 eV. Accordingly, the summed atomic cross sections of Henke *et al*. are fitted by regression to a 4-term polynomial, which is analytically integrated to determine the $S(p)$ in the interval 108.5–400 eV. These $S(p)$ are given in Table 5.7, and the polynomial coefficients appear in Table 5.8.

f   The nitrogen K-edge region, 400–455 eV

As mentioned in Sect. 5.3.1, absolute cross sections in this region are not available. In Fig. 5.7, we have attempted to normalize the relative cross sections of Bianconi *et al*. (1978) at 452.2 eV (approximately 40 eV above the edge), using the summed atomic cross sections of Henke *et al*. at this energy. Some guidance was also sought from the inelastic scattering spectrum of Wight and Brion (1974b). We report the inferred oscillator strengths of the two prominent peaks at 401.16 and 404.92 eV separately. Their large oscillator strengths are attributed to $N(1s) \rightarrow \pi^*$ resonances from N(terminal) and N(central), respectively. The contributions to $S(p)$ of the higher energy structure, 405–455 eV, have been calculated by trapezoidal integration, and given in Table 5.7. At this energy, the most sensitive sum rules are $S(0)$ and $S(+1)$.

g   The inter-edge region, 455–506 eV

The oxygen K-edge in N_2O occurs at 541.4 eV (Jolly *et al*., 1984). Prominent structure appears prior to that edge, at 534.6 eV (Wight and Brion, 1974b).

Table 5.8 Coefficients of the polynomial $\mathrm{d}f/\mathrm{d}E = ay^2 + by^3 + cy^4 + dy^5$ fitted to data at various energies[a]

Energy range, eV	a	b	c	d
108.5–400.0	−6.751 02	858.0453	−7421.56	22 493.22
603.0–2042.4	−12.9042	25 292.21	−776 713	10 465 814
2042.4–10 000	−12.083	27 903.97	−1 631 852	81 315 047

[a]$\mathrm{d}f/\mathrm{d}E$ in Ry units, $y = B/E$, $B = \mathrm{IP} = 12.8900\,\mathrm{eV}$.

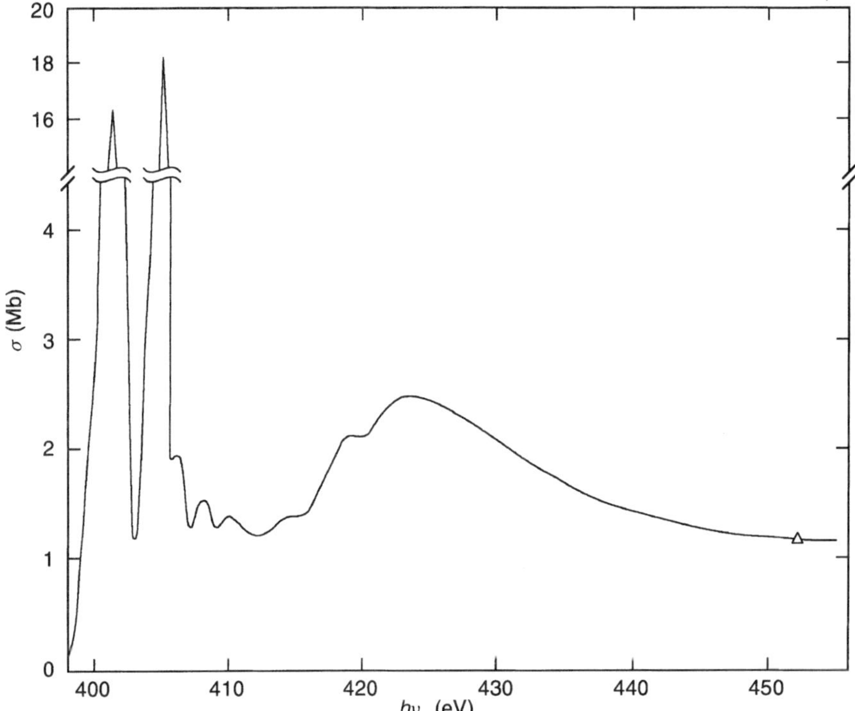

Fig. 5.7 Absolute photoabsorption spectrum of N_2O – nitrogen K-edge. Relative intensities from Bianconi *et al.* (1978) and Wight and Brion (1974b) normalization; ($\triangle$) from Henke *et al.* (1993)

Barrus *et al.* (1979) have measured photoabsorption cross sections in the oxygen K-edge region, which we discuss below. Here, we note that at their lowest energy (506 eV), they obtain $\sigma = 0.925\,\mathrm{Mb}$, while our calibration of Fig. 5.7 gives $\sigma = 1.175\,\mathrm{Mb}$ at 455 eV. We negotiate the transition region with the linear interpolation

$$\sigma = -0.004\,90E + 3.4054$$

with σ in Mb, E in eV, and record the integrated $S(p)$ in Table 5.7.

h   The oxygen K-edge region

Absolute photoabsorption cross sections are given by Barrus *et al.* from 506–603 eV. At their extremities, they agree fairly well with the summed atomic cross sections of Henke *et al.* (see Fig. 5.8). However, they are poor in the region of intense structure, 533–543 eV. We supplement their data by normalizing the inelastic electron scattering relative intensities of Wight and Brion to the Barrus cross sections, as shown in Fig. 5.8. After graphical integration, the contributions to $S(p)$ of the Barrus *et al.* (1979) data and the normalized Wight/Brion data are shown separately in Table 5.7.

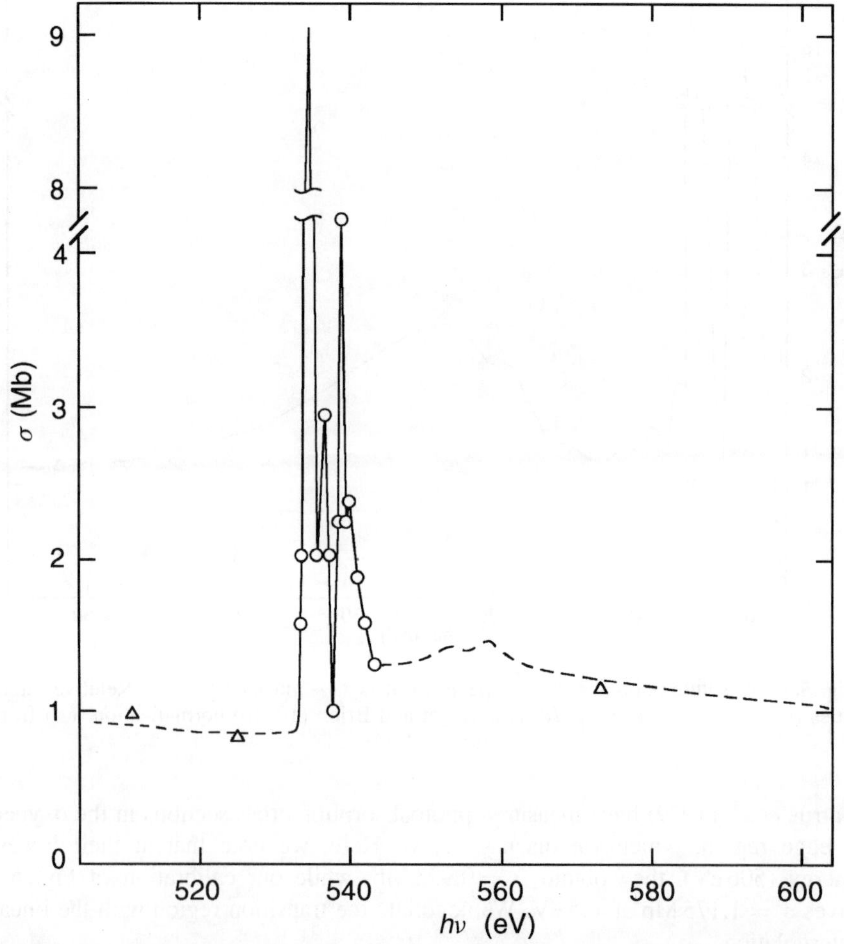

Fig. 5.8   Absolute photoabsorption spectrum of N_2O – oxygen K-edge. Relative intensities from Barrus *et al.* (1979) ——, Wight and Brion (1974b) (•); normalization (△) from Henke *et al.* (1993)

i Post K-edges, 603–10 000 eV

The summed atomic cross sections of Henke *et al.* are fitted to two 4-term polynomials, one between 603–2042.4 eV, the other from 2 042.4–10 000 eV. The coefficients of the polynomials are given in Table 5.8, the corresponding $S(p)$ obtained by analytic integration in Table 5.7. This domain contributes 62% to $S(+1)$ and 48% to $S(+2)$, but is almost negligible for the negative sum rules.

j Post K-edges, 10^4–10^5 eV

The calculated atomic cross sections of nitrogen and oxygen (Chantler, 1995) are stoichiometrically summed, and used to evaluate $S(p)$. This region contributes $\sim$29% to $S(+2)$, $\sim$3% to $S(+1)$, and negligibly to the other $S(p)$.

5.3.3 The analysis

Kirouac and Bose (1973) measured the dielectric constant of N_2O at various temperatures and pressures, and from the Debye equation deduced an electric dipole moment of, $\mu = 0.18$ D and a polarizability, $\alpha = 3.03 \times 10^{-24}$ cm^3. Since in this case the dipole moment contributes the minor component to the molar polarization, it is possible to improve the derived polarizability by introducing a more accurate value of μ. Thus, with a molar polarization $P = 8314(2)$ cm^3/mol at 30.1°C (Kirouac and Bose, 1973) and $\mu = 0.160\,880\,(23)$ D from Reinartz *et al.* (1978), we infer a molar refractivity $R = 7.794(2)$ cm^3/mol, equivalent to $\alpha = 3.090(1) \times 10^{-24}$ cm^3, or $S(-2) = 5.213(2)$ Ry units. Nitrous oxide has a non-negligible infrared contribution. Using infrared intensities from Kagann (1982), we obtain $S(-2)_{ir} = 0.28$. Bishop and Cheung (1982) had previously calculated $S(-2)_{ir} = 0.27$, based on earlier data. Consequently, the ultraviolet, or electronic component, $S(-2)_{uv} = 4.93(1)$ Ry units.

Zeiss *et al.* (1977b) have obtained $S(-2) = 4.925$ Ry units from a Cauchy analysis of archaic, but presumably accurate refractive index measurements at visible and ultraviolet wavelengths. Höhm (1994) has reported more modern measurements which tend to increase this value slightly, to $\sim$4.927 Ry units. Thus, excellent agreement is found for $S(-2)$ from both dielectric constant and refractive index data.

Our spectral sum for $S(-2)$, 4.837 Ry units, is lower by 1.8% while our $S(0)$ value exceeds that required by the TRK sum rule by $\sim$1%. For comparison, Shaw *et al.* obtain almost exact agreement with expectation for $S(0)$, but their value of $S(-2)$ falls shy by $\sim$3.4%. Some of the differences between the results of Shaw *et al.* and the present ones may reflect our reading of their graphical data. Zeiss *et al.* used the expectation values of $S(-2)$ and $S(0)$ in their constraint procedure. They did not present a detailed distribution of oscillator strengths for comparison with the currently used data. The experimental measurements available to them did not include those of Lee and Suto, Shaw *et al.*, and Chan *et al.* Their value of $S(-1)$ is 1.8% higher than the present one, but a precise expectation value is unavailable.

Hartree–Fock calculations of electron densities for atomic nitrogen and oxygen can be summed to provide plausible estimates for $S(+2)$. Thus, from Fraga et al. (1976) we obtain $S(+2) = 90\,089.5$, or alternatively from Bunge et al. (1993), $90\,089.1$. Our spectral sum for $S(+2)$ is 2.0% higher, while the value of Zeiss et al. is marginally closer to the 'expectation' value, but still 1.7% higher.

Atomic additivity would imply $S(+1) = 457.1$ (Fraga et al., 1976). Correlation is expected to increase this value. Zeiss et al. obtain $S(+1) = 462.6\,\text{Ry}$ units, while the current spectral distribution yields 483.0. Since Zeiss et al. use the correct value of $S(0)$ as one constraint, and their $S(+2)$ is slightly closer to expectation than the present calculation, while our numbers are both higher, their value of $S(+1)$ may be preferable, but the expectation value probably lies between them.

The quantum yield of ionization appears to be unity above 20.19 eV, according to Shaw et al. Between IP–20.19 eV, they present measurements of this quantity, which can be combined with their photoabsorption cross sections to infer photoionization cross sections. With this information, we compute the ionized component of $S(-1)$, i.e. $S_i(-1) = 5.541$. Shaw et al. obtain $S_i(-1) = 5.41$, using their data and choices for higher energy. This quantity is equivalent to the dipole matrix element for total ionization, but a direct measurement is unavailable for N_2O. In the autoionization region IP–20.19 eV, about 77.5% of photoabsorption leads to photoionization, a figure comparable to that found for other first row triatomic molecules, but higher than that found for diatomic molecules.

5.4 Nitrogen Dioxide (NO₂)

5.4.1 Preamble

Nitrogen dioxide (O–N–O structure) is bent in its neutral ground state, at an angle of 133.85° (Morino and Tanimoto, 1984). In the independent particle approximation, its orbital configuration in C_{2v} symmetry may be written as

$$(1a_1)^2(1b_2)^2(2a_1)^2(3a_1)^2(2b_2)^2(4a_1)^2(3b_2)^2(1b_1)^2(5a_1)^2(1a_2)^2(4b_2)^2(6a_1)^1 \; {}^2\tilde{A}_1$$

(see, for example, Baltzer et al. (1998); Edqvist et al. (1970)).

The $1a_1$ and $1b_2$ orbitals are essentially oxygen 1s orbitals, and the $2a_1$ is nitrogen 1s. The $3a_1$, $2b_2$ and $4a_1$ orbitals may be expected to be various combinations of oxygen 2s and nitrogen 2s orbitals, but ionization from $3a_1$ and $2b_2$ gives rise to a multiplicity of states, indicating departure from the independent particle model, as is frequently encountered for inner valence orbitals (Baltzer et al., 1998). This splitting is less evident for $(4a_1)^{-1}$. The outermost orbital $(6a_1)$ is singly occupied, and is responsible for the bent neutral structure. Emission of this electron leaves 16 valence electrons, isoelectronic with CO_2, and indeed the ground state of $NO_2{}^+$ is linear. Single-electron emission from all other occupied orbitals generates both singlet and triplet states, contributing to a complex photo-electron spectrum. The lowest-lying unoccupied orbitals are $2b_1$, $7a_1$ and $5b_2$. Valence and core transitions are observed to occur to these orbitals, but also to the half-filled $6a_1$.

Photoabsorption measurements have been stimulated in recent years by the awareness that NO_2 plays an important role in the coupling of NO_x and ClO_x reactions which control the amount of ozone in the stratosphere. Most recent investigations have concentrated on the visible and near ultraviolet. Vandaele *et al*. (1998) have recently reported high resolution measurements up to 5.2 eV, and compared with earlier work. Their estimated error is <3% for most of this range. Higher-energy absolute photoabsorption measurements were carried out earlier by Nakayama *et al*. (1959) between 4.6–11.5 eV, and by Morioka *et al*. (1978) between 11.3–24.8 eV. In all of these measurements of the Beer–Lambert type, in addition to the usual concerns about impurities, care had to be exercised to minimize or take into account the equilibrium involving the dimer, N_2O_4. This was also a problem in measurements of the dielectric constant and refractive index.

This is less problematic in inelastic electron scattering (e,e) measurements, where the molecular beams are at relatively low pressure. Au and Brion (1997) have recently provided such pseudo-photoabsorption data covering the range 2–200 eV, which not only overlaps true photoabsorption measurements, but extends them to significantly higher energy. Earlier, Zhang *et al*. (1990) obtained (e,e) data in the vicinity of the nitrogen and oxygen K-edges. Sum rule analyses appear to be scarce. Au and Brion proffered a truncated analysis of $S(-2)$.

5.4.2 The data

Transitions between the bent neutral ground state of NO_2 and the linear ionic ground state might be expected to generate a vibrational progression in the bending mode, and for many years this was thought to be the case (Brundle *et al*., 1970). Recently, Baltzer *et al*. obtained a higher-resolution He I photoelectron spectrum, and showed that the vibrational structure surmounting the broad first peak had irregular spacings, typically $\sim$1/3 of the bending frequency. The photo-electron intensity dwindles to the base line, implying very low Franck–Condon factors, at $\sim$10.2 eV. An early photoionization measurement by Nakayama *et al*. gave AIP = 9.76 eV; at this energy the quantum yield of ionization was $\sim$2 × 10^{-4}, and dropping rapidly. Subsequently, Killgoar *et al*. (1973) used photoion-ization mass spectrometry and analysis to conclude that AIP (NO_2) $\leq$ 9.62 eV. More recently, Bryant *et al*. (1994) circumvented the Franck–Condon problem by using triple-resonant ZEKE, achieved rotational resolution, and arrived at AIP (NO_2) = 77 315.9 $\pm$1.0 cm^{-1}, or 9.585 95(2) eV.

a   The discrete spectrum, and transitions below the IP

Vandaele *et al*. have presented an absolute photoabsorption spectrum between 1.24–5.20 eV, at a resolution of 0.000 25 eV, with a stated accuracy of <3%,

Table 5.9 Spectral sums and comparison with expectation values for NO_2[a]

Energy, eV	$S(-2)$	$S(-1)$	$S(0)$	$S(+1)$	$S(+2)$
1.24–4.86[b]	0.1496	0.0333	0.0076 (0.0095)[c]	0.0018	0.0004
4.80–6.30[d]	0.0260	0.0109	0.0046 (0.0068)[c]	0.0020	0.0008
6.30–7.85[d]	0.4603	0.2481	0.1339 (0.1287)[c]	0.0724	0.0393
7.85–8.25[d]	0.1491	0.0881	0.0521 (0.0502)[c]	0.0308	0.0182
8.25–9.38[d]	0.2853	0.1835	0.1181 (0.1228)[c]	0.0762	0.0492
9.38–AIP[d]	0.0530	0.0371	0.0259 (0.0259)[c]	0.0181	0.0127
AIP–11.24[d]	0.3910	0.2920	0.2186 (0.2205)[c]	0.1641	0.1234
11.24–22.00[e]	1.3539 (1.9143)[c]	1.6258 (2.3124)[c]	2.0210 (2.8926)[c]	2.5937 (3.7348)[c]	3.4246 (4.9582)[c]
22.0–90.0[c]	1.4144	3.5684	10.2753	34.3780	133.0634
90.0–200.0[c]	0.0299 (0.0313)[f]	0.2569 (0.2706)[f]	2.3090 (2.4462)[f]	21.8237 (23.2742)[f]	217.5248 (233.296)[f]
200.0–437.5[f]	0.0021	0.0395	0.7751	15.9481	344.91
400–437.5[g]	0.0005	0.0165	0.5011	15.2726	465.715
437.5–566.3[f]	0.0005	0.0171	0.6180	22.4681	821.127
528.0–566.3[g]	0.0004	0.0146	0.5863	23.4916	941.627
566.3–2293.2[f]	0.0011	0.0644	3.9888	279.60	22 792.2
2293.2–10 000[f]	–	0.0014	0.3312	90.75	29 104.7
10^4–10^5	–	–	0.0148	18.76	32 092.3
10^5–∞	–	–	–	0.997	23 222.5
Total[h]	4.3171	6.4976	21.9814	526.448	110 139.3
Total[i]	4.8775	7.1842	22.8530	527.589	110 140.9
Total[j]	4.8789	7.1979	22.9902	529.040	110 156.6
Expectation values	4.86 ±0.07[k]	–	23.0	–	107 710.2[l]
Other values	4.914[c]	–	–	–	107 707.7[m]

[a] In Ry units.
[b] Vandaele *et al.* (1998).
[c] Au and Brion (1997).
[d] Nakayama *et al.* (1959).
[e] Morioka *et al.* (1978).
[f] Summed atomic cross sections from Henke *et al.* (1993).
[g] From Zhang *et al.* (1990), but with Bethe–Born correction. See text.
[h] Using data of ref. e, 11.24–22.00 eV.
[i] Using data of ref. c, 11.24–22.00 eV.
[j] Using data of ref. f, rather than ref. c, 90.0–200.0 eV.
[k] See Sect. 5.4.3, text.
[l] Atomic additivity, using Fraga *et al.* (1976).
[m] Atomic additivity, using Bunge *et al.* (1993).

and shown to be in good agreement with earlier data by Mérienne *et al.* (1995), Jenouvrier *et al.* (1996), Yoshino *et al.* (1997) and Bass *et al.* (1976). Their spectrum has been scanned, digitized and trapezoidally integrated to evaluate the $S(p)$ given in Table 5.9. Also shown for comparison is the integrated oscillator strength between 1.58–4.80 eV obtained by Au and Brion using inelastic electron scattering with a resolution of 0.05 eV. The latter is ∼25% larger, but the absolute values are very small. Between 4.80 eV and the AIP, the photoabsorption spectra of Nakayama *et al.* have been scanned, digitized and trapezoidally integrated in segments corresponding to the partitioning used by Au and Brion. Although the contributions within segments vary somewhat between these two groups (see Table 5.9), their integrated oscillator strengths from 4.80 eV to the AIP are almost identical (0.3346, Nakayama *et al*; 0.3344, Au and Brion). The major discrepancy occurs between ∼5.6–6.2 eV, where the cross sections differ by as much as a factor 2 (see Fig. 5.9). Au and Brion suggest that this indicates 'the presence of an additional transition at ≈6.1 eV not observed in the optical study'. Also shown in Fig. 5.9 are data from Bass *et al.* which display even lower cross sections than those of Nakayama *et al.* in the disputed region. It is possible that a non-dipole transition contributes to the (e,e) data of Au and Brion in this region.

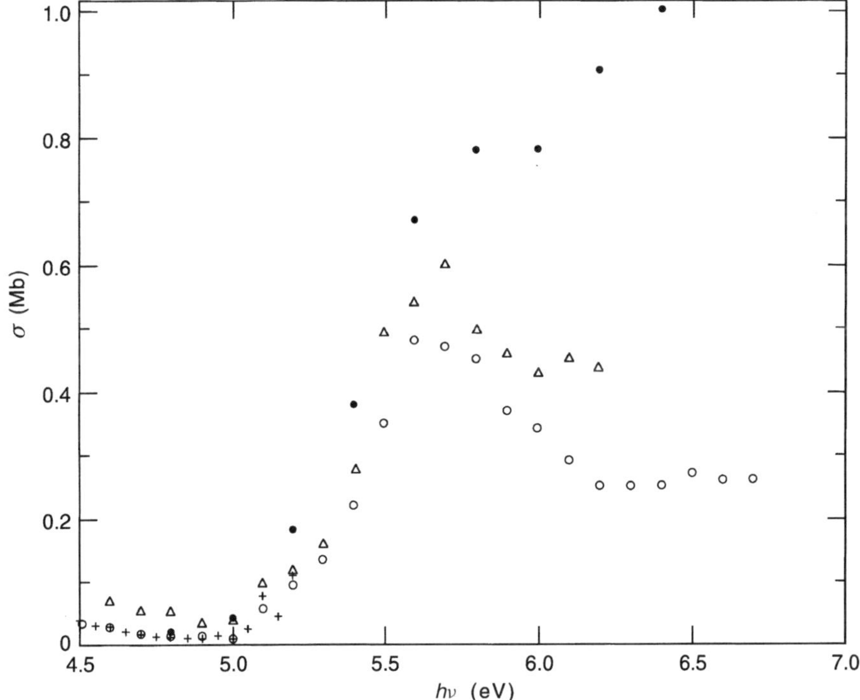

Fig. 5.9 Absolute photoabsorption spectrum of NO_2, 4.5–6.5 eV. • Au and Brion (1997); ○ Bass *et al.* (1976); △ Nakayama *et al.* (1959); + Vandaele *et al.* (1998)

Although locally significant, the effect on the integrated oscillator strength is small.

b The continuum

b.1 IP–11.24 eV In this interval, the photoabsorption measurements of Nakayama *et al.* continue to provide integrated oscillator strengths within 1% of those given in the (e,e) data of Au and Brion, as seen in Table 5.9, although there are local differences of larger magnitude.

b.2 11.24–22.00 eV In contrast to the generally good agreement between photoabsorption cross sections and those based on inelastic scattering below 11.24 eV, the discrepancy between 11.24–22.00 eV is systematically larger, between 25.6–36.6%. Here, the comparison is with the photoabsorption measurements of Morioka *et al.* These latter authors acquired their data on photographic film, which had to be converted to absorption coefficients. They acknowledged an experimental error of about 15%, but it appears that their systematic error was approximately double that value. In Table 5.9, we record the $S(p)$ derived from their work, and the corresponding quantities obtained from Au and Brion. Au and Brion provide increments in $S(0)$; we have scaled the ratios of ($S(0)$, Au-Brion : $S(0)$, Morioka) within the small energy increments to arrive at the other $S(p)$, and record the overall results between 11.24–22.00 eV in Table 5.9.

b.3 22.0–90.0 eV This is a critical region for sum rule analysis. It will be shown to contribute 29%, 50% and 45% to $S(-2)$, $S(-1)$ and $S(0)$, respectively. We are dependent on the (e,e) data of Au and Brion for absorption cross sections in this interval. The summed atomic cross sections of Henke *et al.* (1993) are in excellent agreement with the Au/Brion data at 90 eV, but dip below the (e,e) values at lower energy, where molecular effects can be expected.

b.4 90.0–200.0 eV In this interval, atomic additivity is expected to provide a good approximation to the molecular cross sections. The (e,e) data of Au and Brion are actually ∼6% lower than the summed atomic cross sections of Henke *et al.*, as indicated in Table 5.9. In absolute values of $S(p)$, the differences are small.

b.5 200.0–437.5 eV The nitrogen K-edge in NO_2 occurs at 412.6 eV (Jolly *et al.*, 1984). Pre-edge structure begins at ∼400 eV. Zhang *et al.* present their pseudo-photoabsorption spectrum with both background and continuum subtracted, such that the cross section is zero for $E < 400$ eV. To compensate for this missing continuum, we fit the summed atomic cross sections of Henke *et al.* between 192.6–392.4 eV by regression to a 4-term polynomial, and utilize this function to 437.5 eV, which is approximately 25 eV above the K-edge, where structure is assumed to be insignificant. This function is analytically integrated

Table 5.10   Coefficients of the polynomial $\mathrm{d}f/\mathrm{d}E = ay^2 + by^3 + cy^4 + dy^5$ fitted to data at various energies[a]

Energy range, eV	a	b	c	d
200–437.5	-18.1186	3 745.083	$-84\,005.1$	758 872.2
566.3–2293.2	$-6.401\,15$	67 124.61	$-2\,609\,476$	42 470 841
2293.2–10 000	-10.4969	63 942.04	221 907.1	402 718 826

[a] $\mathrm{d}f/\mathrm{d}E$ in Ry units, $y = B/E$, $B = 9.585\,95\,\mathrm{eV}$.

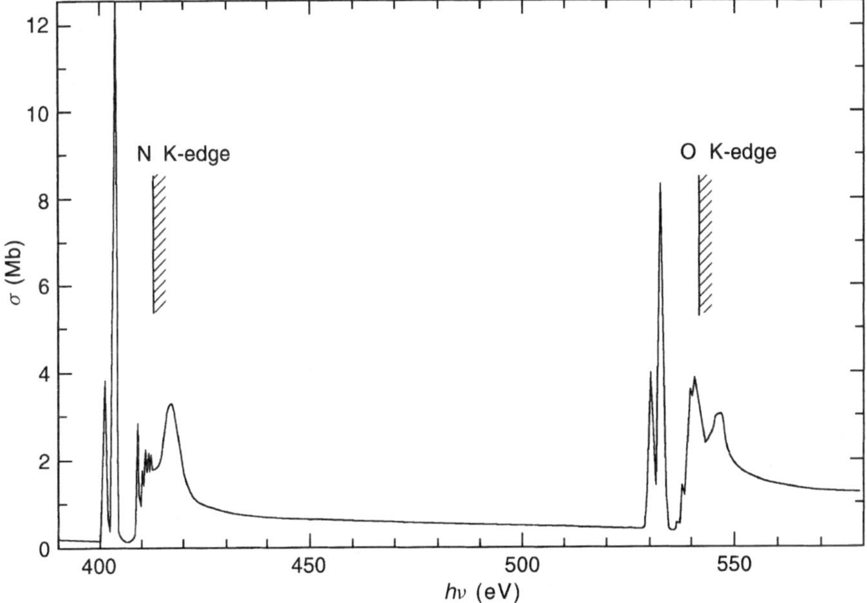

Fig. 5.10   Absolute photoabsorption spectrum of NO_2, K-edge regions. Based on Zhang *et al.* (1990), with corrections described in text

to evaluate the contributions to $S(p)$ shown in Table 5.9. The coefficients of the polynomial are given in Table 5.10.

b.6   400.0–437.5 eV   Zhang *et al.* present this structured region in their Figs. 2 and 3 in absolute units, but explicitly note that the spectrum is not Bethe–Born corrected. We have digitized these spectra, and subsequently performed the Bethe–Born correction. The $S(p)$ resulting from trapezoidal integration are listed in Table 5.9. In Fig. 5.10, this spectrum is shown after Bethe–Born correction, and including the underlying continuum. At 437.5 eV, we obtain $\sigma = 0.659\,\mathrm{Mb}$, in excellent agreement with the summed atomic cross sections of Henke *et al.* between nitrogen and oxygen K-edges extrapolated to 437.5 eV (see below), which yields $\sigma(437.5\,\mathrm{eV}) = 0.644\,\mathrm{Mb}$.

b.7   437.5–566.3 eV   Henke *et al.* provide only three data points between the nitrogen and oxygen K-edges. We approximate this region by the linear expression

$$\sigma(\text{Mb}) = -0.0028E(\text{eV}) + 1.932.$$

The oxygen K-edge occurs at 541.3 eV (Jolly *et al.*, 1984). Once again, we assume that summed atomic cross sections can accurately represent the molecular cross section 25 eV above the edge. To compensate for the missing continuum in the structured oxygen K-edge region given in Figs. 6 and 7 of Zhang *et al.*, we integrate the linear expression across the indicated interval to determine the $S(p)$.

b.8   528.0–566.3 eV   The structured, pre-K-edge region begins at $\sim$528 eV, and we terminate it at 25 eV above that edge. As before, the figures given by Zhang *et al.* covering this region are not Bethe–Born corrected. The figures have been digitized, Bethe–Born corrected, and trapezoidally integrated. Figure 5.10 displays this spectrum after Bethe–Born correction, and including the underlying continuum. At 566.3 eV, we obtain $\sigma = 1.27$ Mb, whereas extrapolation of the post-edge continuum (see below) yields $\sigma(566.3\,\text{eV}) = 1.36$ Mb.

b.9   566.3–10 000 eV   The experimentally based atomic cross sections of Henke *et al.* are stoichiometrically summed and fitted by regression to one 4-term polynomial between 566.3–2293.2 eV, and another from 2293.2–10 000 eV. In Table 5.9, we see that this region contributes the 4 oxygen K-shell electrons, 70% of $S(+1)$ and 47% of $S(+2)$. The coefficients of these polynomials can be found in Table 5.10.

5.4.3   The analysis

Au and Brion cite three independent dielectric constant measurements (Zahn, 1933, Williams *et al.*, 1936; Schulz, 1938) as reporting strikingly similar electric dipole polarizabilities, $3.025 \times 10^{-24}\text{cm}^3$, $3.021 \times 10^{-24}\text{cm}^3$ and $3.021 \times 10^{-24}\text{cm}^3$, respectively. A careful reading of these papers reveals that none of them measured this quantity. In dielectric constant measurements, investigators make use of the Debye equation

$$P = \frac{\varepsilon - 1}{\varepsilon + 2} \cdot \frac{M}{\rho} = \frac{4\pi N_{\text{A}}}{3}(\alpha_0 + \mu^2/3kT),$$

where P is molar polarization, ε is the dielectric constant, M is the molecular weight, ρ is the density, μ is the electric dipole moment, T is the temperature, and N_{A}, k are Avogadro's number and the Boltzmann constant, respectively. The polarizability α_0 is largely due to electronic motion (α_{vis}) but also has a small contribution due to vibrational excitation (α_{ir}). The three groups cited above were primarily interested in extracting a value of μ from their measurements. From a plot of $A + B/T$ versus $1/T$, they could have extracted α_0 as intercept and B (hence μ^2) from the slope. However, they chose to impose α_0 from refractive index measurements, performed earlier by Cuthbertson and Cuthbertson (1913b). Such a measurement gives α_{vis}, in fact α (6438 Å), since that was the only wavelength used for careful measurement. That value is the one which Au and Brion attributed to the three dielectric constant measurements. Perhaps the most extensive of the three studies was performed by Schulz (1938). He estimated α_{ir} as 6% of α_{vis}, by analogy with N_2O. Subsequent absolute intensity measurements of the vibrational transitions in NO_2 have essentially validated this assumption. The most significant contributor is ν_3, for which Malathy Devi $et\ al.$ (1982) give an integrated intensity of 1419 ± 58 cm^{-2} atm^{-1}, compared to 1430 ± 300 cm^{-2} atm^{-1} obtained earlier by Goldman $et\ al.$ (1975). A much smaller intensity for ν_2 is given by Malathy Devi $et\ al.$ (1981), and a still smaller value for ν_1 by Perrin $et\ al.$ (1984). From these sources, we calculate $\alpha_{ir} = 0.186 \times 10^{-24}$ cm^3, which is about 6.2% of α_{vis}.

Schulz subtracted the contributions of $\alpha_{vis} + \alpha_{ir}$ from his measured P, then used the residual to evaluate $\mu = 0.286$ D. The currently accepted value, $\mu = 0.316$ D, is based on microwave measurements by Hodgeson $et\ al.$ (1963). Schulz also showed that if he ignored α_{ir}, he would have obtained $\mu = 0.320$ D, almost identical to that of Hodgeson $et\ al.$ A reduction of α_{vis} by 6%, and inclusion of α_{ir}, would achieve the same result. It will be recalled that α_{vis} used by Schulz and others was really $\alpha(6438$ Å$)$, but the desired value of α_{vis} (the static polarizability) is at $\lambda = \infty$. This analysis suggests that $\alpha(\lambda = \infty)$ should be $\sim$6% lower than $\alpha(6438$ Å$)$, or $\sim 2.84 \times 10^{-24}$ cm^3.

More recently, Goebel $et\ al.$ (1994; 1995) have employed dispersive Fourier transform spectroscopy to study the refractive index of NO_2. Their results are quite close to the value of Cuthbertson and Cuthbertson. At a slightly shorter wavelength, $\lambda = 6329.9$ Å, they explicitly give $\alpha = 3.0233(37) \times 10^{-24}$ cm^3. They do not report $\alpha(\lambda = \infty)$, but their polarizability curve decreases with increasing λ, as expected from the Cauchy relation, to $\sim 2.92 \times 10^{-24}$ cm^3 at $\sim$9090 Å, already a reduction of 3.4%. Here, we adopt a mean of the polarizability from the dielectric constant measurements and the upper limit from the modern refractive index data, or $\alpha_{vis}(\lambda = \infty) = 2.88 \pm 0.04 \times 10^{-24}$ cm^3, which implies $S(-2) = 4.86 \pm 0.07$ Ry units.

In Table 5.9, two alternative spectral sums are shown for $S(-2)$, 4.317 Ry units using Morioka $et\ al.$ between 11.24–22.00 eV, and 4.877 Ry units, based on the values of Au and Brion in this interval. Clearly, the latter is in much better agreement with the expectation value. This observation confirms the earlier surmise that the data of Morioka $et\ al.$ are plagued with a systematic error which diminishes their cross sections by $\sim$30%. The spectral sum for $S(0)$ is 22.85, compared to the TRK expectation value of 23. If we substitute values based on atomic additivity for the Au/Brion cross sections in the 90.0–200.0 eV region, $S(0)$ improves to 22.99, while $S(-2)$ increases only slightly to 4.879 Ry units, but this refinement is unwarranted, in view of the experimental uncertainties. The value of $S(-1)$ is inferred to be 7.18–7.20 Ry units.

The spectral sum for $S(+2)$, $\sim$110 150 Ry units, is approximately 2.3% larger than the calculated sum of Hartree–Fock charge densities at the nuclei (Fraga *et al.*, 1976; Bunge *et al.*, 1993). The spectral sum for $S(+1)$ is $\sim$528 Ry units, larger than atomic additivity applied to the Fraga calculations, 504.4 Ry, as expected.

To compute $S_i(-1)$, it is necessary to know the absolute photoionization cross section (or alternatively the quantum yield of ionization, η_i) from the AIP to an energy where $\eta_i = 1.0$. Nakayama *et al.* provide this information between 9.72–11.48 eV, where η_i varies from 4×10^{-5} to 0.28. Au and Brion tabulate η_i between 15.0–24.5 eV, i.e., η_i between 0.48 and 1.00. They also provide information for estimating η_i from 10.5–15.0 eV. These data yield $\Delta S_i(-1)$, IP -24.5 eV $=$ 1.8738 Ry. Together with $\Delta S(-1)$, 24.5 eV $\rightarrow \infty = 3.4559$ Ry units, we arrive at $S_i(-1) \cong 5.33$ Ry. We are unaware of a direct experimental value.

Between AIP–11.02 eV, only 0.4% of photoabsorption leads to photoionization. With increasing energy, the quantum yield sharply increases, such that the integrated fractional ionization is 59% between AIP–22.0 eV, and 67% between AIP–24.5 eV.

5.5 Hydrogen Sulfide (H$_2$S)

5.5.1 Preamble

The H$_2$S molecule, with C_{2v} symmetry, has an electronic ground state configuration which can be described in the independent particle approximation by the following orbital sequence:

$$(1a_1)^2(2a_1)^2(1b_2)^2(3a_1)^2(1b_1)^2(4a_1)^2(2b_2)^2(5a_1)^2(2b_1)^2, \tilde{X}^1 A_1$$

The lowest energy unoccupied orbitals, $6a_1$ and $3b_2$, play a role in both the valence and core absorption spectra. They are antibonding conjugate partners of the $5a_1$(S(3p) + H(1s) hydrogen angular bonding) and $2b_2$(S(3p) + H(1s) hydrogen σ bonding) orbitals. The uppermost occupied orbital, $2b_1$, is essentially a S(3p) lone-pair, while the $1b_1$, $3a_1$ and $1b_2$ orbitals represent S(2p) in C_{2v} symmetry, the $2a_1$ is predominantly S(2s) and the $1a_1$ orbital, S(1s). The inner-valence $4a_1$ orbital is largely S(3s) + H(1s) bonding in character. The photoelectron spectrum corresponding to ionization from this orbital is dispersed among several peaks, manifesting a breakdown of the single particle approximation often encountered in the inner valence region. These conclusions are based on early self-consistent field calculations by Boer and Lipscomb (1969) and later, more extensive ones by Diercksen and Langhoff (1987).

A sum rule analysis was performed on H$_2$S by Pazur *et al.* (1988). These authors utilized early photoabsorption measurements by Clark and Simpson (1965) and Watanabe and Jursa (1964), together with inelastic electron scattering (e,e) measurements by Brion *et al.* (1986). For photon energies in the vicinity of the S(2p), S(2s) and S(1s) excitation energies, Pazur *et al.* estimated photoabsorption cross sections by 'mixture rules'. They optimized the initially available cross sections with the constraints that $S(-2)$ and $S(0)$ should match expectation values. Shortly before their paper was submitted, an extensive set of photoabsorption measurements, covering the range 490–2400 Å (5.17–25.30 eV) was reported by Lee *et al.* (1987). Later, Xia *et al.* (1991) presented data over a more restricted range, $\sim$IP–11.7 eV, where IP $\cong$ 10.45 eV. Recently, Wu and Chen (1998) examined temperature effects on the broad (1600–2400 Å) band, but there was essential agreement with the data of Lee *et al.* at room temperature.

Absolute photoabsorption cross sections in the S(2p) region, 164–171 eV, were obtained by Hayes and Brown (1972). More recently, Reynaud *et al.* (1996) reported the absolute cross sections in the S(1s) region.

As the current analysis was in progress, the Vancouver group (Feng *et al.*, 1999a) presented improved data, with 0.05 eV resolution (FWHM) to 30 eV, and with lower resolution extending to 260 eV.

With these additional data available, a sum rule analysis was undertaken to examine how well the experimental spectral distribution (without adjustment) conformed to the expectation values of the sum rules.

5.5.2 The data

Baltzer *et al.* (1995) have obtained 10.4666 eV for the adiabatic IP of H_2S from He I PES at the rotational resolution level. This result is in excellent agreement with a Rydberg series extrapolation, 10.466 $\pm$0.001 eV, given by Masuko *et al.* (1979). Photoionization threshold values are somewhat lower, while a ZEKE measurement by Wiedmann and White (1992) yielded a higher value, 10.4682 $\pm$0.0002 eV. Here we select the Baltzer value, 10.4666 eV.

a The discrete region, to 10.4666 eV (IP)

The first significant absorption feature in H_2S is a very broad band ($\sim$5–7.75 eV) with some superimposed, diffuse vibrational structure. It has been attributed to $(2b_1)^{-1} \rightarrow 6a_1$ excitation (Masuko *et al.*, 1979; Wu and Chen, 1998). Resolution is not a factor in the various photoabsorption measurements, which are in good agreement on the magnitude of the cross sections. Pazur *et al.* used the data of Watanabe and Jursa and Clark and Simpson, which did not span the entire band. Lee *et al.* and Wu and Chen display the full band, and their cross sections agree to within their stated errors of $\pm$10%. We choose the data of Lee *et al.* for continuity and consistency, since their measurements extend to higher energies.

In the higher energy VUV, the more recent measurements of Xia *et al.*, using more modern technology, claim the highest accuracy (<5% error). Unfortunately, their data are limited to the region 10.45–11.69 eV. In this domain, their cross sections are in good agreement ($\leq$10%) with those of Watanabe and Jursa and Lee *et al.*, tending to be lower than the former and higher than the latter. This comparison provides some support for the use of either data set between 7.8–10.45 eV. However, Watanabe and Jursa caution that they observed considerable pressure dependence for all strong bands in this region, increasing their error bars and (perhaps coincidentally) omitting an absolute scale in their figure. Lee *et al.* used higher resolution (0.4 Å FWHM) and low gas pressures, claiming to avoid satura-

Table 5.11 Spectral sums and comparison with expectation values for H_2S[a]

Energy, eV	$S(-2)$	$S(-1)$	$S(0)$	$S(+1)$	$S(+2)$
5.22–7.80[b]	0.2554	0.1185	0.0552	0.0258	0.0121
7.80–10.446[b]	1.5936	1.0758	0.7300	0.4979	0.3413
10.446–10.4666[c](IP)	0.0103	0.0079	0.0061	0.0047	0.0036
IP–11.69[c]	0.6957	0.5668	0.4622	0.3773	0.3083
11.69–25.48[b]	3.4981	4.1004	5.0052	6.3748	8.4727
11.7–25.5[d]	3.4464	4.0262	4.8930	6.1981	8.1857
25.5–163.5[d]	0.1839	0.4893	1.6068	7.1830	43.9766
163.5–171.1[e]	0.0010	0.0119	0.1464	1.8078	22.3288
163.5–171.2[d]	0.0005	0.0065	0.0800	0.9891	12.2263
171.1–220.1[f]	0.0077	0.1084	1.5326	21.7902	311.45
171.2–220.2[d]	0.0059	0.0844	1.2117	17.4763	253.30
220.1–929.7[g]	0.0096	0.2257	5.8927	177.1546	6 232.43
929.7–2470[g]	0.0001	0.0070	0.6646	67.4550	7 393.82
2470–2510[h]	–	0.0003	0.0477	8.7316	1 597.49
2510–10 000[g]	–	0.0052	1.4111	431.7572	152 973.0
10^4–10^5	–	0.0001	0.1162	153.120	279 879.4
10^5–∞	–	–	0.0009	11.161	276 893.7
Total[i]	6.2554	6.7173	17.6777	887.4409	725 356.7
Total[j]	6.2014	6.6137	17.1782	882.1316	725 288.2
Expectation value	6.193(4)[k]	–	18.0	–	715 627.4[l]
					715 644.7[m]
Other values	(6.178)[n]	6.565[n]	(18.0)[n]	922.8[n]	716 400.0[n]
	5.936[o]	6.249[o]	–	–	–

[a]In Ry units.
[b]Lee *et al.* (1987).
[c]Xia *et al.* (1991).
[d]Feng *et al.* (1992a).
[e]Hayes and Brown (1972).
[f]Relative spectrum of Hudson *et al.* (1994) normalized to ref. d at 228.3 eV peak. See text.
[g]Summed atomic cross sections from Henke *et al.* (1993).
[h]Reynaud *et al.* (1996).
[i] Using photoabsorption data where possible.
[j] Using (e,e) data between 11.7–220.2 eV.
[k]Russell and Spackman (1997).
[l]Fraga *et al.* (1976).
[m]Bunge *et al.* (1993).
[n]Pazur *et al.* (1988).
[o]From ref. d, which extends to 260 eV. When supplemented by data in Table 5.11 above 260 eV, $S(-2) = 5.943$ and $S(-1) = 6.434$.

tion. Consequently, their Fig. 3 has been scanned and digitized to provide absorption cross sections between 7.80–10.446 eV. Trapezoidal integration with a fine mesh yielded the $S(p)$ given in Table 5.11. The gap between 10.446–10.4666 eV (IP) is filled using the ostensibly more accurate data from Fig. 7 of Xia *et al.* Feng *et al.* display excellent agreement with Lee *et al.* below the IP.

b   The continuum

b.1   IP–25.3 eV   This energy range is chosen because it reflects the upper limit of measurements reported by Lee *et al.* We adopt their data in this domain, with two caveats.

(a) Careful measurement of their Figs. 1 and 2 indicates an overlap in wavelength between 1060–1065 Å, but the spectral features do not match. Presumably no overlap was intended, and Fig. 1 should terminate at 1060 Å. This ambiguity introduces an uncertainty of 1% in $\Delta S(0)$, but less in $\Delta S(-2)$.

(b) Alternative data sets in this region are available, including photoabsorption measurements by Ibuki *et al.* (1985) and (e,e) experiments by Feng *et al.* In the relatively smooth continuum between $\sim$15–22 eV, the cross sections tabulated by Feng *et al.* are in excellent agreement with the photoabsorption measurements of Lee *et al.* Between $\sim$22–25 eV, they diverge, the Lee data becoming $\sim$10% larger. The photoabsorption curve of Ibuki *et al.* crosses that of Lee *et al.* being larger below 20 eV and smaller than Lee *et al.* (and also of Feng *et al.*) above 20 eV. The data of Lee *et al.* are preferred over those of Ibuki *et al.* because their estimated errors are smaller (10% versus 20%), they corrected for second-order radiation, and they span the indicated spectral range (Ibuki *et al.* start at 12.16 eV). In comparison with inelastic scattering experiments, we prefer photoabsorption because no auxiliary calibration is required, and the energy resolution is superior, although the latter is not a major factor above 15 eV. In Table 5.11, we record the $S(p)$ between 11.7–25.5 eV using alternately the values of Lee *et al.* and Feng *et al.* They are seen to be consistent, differing by 0.11 in $S(0)$ and 0.05 Ry units in $S(-2)$.

b.2   25.3–163.5 eV   Prior to the new data of Feng *et al.*, this region was largely unexplored. Fig. 5.11 displays the tabulated data of Feng *et al.*, together with summed atomic cross sections from Henke *et al.* (1993) above 91.5 eV. The latter are seen to be slightly larger.

b.3   163.5–171.1 eV   The sulfur $L_{II,III}$ edges in H_2S are given by Coville and Thomas (1995) as 170.37 eV for $2p_{3/2}$, and 171.57 eV for $2p_{1/2}$. These values are very close to those arrived at by Hudson *et al.* (1994) from Rydberg analysis.

Four research groups have reported photoabsorption spectra in the region approaching these ionization edges and beyond. They include Vinogradov and Zimkina (1971); Hayes and Brown (1972); Thomas *et al.* (1992) and Hudson *et al.* The new (e,e) data of Feng *et al.* also cover this range, but with much lower ($\sim$1 eV) resolution.

The spectrum of Hayes and Brown, with a resolution of $\sim$0.16 eV, is the only photoabsorption measurement with an absolute intensity scale. However, it only covers the structured region ($\sim$164–171.1 eV). Their pre-structure continuum has a cross section of about 0.5 Mb, in fair agreement with the extrapolated continuum based on the summed atomic cross sections of Henke *et al.*, 0.4 Mb. Thomas *et al.*

display a spectrum covering 164–172.6 eV which has better resolution than the spectrum of Hayes and Brown; the relative intensities agree, and hence we adopt the absolute data of Hayes and Brown in this interval. Although the absolute scale cannot be verified by other data, and may be too high, the contributions to $S(p)$ in this short interval are small. We note parenthetically that Hudson *et al.* interpret the first, broad peak at 165.4 eV to overlapping transitions from sulfur $(2p_{3/2,1/2})$ to the low-lying unoccupied orbitals $6a_1$ and $3b_2$. The partially resolved higher energy peaks are assigned to various s, p and d-like Rydberg series.

b.4   171.1–220.1 eV   The spectrum of Hudson *et al.* has the highest resolution (~0.030 eV) of the groups cited above, and spans the range, 160–240 eV. Normal operating procedure would be to scale this relative intensity spectrum to atomic additivity at ≥ 25 eV above the sulfur (2p) edge, after a suggestion of Hitchcock and Mancini (1994). Here we have scaled the data of Hudson *et al.* at 228.3 eV to a corresponding peak in the data of Feng *et al.*, which appears at 230 eV. However, there appear to be some inconsistencies in the relative spectrum of Hudson *et al.*, compared to other sources.

(a). The intensity of the first, broad peak at 165.4 eV is much higher, relative to the onset of the continuum in the data of Hudson *et al.* than in the other three photoabsorption spectra. The ratio is approximately 0.66 (Hayes and Brown; Thomas *et al.*), 0.74 (Vinogradov and Zimkina) and 0.9 (Hudson *et al.*).

(b). The post edge spectra differ significantly (see Fig. 5.11). The spectrum of Hudson *et al.* increases slightly from 171–176 eV, then decreases sharply to 240 eV. The (e,e) cross sections of Feng *et al.* continue to increase from 170–190 eV, then gradually begin their descent. The figure of Vinogradov and Zimkina (not shown in Fig. 5.11) displays an intermediate behavior, increasing between 170–180 eV, then remaining essentially flat to 206 eV. Values based on atomic additivity appear to follow the trend shown by Hudson *et al.*, although the decline in cross section above 180 eV is not as precipitous. Molecular effects could conceivably influence this region. Our modus operandi is to compute $S(p)$ from both Feng *et al.* and the normalized data of Hudson *et al.* and to appeal to the sum rules for adjudication.

b.5   220.1–2470 eV   The S(2s) edge occurs at 234.5 eV (Siegbahn *et al.*, 1969). The spectrum of Hudson *et al.* displays some weak features approaching this edge, at ~228.3 and 231.5 eV. That of Feng *et al.* also has a peak (230 eV), but otherwise declines between 220–260 eV parallel to atomic additivity, though ~15% lower. We assume a smooth, monotonic decline which can be represented by summed atomic cross sections until structure heralding the sulfur K-edge begins to appear at ~2470 eV. Accordingly, we fit the atomic cross sections of Henke *et al.* by regression to 4-term polynomials in two sections, 220.1–929.7 eV and 929.7–2470 eV. The functions are analytically integrated to

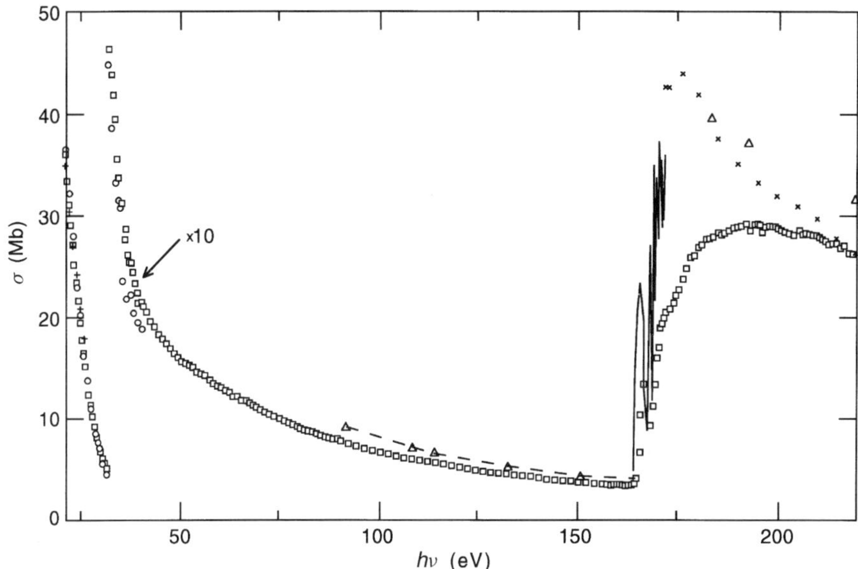

Fig. 5.11   Absolute photoabsorption spectrum of H$_2$S, 20–220 eV. + Lee *et al.* (1987); o Brion *et al.* (1986); △ Henke *et al.* (1993) + additivity; —— Hayes and Brown (1972); × Hudson *et al.* (1994); □ Feng *et al.* (1999a)

yield the $S(p)$ shown in Table 5.11, where one notes very significant contributions to $S(0)$ and $S(+1)$. The coefficients of the polynomials are given in Table 5.12.

b.6   2470–2510 eV   Reynaud *et al.* have presented an absolute photoabsorption spectrum that spans this region, and indeed continues to 2800 eV. It merges reasonably well at its extremities with the experimental atomic sulfur cross sections of Henke *et al.* The K-edge in H$_2$S occurs at 2478.5 eV (Jolly *et al.*, 1984). The first broad absorption peak, at 2472.7 eV, is again assigned to overlapping transitions to $6a_1$ and $3b_2$, the low-lying unoccupied orbitals. One, and perhaps two higher energy excitations are attributed to Rydberg transitions. The

Table 5.12   Coefficients of the polynomial it $df/dE = ay^2 + by^3 + cy^4 + dy^5$ fitted to data at various energies[a]

Energy range, eV	a	b	c	d
220.1–929.7	58.100 11	12 269.37	−232 016	519 689.7
929.7–2470.0	25.470 02	13 855.41	−105 920	−5 299 876
2510–10 000	28.618 72	269 609.1	−14 385 889	−1 451 559 820

[a]df/dE in Ry units, $y = B/E$, $B = $ IP $= 10.4666$ eV.

spectrum in this interval has been digitized and trapezoidally integrated. It contributes only 1% to $S(+1)$, and less to the other $S(p)$.

b.7 2510–10 000 eV   The atomic sulfur cross sections given by Henke *et al.* are fitted by regression to a 4-term polynomial, as before. About 50% of $S(+1)$ derives from this region, but significant quantities accrue also to $S(0)$ and $S(+2)$.

5.5.3   The analysis

The most recent analysis of old refractivity data, by Russell and Spackman (1997) yields $S(-2) = 24.77(2)$ a.u., or $6.139(4)$ Ry units. These authors find the older refractivity measurements of Cuthbertson and Cuthbertson (1909) to be 'remarkably reliable with almost exact agreement' with their theoretical estimates. The current spectral sum is 1.0% larger when using primarily photoabsorption data, and only 0.1% larger when using (e,e) data between 11.7–220.2 eV. However, the $S(0)$ sum rule leads to a more decisive conclusion. Here, the selection of (e,e) data falls 4.6% shy of the Thomas–Reiche–Kuhn value, whereas the dominantly photoabsorption data are only 1.8% lower. Had we continued to use the (e,e) data to 260 eV, the discrepancy would increase to 5.5%. It is noteworthy that Feng *et al.*, using only their data, had a shortfall of 4.0% even for $S(-2)$. Another tell-tale sign is the spectral sum for $S(+1)$. Atomic additivity, using Hartree–Fock calculations for atomic sulfur from (Fraga *et al.*, 1976), yields $S(+1) = 881.58$ Ry units, very slightly lower than the value (882.13) forthcoming from the spectral sum using (e,e) data. If the (e,e) data were extended to 260 eV, the spectral sum would actually fall 2.3 Ry units below atomic additivity. Correlation effects typically enhance the molecular value above atomic additivity, and the spectral sum based on photoabsorption data does achieve this goal. The primary difference in the data sets occurs in the immediate post $L_{\text{II,III}}$ edge region, as seen in Fig. 5.11.

In contrast to $S(+1)$, additivity is expected to be a satisfactory estimate for $S(+2)$. The current spectral sum is only 1.3% larger than additivity; the value deduced by Pazur *et al.* is within 0.1% of atomic sums. Pazur *et al.* fix the values of $S(-2)$ and $S(0)$ in the optimization procedure for treating input data. Consequently, we need only compare $S(+1)$ and $S(-1)$. Surprisingly, their value of $S(+1)$ is about 35 Ry units above the current spectral sum, and well above atomic additivity. The current spectral sum for $S(-1)$ is 2.3% higher than that of Pazur *et al.* The corresponding value reported by Feng *et al.* is still lower, but does not include contributions above 260 eV. When supplemented by higher energy data from Table 5.11, their value becomes 6.434 Ry units. If we enhance their value by 4%, their per cent shortfall for $S(-2)$, we arrive at $S(-1) = 6.691$ Ry units, very close to the current spectral sum.

We estimate $S_i(-1)$ by replacing the photoabsorption cross section (σ_a) with the photoionization cross section (σ_i) between the IP and 18.0 eV, beyond which

$\sigma_i \cong \sigma_a$, according to Brion *et al.* Both Xia *et al.* and Watanabe and Jursa provide values of σ_i from IP-11.69 eV that are similar. We use the more recent data of Xia *et al.* Above 11.69 eV, we must resort to the relatively coarse plot of η_i versus E given by Brion *et al.* Combining η_i with σ_a from Lee *et al.* yields σ_i .in the range 11.69–18.0 eV. The resulting $\Delta S_i(-1) = 2.4554$ from IP-18.0 eV, supplemented by $\Delta S(-1) = 2.0881$ from 18.0 eV$-\infty$, results in $S_i(-1) = 4.544$. Rieke and Prepejchal (1972) have obtained $M_i^2 = 5.03 \pm 0.27$ from high energy electron impact ionization. Their value is $\sim 10\%$ higher, as has often been found in such comparisons.

5.6 Sulfur Dioxide (SO₂)

5.6.1 Preamble

In the independent particle approximation, the electron configuration of SO_2 in its ground state (C_{2v} symmetry) may be written as

$$(1a_1)^2(1b_2)^2(2a_1)^2(3a_1)^2(2b_2)^2(1b_1)^2(4a_1)^2\} \text{ for the core, and}$$

$$(5a_1)^2(3b_2)^2(6a_1)^2(2b_1)^2(7a_1)^2(4b_2)^2(5b_2)^2(1a_2)^2(8a_1)^2\} \text{ for the valence shells}$$

The high resolution He I spectra of Holland *et al.* (1994) have clarified the ordering of the overlapping $(1a_2)^{-1}$ and $(5b_2)^{-1}$ states, as well as the relative energy of $(4b_2)^{-1}$. The core orbitals have the following properties: $1a_1$ is essentially S(1s), $1b_2$ and $2a_1$ are essentially O(1s), $3a_1$ is predominantly S(2s) while $2b_2$, $1b_1$, and $4a_1$ represent S(2p) in C_{2v} symmetry. The $5a_1$ and $3b_2$ inner valence orbitals, having S(3s)/O(2s) character and S(3p)/O(2s) character, respectively, are observed as broad features in the photoelectron spectrum and calculations indicate that the independent particle model is invalid for these excitations (Holland *et al.*, 1994). According to Hillier and Saunders (1971) and Guest and Saunders (1975), the remaining valence orbitals may be described as follows:

$6a_1$, S(3s)/O(2s,2p), non-bonding; $2b_1$, S(3p)/O(2p), π-bonding; $7a_1$, S(3p)/ O(2p), σ-bonding; $4b_2$, primarily O(2p), weakly S–O bonding; $5b_2$, primarily O(2p), O–O antibonding, weakly S–O bonding; $1a_2$, S(3d)/O(2p), π-bonding; and $8a_1$, S–O antibonding (see also Roos and Siegbahn, 1971).

The three lowest energy unoccupied orbitals are (in order of increasing excitation) $3b_1$, S(3p)/O(2p); $9a_1$, S(3s,3p)/O(2s,2p); and $6b_2$, S(3p)/O(2s,2p), according to Mazalov *et al.* (1972). They play a role not only in low-energy vacuum ultraviolet excitation, but also in pre-edge, inner shell excitation.

Kumar and Meath (1985b) performed a sum-rule analysis of SO_2, using photoabsorption data available to them up to 70.9 eV, and 'mixture rules' at higher energy. They optimized existing data to match expectation values for $S(-2)$ and

$S(0)$. Since that time, significant new information has been forthcoming. Hamdy *et al.* (1991), using line sources, have measured photoabsorption cross sections between 12.19–84.34 eV with stated accuracies of 1–2%. Holland *et al.* (1995a) used synchrotron radiation to determine both photoabsorption and photoionization cross sections in the structured region between the IP ($\sim$12.35 eV) and 16.5 eV, and beyond to $\sim$31.0 eV. Feng *et al.* (1999b) performed electron energy loss (EELS) measurements, and converted their data to pseudo-photoabsorption cross sections between 3.5–260 eV, correcting earlier data from this group (see Cooper *et al.*, 1991).

Photoabsorption near S_K, S_L and O_K edges has also been explored. Reynaud *et al.* (1996) have reported absolute cross sections in the structured region near S(1s), while Sze *et al.* (1987), using EELS, obtained relative cross sections near S(2p), S(2s) and O(1s). Gedat *et al.* (1998) have recently presented relative photoabsorption cross sections in the S(2p) region at high resolution. These aforementioned data sets offer the prospect of more accurate local cross sections above the IP. In the discrete spectrum, Manatt and Lane (1993) tried to place prior photoabsorption measurements in the range 3.08–11.70 eV on a consistent basis. Digital information from this compilation was made available to the author.

5.6.2 The data

The most accurate adiabatic IP (SO_2) currently available is based on He I photoelectron spectroscopy. Wang *et al.* (1987) proffered 12.3494(2) eV. More recently, Holland *et al.* obtained 12.3482 eV. We choose the latter.

a The discrete spectrum, and transitions below the IP

In the UV-VUV region, one finds three broad and relatively weak bands with maxima at $\sim$4.3, 6.25 and 8.25 eV (see, for example, Manatt and Lane, 1993; Feng *et al.*, 1999b). The lowest energy band has been attributed to singlet excitation to the lowest energy unoccupied orbital, $3b_1$ (Hillier and Saunders, 1971). The 6.25 and 8.25 eV bands likely involve excitation to the $9a_1$ and $6b_2$ unoccupied orbitals. Between $\sim$9.2 eV and the IP, more intense, probably overlapping Rydberg excitations are observed. The oscillator strengths of the first three bands and also the 9.20–11.65 eV region, have been arrived at in two ways, both of which are listed in Table 5.13. The first involves scanning and digitizing the figures of Warneck *et al.* (1964) and Golomb *et al.* (1962). The second uses the digital compilation of Manatt and Lane. Fine-grained trapezoidal integration was performed with each set. The two sources are seen to agree to $\sim$1%. Manatt and Lane stated that they blue-shifted the data of Golomb *et al.* by $\sim$3 Å in their compilation, as verified currently by matching peaks. In Fig. 4a of Feng *et al.*, it is shown red-shifted. Here, we shall accept the original calibration of Golomb *et al.*

Our practice has been to use photoabsorption data, rather than EELS data, when line saturation is not expected to be a problem. Usually the energy resolution is

Table 5.13 Spectral sums and comparison with expectation values for SO_2[a]

Energy, eV	$S(-2)$	$S(-1)$	$S(0)$	$S(+1)$	$S(+2)$
3.07–5.27[b]	0.0473	0.0152	0.0049	0.0016	0.0005
3.07–5.27[c]	0.0467	0.0150	0.0048	0.0016	0.0005
5.27–7.45[d]	0.3065	0.1416	0.0656	0.0305	0.0142
5.27–7.45[c]	0.3093	0.1428	0.0662	0.0307	0.0144
7.45–9.20[d]	0.1375	0.0850	0.0526	0.0327	0.0204
7.45–9.20[c]	0.1391	0.0861	0.0534	0.0332	0.0207
9.20–11.65[d]	1.1166	0.8373	0.6300	0.4757	0.3604
9.20–11.65[c]	1.1306	0.8497	0.6406	0.4847	0.3680
11.65–12.35(IP)[e]	0.1840	0.1626	0.1437	0.1271	0.1124
IP–16.67[f]	1.6072	1.6934	1.7980	2.2417	2.0738
16.67–84.11[g]	2.7708	5.2721	11.8438	32.7994	111.9126
84.11–163.0[e]	0.0340	0.2666	2.1571	18.0705	156.9212
163.0–177.0[h]	0.0022	0.0273	0.3430	4.3100	54.1871
177.0–260.0[e]	0.0116	0.1801	2.8235	44.7948	719.1327
260.0–525.0[i]	0.0062	0.1524	3.8955	103.558	2866.44
525.0–565.0[j]	0.0004	0.0149	0.5966	23.925	959.91
565–2470[i]	0.0014	0.0770	4.7979	342.218	28 874.8
2470–2510[k]	–	0.0003	0.0600	10.978	2009.0
2510–10 000[i]	–	0.0060	1.6251	494.734	174 285.0
10^4–10^5	–	0.0001	0.1278	167.879	305 162.7
10^5–∞	–	–	0.0009	11.955	295 400.2
Total	6.2257	8.9319	30.9660	1258.13	810 602.8
Expectation values	6.395(13)[l]	–	32.0	–	799 170.6[m]
	–	–	–	–	799 184.9[n]
Other values	(6.403)[o]	9.175[o]	(32.0)[o]	1281.4[o]	800 800[o]
	6.363[f]	9.200[f]	32.121[f]	–	–
	6.363[e]	8.909[e]	–	–	–

[a]In Ry units.
[b]Warneck *et al.* (1964).
[c]Manatt and Lane (1993).
[d]Golomb *et al.* (1962).
[e]Feng *et al.* (1999b).
[f]Holland *et al.* (1995a).
[g]Hamdy *et al.* (1991).
[h]Gedat *et al.* (1998), normalized. See text.
[i]Henke *et al.* (1993).
[j]Sze *et al.* (1987).
[k]Reynaud *et al.* (1996).
[l]See text, Sect. 5.6.3.
[m]Fraga *et al.* (1976).
[n]Bunge *et al.* (1993).
[o]Kumar and Meath (1985b).

substantially better, providing more accurate cross sections in structured regions. Also, the calibration of cross section is primary, rather than being itself based on sum rules. However, there is a gap in the photoabsorption data between ~11.65 eV and the IP (~12.35 eV) which we have filled by utilizing the high resolution (0.05 eV FWHM) EELS data of Feng *et al.*

b The continuum

b.1 Autoionization, IP–16.67 eV Here, the line source photoabsorption measurements of Hamdy *et al*. and the synchrotron-based data of Holland *et al*. come into consideration. Although the cross sections of Hamdy *et al*. are stated to be somewhat more accurate, line sources are a haphazard match for autoionization structure. Consequently, Figs. 2 and 3 from Holland *et al*. have been scanned, digitized and trapezoidally integrated to evaluate the $S(p)$ given in Table 5.13.

b.2 16.67–84.11 eV The cross sections of Holland *et al*. and Hamdy *et al*. are in excellent agreement between $\sim$16.5–17.5 eV, but at higher energy the Hamdy values are systematically lower. This is a region where no sharp structure is evident. Synchrotron sources are more likely suspect for such systematic discrepancies (incomplete corrections for order-sorting, scattered light) than are line sources. We utilize the tabulated values of Hamdy *et al*. for this region, although the EELS data of Feng *et al*. favor slightly higher cross sections between $\sim$20–40 eV.

b.3 84.11–163.0 eV The continuum cross section declines monotonically just prior to the pre-edge structure involving S(2p). Feng *et al*. (1999b) present tabulated EELS data, which have somewhat lower (1–15%) cross sections than those obtainable by summing the atomic cross sections of Henke *et al*. (1993). We adopt the EELS data, and thus have chosen the lower possibility between 16.67–84.11 eV and 84.11–163.0 eV. We shall estimate the effect of these choices in Sect. 5.6.3.

b.4 163.0–177.0 eV The structure near the S(2p) edge in SO_2 has been observed with successively higher resolution in photoabsorption studies by Zimkina and Vinogradov (1971), Krasnoperova *et al*. (1976) and Gedat *et al*. Only relative cross sections were reported. Corresponding EELS measurements were obtained by Sze *et al*. and Feng *et al*. The latter provided absolute cross sections, but with relatively poor resolution.

Coville and Thomas (1995) have measured 174.78 eV for the $S(2p_{3/2})$ edge, and 175.99 eV for the $S(2p_{1/2})$ edge. These values are in substantial agreement with the Rydberg analysis of Gedat *et al*., though the latter group observes additional ligand field splitting of $S(2p_{3/2})$. Term values relative to these IPs of $\sim$10, 5.6 and 4.45 eV can be deduced from the spectra, which roughly correspond to excitations from S(2p) to the unoccupied $3b_1$, $9a_1$ and $6b_2$ orbitals, respectively. Ab initio calculations have supported these assignments (see Kondratenko *et al*. (1980), who performed Hartree–Fock calculations, and Sze *et al*., who used multichannel quantum defect theory). Both groups of authors concluded that the oscillator strength for S(2p) $\rightarrow$ $3b_1$ was of order 4×10^{-4}, whereas transitions to $9a_1$ and $6b_2$ were $\sim$$10^{-2}$. A simplistic interpretation utilizing the atomic orbital description of the unoccupied orbitals given in Sect. 5.6.1 can rationalize this conclusion, since $3b_1$ has p composition, whereas $9a_1$ and $6b_2$ have some S(3s) or O(2s) component (see, for example, Sze *et al*.). The detailed spectrum, which involves spin-orbit splitting, ligand field splitting and vibrational fine structure, is far more complex. Rydberg states have also been identified at slightly higher energies.

We have chosen to use the high resolution (0.03 eV, FWHM) spectrum of Gedat *et al.*, normalized at 177 eV to the absolute cross section (3.90 Mb) extracted from Feng *et al.*, to evaluate the contribution to oscillator strength in this region. Since it amounts to only $\sim$0.34, and the relative contributions to the other $S(p)$ are smaller, the uncertainty in calibration has little effect on the sum rule analysis.

b.5   177.0–260.0 eV   Above the S(2p) edge, very broad features have been observed at $\sim$179 and 195 eV (Zimkina and Vinogradov 1971; Sze *et al.*, 1987; Feng *et al.*, 1999b) and tentatively assigned to shape resonances. Weaker features at $\sim$229.1 and 233.7 eV have been attributed to S(2s) $\to$ $3b_1$ and S(2s) $\to$ $9a_1$, $6b_2$ by Sze *et al.* and Feng *et al.*, who estimate the S(2s) edge at 239.1 eV. The absolute cross sections given by Feng *et al.* are about 25% lower than summed atomic cross sections at $\sim$180 eV, but converge at higher energy (1% difference at 220 eV). We utilize the data in Table 1 of Feng *et al.* to compute $S(p)$ for this region.

b.6   260.0–525.0 eV   The extremities of this region are dictated by the terminus of the data of Feng *et al.* and the onset of structure presaging the O(1s) edge. The good agreement between atomic additivity and the upper energies of the Feng data provide support for traversing this span with summed atomic cross sections from Henke *et al.* These sparse data have been fitted by regression to a 4-term polynomial. The coefficients of this polynomial are given in Table 5.14; the analytically integrated contributions to $S(p)$ can be found in Table 5.13.

b.7   525.0–565.0 eV   The O(1s) edge occurs at 539.84 eV (Jolly *et al.*, 1984). The structure near the O(1s) edge has been observed in photoabsorption by Akimov *et al.* (1982) and by Sze *et al.* using EELS, at higher resolution and over a more extended energy range. The spectrum consists of a relatively strong O(1s) $\to$ $3b_1$ transition, with partially resolved transitions to $9a_1$, $6b_2$ and Rydberg states. Both spectra are given in relative intensity units. We attempt a calibration of Figs. 6(a) and 8 of Sze *et al.* by using summed atomic cross sections from Henke *et al.*, yielding $\sigma = 0.76$ Mb at 525 eV, and $\sim$1.61 Mb at 565 eV. This procedure

Table 5.14   Coefficients of the polynomial $\mathrm{d}f/\mathrm{d}E = ay^2 + by^3 + cy^4 + dy^5$ fitted to data at various energies[a]

Energy range, eV	a	b	c	d
260.0–525.0	114.000 4	2 625.064	10 890.22	−850 569
565.0–2470.0	19.679 7	31 772.13	−745 916	5 771 544
2510–10 000	14.369 06	188 727.9	−7 631 224	−714 781 813

[a]$\mathrm{d}f/\mathrm{d}E$ in Ry units, $y = B/E$, $B = \mathrm{IP} = 12.3482$ eV.

implies that Fig. 6(a) requires a background subtraction of $\sim$0.5 Mb. The manually extracted cross sections are integrated trapezoidally and the $S(p)$ recorded in Table 5.13. There is substantial margin for error here, perhaps 25%, but the contribution of this region to $S(0)$ and $S(+1)$ amounts to <2%, and much less for the other $S(p)$.

b.8   565–2470 eV   This energy range goes from beyond the O(1s) edge to the onset of structure approaching the S(1s) edge. The summed atomic cross sections of Henke *et al.* are fitted by regression to a 4-term polynomial, whose coefficients are given in Table 5.14. The function is analytically integrated to obtain the $S(p)$ values entered in Table 5.13. Significant contributions of 15% to $S(0)$ and 27% to $S(+1)$ accrue here.

b.9   2470–2510 eV   The S(1s) edge occurs at 2483.7 eV (Jolly *et al.*, 1984). Reynaud *et al.* have presented a photoabsorption spectrum in this region which is slightly better resolved than an earlier one by Bodeur and Esteva (1985), and also provided absolute cross sections which agree well at their termini with summed atomic cross sections extrapolated from Henke *et al.* The spectrum is similar to that preceding the O(1s) – a strong S(1s) $\rightarrow 3b_1$ excitation, followed by partially resolved transitions to $9a_1$ and $6b_2$, some Rydberg structure, and a broad (shape?) resonance at $\sim$2496 eV. Cross sections have been manually extracted from Fig. 4 of Reynaud *et al.*, and integrated trapezoidally.

b.10   2510–10 000 eV   As noted above, there is a smooth transition from the cross section of Reynaud *et al.* and the extrapolated sum of atomic cross sections (1993). The latter have been fitted by regression to the familiar 4-term polynomial, whose coefficients appear in Table 5.14. The analytically integrated contributions to $S(p)$ reveal a plausible increment to $S(0)$, less than 2.0 because of Pauli exclusion effects, and major additions to $S(+1)$ and $S(+2)$.

5.6.3   The analysis

Both Feng *et al.* and Holland *et al.* cite Bridge and Buckingham (1966) as providing a 'direct experimental value' of the polarizability α, or equivalently, $S(-2)$. Actually, Bridge and Buckingham derived $\alpha = 3.89 \times 10^{-24}$ cm^3 from earlier work, and it is not the static electric dipole polarizability, but refers to $\lambda = 6328$ Å. Kumar and Meath have examined the wavelength dependence of molar polarization (or equivalently, refractive index or α) and deduced $S(-2) = 25.61$ a.u., or a static polarizability of 3.795×10^{-24} cm^3. In their data set, $\alpha(6328$ Å$) = 3.89 \times 10^{-24}$ cm^3. Hence, the reference to Bridge and Buckingham is inappropriate in the present context. Feng *et al.* also cite the compilation of Maryott and Buckley (1953b) as a source of dielectric constants. These latter authors give 10.8 ±0.8 as the total molar polarization, 9.54 at $\lambda = \infty$, obtained from LeFèvre *et al.* (1950). A careful reading of LeFèvre *et al.* reveals that the value 9.54 comes from refractive index measurements as used by Kumar and Meath, not from dielectric constant measurements. The dielectric constant measurements yield a total molar polarization of 10.9 ±0.6, from which LeFèvre *et al.* deduce an atomic polarization (the infrared contribution) of 1.4 ±0.6. A direct calculation of the infrared contribution (Bishop and Cheung, 1982) gives 0.75. Therefore, a modern interpretation of the data of LeFèvre *et al.* would yield an electronic molar polarization of 10.1 ±0.6, or $\alpha = 4.00$ ±0.24 × 10^{-24} cm^3, clearly not competitive with the accuracy inferred from refractive index measurements. From their analysis of earlier refractive index measurements, LeFèvre *et al.* inferred electronic molar polarization values of 9.54 and 9.58, or $S(-2) = 25.52$ and 25.63 a.u. We take $S(-2) = 25.58$ ±0.05 a.u., or 6.395(13) Ry units, which encompasses the value of Kumar and Meath, with a generous uncertainty.

The present spectral sum for $S(-2)$, 6.2257 Ry units, is approximately 2.6% lower, and $S(0)$ is 3.2% lower than the expectation values. If both deficits are to be attributed to a common spectral region, it is unlikely to be below the IP, since the total oscillator strength is only $\sim$0.9. A large contribution to both $S(-2)$ and $S(0)$ occurs between 16.67–84.11 eV, where the data of Hamdy *et al.* (1991) (1–2% accuracy) were used. It will be recalled that both Feng *et al.* (1999b) and Holland *et al.* (1995a) displayed higher cross sections in a portion of this region, between 20–40 eV. Using the data of Feng *et al.* instead of Hamdy *et al.* would increase $S(-2)$ by 0.07 Ry units, $S(-1)$ by 0.13 Ry units, and $S(0)$ by 0.25, leaving us shy by 1.5% for $S(-2)$ and by 2.5% for $S(0)$. By comparison, the analysis of Holland *et al.* was only $\sim$0.5% low for $S(-2)$, and 0.3% high for $S(0)$. Unfortunately, part of the input to their analysis included the data of Cooper *et al.* The Vancouver group (Feng *et al.*, 1999b) has shown that this earlier data was flawed between 60–260 eV, and they presented corrected values. Direct evaluation in this interval reveals a diminution of 1.194 for $S(0)$, 0.100 for $S(-1)$ and 0.0097 for $S(-2)$. Applying this revision changes the Holland value of $S(0)$ to 30.927, comparable to the current value, 30.966; similarly, their $S(-1)$ becomes 9.100, close to the proposed upward revision in the present evaluation, which becomes 9.06.

Feng *et al.* determined oscillator strengths and absorption cross sections in their EELS measurements (normalized by the TRK sum rule) only to 260 eV. Hence, they confined their analysis of $S(p)$ to $p \leq -1$. Their partial $S(-1)$, if augmented by the currently determined contributions above 260 eV, would yield $S(-1) = 9.16$ Ry units.

Kumar and Meath, who were limited to earlier data, optimized the available distribution of oscillator strengths subject to the required expectation values for $S(-2)$ and $S(0)$. Their inferred $S(-1)$, 9.175 Ry units, is very close to the augmented value from Feng *et al.*, but approximately 1% higher than that of Holland *et al.* (revised) and the present study. This difference may reflect the fact that the latter two studies lack $\sim$1 in $S(0)$, whereas the correct value is imposed by Kumar and Meath. Comparison of their distribution of oscillator strength with the current one reveals that most of the discrepancy occurs above 164 eV, where

Kumar and Meath have used mixture rules, and the bulk of this contribution in the present evaluation has used a different mixture rule, namely summed atomic cross sections from Henke *et al*. This may also account for the larger value of $S(+1)$, 1281.4 Ry units, obtained by Kumar and Meath compared to the current spectral sum, 1258.1 Ry units. Both are larger than the summed atomic values, which yield 1248.0 Ry units (Fraga *et al*., 1976). This is the expected direction of deviation from additivity, but the magnitude is uncertain. By contrast, the present value of $S(+2)$ is 1.4% larger than the summation of atomic values, whereas Kumar and Meath obtain almost the exact 'expectation value'. For the very high energies which contribute most significantly to $S(+2)$, where direct experimental evidence is lacking, both must be considered satisfactory.

To estimate $S_i(-1)$, we utilize the photoionization cross sections of Holland *et al*. from IP–20.66 eV, beyond which the quantum yield of ionization is unity, and the values of $S(-1)$ given in Table 5.13 can be used. To compare with Holland *et al*., who list $\Delta S_i(-1) = 4.983$, we obtain 4.961 between IP and 31 eV. Between 31 eV and a high energy limit, the values are 2.563 (Holland *et al*.) and 2.491 (present result). We note parenthetically that the Holland value has been misplaced in their Table 1, and that it should be revised downward by 0.100, to account for the corrections of Feng *et al*. to the data of Cooper *et al*. used by Holland *et al*. With this correction, $S_i(-1) = 7.446$ (Holland *et al*.) and 7.452 (present result).

<h2 style="text-align:center">5.7 Ozone (O$_3$)</h2>

<h3 style="text-align:center">5.7.1 Preamble</h3>

Ozone (O_3) is bent in its ground state, at an angle of 117° (Colmont *et al*., 1995). In C_{2v} symmetry, the molecular orbital sequence of the dominant configuration is

$$(1a_1)^2(2a_1)^2(1b_2)^2(3a_1)^2(2b_2)^2(4a_1)^2(5a_1)^2(3b_2)^2(1b_1)^2(6a_1)^2(4b_2)^2(1a_2)^2, \tilde{X}^1A_1$$

The lowest vacant orbitals are $2b_1$, $7a_1$, and $5b_2$. (Mason *et al*., 1996; Gejo *et al*., 1999). The three deepest orbitals represent the O(1s) core, where $1a_1$ corresponds to the central atom. The next three are combinations of O(2s), and the upper six of O(2p) orbitals. The a_1 and b_2 are σ-like orbitals (in plane) while a_2 and b_1 are π-like orbitals.

The lowest energy absorption (Wulf band) corresponds to the transition

$$\ldots (6a_1)^2(4b_2)^2(1a_2)^2, \tilde{X}^1A_1 + h\nu \rightarrow \ldots (6a_1)^2(4b_2)^1(1a_2)^2(2b_1)^1, \, ^3A_2$$

It is forbidden by both its triplet character and its A_2 symmetry. It has a very weak photoabsorption cross section, with an onset at 1.18 eV. The next higher absorption (Chappuis band) has been attributed to both $^1A_2$ and $^1B_1$, which undergo an avoided crossing (conical intersection). See Bacis *et al*. (1998). The $^1B_1$ results from the transition

$$\ldots (6a_1)^2(4b_2)^2(1a_2)^2, \tilde{X}\,^1A_1 + h\nu \rightarrow \ldots (6a_1)^1(4b_2)^2(1a_2)^2(2b_1)^1, \, ^1B_1$$

Although this transition is optically allowed, and about 20 times stronger than the Wulf band, its maximum cross section at 2.06 eV is only ~0.005 Mb. The first truly significant absorption (Hartley band) covers the 2100–3000 Å region, with a maximum cross section of ~11.6 Mb at 2550 Å (4.86 eV). This is the absorption band that protects our biosphere from harmful solar ultraviolet radiation. Ironically, this photoabsorption contributes to the destruction of the ozone layer, since it results in photodissociation, primarily to $O(^1D) + O_2(a^1\Delta_g)$ (Bacis *et al.*, 1998). The transition involved is

$$\ldots (6a_1)^2(4b_2)^2(1a_2)^2, \tilde{X}^1A_1 + h\nu \to \ldots (6a_1)^2(4b_2)^2(1a_2)^1(2b_1)^1, {}^1B_2$$

Its cross section has been studied extensively, due to its biological and environmental impact. However, we shall see that it contributes only ~0.09 to $S(0)$, and amounts to only ~1/7 of $S(-2)$.

Figure 5.12 displays currently available knowledge of the photoabsorption cross sections of ozone in the UV and VUV. Between 6.7–11.24 eV, the recent synchrotron-based data of Mason *et al.* (1996) are preferred to earlier data.

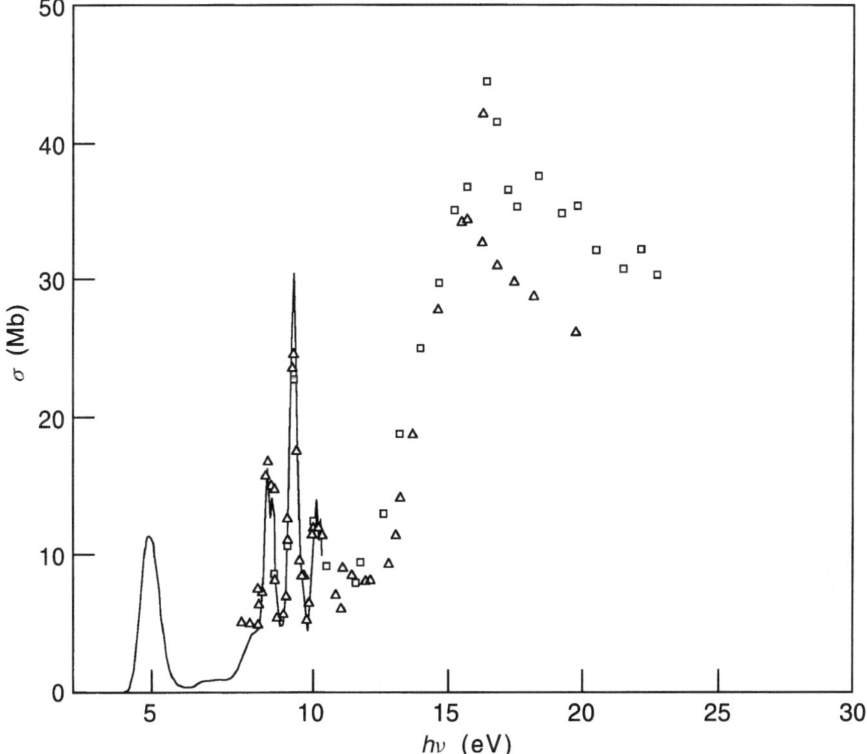

Fig. 5.12   Absolute photoabsorption spectrum of O_3, 4–25 eV. —— Mason *et al.* (1996); □ Ogawa and Cook (1958a); △ Cook (1970)

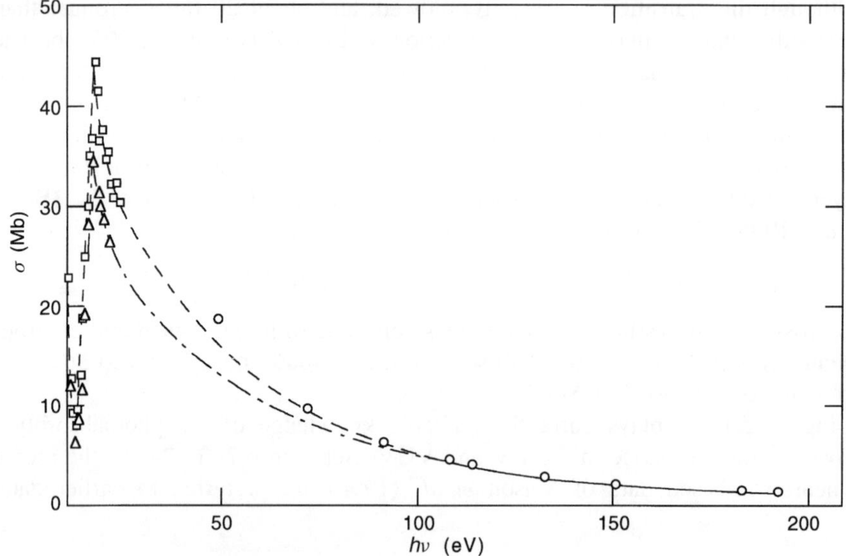

Fig. 5.13 Absolute photoabsorption spectrum of O_3, 10–200 eV. □ Ogawa and Cook (1958a); △ Cook (1970); o Henke *et al.* (1993) + additivity

They have been normalized at 2537 Å (4.887 eV) to 11.37 Mb, a value given by Mauersberger *et al.* (1987) with an rms uncertainty of 0.5%. Although Mason and Pathak (1997) comment that ozone photoabsorption spectra beyond 11.05 eV have not been measured, early data by Ogawa and Cook (1958a) and by Cook (1970) exist, and have been plotted in Fig. 5.12. They are relatively crude (Ogawa and Cook estimate an error ≤20%) but they are nonetheless of some value, since the 11–50 eV region contributes heavily to $S(-2)$.

Even availing ourselves of the data of Ogawa and Cook (1958a) or Cook, there remains a gap between $\sim$20–100 eV, before one can apply atomic additivity with some confidence. We interpolate this domain with a smooth curve, as shown in Fig. 5.13. The region between 108.5–535 eV has been fitted by regression to three times the atomic cross sections given at isolated points by Henke *et al.* (1993).

The photoabsorption cross section of ozone in the K-shell region (525–555 eV) has been measured recently by Gejo *et al.* (1997). We tentatively utilize their values, although post-edge estimates based on atomic additivity suggest that their absolute values may be low by a factor $\sim$3.

5.7.2 The data

Two values of the adiabatic ionization potential have been proffered from PES measurements, a 'low value' of 12.43–12.44 eV (Katsumata *et al.*, 1984; Dyke *et al.*, 1970) and a 'high value' of 12.52–12.53 eV based on PES (Frost *et al.*, 1970) and PIMS (Weiss *et al.*, 1977). The conflict hinges upon whether a weak vibrational peak at onset is or is not a hot band. Moseley *et al.* (1981) measured a threshold for the process

$$O_3{}^+ + h\nu \rightarrow O^+ + O_2$$

of 2.16 ±0.02 eV. A thermochemical cycle using $D_0(O_2-O)$ and IP(O) then yields IP(O_3) = 12.51 ±0.02 eV. Also, they concluded from their data that the dissociation limit for the process

$$O_3{}^+ + h\nu \rightarrow O_2{}^+ + O$$

was 0.59 eV. Together with IP(O_2) and $D_0(O_2-O)$, this implies IP(O_3) = 12.53 eV. They concluded that their results provided 'strong support for an AIP near 12.52 eV'. The more accurate value of $D_0(O_2-O) = 1.0621$ ±0.0004 eV obtained recently by Taniguchi *et al.* (1999) increases the calculated AIP to 12.52 ±0.02 eV and 12.54 eV. We therefore adopt the PIMS value, AIP(O_3) = 12.519 ±0.004 eV.

a   1.31–3.54 eV (Wulf and Chappuis bands, approach to Hartley band)

Burrows *et al.* (1999) and Brion *et al.* (1998) present overview spectra showing the deep minimum between the Chappuis and Hartley bands, ~2.75–3.54 eV. Anderson (1993) presents modern cross sections for the Wulf band (1.31–1.65 eV), while Brion *et al.* or Burkholder and Talukdar (1994), whose data are in good agreement, are used for the Chappuis band (1.65–2.75 eV). These contributions to $S(p)$ are listed separately in Table 5.15, and confirm quantitatively their insignificance for the present purposes.

b   3.54–6.70 eV (Hartley band)

Molina and Molina (1986) provide tabulated data in this interval, with an estimated accuracy in cross section of 1–2%. More recently, Yoshino *et al.* (1993) have presented their own tabulation between 4.89–6.70 eV, which they find to be 2–5% smaller. Burrows *et al.* find a value exactly mid-way at 2356.5 Å (Hg line), which is near the peak of the Hartley band. Here, we utilize the data of Molina and Molina, since they cover the entire band, and their estimated accuracy seems to be substantiated by subsequent investigations.

c   6.70–11.24 eV

Figure 5.12 displays the modern synchrotron data of Mason *et al.* (1996) in this interval, together with the lower energy Hartley band taken from Molina and Molina. The graphs from Mason *et al.* have been manually digitized for transfer to Fig. 5.12, and also for evaluation of the corresponding $S(p)$, which appear in Table 5.15. The contribution to $\Delta S(0)$ is just 2.5% larger than that given directly by Mason *et al.* Up to this point, the photoabsorption cross sections are rather well established, but they contribute only ~25% to $S(-2)$ and ~1.6% to $S(0)$.

Table 5.15 Spectral sums, and comparison with expectation values for O_3[a]

Energy, eV	$S(-2)$	$S(-1)$	$S(0)$	$S(+1)$	$S(+2)$
1.31–1.65 (Wulf)[b]	3.3×10^{-5}	–	–	–	–
1.65–2.75$_5$ (Chappuis)[c]	0.0008_9	0.0001_4	0.0000_2	–	–
2.75$_5$–3.54[d]	1.1×10^{-5}	–	–	–	–
3.54–6.70 (Hartley)[e]	0.706 3	0.2555	0.0930	0.0341	0.0126
6.70–11.24[f]	0.559 9	0.3966	0.2840	0.2053	0.1497
11.24–20.66[g]	1.582 1	1.8970	2.3250	2.9073	3.7011
11.24–20.66[h]	1.316 7	1.5684	1.9104	2.3747	3.0058
20.66–108.5[g]	1.428 9	3.6390	11.0478	41.0873	184.7834
20.66–108.5[h]	1.123 3	2.9145	9.1018	35.0441	163.0503
108.5–535[i]	0.021 6	0.2462	2.9875	43.5581	776.3581
527.5–555.9	0.000 14	0.0055	0.2172	8.6292	342.8984
555.9–2622.4[i]	0.001 4	0.0774	4.7873	342.4822	29 425.455
2622.4–10 000[i]	–	0.0011	0.2928	89.1439	30 985.535
10^4–10^5	–	–	0.0174	22.1379	37 924.89
10^5–10^6	–	–	–	1.1478	18 303.03
10^6–10^7	–	–	–	0.0408	6 392.19
10^7–10^8	–	–	–	0.0012	2 088.05
10^8–10^9	–	–	–	–	667.19
10^9–∞	–	–	–	–	309.69
Total[g]	4.3013	6.5184	22.0520	551.3751	127 403.9
Total[h]	3.7303	5.4653	19.6914	544.7993	127 381.5
Expectation values	4.68$_5$ (avg)[j]	–	24.0	–	125 774.25[n]
	4.737[k]	–	–	–	125 326.27[o]
	4.92$_6$ ±0.05[l]	–	–	–	–
	5.04 ±0.10[m]	–	–	–	–

[a] $S(p)$ in Ry units.
[b] Anderson (1993).
[c] Burkholder and Talukdar (1994).
[d] Burrows *et al*. (1999).
[e] Molina and Molina (1986).
[f] Mason *et al*. (1996).
[g] Ogawa and Cook (1958a). The region 20.66–108.5 eV is a smooth interpolation (Fig. 5.13) between their data and atomic additivity.
[h] Cook (1970). The 20.66–108.5 eV region is interpolated as shown in Fig. 5.13.
[i] Summed atomic cross sections given by Henke (1993).
[j] Selected 'best average' from numerous calculations with correlated wavefunctions. See Andersson *et al*. (1992b).
[k] Coupled cluster calculation, Maroulis (1994).
[l] From refractive index measurement, Cuthbertson and Cuthbertson (1913a).
[m] From dielectric constant measurement, Epprecht (1950), and auxiliary information, see Sect. 5.7.3, text.
[n] Hartree–Fock results and atomic additivity, Fraga *et al*. (1976).
[o] Hartree–Fock results and atomic additivity, Bunge *et al*. (1993).

d 11.24–20.66 eV

Figure 5.12 also shows the tabulated data of Ogawa and Cook and the manually digitized cross sections from the graph given by Cook. Both are in fair agreement with the more accurate values of Mason *et al*. in the region of overlap, but they differ from one another, especially above 16 eV. However, Cook cites an earlier presentation (Cook, 1968), ostensibly the source of his 1970 figure, in which it is explicitly stated that the maximum cross section is 42 Mb at 725 Å (17.1 eV), which would be in very good agreement with Ogawa and Cook, instead of 32.6 Mb extracted from his 1970 figure. An alternative method of scaling the data of Cook (1970) is to examine the overlap with Mason *et al*. There is a peak at 10–10.5 eV which displays structure in the Mason data, but not in the Cook data, although at his stated resolution (0.5 Å $\cong$ 0.004 eV) it should. The maximum appears to be very close to Lyman-α (1215.7 Å $\equiv$ 10.199 eV) in both spectra. Normalizing the Cook cross section (24.6 Mb) to the Mason cross section (30.3 Mb) at 10.2 eV results in a Cook cross section of 40.2 Mb at 725 Å, close to his stated value in 1968, 42 Mb. Thus, although we cannot trace the cause of the change between Cook (1968) and Cook (1970), there is some reason to believe that the earlier, higher cross section may be preferable. For the moment, we evaluate contributions to $S(p)$ from Ogawa and Cook and Cook (1970), and record both in Table 5.15. Ogawa and Cook used emission lines for their light source, one of which is listed at 1216 Å. They chose to normalize their relative absorption coefficients to 22.85 Mb at Lyman-α, a value estimated from Tanaka *et al*. (1953). The alert reader will note that this value differs from that of Mason *et al*., 30.3 Mb. However, four other points from Ogawa and Cook are more nearly consistent with Mason, *et al*. Hence, we retain the listed values of Ogawa and Cook, pending the examination of the sum rules.

e 20.66–108.5 eV

Based on many examples, atomic additivity can be assumed to be a close approximation to the true molecular cross section for $h\nu \gtrsim 100$ eV. The summed cross sections, from Henke, *et al*., are plotted in Fig. 5.13, together with data from Ogawa and Cook (1958a) and Cook (1970). The plausible inference is that the cross section diminishes monotonically between 20–100 eV, but it is difficult to find absolute, or even relative cross section data for verification. Padial *et al*. (1981) calculated absolute partial cross sections using the Stieltjes-Tchebycheff procedure and the separated channel static-exchange approximation. Upon summing their partial cross sections, one finds a peak at $\sim$23.5 eV, then a rapid decline to 40 eV. However, these summed cross sections are about a factor 2 lower than expected. Celotta *et al*. (1974) present electron energy loss (EELS) data to 30.5 eV for O_3, as well as O_2 obtained concomitantly. In principle, they can be converted to relative photoabsorption cross sections by applying an E^3 correction factor. Alternatively, the energy dependence of the O_3 photoabsorption cross section can be related to the known energy dependence of the O_2 cross section. Both approaches were found to yield unrealistic results. Hence, we

Table 5.16 Coefficients of the polynomial $df/dE = ay^2 + by^3 + cy^4 + dy^5$ fitted to data at various energies[a]

Energy range, eV	a	b	c	d
108.5–535	−6.588 49	1 281.059	−11 837.6	38 132.3
555.9–2622.4	3.462 726	33 277.19	−964 529	11 207 822
2622.4–10 000	−9.358 04	35 732.85	−465 170	−88 758 750

[a] df/dE in Ry units, $y = B/E$, $B = $ IP $= 12.519$ eV.

must resort to smooth curve interpolation between the VUV data of Ogawa and Cook (1958a) or Cook (1970) and atomic additivity at ~ 100 eV, as indicated in Fig. 5.13. Again, we shall appeal to the sum rules for a judgment between these alternatives, or beyond them. The $S(p)$ entries in Table 5.15 are obtained by trapezoidal integration.

f   108.5–535 eV

We assume $\sigma(O_3) = 3\sigma(O)$, where $\sigma(O)$ is taken from Henke *et al*. The sparse points are fitted by regression to a 4-term polynomial, whose coefficients appear in Table 5.16. The $S(p)$ are obtained by analytical integration.

g   527.5–555.9 eV

Gejo *et al*. (1997) obtained an absolute photoabsorption spectrum in this region. Two prominent pre-edge peaks are observed, and assigned primarily to $O_T(1s) \rightarrow \pi^*(2b_1)$ and $O_C(1s) \rightarrow \pi^*(2b_1)$. We utilize the data as presented, and determine $S(p)$ by trapezoidal integration. However, the absolute scale is suspect. At 555.9 eV, $\sigma(O_3) \cong 0.55$ Mb. This energy is only ~ 10 eV above the K edge ($O_T^* $K edge $= 541.5$ eV, $O_C = 546.2$ eV), according to Banna *et al*. (1977). The rule-of-thumb (Hitchcock and Mancini, 1994) is that the cross section is close to atomic additivity 25 eV above the edge. The fitted data based on Henke *et al*. yield $\sigma(O_3) = 1.51$ Mb at 570 eV, and 1.60 Mb at 556 eV. This is about a factor 3 higher than the measurements by Gejo *et al*. indicate. Of the $S(p)$, only $S(0)$ and $S(+1)$ are sensitive to this deviation.

h   555.9–2622.4; 2622.4–10 000 eV

The tripled atomic oxygen cross sections of Henke *et al*. are fitted by regression to two 4-term polynomials, partitioned as indicated.

i 10^4 eV–∞

This region is treated analogously to atomic and molecular oxygen.

5.7.3   The analysis

Values of the static electric dipole polarizability of ozone are available from recent calculations using a variety of correlated wavefunctions. Andersson *et al*. (1992b) examined ten methods, and found about five which gave fairly consistent results, i.e., $\alpha = (2.54 – 2.88) \times 10^{-24}$ cm^3. Although they did not specify a best value, their multi-reference configuration interaction ($\alpha = 2.764 \times 10^{-24}\,\text{cm}^3$) and complete active space perturbation theory ($\alpha = 2.790 \times 10^{-24}\,\text{cm}^3$) calculations appear to be most favored. Maroulis (1994) used coupled cluster theory with double substitutions corrected for the effects of single and triple excitations, and arrived at $\alpha = 2.808 \times 10^{-24}\,\text{cm}^3$.

Experimental values are scarce, and old. Cuthbertson and Cuthbertson (1913a) measured the refractive index at 8 wavelengths in the visible spectrum. This is the region of weak absorption by the Chappuis band. They estimated an accuracy of $\sim 1\%$ in $n - 1$. Upon fitting these refractivities to the Cauchy expansion at the quadratic level, we obtain $\alpha = 2.92 \pm 0.03 \times 10^{-24}\,\text{cm}^3$. Epprecht (1950) measured the dielectric constant in the microwave region, using a two-cavity comparative method. His primary goal appeared to be the determination of the electric dipole moment (μ) from the temperature dependence of the dielectric constant ε, and the Debye equation (see Sect. 5.4.3). His result, $\mu = 0.52 \pm 0.03\,\text{D}$, was quite good. Modern molecular beam methods have obtained $\mu = 0.533\,73(7)\,\text{D}$ (Meerts $et\ al.$, 1977) and $\mu = 0.533\,747\,(5)\,\text{D}$ (Mack and Muenter, 1977). Since his measurements of ε were more precise than those of μ, we have used his $\varepsilon = 1.001\,90(2)$ at 1 atm, 0 °C, and the modern value of μ to deduce α. Thus, the total molar refractivity, $R_{\text{tot}} = 14.20 \pm 0.15\,\text{cm}^3$, that due to the permanent dipole moment, $(4\pi N_A \mu^2/9kT) = 6.3516\,\text{cm}^3$, and hence the component due to polarizability, $R(\alpha) = 7.84_8 \pm 0.15\,\text{cm}^3$. This is equivalent to $\alpha = (3.11 \pm 0.06) \times 10^{-24}\,\text{cm}^3$. The infrared component has been calculated by Bishop and Cheung (1982) to be $0.1168 \times 10^{-24}\,\text{cm}^3$, leaving $\alpha_{\text{vis}} = (2.99 \pm 0.06) \times 10^{-24}\,\text{cm}^3$. Thus, within their mutual errors, the two experimental values overlap, and suggest a slightly higher (3–5%) value than the theoretical results. The corresponding values of $S(-2)$, ranging from 4.68_5 to 5.04 Ry units, are recorded in Table 5.15.

A comparison of the spectral sums with expectation values in Table 5.15 immediately reveals that the higher spectral sum is 8.2% lower than the lowest expectation value for $S(-2)$, and also for $S(0)$. The lower spectral sum has a deficit of 20% and 18%, respectively. Clearly, these sum rules favor Ogawa and Cook over Cook (1970), and imply that the maximum cross section (42 Mb) given by Cook (1968) is closer to reality. Atomic additivity is generally a good approximation for $S(+2)$, and indeed the spectral sum is just 1.5% larger than the expectation value based on Hartree–Fock calculations. But $S(+2)$ is insensitive to the regions of experimental controversy encountered here. However, $S(+1)$ typically results in a value larger than additivity. From Fraga $et\ al.$ (1976) and additivity we obtain $S(+1) = 551.649$, essentially the same value as given by the higher spectral sum, again suggesting a spectral deficit. Let us explore possible improvements in the experimental data suggested by the sum rules.

As indicated by the data presented in Sect. 5.7.2, the cross sections for $h\nu < 11\,\text{eV}$ and $h\nu > 570\,\text{eV}$ are fairly well established. In fact, $S(-2)$ receives very little contribution above 108.5 eV. If we therefore confine ourselves to the region $11.24\,\text{eV} \leq h\nu \leq 108.5\,\text{eV}$, the simplest assumption is to maintain the same shape of the spectral distribution, but to increase its scale. A 12% enhancement in this domain leads to $S(-2) = 4.663$, $S(0) = 23.657$, and $S(+1) = 556.77$, which comes quite close to satisfying these three sum rules. We had also questioned the absolute scale of the experimental cross sections in the K-shell region. Tripling this scale has little effect on $S(-2)$ and $S(-1)$, but further increases $S(0)$ to 24.091 and $S(+1)$ to 574.03, which remains plausible. A constraint in the opposite direction is posed by $S(-1)$. The molecular value is typically lower than that based on atomic additivity. From Fraga *et al.*, we obtain an atomic additivity value of 7.559, compared to a spectral sum (enhanced by 12%) of 7.183, which is acceptable. Thus, the sum rules have pointed the way to a plausible description of a very important region of the ozone photoabsorption spectrum, which was at best poorly known, and was regarded by some as terra incognita.

For completeness, we attempt to estimate $S_i(-1)$. Cook (1970) displays this photoionization spectrum, but since we have concluded that his photoabsorption cross sections are low, we assume that his photoionization cross sections are low by the same factor (~ 1.37). Trapezoidal integration gives $\Delta S_i(-1) = 1.560$ between IP–20.66 eV. We make the usual assumption that the quantum yield of ionization above 20.66 eV is unity. Cook's spectrum appears to approach this behavior. Then, for $h\nu > 20.66$ eV, $\Delta S_i(-1) = \Delta S(-1) = 4.406$, or $S_i(-1) = 5.97$.

6

Polyatomic Molecules

6.1 Ammonia (NH$_3$)

6.1.1 Preamble

The electronic ground state of NH$_3$ has the electronic configuration

$$(1a_1)^2(2a_1)^2(1e)^4(3a_1)^2,\ \tilde{X}^1A_1$$

It is pyramidal (C_{3v} symmetry) but photoabsorption in the UV-VUV is to planar (D_{3h}) Rydberg states leading to NH$_3{}^+$, which is also planar. This type of transition is apparent in the long vibrational progression of the ν_2 mode (umbrella, or out-of-plane bend) in both Rydberg and photoelectron spectra. The most intense transitions below the IP are $\tilde{X}^1A_1 \to \tilde{A}^1A_1$, $3a_1 \to 3$s at $\sim$6.5 eV and $\tilde{X}^1A_1 \to \tilde{D}^1E''$, $3a_1 \to 3$d at $\sim$9.2 eV (for the latter assignment, see Glownia et al. (1980)). Much weaker absorptions occur to the $\tilde{B}$ and $\tilde{C}$ states, assigned to $3a_1 \to 3$p transitions at $\sim$8.0 and $\sim$8.5 eV. These latter states may be connected by a dynamic Jahn–Teller effect (Glownia et al., 1980).

The oscillator strength distribution for NH$_3$ has been addressed recently by experimental methods (Burton et al. (1993a)) and by sum rule analysis (Burton et al., 1993b). Our emphasis here is slightly different. Where possible, we prefer to use photoabsorption studies as our sources, since absolute cross sections are forthcoming from the Beer–Lambert law, whereas inelastic electron scattering methods, such as Burton et al. (1993a), require some auxiliary normalization. The latter authors have made the point that photoabsorption can result in saturation for narrow features, but dissociation or pre-dissociation broadens the aforementioned Rydberg states sufficiently to minimize this concern. As an example, the vibrational band widths for the $\tilde{X} \to \tilde{A}$ transition have been measured to be 34–293 cm^{-1} FWHM (Vaida et al., 1987). The instrumental resolution used by Watanabe (1954) was $\sim$27 cm^{-1}. Also, saturation should result in reduced oscillator strengths, but we shall see that in most cases, photoabsorption in NH$_3$ leads to larger oscillator strengths than the electron scattering data. A possible exception is the $\tilde{C}$ state, from which emission may occur (Vaida et al., 1987), implying a longer-lived, sharper feature.

The sum rule analysis of Burton et al. (1993b) imposes constraints on $S(-2)$, from refractivity measurements, and on $S(0)$ from the TRK sum rule, to arrive at a 'global' distribution of oscillator strengths by modifying available data. This approach is not intended to provide accurate oscillator strengths at specific energies. Our goal is to distinguish, from available data, the best selection of oscillator strengths at all energies, using the sum rules as a guide. There is no dearth of photoabsorption measurements from the onset of absorption (~ 5.7 eV) to ~ 20 eV. Our task is to arrive at plausible choices, to provide rationales for these choices, and to test these selections by sum rule analysis. One source of photoabsorption data (Samson et $al.$, 1987b) uses line sources and extends to ~ 150 eV. Another (Edvardsson et $al.$, 1999) uses synchrotron radiation, and spans IP–24.8 eV. Absolute cross section measurements in the nitrogen K-edge region (400–425 eV) also exist. Other high-energy regions, presumably devoid of structure, can be adequately estimated by summed atomic cross sections, which for NH_3 is predominantly atomic nitrogen.

6.1.2 The data

The adiabatic ionization potential of NH_3, obtained by non-resonant, two-photon ZEKE spectroscopy (Reiser et $al.$, 1993) is $82\,159 \pm 1\,\mathrm{cm}^{-1} \equiv 10.1864(1)$ eV. This value confirms an earlier result by Habenicht et $al.$ (1991). Electric field effects were taken into account, but since only $v_2 = 1$ of NH_3^+ was accessed, it was necessary to subtract the term value $G(v_2 = 1) - G(v_2 = 0)$ given by Lee and Oka (1991). More recently, Dickinson et $al.$ (1997) obtained $82\,153.6$ and $82\,151.3\,\mathrm{cm}^{-1}$ from interpretations of features in their spectrum. The small discrepancy has not been stressed by these latter authors, and is of minor consequence in the apportionment of oscillator strengths, so we shall retain the value of Reiser et $al.$ (1993).

a The discrete spectrum and transitions below the IP

One criterion we shall invoke in selecting among various photoabsorption measurements in the VUV is the declared error estimates of the authors. Thus, Watanabe (1954), Watanabe and Sood (1965) and Suto and Lee (1983) estimate $\pm 10\%$, Xia et $al.$ (1991) assert $\pm 5\%$, Samson et $al.$ (1987b) state $\pm 3\%$ and Syage et $al.$ (1992) offer $\pm 20\%$. On this basis, we eliminate the last group, a decision which is bolstered by the characteristically larger cross sections found by Syage et $al.$ than by others (see the tables of Burton et $al.$, 1993a). The lowest error estimates are from Samson et $al.$, and their track record is very good. They use a discrete line light source, which is a haphazard match for regions containing sharp absorption features. Their data do not extend below 11.07 eV, but we shall use their results as a secondary standard in regions of overlap. Thus, in Fig. 6.1 we compare the results of various authors between 11–13.5 eV. We find that the cross sections of Samson et $al.$ are in good agreement with those of Xia et $al.$, the group with the second lowest error estimate. The values of Watanabe and Sood are also in fair agreement below 11.8 eV, but appear to be too high between 11.8–12.8 eV. Since sharp absorption features do not occur above 11.8 eV we opt for the data of Samson et $al.$ at higher energies. The two representative points of Suto and

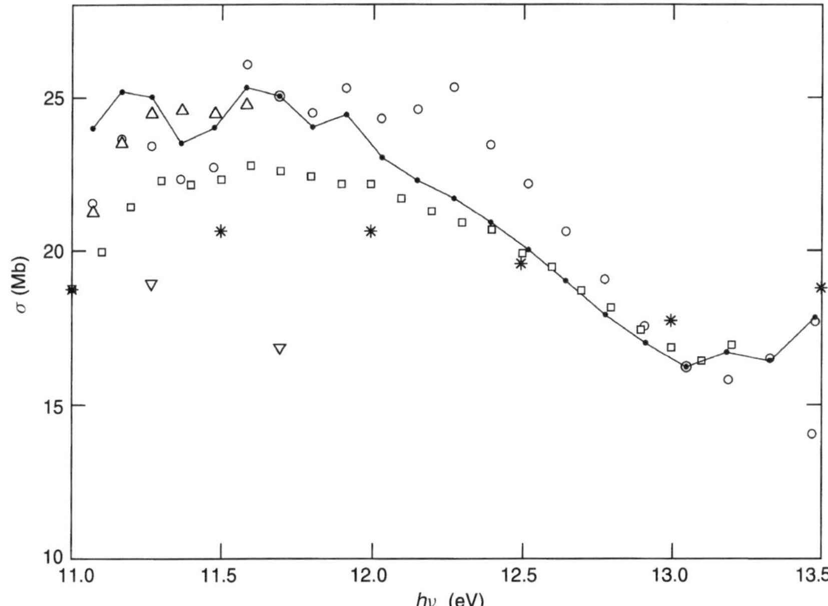

Fig. 6.1 Absolute photoabsorption spectrum of NH_3, 11.0–13.5 eV. ● Samson *et al.* (1987b); ○ Watanabe and Sood (1965); △ Xia *et al.* (1991); □ Burton *et al.* (1993a) 'hi-resol'; ∗ Burton *et al.* (1993a) 'lo-resol'; ▽ Suto and Lee (1983)

Lee are seen to be 25–30% lower than those of Samson *et al.* Burton *et al.* (1993a) also find that the oscillator strengths of Suto and Lee are often (though not always) low, compared to other measurements. However, they employ coarser resolution, which detracts from our goal of local accuracy, when possible.

The inelastic electron scattering results of Burton *et al.* (1993a) are seen to be lower than those of Samson *et al.* below 12.4 eV, more so for their low resolution data than that at higher resolution. Burton *et al.* have noted their discrepancy with Samson *et al.* and with Watanabe and Sood in this region, but find support for their values from earlier electron scattering results. We can find no plausible reason why the measurements of Samson *et al.* should be '10–15% too high' around 12 eV, but otherwise acceptable. The most recent synchrotron-based data of Edvardsson *et al.* (1999), not shown in Fig. 6.1, 'strongly support' those of Samson *et al.* between 11.3 and 11.8 eV. By elimination, we find the data of Xia *et al.* and Watanabe and Sood, which use continuum light sources, as the best choices between the IP (10.186 eV) and 11.8 eV. They are also in excellent agreement with one another (1.2% for $\Delta S(0)$, 1.5% for $\Delta S(-2)$).

The data of Xia *et al.* initiate at the IP, but Watanabe and Sood continue to a lower limit of ~7.52 eV. We tentatively (pending sum rule analysis) utilize the cross sections of Watanabe and Sood (scanned and digitized) from 7.52 eV to the IP, and continue with Watanabe (1954) to lower energies, essentially the

$\tilde{X} \to \tilde{A}$ band. In Table 6.1, we list our incremental sums for the various vibronic transitions below the IP, in a manner which can be compared with the data of Burton *et al*. (1993a). Our summed oscillator strength for the $\tilde{A}$ band, from Watanabe, is 0.0819. Burton *et al*. (1993a) obtain 0.074 49 from their (e,e) data, and 0.0781 from their integration of Watanabe. Interestingly, the constrained sum rule adjustment shown by Burton *et al*. (1993b) seems to prefer a higher value of $\sim$0.082, in better agreement with our result than with Burton *et al*. (1993a). Parenthetically, we note that different authors, digitizing graphical data from semilog plots, do not necessarily get the same result. Using the data of Watanabe between 6.00–7.29 eV, Burton *et al*. (1993a) obtain 0.072 99, Lee and Suto report 0.088, while the present integration yields 0.076 44.

Table 6.1 Contributions to $S(p)$ of transitions below the IP in NH_3[a]

v'	ΔE, eV	$S(-2)$	$S(-1)$	$S(0)$
a. $\tilde{X}^1A_1 \to \tilde{A}^1A_1$, 5.67–7.42 eV[b]				
0	5.67–5.78	0.1316	0.0554	0.0233
1	5.78–5.89	0.5291	0.2271	0.0974
2	5.89–6.00	1.4716	0.6427	0.2807
3	6.00–6.12	2.3415	1.0425	0.4642
4	6.12–6.24	3.5749	1.6217	0.7357
5	6.24–6.35	4.2812	1.9789	0.9147
6	6.35–6.46	4.7033	2.2132	1.0415
7	6.46–6.58	4.3221	2.0714	0.9927
8	6.58–6.70	4.1995	2.0482	0.9990
9	6.70–6.82	3.5165	1.7458	0.8667
10	6.82–6.94	2.6117	1.3198	0.6670
11	6.94–7.06	1.7242	0.8862	0.4555
12	7.06–7.18	1.1375	0.5945	0.3107
13	7.18–7.29	0.7049	0.3747	0.1992
14	7.29–7.42	0.4729	0.2556	0.1381
Total		35.7225	17.0777	8.1864
b. $\tilde{B}$ and $\tilde{C}$ states, 7.42–8.62 eV[c]				
	ΔE, eV	$S(-2)$	$S(-1)$	$S(0)$
	7.42–7.52[b]	0.2592	0.1423	0.0781
	7.52–7.65	0.2197	0.1224	0.0682
	7.65–7.75	0.2022	0.1143	0.0646
	7.75–7.87	0.3050	0.1750	0.1004
	7.87–8.00	0.5215	0.3040	0.1772
	8.00–8.11	0.5451	0.3226	0.1910
	8.11–8.25	0.5529	0.3325	0.1999
	8.25–8.36	0.2662	0.1625	0.0992
	8.36–8.49	0.3869	0.2398	0.1486
	8.49–8.62	0.4511	0.2839	0.1787
	Total	3.7098	2.1993	1.3059

Table 6.1 *(Continued)*

c. 8.62–9.32 eV[c]

ΔE, eV	$S(-2)$	$S(-1)$	$S(0)$
8.62–8.68	0.3376	0.2146	0.1364
8.68–8.83	1.4392	0.9265	0.5964
8.83–8.95	2.1616	1.4111	0.9211
8.95–9.07	2.9210	1.9326	1.2787
9.07–9.20	3.0608	2.0532	1.3773
9.20–9.32	3.5427	2.4078	1.6364
Total	13.4629	8.9458	5.9463

d. 9.32–10.186 eV (IP)[c]

ΔE, eV	$S(-2)$	$S(-1)$	$S(0)$
9.32–9.45	3.3253	2.2919	1.5797
9.45–9.59	3.0145	2.1078	1.4738
9.59–9.72	3.1953	2.2654	1.6062
9.72–9.84	2.8317	2.0343	1.4614
9.84–9.97	2.6895	1.9581	1.4255
9.97–10.09	2.2944	1.6907	1.2459
10.09–10.186	2.0379	1.5191	1.1324
Total	19.3886	13.8673	9.9249
$\sum$, 5.67–IP	72.2838	42.0901	25.3635

[a] $S(p)$ in Ry units. The numbers shown should be divided by 10^2.
[b] Digitized and integrated from Watanabe (1954).
[c] Digitized and integrated from Watanabe and Sood (1965), except for 7.42–7.52 eV region.

Our integrated oscillator strengths, based on the data of Watanabe and Watanabe and Sood, are typically larger than the (e,e) data of Burton *et al.* (1993a), averaging 10% larger in the sub-ionization region. The exception is the weak region involving the $\tilde{C}$ state, where sharp structure may indeed lead to saturation in the photoabsorption measurements. The difference here (0.002) is slight compared to the total oscillator strength (0.194) in the sub ionization region.

b The 'autoionization' region, IP–11.65 eV

The photoabsorption curve of NH_3 continues to evidence structure above the IP, to about 11.6 eV, but the structural features appear predominantly as predissociation, rather than autoionization. The discrete line source used by Samson *et al.* does not track this structure. Hence, we opt for the synchrotron-based data of Xia *et al.*, which are in good agreement with the cross sections of Watanabe and Sood in this interval, but were obtained with higher resolution and a smaller stated uncertainty. The recent data of Edvardsson *et al.* would have been preferred in this region, but their absolute cross sections are presented in a very compressed figure, and their more expanded figure, obtained in a different manner, is apparently not absolute. The digitized data have been integrated trapezoidally and are listed in Table 6.2.

Table 6.2 Spectral sums, and comparison with expectation values for NH_3[a]

Energy, eV	$S(-2)$	$S(-1)$	$S(0)$	$S(+1)$	$S(+2)$
5.67–IP[b]	0.7228	0.4209	0.2536	0.1578	0.1009
IP–11.65[c]	0.3846	0.3107	0.2513	0.2036	0.1652
11.65–112.71[d]	2.5086	3.7825	6.9722	17.0695	57.8350
112.71–400.0[e]	0.0035	0.0392	0.4809	6.5714	101.475
400.66[f]	–	0.0003	0.009	0.2650	7.805
402.33[f]	–	0.0009	0.027	0.7984	23.610
400–425[f]	0.0003	0.0090	0.2723	8.2481	249.91
425–2042.4[e]	0.0008	0.0325	1.5480	85.3463	5668.22
2042.4–10 000	–	0.0004	0.0932	23.0813	6834.62
10^4–10^5[g]	–	–	0.0032	4.0058	6809.03
10^5–10^6[h]	–	–	–	0.1965	3124.57
10^6–10^7[h]	–	–	–	0.0067	1077.18
10^7–10^8[h]	–	–	–	0.0002	350.34
10^8–10^9[h]	–	–	–	–	111.79
$10^9 \to \infty$[h]	–	–	–	–	51.77
Total	3.6206	4.5964	9.9107	145.95	24 468.4
Expectation values	3.642[i]		10.0		24 172.3[j]
					24 172.8[k]
Other values	(3.640)[l]	4.655[l]	(10.0)[l]	146.58[l]	24 772[l]
	3.596[m]	4.625[m]	9.935[m]		

[a] $S(p)$ in Ry units.
[b] See Table 6.1 and text.
[c] Xia *et al.* (1991).
[d] Samson *et al.* (1987b).
[e] Henke *et al.* (1993).
[f] Akimov *et al.* (1988).
[g] Chantler (1995).
[h] Assuming hydrogen-like behavior, bare nitrogen atom with screening, K shell only.
[i] Calculated from wavelength dependent polarizabilities given by Hohm (1994).
[j] Fraga *et al.* (1976).
[k] Bunge *et al.* (1993).
[l] Burton *et al.* (1993b).
[m] Edvardsson *et al.* (1999).

c   The continuum, 11.65–112.71 eV

Above the 'autoionization' region, the photoabsorption cross section initially declines to $\sim$13.2 eV, then ascends to a broad peak at $\sim$16 eV, doubling its value. This is the region corresponding to excitation and ionization from the $1e$ orbital, which manifests a broad photoelectron band from $\sim$14.8–18.6 eV. Beyond $\sim$17 eV, the cross section declines monotonically toward the nitrogen K-edge $(1a_1)^{-1}$ with very little indication of $2a_1$ excitation at $\sim$26 eV. Edvardsson *et al.* present an expanded view of the absolute cross sections between 13.2–17.7 eV, which we have compared with the data of Samson *et al.* The synchrotron-based data reveal weak oscillations, attributed to Rydberg series converging to the $\tilde{A}^2E$ ionization threshold, but essentially track the line source data of Samson *et al.*,

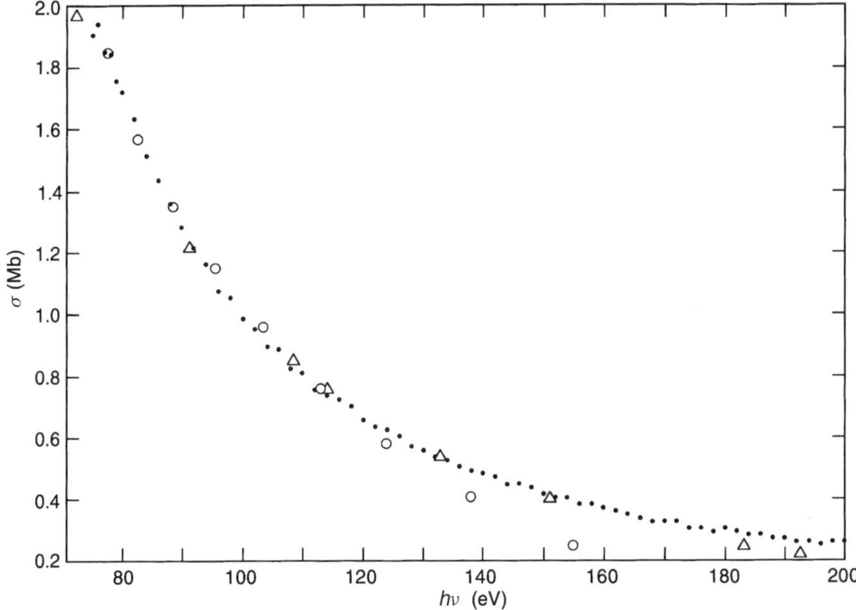

Fig. 6.2 Absolute photoabsorption spectrum of NH_3, 70–200 eV. ● Burton *et al.* (1993a);
△ Samson *et al.* (1987b); ○ Henke *et al.* (1993) + additivity

lying 3.3% higher on average. This is about the accuracy claimed by Samson *et al*; Edvardsson *et al*. do not explicitly estimate their accuracy for this molecule. In Fig. 6.2, we examine the segment between 75–200 eV. The data set of Samson *et al*. (1987b), which is the basis for our description above 11.65 eV, is in good agreement with the (e,e) values of Burton *et al*. (1993a) to 112.7 eV, but falls lower at higher energies. In support of the (e,e) data, the atomic nitrogen cross sections of Henke *et al*. (1993) also lie above those of Samson *et al*. beyond 112.7 eV.

At this high energy, atomic additivity (in this case, just nitrogen) is generally reliable for molecules consisting of first row atoms. Hence, this interval, utilizing the cross sections of Samson *et al*., will terminate at 112.7 eV. The contributions to $S(p)$, determined by trapezoidal integration, are reported in Table 6.2.

d The continuum, 112.71–400.0 eV

The nitrogen K-edge of NH_3 occurs at 405.6 eV (Jolly *et al*., 1984), but prominent structure begins to appear at ~400 eV (Akimov *et al*., 1988). The cross sections in the interval 112.7–400 eV are estimated by fitting the atomic nitrogen cross sections of Henke *et al*. to a 4-term polynomial. Analytical integration of the polynomial yields the $S(p)$ given in Table 6.2. The coefficients of the polynomial appear in Table 6.3.

Table 6.3 Coefficients of the polynomial $df/dE = ay^2 + by^3 + cy^4 + dy^5$ fitted to data at various energies[a]

Energy range, eV	a	b	c	d
112.71–400.0	−1.487 72	370.6193	−3527.93	11 717.53
425.0–2042.4	1.376 059	3899.576	−65 606.8	284 359.2
2042.4–10 000	−3.820 36	13 277.33	−547 111	24 423 190

[a] df/dE in Ry units, $y = B/E$, $B = IP = 10.1864$ eV.

e The nitrogen K-edge region, 400–425 eV

Akimov *et al*. (1988) present an experimental absolute photoabsorption spectrum of NH_3 in this region. Schirmer *et al*. (1993) have obtained a spectrum which displays somewhat better resolution, but it is in relative intensity units. There are two major peaks, at 400.66 eV and 402.33 eV, which have been assigned to excitations to the lowest unoccupied orbitals, 3s (a_1) and 3p (e). Akimov *et al*. give the oscillator strengths for these peaks explicitly, 0.009 and 0.027, respectively. For the weaker peaks and the underlying continuum, the spectrum of Akimov *et al*., has been digitized and integrated trapezoidally. The results, given in Table 6.2, yield a slightly larger integrated oscillator strength than can be gleaned from a curve of running oscillator strength given by Akimov *et al*.

f Post K-edge, 425 eV–10 keV

The atomic nitrogen cross section is essentially the only contributor here. The cross sections of Henke *et al*. are fitted to two 4-term polynomials, spanning 425–2042.4 eV and 2042.4–10 000 eV. The contributions to $S(p)$, obtained by analytical integration, are given in Table 6.2, while the coefficients of the polynomials can be found in Table 6.3. This interval contributes $\sim$74% to $S(+1)$.

g Post K-edge, 10^4–10^5 eV

This section is identical to that for atomic nitrogen. This region contributes $\sim$27.8% to $S(+2)$, $\sim$2.7% to $S(+1)$ and negligibly to the other $S(p)$.

6.1.3 The analysis

Burton *et al*. (1993b) used $S(-2) = 14.56$ a.u, or 3.640 Ry units, a value which had been deduced previously by Zeiss *et al*. (1977) by fitting earlier refractivity data to a Cauchy expansion. Edvardsson *et al*. (1999) cite $\alpha = 2.22 \times 10^{-24}$ cm^3, equivalent to $S(-2) = 3.745$ Ry units, attributed to Bridge and Buckingham (1966). The value listed by Bridge and Buckingham had been derived from earlier data, and refers to $\lambda = 6328$ Å, not the static polarizability. Hohm (1994) has re-determined the refractivity of ammonia at four wavelengths. His values are in substantial agreement with earlier work. Our fit to his data yields $\alpha_0 = 2.159 \times 10^{-24}$ cm^3, and $S(-2) = 3.642$ Ry units, almost identical to that reported by Burton *et al*. (1993b). The latter authors used their derived value of $S(-2)$, and the TRK sum for $S(0)$, as fixed inputs to their optimization procedure for treating input data. The current spectral sum for $S(-2)$, 3.6206 Ry units, is 0.6% lower than the expectation value. Edvardsson *et al.* obtain a still lower value. A comparison of their partial sum rule analysis with the present one reveals that their selection of sub-ionization contributions is substantially lower than the current one (0.59 versus 0.72 Ry units), partially compensated by higher values in their experimental range, 500–1217 Å (2.50 versus 2.39 Ry units). Their value of $S(0)$ is slightly closer to the TRK expectation value than the current spectral sum, though both are low by $<0.9\%$. A detailed comparison shows that their contribution between 500–1217 Å exceeds the present one by 3.6%. This is approximately the difference found earlier between the data of Samson *et al.* and that of Edvardsson *et al.* Below and above their experimental range, the current contributions to $S(0)$ are larger, making the total values of $S(0)$ much closer.

The values of $S(-1)$ span the range 4.596 (ours), 4.625 (Edvardsson *et al.*) and 4.655 (Burton *et al.*, 1993b), all in Ry units. The current value may be slightly too low, judging by the larger deficit in $S(0)$. Our spectral sum for $S(+2)$ is $\sim 1.2\%$ higher than that based on the sum of electron densities at the nuclei (essentially that of atomic nitrogen) forthcoming from Hartree–Fock calculations (Fraga *et al.*, 1976; Bunge *et al.*, 1993). The value of $S(+2)$ given by Burton *et al.* is approximately 2.5% higher than the 'expectation' value, and their $S(+1)$, 146.58 Ry units, is slightly higher than our spectral sum, 145.95 Ry units. The value based on atomic additivity (Fraga *et al.*, 1976) is 140.6 Ry units.

$S_i(-1)$ has been determined by using the photoionization cross sections of Xia *et al.* between 1060 Å and the IP, and values from Samson *et al.* between 690–1060 Å. Below 690 Å, the quantum yield of ionization (η_i) is unity, according to Samson *et al.*, Watanabe and Sood, and Edvardsson *et al.* The result, $S_i(-1) = 3.395$, is lower than $M_i^2 = S_i(-1) = 3.58 \pm 0.35$ measured by Rieke and Prepejchal (1972), but this time within their error limit. Edvardsson *et al.* find $S_i(-1) = 3.621$, uncharacteristically higher than the value measured by Rieke and Prepejchal. Their values of η_i closely follow those of Samson *et al.* from 18 to ~ 15 eV, but they begin to deviate to lower energy. Below 13.4 eV, Samson *et al.* adopt η_i from Watanabe and Sood, and indeed the two sets of data appear to smoothly merge. There is also very good agreement in $\Delta S_i(-1)$ from IP–11.7 eV between Xia *et al.* and Watanabe and Sood. By contrast, the η_i from Edvardsson *et al.* deviate more strongly, and there is a sudden jump at 11.8 eV, which the authors attribute to the cut-off in their lithium fluoride filter. This anomalous behavior leads us to prefer the current selections to those of Edvardsson *et al.* in evaluating $S_i(-1)$.

Between 690 Å (18 eV) and the IP, about 57% of absorption results in ionization. This fraction is lower than found for three triatomic species, and closer to that found for five diatomic species. It reflects, in part, the dominant role played by predissociation in the first 3 eV above the IP, where large excursions in the photoabsorption cross section are not mimicked in the photoionization cross section, the latter appearing more like a direct photoionization process.

6.2   Methane (CH$_4$)

6.2.1   Preamble

Methane is the prototypical organic molecule with tetrahedral structure. In tetrahedral (T_d) symmetry, its electronic configuration is

$$(1a_1)^2 (2a_1)^2 (1t_2)^6, \; {}^1A_1$$

Removal of an electron from the valence orbital would give rise to a triply degenerate state of CH$_4{}^+$, but Jahn–Teller interaction distorts the cation from T_d symmetry to C_{2v}, D_{2d} and C_{3v} symmetries (see e.g. Frey and Davidson, 1988a). The corresponding He I photoelectron spectrum displays a complex vibronic structure, ~3 eV wide (see, for example, Turner *et al.*, 1970). The $1t_2$ orbital has dominant C(2p) character. The first prominent excitation, at ~9.5 eV, is to a Rydberg state with 3s character. Photoabsorption to this state is similar in breadth and shape to the photoelectron spectrum, both displaying two broad maxima Narayan (1972). Hence, it appears to have similar distortions to that of the cation. The only additional feature in the photoabsorption cross section (see Fig. 6.3) from this region to the IP is a shoulder at ~11.5–12.0 eV, presumably excitation

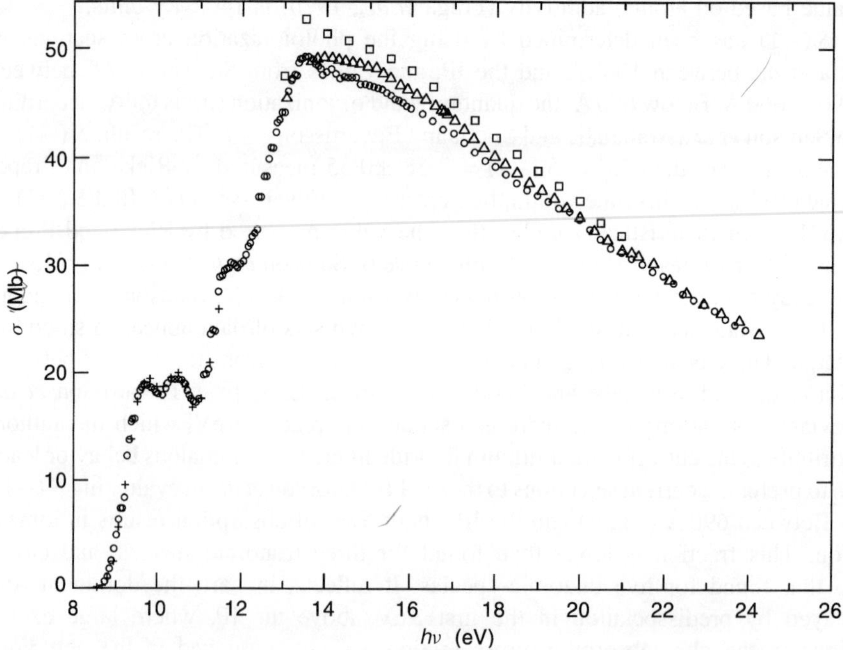

Fig. 6.3   Absolute photoabsorption spectrum of CH$_4$, 8.5–24 eV. + Watanabe *et al.* (1953b); o Au *et al.* (1993a); △ Samson *et al.* (1989); □ Person *et al.* (1975)

to a higher state, which may be 4s or 3d. No sharp structure is evident, so that photoabsorption measurements should not be affected by saturation problems.

The $2a_1$ orbital has dominant C(2s) character. The photoelectron spectrum displays a distinct vibrational progression. However, photoabsorption from such orbitals (2s-like, containing first row atoms) is typically weak. Lee *et al.* (1977) report some barely discernible structure in their photoabsorption curve between 21.2–22.6 eV, which may be an early Rydberg state converging to $(2a_1)^{-1}$.

The $1a_1$ orbital obviously has C(1s) character. Consequently, transition to $3a_1$ (or 3s) is expected to be and is weak. The dominant excitation in the pre-edge region is to $2t_2$, which has p character.

Recent work at higher precision has superseded earlier measurements. The more current cross sections tend to be in good agreement with one another. This will be seen in the figures presented, and in the deduced oscillator strength distributions.

6.2.2 The data

The diffuseness of the absorption spectrum probably accounts for the lack of Rydberg series extrapolations of the adiabatic ionization potential of CH_4. Until quite recently, the best AIP was that based on photoion yield measurements. Chupka and Berkowitz (1971) studied CH_4 at 78 K and 300 K, observed no difference, and concluded that AIP $(CH_4) \leq 12.615 \pm 0.010$ eV. Later, Berkowitz *et al.* (1987a) examined CD_4 and re-investigated CH_4. They found AIP $(CH_4) = 12.61^{+0.01}_{-0.00}$ eV, and AIP $(CD_4) = 12.65_8 \pm 0.01_5$ eV. At this writing, Signorell and Merkt (1999) have reported on their pulsed-field-ionization (PFI) zero-kinetic-energy (ZEKE) photoelectron spectra of CH_4 and CD_4. From their preliminary analysis, AIP $(CH_4) = 12.6182 \pm 0.0043$ eV and AIP $(CD_4) = 12.6724 \pm 0.0031$ eV. Their analysis follows the pseudorotation among C_{2v} states developed by Frey and Davidson.

a   The oscillator strength distribution below the IP

Figure 6.3 displays the absolute photoabsorption cross section of CH_4 from the onset of absorption ($\sim$8.8 eV) to 24 eV. Samson and Yin (1989) estimate an overall accuracy of 1–2% for their data. We shall regard this data set as a secondary standard in assessing other measurements. The reader should be fore-warned that Samson and Yin (1989) supersedes the data of Samson and Haddad (1984). The latter display prominent dips between $\sim$13.25–13.75 eV which are not observed in other measurements.

The inelastic electron scattering data of Au *et al.* (1993) obtained by digitizing their figure, span the region from $\sim$8.8–24 eV. They are seen to fall below the data of Samson *et al.* between 14.25–19.5 eV by $\sim$4%, but are in good agreement at 13 eV and above 20 eV. These measurements were originally normalized by the TRK sum rule, but subsequently it was suggested that they be increased by $\sim$4% to conform to refractivity data and the $S(-2)$ sum rule (see Olney *et al.*, 1997). At lower energies (8.8–11.5 eV), the cross sections of Au *et al.* are seen to be in very good agreement with early photoabsorption measurements by Watanabe *et al.* (1953b). A photoabsorption spectrum presented by Koch and Skibowski (1971) has the same general shape below the IP as that seen in Fig. 6.3, but their estimated error in absolute cross sections is ±20%, and hence it is not shown.

We choose the absolute cross sections of Watanabe *et al.* to compute $S(p)$ between 8.61–11.27 eV by trapezoidal integration. In this interval, the $S(p)$ based on the data of Au *et al.* are almost identical. For the following segment, 11.27 eV–IP, we utilize the data of Au *et al.* The results are recorded in Table 6.4. These values may be ∼4% too low, as indicated by the above comments.

b IP–150 eV

Several observations must be made before proceeding with selections and computations in this region. The data of Au *et al.* seen in Fig. 6.3, and those of Koch and Skibowski, display a monotonic increase in cross sections from the IP to a broad maximum at ∼13.6 eV. The tabulated data of Samson *et al.* (1989) commence at 13.05 eV, and skip to 13.62 eV, thereby avoiding the dubious dips of Samson and Haddad (1984). To bridge the gap between 12.61–13.62 eV, we use the cross sections extracted from Au *et al.*

Samson *et al.* considered some prior work in the energy domain up to 24 eV, and found satisfactory agreement with the results of Rustgi (1964) and with Lee *et al.* between 18–24 eV, but less so with the cross sections of Metzger and Cook (1964). We had already noted that the data extracted from Au *et al.* were lower than Samson's, but converged for $20\,\text{eV} < h\nu < 24\,\text{eV}$. In Fig. 6.3, we also present the data of Person *et al.* (1975) which, though less accessible, deserve attention. Their cross section measurements have very nearly the same spectral shape as those of Samson *et al.*, but are about 4% higher. The Samson *et al.* (1989) measurements offer the highest estimated accuracy, and represent a rough mean of the cross sections from Au *et al.* and Person *et al.* We shall utilize the Samson *et al.* (1989) cross sections between 13.62–24.0 eV.

Figure 6.4 compares various sources of absorption cross sections between 25–200 eV. The agreement between the photoabsorption measurements of Samson *et al.* and the (e,e) data of Au *et al.* is excellent. We elect to use the Samson cross sections between 24.0–112.7 eV to evaluate the $S(p)$, but the (e,e) data yield essentially the same results. Above 112.7 eV, there is only one data point from Samson *et al.* We make a smooth transition to the values of Au *et al.* for the 112.7–150.0 eV interval. Trapezoidal integration is used throughout this domain. The values of $S(p)$ in the stated intervals are listed in Table 6.4.

c The continuum, 150.0–285.0 eV

Above 150 eV, Fig. 6.4 displays the (e,e) data of Au *et al.*, a few accurate photoabsorption measurements of Denne (1970) and atomic carbon cross sections from Henke *et al.* (1993). Denne's data points are somewhat lower that those of Au *et al.* Those of Henke *et al.* lie closer to Denne's with increasing photon energy.

Table 6.4 Spectral sums, and comparison with expectation values for CH_4[a]

Energy, eV	$S(-2)$	$S(-1)$	$S(0)$	$S(+1)$	$S(+2)$
$8.61-11.27$[b]	0.6200	0.4631	0.3472	0.2612	0.1972
$11.27-IP$[c]	0.4856	0.4270	0.3758	0.3311	0.2920
$IP-13.62$[c]	0.4510	0.4351	0.4200	0.4056	0.3918
$13.62-24.0$[d]	2.1529	2.7350	3.5624	4.7608	6.5266
$24.0-112.7$[d]	0.4772	1.1409	3.0129	9.1857	33.6401
$112.7-150.0$[c]	0.0011	0.0104	0.0982	0.9342	8.9453
$150.0-285.0$[e]	0.0006	0.0086	0.1238	1.7938	27.3058
$285.0-340.0$[f]	0.0011	0.0240	0.5436	12.3369	280.6989
$340.0-1740.0$[e]	0.0010	0.0335	1.2594	55.3115	3033.196
$1740.0-10\,000$[e]	–	0.0004	0.0700	15.1332	4026.225
10^4-10^5[g]	–	–	0.0016	1.9577	3295.040
10^5-10^6[h]	–	–	–	0.0900	1426.220
10^6-10^7[h]	–	–	–	0.0031	485.436
10^7-10^8[h]	–	–	–	0.0001	157.210
10^8-10^9[h]	–	–	–	–	50.093
$10^9-\infty$[h]	–	–	–	–	23.187
Total	4.1905	5.2780	9.8149	102.50	12 854.6
Expectation values	4.314$_3$[i]		10.0		12 834.1[j]
Other values	(4.318)[k]	5.43[k]	(10.0)[k]	101.44[l]	12 824[l]
		5.38[m]			

[a]In Ry units.
[b]Watanabe *et al.* (1953b).
[c]Au *et al.* (1993).
[d]Samson *et al.* (1989).
[e]Atomic carbon cross sections from Henke *et al.* (1993).
[f]Kivimäki *et al.* (1996).
[g]Atomic carbon cross sections from Chantler (1995).
[h]Assuming hydrogen-like behavior, bare carbon atom with screening, K shell only.
[i]Hohm (1993).
[j]Atomic additivity using Fraga *et al.* (1976); same result using Bunge *et al.* (1993).
[k]Thomas and Meath (1977).
[l]Mulder and Meath (1981).
[m]Olney *et al.* (1997).

The carbon K-edge in CH_4 occurs at 290.707 eV (Asplund *et al.*, 1985), but significant pre-edge structure exists (vide infra). We span the smooth continuum between 150–285 eV by fitting the data points of Denne to a 4-term polynomial. Values of $S(p)$ are determined by analytical integration of the polynomial, and appear in Table 6.4. The coefficients of the polynomial are given in Table 6.5.

d The carbon K-edge region, 285.0–340.0 eV

The absolute photoabsorption cross section of CH_4 in the vicinity of the carbon K-edge has been measured by Sivkov *et al.* (1986) and more recently by Kivimäki *et al.* (1996). The latter displays more features and a much higher intensity for the dominant peak at 288.0 eV, implying higher resolution. This peak, alluded to in Sect. 6.2.1, has been assigned to the transition $1a_1 \to 2t_2$, or $C(1s) \to 3p$,

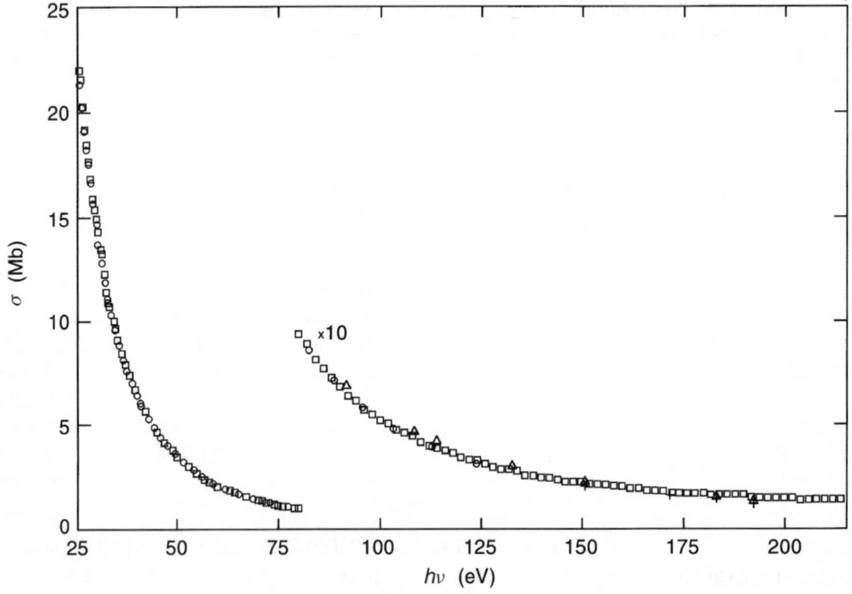

Fig. 6.4 Absolute photoabsorption spectrum of CH_4, 25–210 eV. □ Au *et al*. (1993a); o Samson *et al*. (1989); △ Henke *et al*. (1993) + additivity; + Denne (1970)

Table 6.5 Coefficients of the polynomial $df/dE = ay^2 + by^3 + cy^4 + dy^5$ fitted to data at various energies[a]

Energy range, eV	a	b	c	d
150.0–285.0	−4.031 33	261.0982	−3 005.43	11 572.96
340.0–1740.0	−13.022 5	4303.263	−77 622.4	446 724.9
1740.0–10 000	−1.116 77	3478.699	−48 076.5	−364 234

[a]df/dE in Ry units, $y = B/E$, $B = IP = 12.61$ eV.

the upper state having mixed valence-Rydberg character (Koch and Peyerimhoff, 1992; Ueda *et al*., 1995). We have digitized and integrated the data of Kivimäki *et al*., and entered the values of $S(p)$ in Table 6.4. Sivkov *et al*. present integrated oscillator strengths up to ∼320.7 eV (see also Sivkov *et al*., 1987) which we have compared with our integration of the data of Kivimäki *et al*., and there is good agreement in the region of overlap.

e Post K-edge, 340.0–10 000 eV

The atomic carbon cross section is essentially the only contributor here. The cross sections of Henke *et al*. are fitted, in two segments (340.0–1740.0 eV; 1740.0–10 000 eV), to 4-term polynomials. Each polynomial is analytically integrated to arrive at the corresponding $S(p)$, shown in Table 6.4. The coefficients of the polynomials are given in Table 6.5. At 340.0 eV, the fitted function yields $\sigma = 0.695\,\text{Mb}$, whereas the digitized data of Kivimäki *et al.* provide $\sigma = 0.72\,\text{Mb}$, i.e. the juncture is quite satisfactory.

f Post K-edge, 10^4–10^5 eV

The calculated atomic carbon cross sections of Chantler (1995) are used to evaluate $S(p)$. This region contributes $\sim26\%$ to $S(+2)$, $\sim2.0\%$ to $S(+1)$, and insignificantly to the other $S(p)$.

6.2.3 The analysis

Hohm (1993) has recently re-determined the molar refractivity of CH_4 at several wavelengths in the visible and UV regions, and from a Cauchy moment analysis deduced $S(-2) = 17.257\,\text{a.u.} \equiv 4.314_3\,\text{Ry}$ units. This is very slightly lower than that previously obtained by Thomas and Meath (1977) using older data. The selected polarizability of Miller (1999), $\alpha = 2.593 \times 10^{-24}\,\text{cm}^3$, is equivalent to $S(-2) = 4.375\,\text{Ry}$ units. The difference is attributable to infrared intensities (Thomas and Meath, 1977) which should be subtracted for the present comparison.

Our spectral sum, $S(-2) = 4.1905$, is 2.9% lower than the Hohm value, and our $S(0)$ is 1.9% lower than the TRK sum. Since the highest accuracy claimed in the data sets used is $\sim2\%$ by Samson *et al.*, fine tuning may be a luxury not worth undertaking. However, the negative deviation of both $S(-2)$ and $S(0)$ does indicate a direction of improvement. In Sect. 6.2.2.a, evidence was presented for increasing the cross section of Au *et al.* by $\sim4\%$. Such an increase for the relevant data of Au *et al.* (essentially 8.6–13.6 eV) would increase $S(-2)$ to 4.253 and $S(0)$ to 9.861, both now $\sim1.4\%$ lower than anticipated. The residual deficit suggests that the cross sections of Samson *et al.*, which contribute very significantly to both $S(-2)$ and $S(0)$, could be increased by 1.4%, which is within their margin of error and in the direction of Person *et al.* Applying the same corrections, $S(-1)$ would increase to 5.3853 Ry units.

Thomas and Meath use $S(-2)$ and $S(0)$ as constraints in their least-squares procedure for estimating the distribution of oscillator strengths. Their value for $S(-1)$, 5.43, is 0.8% higher than our best estimate. Olney *et al.* (1997) obtained $S(-1) = 5.38\,\text{Ry}$ units, after increasing the original cross sections of Au *et al.*

Our spectral sum for $S(+2)$ is 0.2% higher than a value based on the charge density at the nucleii (predominantly carbon) calculated at the Hartree–Fock level. The values for the latter, obtained from Fraga *et al.* (1976) and Bunge *et al.* (1993) are identical. Mulder and Meath (1981) have reported $S(+2) = 12\,824\,\text{Ry}$ units, based on the earlier work of Thomas and Meath. Their result, also based on a hydrogenic approximation, is actually 0.1% less than the 'expectation' value.

For $S(+1)$, additivity and the calculation of Fraga *et al.* yield 102.12 Ry units, compared to 101.44 given by Mulder and Meath and 102.50 from the current spectral sum. Correlation is expected to increase $S(+1)$ above additivity, but in this instance it is very slight, at best. The major contribution in our spectral sum derives from the experimentally based atomic carbon cross sections of Henke *et al.*

To arrive at $S_i(-1)$, we utilize the photoionization cross section Samson *et al.* with some interpolation near the IP from their Fig. 1. The ionization yield reaches unity below $750\,\text{Å} \equiv 16.53\,\text{eV}$. The resulting $S_i(-1) = 3.756$. Rieke and Prepejchal (1972), using a mixture of $CH_4 + CD_4$, obtained two values for $M_i^2 \equiv S_i(-1)$:4.23 ± 0.13 using electrons, and 3.69 ± 0.74 using positrons. Interestingly, the latter, with much larger uncertainty, is in good agreement with the present result.

Between the IP and $750\,\text{Å}$, about 62% of photoabsorption leads to photoionization, compared to 73% in H_2O and 57% in NH_3, all encompassing domains where the ionization yield is less than, or approaches unity. There is no obvious correlation with IP, or the number of hydrogens.

6.3 Acetylene (C_2H_2)

6.3.1 Preamble

Acetylene (C_2H_2) is linear in its ground state, with the electron configuration

$$(1\sigma_g)^2(1\sigma_u)^2(2\sigma_g)^2(2\sigma_u)^2(3\sigma_g)^2(1\pi_u)^4, \tilde{X}^1\Sigma_g^+$$

Isoelectronic with N_2, acetylene has two uppermost occupied orbitals responsible for the strong C–C triple bond. It also displays sharp absorption features below the IP, characteristic of small molecules. The inner valence $2\sigma_g$ and $2\sigma_u$ orbitals, bonding and antibonding combinations of C(2s) orbitals, have binding energies $\sim$20 eV and display weak absorption. The $1\sigma_g$ and $1\sigma_u$ are essentially the carbon K-shell.

Significant sources of new data have become available since our earlier effort (Berkowitz, 1979), and indeed since Kumar and Meath (1992) made a sum-rule analysis. The latter authors considered data published up to 1988. Rather recently, Cooper *et al.* (1995a) presented cross section measurements using dipole (e,e) spectroscopy in the valence region with $\sim$0.05 eV resolution, and to 200 eV with $\sim$1 eV resolution. In the sub-ionization region, photoabsorption measurements by Wu *et al.* (1989), Chen *et al.* (1991) and Smith *et al.* (1991) are available for comparison. Just above the IP (11.4–11.7 eV), there are newer data by Xia *et al.* (1991), and also from $\sim$13.25–22.5 eV, Ukai *et al.* (1991) proffer more recent measurements. Finally, absolute photoabsorption measurements in the vicinity of the carbon K-shell have been presented by Kivimäki *et al.* (1997). As might be expected, these more recent data sets help to improve agreement with the sum rules.

6.3.2 The data

The best current adiabatic ionization potential (AIP) of C_2H_2, obtained from a Rydberg series limit, is $91\,956 \pm 1\,\text{cm}^{-1} \equiv 11.4011 \pm 0.0001\,\text{eV}$ (Shafizadeh *et al.*,

1997). This is about $4\,cm^{-1}$ higher than a ZEKE measurement by Pratt *et al.* (1993), which resulted in AIP $= 11.4006 \pm 0.0002\,eV$.

a *The oscillator strength distribution below the IP*

The (e,e) spectrum of Cooper *et al.* clearly shows weak absorption to the valence $\tilde{A}$, $\tilde{B}$ and $\tilde{E}$ states, and much stronger absorption to Rydberg states. The $\tilde{X} \to \tilde{A}$ and $\tilde{X} \to \tilde{B}$ vibronic bands are only slightly resolved by Cooper *et al.* The photoabsorption measurements of Wu *et al.* display better resolution, and they explicitly present $S(0)$, $S(-1)$ and $S(-2)$ resulting from the $\tilde{X} \to \tilde{B}$ transitions. We choose this source, and Cooper *et al.* for the $\tilde{X} \to \tilde{A}$ transitions, which are still weaker by a factor 10. The distinction between Wu *et al.* and Cooper *et al.* here is moot, since there is good agreement between the two, and the contributions to the $S(p)$ are small.

For the $\tilde{X} \to$ Rydberg region (8.161–11.126 eV), Cooper *et al.* (1995a) make a reasonable case that saturation occurs for some of the stronger peaks in the data of Suto and Lee (1984). Unfortunately, most of the newer photoabsorption measurements do not cover this region, but there is some support from Smith *et al.* for the oscillator strength of the 8.161 eV strong peak given by Cooper *et al.* The resolution of the (e,e) measurements is insufficient to separate the vibronic bands, especially for the higher Rydbergs. We assume that their integrated oscillator strength for the entire set of vibronic bands is more accurate then their deconvolution of the individual bands. This integrated oscillator strength is approximately 30% larger than that deduced from the data of Suto and Lee, and makes a significant contribution to $S(-2)$. We choose the values of Cooper *et al.* here, noting that use of the data of Suto and Lee would lower $S(-2)$ by about 0.4 Ry units which, in the final analysis, would be unacceptable.

Cooper *et al.* have presented oscillator strengths for the $\tilde{X} \to \tilde{E}$ vibronic bands. In principle, equivalent photoabsorption data are contained in Fig. 1 of Suto and Lee, but accurate oscillator strengths are difficult to extract. The two lowest energy peaks in the spectrum of Suto and Lee appear to have 1/2 or less of the oscillator strengths given by Cooper *et al.*, but the total oscillator strength for the $\tilde{X} \to \tilde{E}$ transitions, which is given as 0.12, may be comparable. We choose the default values of Cooper *et al.*, with the proviso that they may be high. They are also the result of deconvolutions.

There is a gap in tabulated oscillator strengths between 11.126 eV and the IP. From the compressed Fig. 3 of Cooper *et al.* we extract $f = 0.0789$. From Suto and Lee, Fig. 1, which is more clearly resolved, we obtain 0.0718. We choose the latter. The selected oscillator strengths in the sub-ionization region are summarized in Table 6.6.

b *The autoionization region, IP–11.70 eV*

In Fig. 6.5, we display the photoabsorption cross sections of several investigators from the IP to 13.78 eV (1087.5–900 Å). The high resolution data of Xia *et al.* reveal relatively sharp autoionization structure between 1087.5–1060 Å

Table 6.6 Contributions to $S(p)$ of transitions below the IP in C_2H_2[a]

Energy, eV	$S(-2)$	$S(-1)$	$S(0)$
$\tilde{X} \to \tilde{A}$,			
5.865–6.640[b]	0.0049	0.0023	0.0011
$\tilde{X} \to \tilde{B}$,			
6.68–8.00[c]	0.0379	0.0202	0.0108
$\tilde{X} \to \tilde{R}$,			
8.161–11.126[b]	1.7400	1.2155	0.8571
$\tilde{X} \to \tilde{E}$[b],			
9.102–9.619	0.2580	0.1783	0.1233
11.126–11.400[d]	0.1048	0.0868	0.0718
Total	2.1456	1.5031	1.0641

[a] $S(p)$ in Ry units.
[b] Cooper *et al*. (1995a).
[c] Wu *et al*. (1989).
[d] Suto and Lee (1984).

(IP–11.7 eV). This confirms (and improves upon) earlier work by Person and Nicole (1970). The structure here is attributed to Rydberg series converging to vibrationally excited levels of the ground state ($\tilde{X}^2\Pi_u$) of $C_2H_2^+$. At higher energies (13.5–17.5 eV), Ukai *et al*. have observed broader, more diffuse features attributed to Rydberg series converging to $(3\sigma_g)^{-1}$ and $(2\sigma_u)^{-1}$. Between these regions, high-resolution photoabsorption data are unavailable, but the prevailing evidence from lower-resolution measurements is that the cross section increases smoothly with increasing energy. Here, we rely primarily on Cooper *et al*., which connects these regions. Earlier photoabsorption measurements by Metzger and Cook (1964) and Han *et al*. (1989) also display no structure, but their absolute cross sections lie considerably lower than more recent data, and hence are rejected.

The cross sections of Xia *et al*., estimated to be accurate to $\leq$5%, have been used to evaluate $S(p)$ between IP–11.7 eV by trapezoidal integration.

c   11.70–13.26 eV (1060–935 Å)

For reasons given in the previous section, a smooth curve is drawn joining the data of Xia *et al*. and Ukai *et al*. Some points from Cooper *et al*. lie on this curve. The contributions to $S(p)$ are determined by trapezoidal integration.

d   13.26–20.66 eV (935–600 Å)

The cross sections of Ukai *et al*., described as having a maximum error of 'a few per cent', are favored in this interval. The data of Cooper *et al*. agree with Ukai *et al*., but the latter provide more detailed structure. Earlier photoabsorption measurements by Wu and Judge (1985) covering some of this region lie ~20%

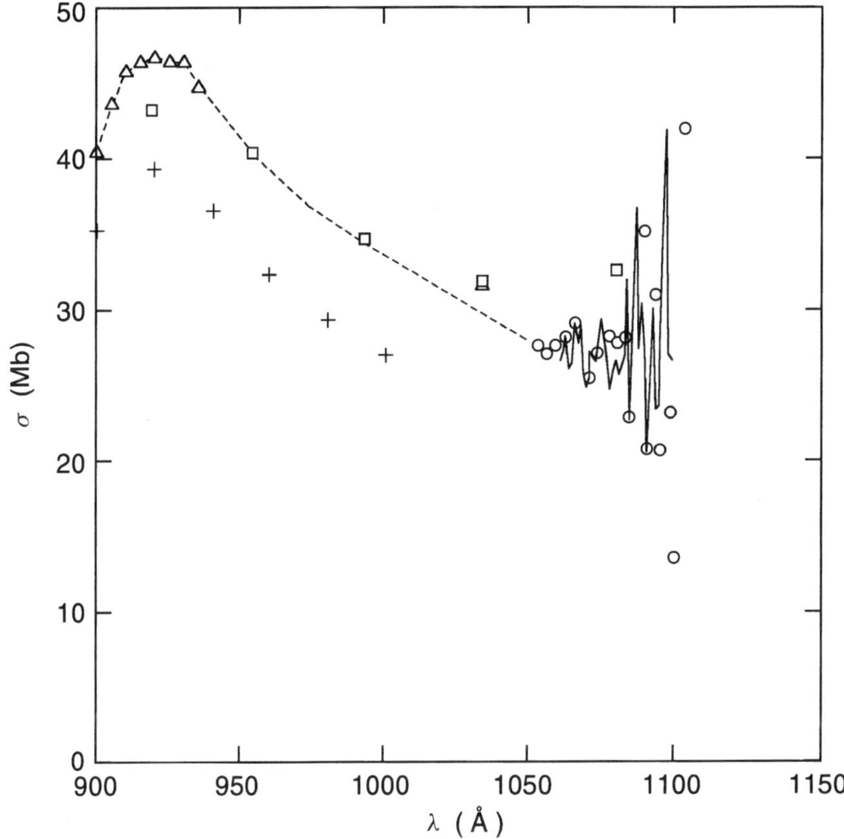

Fig. 6.5   Absolute photoabsorption spectrum of C_2H_2, 900–1100 Å. — Xia *et al*. (1991); o Person and Nicole (1970); △ Ukai *et al*. (1991); □ Cooper *et al*. (1995a); + Metzger and Cook (1964)

lower in cross section. The graphical data of Ukai *et al*. have been electronically scanned and the $S(p)$ determined by fine-grid trapezoidal integration.

e   20.66–61.99 eV (600–200 Å)

The data of Ukai *et al*. extend only to $\sim$23.5 eV, and here tend to be $\sim$5% higher than the values of Cooper *et al*. The cross sections of Wu and Judge (1985) remain lower than those of Cooper *et al*. for most of this interval, but cross at $\sim$50 eV and actually become higher at 62 eV. Clearly, the shapes of the absorption curves differ. The synchrotron-based data of Wu and Judge may have been affected by higher-order radiation, although filters were used. The values of Cooper *et al*. bridge the region of directly measured photoabsorption cross sections of Ukai *et al*. and (see below) the summed atomic cross sections

(Henke *et al.*, 1993). Using the tabulated cross sections of Cooper *et al.*, the $S(p)$ in this interval were evaluated by trapezoidal integration.

f 62–90 eV; 90–285 eV

In Fig. 6.6, we compare the cross sections of Wu and Judge, which terminate at 68.9 eV, the tabulated values of Cooper *et al.*, which extend to 200 eV, and summed atomic cross sections from Henke *et al.*. Only the highest 7 eV of Wu and Judge are relevant to this interval, and the shape of their photoabsorption curve has been questioned. Of more concern is the observation that the cross sections of Cooper *et al.* dip below atomic additivity at ∼90 eV, and remain lower to 200 eV. In the case of CH_4, where accurate, directly measured photoabsorption values were available, they agreed with atomic additivity in this interval to about the experimental error. Here, the difference is about 25%. Our modus operandi is to divide the 62–285 eV region (upper limit set by the approach to the K-edge) into two regions, in two separate ways, and to compare them. Mode 1 utilizes the data of Cooper *et al.* from 62–90 eV (the approximate crossing point) and summed atomic cross sections to 285 eV. $S(0)$ for the total interval is 1.143. Mode 2 retains the values of Cooper *et al.* from 62–150 eV, then continues with summed atomic cross section to 285 eV. $S(0)$ for the interval becomes 1.092, reflecting the lower cross sections of Cooper *et al.*. The difference in $S(0)$ is very small (0.05 cf. 14.0), and for the other $S(p)$ it is relatively less important. The

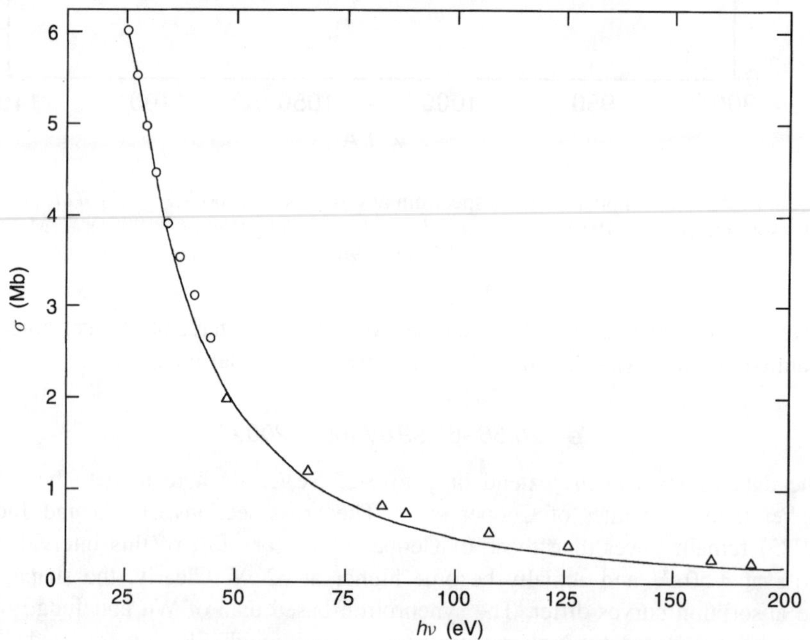

Fig. 6.6 Absolute photoabsorption spectrum of C_2H_2, 25–200 eV. — Cooper *et al.* (1995a); △ Henke *et al.* (1993); ○ Wu and Judge (1985)

Table 6.7 Spectral sums, and comparison with expectation values for C_2H_2[a]

Energy, eV	$S(-2)$	$S(-1)$	$S(0)$	$S(+1)$	$S(+2)$
5.865–IP[b]	2.1456	1.5031	1.0641	0.7609	0.5492
	(2.0980)[c]	(1.4699)[c]	(1.0408)[c]		
IP–11.70[d]	0.1026	0.0870	0.0739	0.0627	0.0532
11.70–13.26[e]	0.5863	0.5391	0.4964	0.4576	0.4225
13.26–20.66[f]	2.0263	2.4272	2.9513	3.6427	4.5634
20.66–61.99[g]	0.9218	1.9396	4.4146	11.0140	30.2768
	(0.8999)[c]	(1.8934)[c]	(4.3095)[c]	(10.7518)[c]	(29.5559)[c]
62–285[h]	0.0252	0.1602	1.1429	9.5356	95.4472
285.9[i]	0.0003	0.0072	0.1508	3.1688	66.5868
286.5–360[i]	0.0021	0.0500	1.1708	27.5301	649.8762
360–1740[h]	0.0016	0.0596	2.3946	111.5772	6254.826
1740–10 000[h]	–	0.0008	0.1400	30.2664	8052.450
10^4–10^5[j]	–	–	0.0032	3.9154	6590.080
10^5–10^6[k]	–	–	–	0.1800	2852.440
10^6–10^7[k]	–	–	–	0.0062	970.872
10^7–10^8[k]	–	–	–	0.0002	314.420
10^8–10^9[k]	–	–	–	–	100.186
10^9–∞[k]	–	–	–	–	46.374
Total	5.8118	6.7738	14.0026	202.12	26 029.4
Total (adj)	(5.7423)[c]	(6.6944)[c]	(13.8742)[c]	(201.86)[c]	(26 028.7)[c]
Expectation values	5.740[l]		14.0		25 636.3[m]
Other values	(5.740)[l]	6.700[l]	(14.0)[l]	201.0[l]	26 888[l]
	5.738[n]	6.52[n]			
	5.07[o]	5.88[o]	13.01[o]		
			13.55[f]		

[a]In Ry units.
[b]See Table 6.7 and text.
[c]Data of Cooper *et al.* (1995a), adjusted downward by 2.4%.
[d]Xia *et al.* (1991).
[e]Interpolated values. See Fig. 6.5 and text.
[f]Ukai *et al.* (1991).
[g]Cooper *et al.* (1995a).
[h]Summed atomic cross sections, from Henke *et al.* (1993).
[i]Kivimäki *et al.* (1997); Kempgens *et al.* (1997b).
[j]Atomic carbon cross sections from Chantler (1995).
[k]Assuming hydrogen-like behavior, bare carbon atom with screening, K-shell only. See text.
[l]Kumar and Meath (1992).
[m]Using atomic additivity and Hartree–Fock calculations for atomic carbon. Identical results are obtained from Fraga *et al.* (1976) and Bunge *et al.* (1993).
[n]Olney *et al.* (1997).
[o]Wu *et al.* (1989).

absolute cross sections are very low in this region, and the sum rules are not very sensitive, and hence cannot distinguish between these alternative cross sections. We select mode 1 for inclusion in Table 6.7. The summed atomic cross sections have been fitted by regression to a 4-term polynomial. The coefficients of the polynomial are given in Table 6.8.

Table 6.8   Coefficients of the polynomial $df/dE = ay^2 + by^3 + cy^4 + dy^5$ fitted to data at various energies[a]

Energy range, eV	a	b	c	d
90–285	−4.095 82	371.7201	−3 488.22	11 226.67
360–1740	14.123 06	7389.95	−105 703	259 751.7
1740–10 000	−2.732 31	9413.515	−143 892	−1 205 737

[a] df/dE in Ry units, $y = B/E$, $B = \text{IP} = 11.4011\,\text{eV}$.

g   Carbon K-shell region, 285–360 eV

The absolute photoabsorption cross section of C_2H_2 between 280–340 eV has been reported recently by Kivimäki *et al*. We extend this information to 360 eV, from additional data given by Kempgens *et al*. (1997b). The K-edge in C_2H_2 occurs at $\sim$291.1 eV (Jolly *et al*., 1984). However, the strong $1s \rightarrow \pi^*$ resonance peak occurs at 285.9 eV, and there is additional pre-edge structure. The $1s \rightarrow \pi^*$ resonance is treated separately. We extract $f = 0.1508$ for this resonance. The remaining cross sections to 360 eV have been extracted from figures in the indicated references, and trapezoidally integrated to evaluate the $S(p)$.

h   Post K-edge; 360–10 000 eV

We use atomic additivity, but the atomic carbon cross section is essentially the only contributor here. The cross sections of Henke *et al*. are partitioned into two regions, 360–1740 eV and 1740–10 000 eV. Each region is fitted to a 4-term polynomial, which is then integrated to yield $S(p)$. The coefficients of the polynomials are given in Table 6.8. At 360 eV, we calculate $\sigma = 1.217\,\text{Mb}$, in fortuitous agreement with the extracted value from Kivimäki *et al*. and Kempgens *et al*., 1.21 Mb.

i   Post K-edge, 10^4–10^5 eV

The calculated atomic carbon cross sections of Chantler (1995) are used to evaluate $S(p)$, which are recorded in Table 6.7. This region contributes $\sim$25% to $S(+2)$, $\sim$2% to $S(+1)$.

6.3.3   The analysis

Expectation values for $S(-2)$ can usually be inferred from molar refractivity or dielectric constant measurements. Acetylene is one of those cases where the static polarizability (as measured by the dielectric constant) is larger than that deduced from refractivity measurements in the visible and ultraviolet. Thus, Miller (1997) selects a polarizability (α) of $3.93 \times 10^{-24}\,\text{cm}^3$, based on dielectric constant measurements, which translates to $S(-2) = 6.63$. Kumar and Meath (1992) have analyzed molar refractivity measurements of 1936 vintage to extract $S(-2) = 5.74$. In order to rationalize the difference, we have examined the possible role of infrared contributions. Acetylene, a linear molecule, has five fundamental normal vibrations. Of these, three are Raman active, two are infrared active. One of the two latter vibrations, a Π_u bending mode at $730\,\mathrm{cm}^{-1}$, makes the dominant contribution to $S(-2)$. Its integrated intensity has been measured by Kim and King (1979), and they report $A = 177.1 \pm 2.1\,\mathrm{km/mol}$. The contribution of this mode to $S(-2)$ is 0.751. An additional contribution of 0.015 can be deduced for the other infrared active mode. The sum, 0.766, is close to the difference (0.89) of $S(-2)$ values inferred from dielectric constant and molar refractivity measurements. Bishop and Cheung (1982) had earlier obtained a result equivalent to $S(-2)_{\mathrm{vib}} = 0.758$. Russell and Spackman (1996) have obtained good agreement with the experimental values of both $S(-2)_{\mathrm{elec}}$ and $S(-2)_{\mathrm{vib}}$, using correlated wave functions. Since the sum rules refer to electronic transitions rather than nuclear motions, the value based on molar refractivity measurements is the appropriate one to compare with our spectral sum.

The current spectral sum for $S(-2)$, 5.8118 Ry units, is 1.25% higher than the expectation value. This is probably fortuitously close, and normally would not warrant further correction. Cooper *et al.* had obtained a slightly higher value, $S(-2) = 5.88$ Ry units. However, Olney *et al.* (1997) subsequently adjusted their earlier data (Cooper *et al.*, 1995a) to make it agree with their expectation value, $S(-2) = 5.74$ Ry units. If we apply the same correction factor only to the data of Cooper *et al.*, it coincidentally leads to a spectral sum $S(-2) = 5.74$ Ry units. However, this reduces our spectral sum for $S(0)$ from the extraordinarily good 14.0026 to 13.874, which is now $\sim1\%$ too low. The adjusted $S(p)$ are recorded separately in Table 6.7. The uncertainty in measurement clearly does not enable us to choose between these alternatives.

The expectation value for $S(+2)$ is taken as the sum of corresponding expectation values for the constituent atoms, based on Hartree–Fock values of the charge densities at the nucleii. Identical values for this quantity are obtained from Fraga *et al.* (1976) and from Bunge *et al.* (1993). Our spectral sum for this quantity, 26 029.4 Ry units, is 1.5% higher than the expectation value. Kumar and Meath obtain a value of $S(+2)$ which is 4.9% higher.

As mentioned in Sect. 6.3.1, Kumar and Meath used earlier input data than was available for the current sum rule analysis, and adjusted it to conform to two constraints namely $S(-2) = 5.740$ and $S(0) = 14.000$. Their inferred value of $S(-1)$, 6.700, is about 1% lower than our unadjusted spectral sum, but almost identical to the adjusted one. Other values of $S(-1)$ listed in Table 6.7 which do not satisfy the constraints for $S(-2)$ and $S(0)$ are not likely to be as accurate.

The value of $S(+1)$ deduced from atomic additivity (Fraga *et al.*, 1976) is 196.2, which is lower than the current spectral sum (202.1 or 201.9) and that of Kumar and Meath, 201.0. This is the expected direction of deviation from additivity, which can be attributed to correlation effects.

To arrive at $S_i(-1)$, we proceed as follows:

1. Xia *et al.* present the photoionization cross section between the IP and 11.70 eV.
2. Between 11.70–13.26 eV, the currently interpolated photoabsorption cross sections (see Fig. 6.5) are combined with quantum yields of ionization from

Metzger and Cook (1964). Although we earlier rejected the photoabsorption cross sections of Metzger and Cook, their quantum yields of ionization appear reasonable, according to Fig. 2 of Ukai *et al*. In that figure, the quantum yield approaches unity at $\sim$600 Å (20.66 eV).

3. The photoionization cross sections of Ukai *et al*. are used from 13.26–20.66 eV.

The contribution to $S_i(-1)$ from IP–20.66 eV is 2.5379. Above 20.66 eV, the quantum yield of ionization is assumed to be unity, and hence from Table 6.7, the contribution to $S_i(-1)$ is 2.2174 Ry units, or 2.1712 Ry units if we adjust the data of Cooper *et al*. Hence, $S_i(-1)_{\text{total}} = 4.755$ Ry units, or 4.709 Ry units (adjusted). Both Wu and Judge (1985) and Wu *et al*. (1989) have obtained 4.08 for this quantity. Rieke and Prepejchal (1972) measured $M_i^2 = 5.21 \pm 0.086$, about 10% larger than our optical $S_i(-1)$.

Between the IP and 20.66 eV, approximately 84% of photoabsorption leads to photoionization. This figure is relatively high, partly due to somewhat arbitrarily extending the range of consideration. Between 700 Å (17.17 eV) and 600 Å (20.66 eV), the quantum yield of ionization is already quite high, but does not reach unity until its upper extremity.

6.4 Ethylene (C_2H_4)

6.4.1 Preamble

Ethylene, C_2H_4, is the prototypical organic molecule with a C=C double bond. In its electronic ground state, it is planar, with D_{2h} symmetry. Within the independent particle approximation, the electron configuration is

$$(1a_g)^2(1b_{1u})^2(2a_g)^2(2b_{1u})^2(1b_{2u})^2(3a_g)^2(1b_{3g})^2(1b_{3u})^2, \tilde{X}^1A_g.$$

The two equivalent carbon atoms give rise to the two almost degenerate core orbitals $1a_g$ and b_{1u}. Single electron ejection from each of the other six occupied orbitals results in states of $C_2H_4^+$ encompassing the range 10.5–24 eV. This is clearly seen in a He II photoelectron spectrum by Holland *et al*. (1997), perhaps the latest of many photoelectron studies.

The electronic ground state of $C_2H_4^+$ (and the Rydberg states converging to it) has a twisted, non-planar structure, with a torsion angle of $\sim$25°. This behavior was initially predicted by Mulliken (1959). It is evident in the photoelectron spectrum by the prominent appearance of vibrational structure corresponding to two-quantum excitation of the torsional frequency (see, for example, Holland *et al*., 1997). Calculational efforts to verify this loss of symmetry have sometimes led to ambiguous results, but the most extensive calculations appear to have verified the point (see, for example, Salhi-Benachenhou *et al*., 1998).

Jhanwar *et al*. (1983) performed a sum-rule analysis of the oscillator strength distribution based on data available to them. Since that time, significant new data have come to light, notably photoabsorption measurements from the IP to 25 eV

(Holland *et al.*, 1997), and in the carbon K-edge region, 284–340 eV (Kempgens *et al.*, 1995; 1998). This information has been complemented by inelastic electron scattering measurements by Cooper *et al.* (1995b) covering the range 6–200 eV. In addition, modern refractivity measurements (Hohm, 1993) have established $S(-2) = 6.947$ Ry units.

6.4.2 The data

Two fairly recent studies have established the adiabatic IP of C_2H_4 to higher precision, and presumably higher accuracy, than earlier work. Williams and Cool (1991) have observed a two-photon allowed nf-Rydberg series with an extrapolated IP of $84\,799 \pm 5\,\mathrm{cm}^{-1} \equiv 10.5137 \pm 0.0006$ eV. Holland *et al.* obtained a well-resolved He I photoelectron spectrum which yields AIP $= 10.5122$ eV 'with an accuracy of the order of ± 1 meV or better'. We adopt an average value of 10.513 eV.

a   The discrete spectrum and transitions below the IP

The alternatives here are early photoabsorption measurements by Zelikoff and Watanabe (1953) and recent (e,e) spectra of Cooper *et al.*. The resolution in the photoabsorption measurements was approximately 5 meV, that in the inelastic electron scattering measurements is about ten times poorer. This manifests itself in more clearly resolved vibronic structure in the photoabsorption experiments, which are much less apparent in the (e,e) spectra. Also, the photoabsorption studies could use Beer's law absolute calibrations, whereas the (e,e) measurements used TRK normalization (and later, $S(-2)$ normalization, see Olney *et al.*, 1997). We shall adopt the procedure of utilizing photoabsorption data when available, subject to a scaling adjustment upon completion of the sum rule analysis. Thus, we tentatively list the $S(p)$ for this sub-ionization region after electronic scanning of the figures given by Zelikoff and Watanabe in Table 6.9. We shall find that the spectral value of $S(-2)$ is sensitive to the oscillator strengths in this region.

b   The autoionization region and beyond: IP–24.8 eV (500 Å)

Here we turn to the recent absolute photoabsorption measurements of Holland *et al.*, which were performed with considerably higher resolution and hence display structure that is largely absent in the (e,e) data of Cooper *et al.*. The structure is broad enough to negate concerns regarding saturation. Earlier photoabsorption experiments studied portions of this range. Person and Nicole (1968) examined the region IP–11.8 eV; their cross sections are very close to those of Holland *et al.*. However, Person and Nicole (1974), using a helium continuum light source for 12.5 eV $\leq hv \leq 20$ eV, observed cross sections about 5% higher than those of Holland *et al.*. The Person/Nicole cross sections are closer to the (e,e) results of Cooper *et al.*. This observation will provide further support for our conclusion, following the sum rule analysis, that the cross sections of Holland *et al.* require upward adjustment. A short region (IP–11.7 eV) has also been

Table 6.9 Spectral sums, and comparison with expectation values for C_2H_4[a]

Energy, eV	$S(-2)$	$S(-1)$	$S(0)$	$S(+1)$	$S(+2)$
6.62–IP[b]	1.8548	1.1507	0.7249	0.4638	0.3013
IP–24.8[c]	4.1691	4.9488	6.1609	8.1306	10.9262
24.8–80.0[d]	0.6670	1.6400	4.3543	12.7279	41.4929
80.0–200.0[e]	0.0107	0.0834	0.6879	6.0656	57.3691
200–285[e]	0.0003$_5$	0.0060	0.1037	1.7792	30.4898
284.5[f]	0.0002	0.0036	0.0762	1.5940	33.3663
286–340[f]	0.0020	0.0445	1.0134	23.1177	528.5613
340–1740[e]	0.0020	0.0688	2.6313	117.6608	6410.0455
1740–10 000[e]	–	0.0008	0.1401	30.2746	8054.38
10^4–10^5[g]	–	–	0.0032	3.9154	6590.08
10^5–10^6[h]	–	–	–	0.1800	2852.44
10^6–10^7[h]	–	–	–	0.0062	970.87
10^7–10^8[h]	–	–	–	0.0002	314.42
10^8–10^9[h]	–	–	–	–	100.19
10^9–∞[h]					46.37
Total	6.7062	7.9466	15.8959	205.916	26 041.3
Revised sums[i]	(6.9380)	(8.1377)	(16.0771)	(206.12)	(26 041.5)
Expectation values	6.947[j]		16.0		25 646.9[k]
Other values	(6.925)[l]	8.155[l]	(16.0)[l]	200.2[l]	26 224.0[l]
	7.036[d]		16.1[d]		
	6.9475[m]	7.95[m]			
	6.653[c]	7.906[c]	15.963[c]		

[a]In Ry units.

[b]Zelikoff and Watanabe (1953b). Recommendation after sum-rule analysis is to increase these cross sections by factor 1.08.

[c]Holland *et al.* (1997).

[d]Cooper *et al.* (1995b).

[e]Summed atomic cross sections from Henke *et al.* (1993).

[f]Kempgens *et al.* (1995; 1998).

[g]Chantler (1995).

[h]Assuming hydrogen-like behavior, bare carbon atom with screening, K-shell only (Bethe and Salpeter, 1977).

[i]Increasing cross sections: 6.62 eV–IP ($\times 1.08$), IP–24.8 eV ($\times 1.02$). See text.

[j]Hohm (1993).

[k]Using atomic additivity and Hartree–Fock calculations for atomic carbon. Identical results are obtained from Fraga *et al.* (1976) and Bunge *et al.* (1993).

[l]Jhanwar *et al.* (1983).

[m]Olney *et al.* (1997).

studied by Xia *et al.* (1991). Their cross sections are ∼8% lower than those of Holland *et al.*

Chen and Wu (1999) have recently re-determined photoabsorption (σ_a) and photoionization (σ_i) cross sections from IP–11.5 eV at higher resolution, with a stated accuracy of ±10%. Good agreement with Holland *et al.* can be seen for σ_a between 11.2–11.5 eV, but as the IP is approached, their cross sections fall lower by ∼10%. Their values of σ_i are consistently below those of Holland *et al.* and Cooper *et al.*

The photoabsorption cross sections of Holland *et al.* have been scanned electronically, and the $S(p)$ values determined by fine-mesh trapezoidal integration. The values of $S(-2)$ and $S(-1)$ recorded in Table 6.9 are in excellent agreement with corresponding values given by Holland *et al.*, but our $S(0)$ is $\sim$4% lower. Our result was reproduced to within 0.2% using alternatively an overview figure and sums of partial spectra. Their result is essentially the value of $S(0)$ for this region which we obtain from the cross sections of Cooper *et al.*, despite the fact that the latter cross sections are $\sim$4% higher.

c   The continuum, 24.8–80.0 eV

The lowest energy of this domain is above the ionization energies of the outer- and inner-valence orbitals. Hence, it is expected to be structureless, and the available evidence confirms this view. The alternative published data include early photoabsorption measurements of Lee *et al.* (1973) and the (e,e) experiments of Cooper *et al.*, although the latter publication presents unpublished data of T. Ibuki and J. W. Taylor as well. The energy dependence of the photoabsorption spectrum of Lee *et al.* differs from that of Cooper *et al.*, displaying a distinctly lower cross section at 19 eV, crossing the Cooper curve at $\sim$23 eV, and remaining above the Cooper curve at higher energies (see Fig. 6.7). This disparity in shape also exists when comparing the data of Lee *et al.* with the photoabsorption cross sections of Ibuki and Taylor shown by Cooper *et al.* and

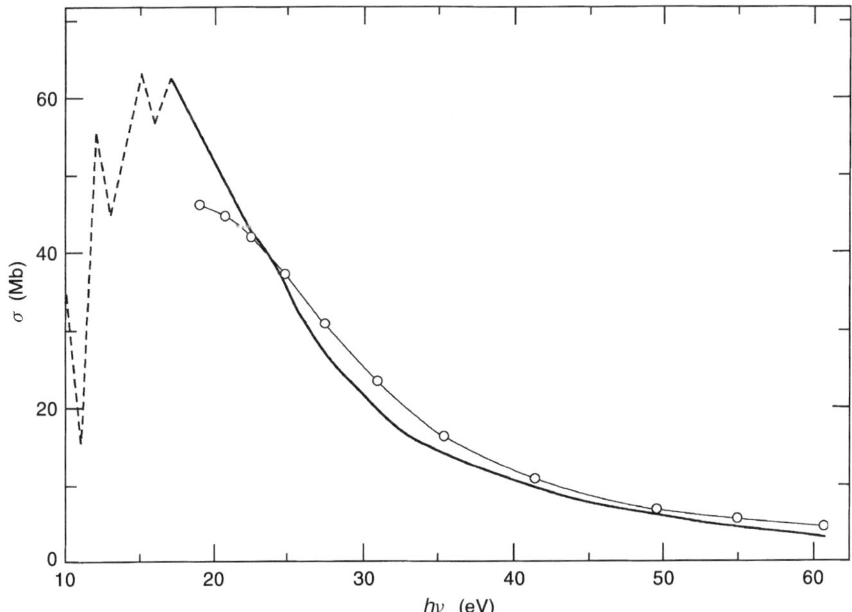

Fig. 6.7   Absolute photoabsorption spectrum of C_2H_4, 10–60 eV. —— Cooper *et al.* (1995b); o Lee *et al.* (1973)

with still earlier photoabsorption studies by Schoen (1962). Thus, the spectral dependence of the photoabsorption curve of Lee *et al.* is suspect, and cannot be corrected by simple scaling. Consequently, we eschew our pattern of preferring photoabsorption data in this instance, and evaluate $S(p)$ by trapezoidal integration of the (e,e) cross sections of Cooper *et al.*

d   The continuum, 80–200 eV; 200–285 eV

The options here are limited to the (e,e) data of Cooper *et al.* and summation of atomic photoabsorption cross sections, taken from Henke *et al.* (1993). In Fig. 6.8, these alternatives are compared. Above 80 eV, the summed atomic cross sections exceed the molecular cross sections given by the (e,e) determinations. We opt for the summed atomic cross sections here since we deem it less likely that the molecular cross sections should be smaller. Cooper *et al.* implicitly seem to accept this view, since they perform sum rule estimates using atomic additivity above 62 eV. The absolute magnitudes are low in this region, and the differences still smaller, making the choice insensitive to any of the sum rules. We fit the summed Henke cross sections to a 4-term polynomial by regression. The $S(p)$ are evaluated analytically from this function, which is continued to 285 eV, where it approaches the structure preceeding the carbon K-edge. The values of $S(p)$ are recorded in Table 6.9, the coefficients of the polynomial in Table 6.10.

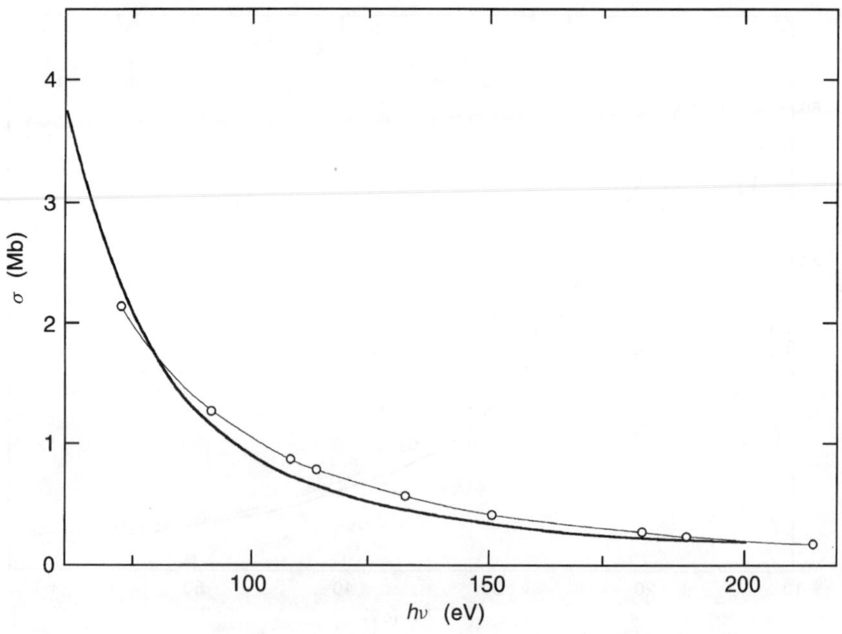

Fig. 6.8   Absolute photoabsorption spectrum of C_2H_4, 35–200 eV. —— Cooper *et al.* (1995b); o Henke *et al.* (1993) + additivity

Table 6.10 Coefficients of the polynomial it $df/dE = ay^2 + by^3 + cy^4 + dy^5$ fitted to data at various energies[a]

Energy range, eV	a	b	c	d
80–285	−0.017 13	283.8744	−2 345.89	6 811.379
340–1740	16.627 49	9 440.229	−146 516	390 663.8
1740–10 000	−3.236 18	12 024.91	−202 311	−1 562 715

[a] df/dE in Ry units, $y = B/E$, $B = \mathrm{IP} = 10.513\,\mathrm{eV}$.

e The carbon K-edge region, $\sim$285–340 eV

Kempgens *et al.* (1998) have recently presented the absolute photoabsorption curve of C_2H_4 between 283–340 eV. It consists of a strong, structured $C(1s) \rightarrow \pi^*$ peak at 284.5 eV, followed by additional structure between $\sim$287–291 eV, and then weaker indulations and a decline to 340 eV. The 284.5 eV peak and 287–291 eV structure had been given in expanded form previously (Kempgens *et al.*, 1995). The cross sections in this region have been extracted from the figures in the indicated references, and integrated trapezoidally to determine the $S(p)$.

f Post K-edge; 340–10 000 eV

We again resort to atomic additivity, using the atomic cross sections of Henke *et al.*, partitioned into two regions, 340–1740 eV and 1740–10 000 eV. Each region is fitted to a 4-term polynomial, which is then integrated to yield $S(p)$. At 340 eV, we calculate $\sigma = 1.387\,\mathrm{Mb}$, to be compared with $\sigma = 1.409\,\mathrm{Mb}$ extracted from Kempgens *et al.* (1998). The juncture is satisfyingly close.

g 10^4–10^5 eV

The calculated atomic cross sections of Chantler (1995) are used to evaluate $S(p)$.

6.4.3 The analysis

The new value for $S(-2)$, obtained by Hohm (6.947 Ry units) essentially confirms the value (6.925) deduced by Jhanwar *et al.* from earlier refractivity data. Our spectral sum, based primarily on existing photoabsorption data, is $\sim$3.5% lower. It is evident from Table 6.9 that the magnitude of $S(-2)$ is influenced primarily from the first two entries, the sub-ionization and autoionization regions. Cooper *et al.* emphasized the larger cross sections they obtained in the sub-ionization region. They explicitly give $\Delta S(0) = 0.805$, whereas we computed $\Delta S(0) = 0.725$ from the photoabsorption data of Zelikoff and Watanabe, lower by 10%. In addition, for the region IP–24.8 eV, we extract $\Delta S(0) = 6.420$ from Cooper *et al.* and $\Delta S(0) = 6.161$ from the photoabsorption data of Holland *et al.*, which is lower by 4% as noted earlier. If we were to use the data of Cooper *et al.* exclusively from 6–80 eV, and maintained our other values above 80 eV, the resulting values would be $S(-2) = 7.151$ (2.9% high) and $S(0) = 16.235$ (1.5% high).

Olney *et al.* (1997) recognized that their value of $S(-2)$ was too high, and re-calibrated to the expectation value. When we scale the data of Cooper *et al.* between 6–80 eV by the factor implied by Olney *et al.*, we obtain $S(-2) = 7.032$ (still about 1.2% high) and $S(0) = 16.043$ (just 0.3% high). This is indeed closer to expectation than our initial assessment. However, it is desirable to regain the higher resolution photoabsorption data, since they offer more accurate local cross sections near peaks and valleys. To achieve this goal, we make use of the comparisons with the (e,e) data to increase the sub-ionization cross sections of Zelikoff and Watanabe by 8% and the values of Holland *et al.* from IP–24.8 eV by 2%. Further support for this adjustment can be found by comparing the cross sections of these groups at the IP. The revised spectral sums now become: $S(-2) = 6.9380$; $S(-1) = 8.1377$; $S(0) = 16.0771$. The modified $S(-2)$ is now 0.1% below the expectation value, while $S(0)$ is 0.5% above the TRK sum rule.

Jhanwar *et al.* imposed constraints on their minimization procedure such that their value of $S(-2)$, and $S(0)$, would be exactly satisfied. Their value of $S(-1)$, 8.155 Ry units, is very close to our currently modified value. The revised value of $S(-1)$ given by Olney *et al.* is $\sim$2.5% lower. The lower values of $S(-2)$, 4.2%, and $S(-1)$, 3%, found by Holland *et al.* reflect their use of the data of Zelikoff and Watanabe, and to a lesser extent, their own data.

The expectation value for $S(+2)$ is taken as the sum of corresponding expectation values for the constituent atoms, based on Hartree–Fock values of the charge densities at the nuclei. Identical values for this quantity are obtained from Fraga *et al.* (1976) and from Bunge *et al.* (1993). Our spectral sum for this quantity, 26 041.5 Ry units, is 1.5% higher than the expectation value. Jhanwar *et al.* obtain a value of $S(+2)$ which is 2.3% higher.

The value of $S(+1)$ deduced from atomic additivity (Fraga *et al.*, 1976) is 198.9. Correlation effects are expected to increase $S(+1)$. Jhanwar *et al.* obtain 200.2, which is slightly lower than this group (see Kumar and Meath, 1992) reported for the smaller acetylene molecule. The current value is 206.1, and reflects enhancement between the IP and the carbon K-edge, relative to acetylene.

The computation of $S_i(-1)$ encounters ambiguities. Holland *et al.* present spectra of photoabsorption and the quantum yield of ionization (η_i) in their Fig. 1. The product of these quantities is the photoionization cross section. Holland *et al.* have performed this operation, and exhibit a photoionization spectrum in their Fig. 9. We have scanned Fig. 9, and performed the indicated operations from Fig. 1, and find differences. Both spectra are displayed in the current Fig. 6.9. The differences appear to be well outside the digitizing uncertainty.

Another point of concern is the dependence of η_i on energy (or wavelength). In their Fig. 1, Holland *et al.* find good agreement with earlier work between IP–11.7 eV (1060 Å), but at higher energy their values of η_i lie above those of earlier workers. This has prompted us to evaluate a third photoionization spectrum, based on a composite η_i extracted from Ibuki *et al.* (1989), Person and Nicole (1968) and Schoen, multiplied by photoabsorption cross sections from

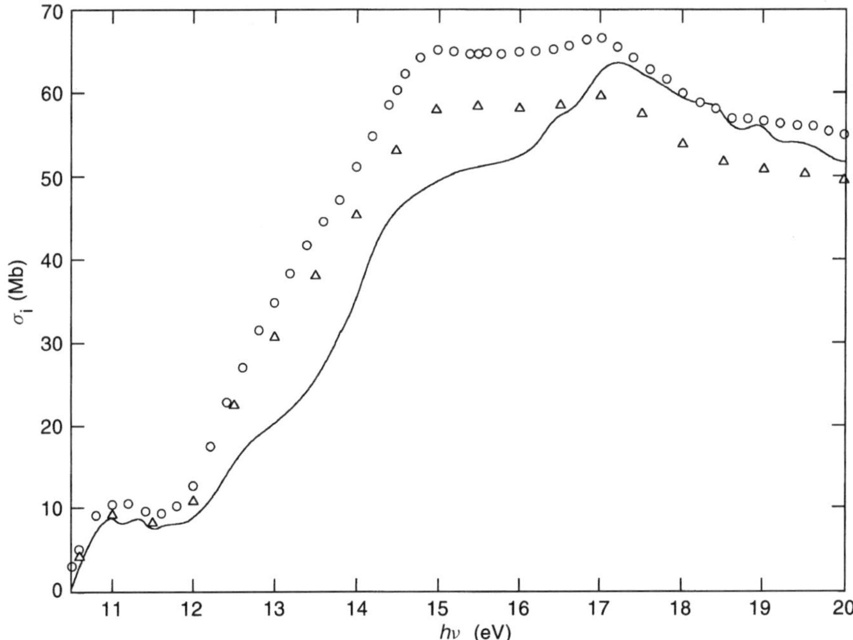

Fig. 6.9 Absolute photoionization spectrum of C_2H_4, IP–20 eV. ○ Holland *et al.* (1997), Fig. 9; △ Holland *et al.* (1997), $\eta^*\sigma_a$, Fig. 1; —— Cooper *et al.* (1995b) and other η_i

Cooper *et al.*, also shown in the current Fig. 6.9. If we assume that $\eta_i = 1$ for $E > 20$ eV, we infer three different values of $S_i(-1)$:

6.502 from Fig. 9, Holland *et al.*
6.116 from Fig. 1, Holland *et al.*
5.890 from our composite photoionization cross section.

The value explicitly given by Holland *et al.* is 6.063, which is close to our inference from their Fig. 1. We take this as the true representation of the data of Holland *et al.*, but additional confirmation of their higher η_i between 1060–700 Å (11.7–17.7 eV) is necessary before the low value (5.890) is ruled out. A value of M_i^2 of 6.75 ±0.10 has been reported by Rieke and Prepejchal (1972). This result is ∼10% higher than our best guess (∼6.1), as it was for acetylene.

6.5 Ethane (C_2H_6)

6.5.1 Preamble

Ethane (C_2H_6) is the prototypical hydrocarbon with a single C–C bond. In the sequence C_2H_2–C_2H_4–C_2H_6, the symmetry reduces from linear to planar to staggered D_{3d}. In photoabsorption, vibronic structure becomes more diffuse and less apparent. In the independent particle approximation, the orbital sequence is

$$(1a_{1g})^2(1a_{2u})^2(2a_{1g})^2(2a_{2u})^2(1e_u)^4(3a_{1g})^2(1e_g)^4, {}^1A_{1g}$$

Electron emission from the $1e_g$ orbital leads to Jahn–Teller distortion. Ionization from the three uppermost occupied orbitals (outer valence) spans the range 11.5–16.5 eV. The C(2s) combinations $2a_{2u}$ and $2a_{1g}$ have ionization energies of $\sim$20.4 and 23.9 eV (inner valence), while $1a_{1g}$ and $1a_{2u}$ account for the carbon K-shell region.

A sum-rule analysis was performed by Jhanwar *et al.* (1981). At that time, experimental photoabsorption cross sections available to them were limited to $h\nu < 70$ eV, and some of these sources were inconsistent. More recent data include photoabsorption measurements from $\sim$9.9–22.5 eV by Kameta *et al.* (1996), inelastic electron scattering (e,e) data from 7.5–220 eV by Au *et al.* (1993), and absolute cross sections in the 285–320 eV (K-edge) region obtained by Ishii *et al.* (1988) using inelastic electron scattering (see also Hitchcock, 1990). A more current value of the refractivity has also been reported by Hohm (1993).

6.5.2 The data

No truly modern determination of the adiabatic IP, such as ZEKE or Rydberg series extrapolation, is known to us. Nicholson (1965) reported 11.521 $\pm$0.007 eV by photoionization. Baker *et al.* (1968) reported a lowest vibronic peak at 11.56 eV, in a He I photoelectron spectrum. This was reproduced by Turner *et al.* (1970). In their compilation, Lias *et al.* (1988) chose 11.52 $\pm$0.01 eV, but later relaxed this to 11.52 $\pm$0.04 eV (Lias, 2000), which covers the two cited values.

a The discrete spectrum and transitions below the IP

In Fig. 6.10, several determinations of the photoabsorption cross section are depicted, from the onset of absorption to 15 eV. Three are truly photoabsorption measurements, the fourth a pseudo-absorption (e,e) determination by Au *et al.* This latter spectrum offers the lowest cross sections. The most recent photo-absorption measurements, by Kameta *et al.*, purport to have errors 'within about 5%' and would be preferred, but they do not extend below 9.9 eV. The spectrum of Raymonda and Simpson (1967) displays the most clearly resolved structure, but the larger-scale spectrum appears to have the wrong shape, with much higher cross sections than Kameta *et al.* and Au *et al.* below 11 eV, then crossing and dipping below Kameta *et al.* above 11 eV. The data of Koch and Skibowski (1971) appear to follow the curve of Kameta *et al.* most closely, but their figure is too compressed for accurate transcription. We opt here for a re-scaled version of the (e,e) data of Au *et al.* Their spectrum, while lower than that of Kameta *et al.*, generally has the same shape until they cross at $\sim$14 eV. It also offers partial resolution of the structure between 8.5–11.5 eV. Detailed comparison of the cross sections of Kameta *et al.* and Au *et al.* between 9.9–11.52 eV suggests

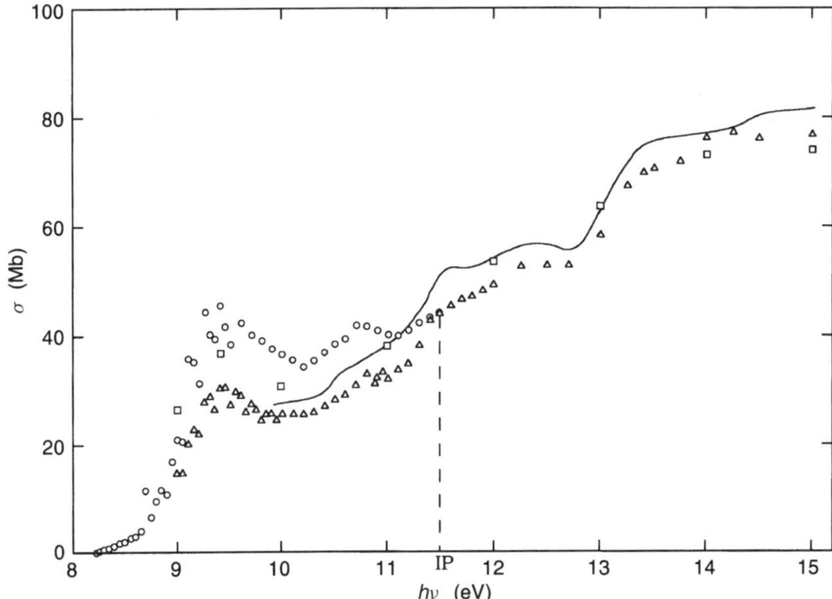

Fig. 6.10 Absolute photoabsorption spectrum of C_2H_6, 8–15 eV. —— Kameta *et al.* (1996); ○ Raymonda and Simpson (1967); △ Au *et al.* (1993a); □ Koch and Skibowski (1971)

an amplification factor of 1.127 for the latter. With this enhancement, the scanned spectrum from Fig. 4 of Au *et al.* is utilized to evaluate the $S(p)$ contributions below the IP, which are recorded in Table 6.11.

b IP–21.5 eV

There is only broad structure suggesting autoionization in this region, in both the (e,e) data of Au *et al.* with 48 meV resolution, and the photoabsorption spectrum of Kameta *et al.*, with 15-meV resolution. We choose the latter, and evaluate $S(p)$ by trapezoidal integration. The cross sections of Au *et al.* are only ∼4% lower in this region.

c 21.5–150 eV

In Fig. 6.11, we compare the synchrotron-based photoabsorption measurements of Lee *et al.* (1973) with photon data by Person and Nicole (1977) using a spark source, and (e,e) data by Au *et al.* Note that the data of Person and Nicole and Au *et al.* are almost superposed, whereas the cross sections of Lee *et al.* are lower below 26 eV and higher above 28 eV. This characteristic behavior of the data of Lee *et al.* has been observed with other molecules studied in that article.

The measurements of Person and Nicole extend only to 50 eV, whereas Au *et al.* report cross sections to 220 eV. Here, we follow Au *et al.*, using their

Table 6.11 Spectral sums, and comparison with expectation values for C_2H_6[a]

Energy, eV	$S(-2)$	$S(-1)$	$S(0)$	$S(+1)$	$S(+2)$
7.86–11.52(IP)[b]	1.4121	1.0556	0.7941	0.6011	0.4577
IP–21.5[c]	4.7087	5.3851	6.3360	7.6662	9.5272
21.5–150[d]	1.2470	2.7230	6.7387	20.4929	83.3119
150–288[e]	0.0013	0.0174	0.2461	3.6010	54.5528
288–320[f]	0.0016	0.0348	0.7696	17.0178	376.65
320–1740[e]	0.0025	0.0798	2.8993	124.1785	6570.54
1740–10 000[e]	–	0.0008	0.1402	30.2825	8056.27
10^4–10^5[g]	–	–	0.0032	3.9154	6590.08
10^5–10^6[h]				0.1800	2852.44
10^6–10^7[h]	–	–	–	0.0062	970.87
10^7–10^8[h]	–	–	–	0.0002	314.42
10^8–10^9[h]	–	–	–	–	100.19
10^9–∞[h]	–	–	–	–	46.37
Total	7.3732	9.2965	17.9272	207.942	26 025.7
Expectation values	7.3895[i]		18.0		25 657.6[j]
Other values	(7.402$_5$)[k]	9.375[k]	(18.0)[k]	203.8[k]	26 228[k]
	7.130[l]		(18.0)[l]		
	7.385[m]	9.225[m]			

[a]In Ry units.
[b]Data from Fig. 4 of Au *et al.* (1993a), normalized to that of c, below. See Fig. 6.10 and text.
[c]Kameta *et al.* (1996).
[d]Ref. b, above, without adjustment. See Fig. 6.11 and text.
[e]Summed atomic cross sections from Henke *et al.* (1993).
[f]Hitchcock (1990). See also Ishii *et al.* (1988a).
[g]Chantler (1995).
[h]Assuming hydrogen-like behavior, bare carbon atom with screening, K-shell only (Bethe and Salpeter, 1977).
[i]Hohm (1993).
[j]Using atomic additivity and Hartree–Fock calculations for atomic carbon. Identical results are obtained from Fraga *et al.* (1976) and Bunge *et al.* (1993).
[k]Jhanwar *et al.* (1981).
[l]Value given by Olney *et al.* (1997), based on data in ref. b and TRK sum rule.
[m]Value given by Olney *et al.* (1997), based on renormalization of data in ref. b to $S(-2)$ sum rule.

Table 8 from 21.5–150 eV. The values of $S(p)$ are determined by trapezoidal integration, and listed in Table 6.11.

d 150–288 eV

Figure 6.12 compares the (e,e) data of Au *et al.* with summed atomic photoabsorption cross sections from Henke *et al.* (1993). The concordance is good; the curves appear to cross at ~150 eV. In the above section, we had retained the cross sections of Au *et al.* to 150 eV. We continue from 150 eV to the onset of structure near the K-edge at ~288 eV, using the summed cross sections of Henke *et al.* These sparse cross sections have been fitted to a 4-term polynomial. The coefficients are given in Table 6.12; the values of $S(p)$, obtained by analytical integration of the polynomial, are listed in Table 6.11.

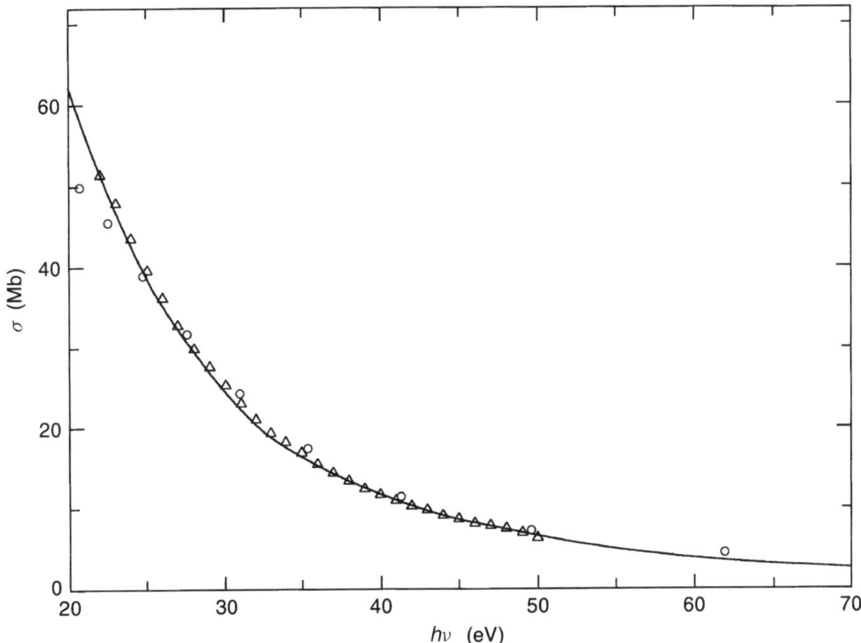

Fig. 6.11   Absolute photoabsorption spectrum of C_2H_6, 20–70 eV. — Au *et al*. (1993a); o Lee *et al*. (1973); △ Person and Nicole (1977)

e   The carbon K-edge, 288–320 eV

The carbon K-edge in ethane occurs at 290.7 eV (Sae *et al*. 1989). Photoabsorption studies at relatively high resolution (0.03–0.04 eV) reveal significant structure below this edge, dominated by a peak at ~287.9 eV (see Ma *et al*., 1991). These authors assign the 287.9 eV peak to a C(1s) → 3p Rydberg transition. Unfortunately, their spectrum covers a limited range, and is given in relative intensity only. Absolute cross sections have been obtained by Ishii *et al*. using inelastic electron scattering, and reproduced by Hitchcock (1990). In these experiments, the resolution was 0.6 eV. The spectrum shown by Hitchcock has been digitized, and the $S(p)$ evaluated by trapezoidal integration. In the stated interval, our extracted cross sections yield $\Delta S(0) = 0.770$, whereas Ishii *et al*. explicitly give 0.14 below the K-edge, 0.61 above the K-edge, or $\Delta S(0) = 0.75$.

f   320–1740 eV; 1740–10 000 eV

The summed atomic photoabsorption cross sections of Henke *et al*. are utilized here between 320–10 000 eV. The region is divided (320–1740; 1740–10 000 eV), and 4-term polynomials are fitted to each region by regression. At 320 eV, the derived function yields $\sigma = 1.58$ Mb, whereas that inferred from the spectrum of Hitchcock is ~1.5 Mb, a reasonable concordance.

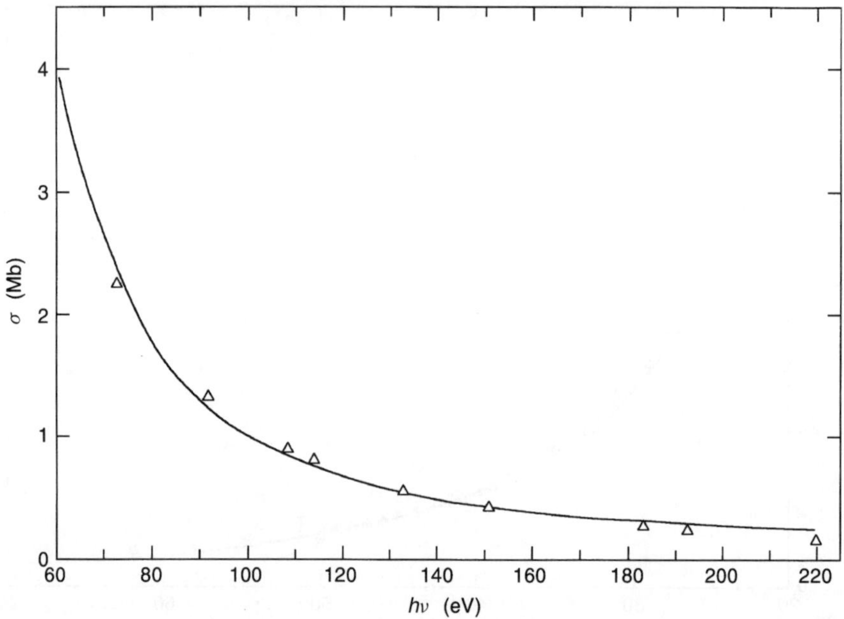

Fig. 6.12 Absolute photoabsorption spectrum of C_2H_6, 60–220 eV. — Au *et al.* (1993a); △ Henke *et al.* (1993) + additivity

Table 6.12 Coefficients of the polynomial $df/dE = ay^2 + by^3 + cy^4 + dy^5$ fitted to data at various energies[a]

Energy range, eV	a	b	c	d
150–288	−12.430 8	746.5164	−8 913.93	37 174.85
320–1740	9.028 358	7916.428	−136 533	745 811.8
1740–10 000	−2.578 45	9040.553	−114 924	−2 946 480

[a]df/dE in Ry units, $y = B/E$, $B = IP = 11.52$ eV.

$$g \quad 10^4 – 10^5 \text{ eV}$$

The calculated atomic cross sections of Chantler (1995) are summed to evaluate $S(p)$.

6.5.3 The analysis

The new value for $S(-2)$, 7.3895 Ry units (Hohm, 1993), is very close to that inferred by Jhanwar *et al.* from older data. Our spectral sum is fortuitously within 0.2% of the expectation value. The major contributors to this quantity are the first three entries in Table 6.11. It will be recalled that the most recent data of Kameta *et al.* were utilized for the range IP–21.5 eV, while the (e,e) data of Au *et al.*, with strong support from earlier photoabsorption measurements by

Person and Nicole, determined the 21.5–150 eV domain. The only significant adjustment performed in this analysis was to enhance the sub-ionization data of Au *et al.* by calibrating to the photoabsorption measurements of Kameta *et al.* In subsequent work, Olney *et al.* (1997) also concluded that their earlier data required upward adjustment, but by a smaller factor and over a larger energy region.

The spectral sum for $S(0)$ is also fortuitously close to TRK sum rule expectations, falling shy by just 0.4%. These results for $S(-2)$ and $S(0)$ suggest that the spectral sum for $S(-1)$ might be marginally improved by an increase of $\sim 0.3\%$, to 9.324 Ry units. Jhanwar *et al.* optimized existing photoabsorption data, supplemented by mixture rules for $h\nu > 70$ eV, and constrained by fixed values for $S(-2)$ and $S(0)$. Their values of $S(-2)$, $S(-1)$ and $S(0)$ are $\sim 0.4\%$ higher than the present ones.

The expectation value for $S(+2)$ is taken as the sum of corresponding expectation values for the constituent atoms, based on Hartree–Fock values of the charge densities at the nuclei. The spectral sum for $S(+2)$ has major contributions above 320 eV from atomic additivity utilizing the experimentally based atomic photoabsorption cross sections of Henke *et al.* from 320–10 000 eV, calculated atomic cross sections of Chantler from 10^4–10^5 eV, and to a lesser extent, hydrogenic behavior above 10^5 eV. The spectrally deduced $S(+2)$ is $\sim 1.4\%$ higher than the 'expectation' value. Jhanwar *et al.* report a slightly higher value, approximately 2.2% above 'expectation'.

The value of $S(+1)$ deduced from atomic additivity (Fraga *et al.*, 1976) is 201.6 Ry units. Correlation is expected to increase this quantity in the molecule. Indeed, Jhanwar *et al.* find $S(+1) = 203.8$, while the present spectral sum yields $S(+1) = 207.9$. Major contributions to this quantity occur in the 320–10 000 eV region (the 'Henke region'), but significant additions can be found in the K-shell region and at lower energies, where the present analysis has the advantage of experimental data unavailable to Jhanwar *et al.*

Although the degree of accord between spectral sums and expectation values for $S(-2)$ and $S(0)$ is exceptionally good, the behavior of $S_i(-1)$, the ionized component of $S(-1)$, is disquieting. If we use the photoionization cross sections of Kameta *et al.* between IP–21.5 eV, to be consistent with current selections, we obtain $S_i(-1) = 7.30$. The directly measured $M_i^2 \equiv S_i(-1)$ obtained by Rieke and Prepejchal (1972) using high-energy electron impact cross sections is 6.80 ± 0.36. In the vast majority of cases, the Rieke/Prepejchal values of M_i^2 have been 10–15% higher than $S_i(-1)$ from spectral sums. Indeed, in Berkowitz (1979), based on earlier data, we had obtained $S_i(-1) \sim 6.1$. The difference can be traced to the cross sections between IP–21.5 eV. (The adjustment of the sub-ionization data of Au *et al.* plays no role here.) The earlier photoabsorption cross sections, taken from Schoen (1962) and Metzger and Cook (1964), were lower than those currently used. In addition, the quantum yield of ionization was also lower, particularly between 13.5–21 eV. In the recent work of Kameta *et al.*, the quantum yield reaches unity essentially at ~ 17 eV, whereas the earlier work, which was inconsistent, was assumed to attain unity at 600 Å (20.66 eV). The photoionization cross sections of Kameta *et al.* receive support from the (e,e) data of Au *et al.* (1993b), which give even a slightly higher contribution to $S_i(-1)$ between IP–21.5 eV.

In Table 3.1, we can compare several determinations of M_i^2 for C_2H_4 and C_2H_6. Ethane has the higher value, by 1.2 (optical) and 1.32 (Schram *et al.*, 1966) and by 2.7 Ry units (Nishimura and Tawara, 1994). The implication from the well-known values of $S(-2)$ and $S(0)$ for C_2H_4 and C_2H_6 is that M_i^2 for C_2H_6 should be approximately 10% larger. However, Rieke and Prepejchal reported almost identical values (6.75 ±0.10 for C_2H_4, 6.80 ±0.36 for C_2H_6). Their larger uncertainty for C_2H_6, taken together with its unusual direction of deviation from the optical value, suggest that M_i^2 for C_2H_6 is more suspect than their value for C_2H_4.

6.6 Methanol (CH₃OH)

6.6.1 Preamble

Methanol (CH_3OH) has C_s symmetry. In the independent particle approximation, its orbital sequence may be written

$$(1a')^2(2a')^2(3a')^2(4a')^2(5a')^2(1a'')^2(6a')^2(7a')^2(2a'')^2$$

The $1a'$ orbital, essentially oxygen K-shell, has a binding energy of $\sim$539 eV; similarly, $2a'$, carbon K-shell, has an ionization energy of $\sim$292.4 eV. The $3a'$ orbital (VIP $\cong$ 32 eV) and $4a'$ (VIP $\cong$ 22.6 eV) are predominantly O(2s) and C(2s), respectively, according to both Robin and Kuebler (1972) and Nordfors *et al.* (1991). The uppermost ($2a''$) orbital is primarily O(2p) lone pair (VIP = 10.94 eV), while the $5a'$ (VIP $\cong$ 17.55 eV), $1a''$ (VIP $\approx$ 15.7 eV), $6a'$ (VIP $\cong$ 15.2 eV) and $7a'$ (VIP $\cong$ 12.64 eV) are various combinations of C(2p), O(2p) and H(1s) atomic orbitals. In the photoabsorption spectrum, sharp structure appears at $\sim$7.72 and 8.32 eV, assigned as low Rydberg members converging to the first IP, but beyond $\sim$9 eV the structure is weak and diffuse. The intensity of the low-lying peaks is sensitive to resolution, but the remainder of the valence spectrum is not.

Jhanwar and Meath (1984) performed a sum rule analysis using data available to them. Most of these data were of 1971 vintage, the latest being 1974. No direct experimental data were available between 21–30 eV or above 100 eV, and hence mixture rules were used. Additional information since that time includes photoabsorption between 6.3–11.7 eV (Nee *et al.*, 1985), inelastic electron scattering data of Burton *et al.* (1992) between 6–360 eV and other (e,e) measurements by Ishii and Hitchcock (1988) providing oscillator strengths in the vicinity of the oxygen and carbon K-edges. Also noteworthy is a photoabsorption study of the 7.7–10.6 eV region by Person and Nicole (1978), which was performed with higher resolution than the experiment of Nee *et al.*

6.6.2 The data

A highly accurate adiabatic ionization potential (AIP) such as might be forth-coming from Rydberg series extrapolation or ZEKE measurements, is not available. MacNeil and Dixon (1977) performed a deconvolution of their He I spectrum and deduced AIP (CH_3OH) = 10.846 ±0.002 eV. Their error bar might be optimistic, since that work also concluded that AIP (O_2) = 12.076 ±0.002 eV, whereas subsequent ZEKE measurements (Merkt *et al.*, 1998) found AIP (O_2) = 12.070 14 (15) eV. Photoionization measurements give AIP (CH_3OH) = 10.84–10.85 eV (see Berkowitz, 1978).

a   *The discrete spectrum and transitions below the IP*

Figure 6.13 is a composite of three data sets describing photoabsorption below the IP. The (e,e) data of Burton *et al.* had a stated resolution of 48 meV, whereas the photoabsorption measurements of Nee *et al.* and Person and Nicole were performed with resolutions of 13 and 6.5 meV, respectively. The agreement among all three groups is quite good in regions devoid of structure, but as anticipated, displays marked influence of resolution near sharp peaks. For clarity, only selected points from the spectrum of Person and Nicole are shown. There is a slight offset in the peak positions from Nee *et al.* and Person and Nicole. The contributions to $S(p)$ have been evaluated by trapezoidal integration. The data of

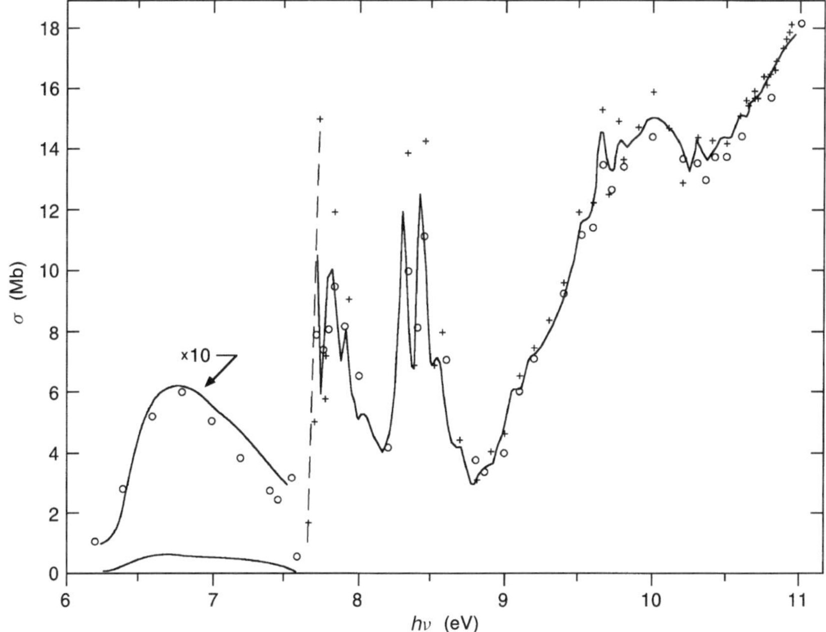

Fig. 6.13   Absolute photoabsorption spectrum of CH_3OH, 6–11 eV. —— Nee *et al.* (1985); o Burton *et al.* (1992); + Person and Nicole (1970; 1978)

Burton *et al.* and Nee *et al.* differ by $\leq 1\%$ in $S(p)$; the corresponding values from the spectrum of Person and Nicole are about 3% higher in the region covered by them, 7.66–10.60 eV. In Table 6.13, we have selected the $S(p)$ from Nee *et al.* since photoabsorption has a well-defined absolute calibration (Beer's law) and their scan covers the entire range under consideration. (Earlier data from Person and Nicole (1970) covers the region 10.0–11.8 eV, and is partially shown in Fig. 6.13.)

Table 6.13 Spectral sums, and comparison with expectation values for CH_3OH[a]

Energy, eV	$S(-2)$	$S(-1)$	$S(0)$	$S(+1)$	$S(+2)$
6.26–10.85 (IP)[b]	0.6194	0.4234	0.2931	0.2052	0.1452
IP–21.21[c]	3.6333	4.0966	4.7707	5.7311	7.0872
21.21–30.0[d]	0.8488	1.5331	2.7959	5.1489	9.5767
30.0–150.0[e]	0.4596	1.4815	5.4692	24.2367	131.8399
150.0–287.0[f]	0.0029	0.0397	0.5643	8.2796	125.7540
288.1[g]	0.0000_6	0.0013	0.0272	0.5751	12.178
289.4[g]	0.0001_2	0.0026	0.0549	1.1686	24.587
287–320[g]	0.0005_8	0.0241	0.5353	11.8940	264.442
320–350[e]	0.0004_2	0.0102	0.2517	6.1884	152.240
350–532[f]	0.0008	0.0243	0.7475	23.3743	741.347
532–568[h]	0.0001_7	0.0070	0.2809	11.3316	457.199
568–2042.4[f]	0.0006	0.0330	2.0157	137.1795	10610.04
2042.4–10 000[f]	–	0.0010	0.2131	53.1475	15851.41
10^4–10^{5}[i]	–	–	0.0074	9.3370	15936.67
10^5–10^{6}[j]	–	–	–	0.4726	7527.23
10^6–10^{7}[j]	–	–	–	0.0167	2616.17
10^7–10^{8}[j]	–	–	–	0.0005	853.22
10^8–10^{9}[j]	–	–	–	–	272.49
10^9–∞[j]	–	–	–	–	126.30
Total	5.5668	7.6778	18.0269	298.29	55719.9
Expectation values	5.491[k]		18.0		54611.0[l]
Other values	(5.485)[m]	7.590[m]	(18.0)[m]	294.0[m]	55880.0[m]
	5.198[n]				
	5.405[o]	7.455[o]			

[a]In Ry units.
[b]Nee *et al.* (1985).
[c]Person and Nicole (1974).
[d]Interpolation from $\sigma(21.21\,eV)$, ref. c, to $\sigma(30.0\,eV)$, ref. e.
[e]Burton *et al.* (1992).
[f]Polynomial fit to summed atomic cross sections from Henke *et al.* (1993).
[g]Wight and Brion (1974), normalized to data of ref. e at 320 eV.
[h]Ishii and Hitchcock (1988), supplemented by valence and carbon K-shell contributions.
[i]Summed atomic cross sections from Chantler (1995).
[j]Summed atomic cross sections calculated from eq. (71.13) of Bethe and Salpeter (1977).
[k]Current evaluation. See Sect. 6.6.3, text.
[l]Summed atomic contributions. See Fraga *et al.* (1976) or Bunge *et al.* (1993).
[m]Jhanwar and Meath (1984).
[n]Olney *et al.* (1997), using data of ref. e.
[o]Reference m, after normalizing data of ref. e to chosen refractivity.

b   IP–21.21 eV

In Fig. 6.14, we compare two older sets of photoabsorption measurements (Person and Nicole, 1974; Ogawa and Cook, 1958b) with recent (e,e) data from Burton *et al*. The two photoabsorption cross section sets are in fair agreement with one another (although the data of Ogawa and Cook display scatter), but on average they appear to be $\sim$10% higher than the (e,e) data. The absolute cross sections are highest in this region ($\sigma_{max} \sim 60$ Mb at 14 eV), and can be anticipated to have a significant impact on $S(-2)$, $S(-1)$ and $S(0)$. We tentatively choose the photoabsorption data of Person and Nicole (1974) in this interval. In the final analysis, we shall consider the effect if we had chosen the cross sections of Burton *et al*.

c   21.21–30 eV

The gap in photoabsorption data in this range, alluded to earlier, still exists. However, Burton *et al*. present (e,e) data in this interval. In Fig. 6.15 we can once again compare the (e,e) cross sections of Burton *et al*. with photoabsorption measurements, this time at the higher energies 38–120 eV (de Reilhac and Damany, 1971). Taken at face value, the latter measurements appear to yield higher ($\sim$6%) cross sections than the (e,e) data at 38–40 eV, then remain rather close up to 120 eV. Hence, there is support for the absolute values of the (e,e) cross sections above 40 eV. If we were to adopt the (e,e) data above 21.21 eV,

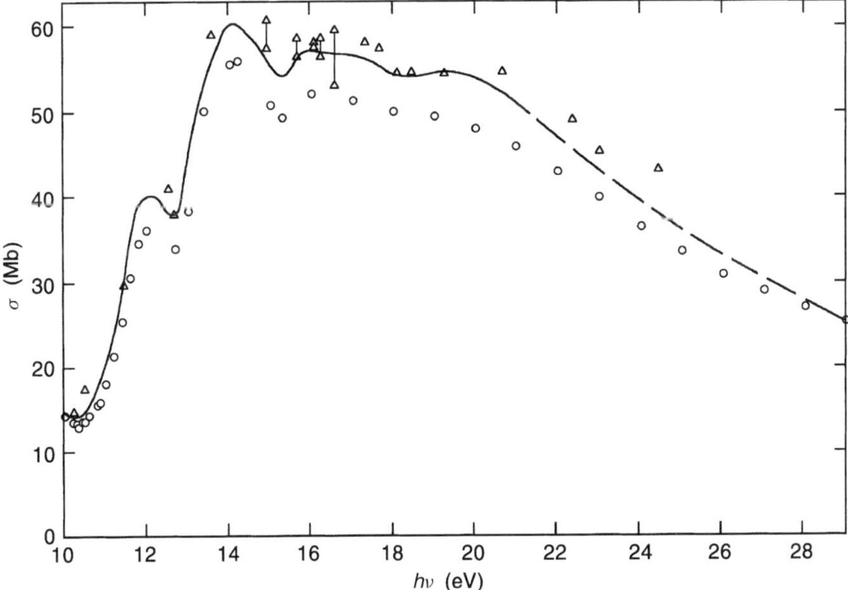

Fig. 6.14   Absolute photoabsorption spectrum of CH_3OH, 10–29 eV. —— Person and Nicole (1974); o Burton *et al*. (1992); $\triangle$ Ogawa and Cook (1958b)

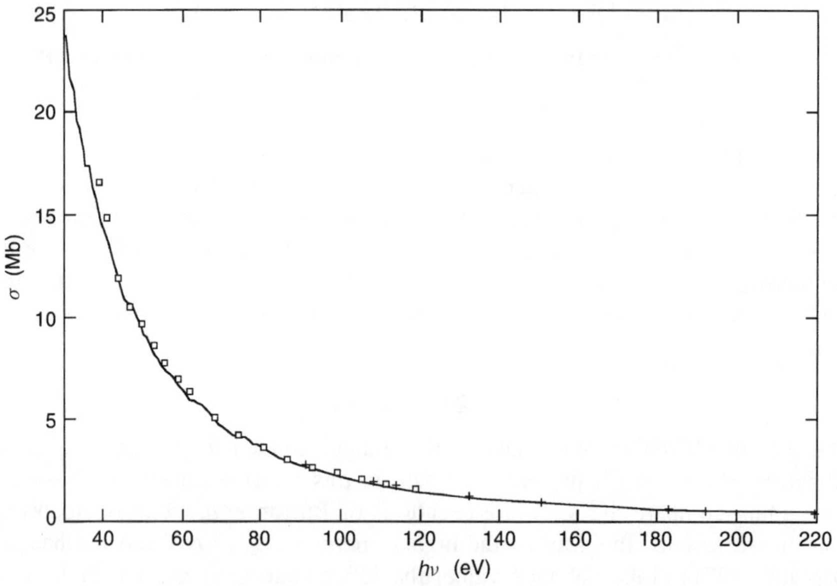

Fig. 6.15 Absolute photoabsorption spectrum of CH_3OH, 30–220 eV. —— Burton *et al.* (1992); □ de Reilhac and Damany (1971); + Henke *et al.* (1993) + additivity

there would be an abrupt discontinuity at that energy unless we retained the lower energy (e,e) cross sections. We shall consider the latter as one option. As an alternative, we smoothly interpolate from the photoabsorption cross section at 21.21 eV (Person and Nicole, 1974) to the (e,e) data of Burton *et al.* at 30 eV. In Table 6.14, we record the $S(p)$ resulting from this interpolation.

d 30–150 eV

Figure 6.15, in addition to comparing the photoabsorption data of de Reilhac and Damany with the (e,e) cross sections of Burton *et al.*, shows that the sum of atomic photoabsorption cross sections taken from Henke *et al.* (1993) is in very good agreement with the (e,e) data above ∼100 eV. Here, we adopt the (e,e) results from 30–150 eV. The importance of this energy region lies in its

Table 6.14 Coefficients of the polynomial $df/dE = ay^2 + by^3 + cy^4 + dy^5$ fitted to data at various energies[a]

Energy range, eV	a	b	c	d
50–287	−20.811 7	1 521.157	−17 539.3	68 087.62
350–532	17.634 33	3 843.691	−41 722.2	–
568–2042.4	1.223 81	22 116.02	−620 107	6 048 279
2042.4–10 000	−11.773	30 317.78	−2 504 940	158 012 313

[a] df/dE in Ry units, $y = B/E$, $B = $ IP $= 10.846$ eV.

contribution to $S(0)$, $\sim$30%, and to $S(-1)$, $\sim$20%, but only $\sim$8% to $S(-2)$ and $S(+1)$. The evaluation of $S(p)$ is performed by trapezoidal integration.

e   150–287 eV

The carbon K-edge occurs at $\sim$292.4 eV (Jolly *et al.*, 1984), but pre-edge structure begins to appear at $\sim$287 eV. Figure 6.15 suggests that the contributions to $S(p)$ from this interval can be evaluated equally well either from the data given by Burton *et al.* or from the summed atomic cross sections of Henke *et al.* In actual execution, the cross sections of Burton *et al.* yield a value of $S(0)$ about 0.11 larger than the Henke values and 1.5 Ry larger for $S(+1)$, with smaller relative influence on the other $S(p)$. The absolute values of Burton *et al.* are determined by normalization to the TRK sum rule, whereas the Henke cross sections are based on photoabsorption measurements. Burton *et al.* present their results in convenient tabular form. As an alternative, we have fitted the sparse cross sections of Henke *et al.* to a 4-term polynomial. Analytical integration of the polynomial yields the $S(p)$ given in Table 6.13.

f   Carbon K-edge, 287–350 eV

Four data sets are available here, three based on inelastic electron scattering, the fourth (Andersen *et al.*, 1997) on true photoabsorption. The latter is much better resolved (35 meV), but the intensity is in relative units, and it encompasses a short (287–293 eV) energy range. Although it could be indirectly scaled, it is of limited use for sum rule purposes. Of the three electron scattering measurements, two, Burton *et al.* and Ishii and Hitchcock provide absolute scales, but the third (Wight and Brion, 1974c) appears to be better resolved. Hence, our approach was to transfer the absolute calibration of Burton *et al.* at 320 eV to the spectrum of Wight and Brion. This cross section, well beyond the K-edge, is in fairly good agreement with that from Ishii and Hitchcock, after subtracting the valence contribution. With this imposed calibration, the cross sections between 287–320 eV were manually extracted from the spectrum of Wight and Brion. Of the three groups, only Burton *et al.* extend their data significantly beyond 320 eV. Hence, we return to their spectrum to encompass the 320–350 eV interval. In Table 6.13, we summarize the contributions to $S(p)$, with the resonant peaks appearing separately.

g   Inter-edge continuum, 350–532 eV

Here, we return to the summed atomic cross sections of Henke *et al.* The data points are fitted to a 4-term polynomial as before. At 350 eV, the calculated cross section from this polynomial is 0.75 Mb, whereas that measured from the data of Burton *et al.* is $\sim$0.85 Mb, which is fair agreement. The coefficients of the polynomial are included in Table 6.14, and the $S(p)$ contributions in Table 6.13.

h   Oxygen K-edge, 532–568 eV

The relative photoabsorption spectrum of Andersen *et al.*, which reveals some undulations that are washed out in the electron energy loss data, unfortunately only covers the range 532–545 eV, a scant 6 eV above the oxygen K-edge. Although spectra are available from both Wight and Brion and Ishii and Hitchcock, only the latter can be considered. Only they present an absolute scale, but more significantly, the shape of the Wight and Brion spectrum, displaying a higher intensity at 532 eV than at 568 eV, is contrary to the expectation from additivity of atomic cross sections. After manually extracting the cross sections from the spectrum of Ishii and Hitchcock, we have added the contributions of valence and carbon K-shell continua, since their figure was constructed by subtracting these contributions. At 568 eV, the supplemented Ishii/Hitchcock cross section is measured to be 0.704 Mb, whereas atomic additivity at this energy (vide infra) is 0.704_6 Mb. This virtual identity is not surprising, since Ishii and Hitchcock normalized their data to atomic additivity at IP +25 eV, or 564 eV. Trapezoidal integration of the modified Ishii/Hitchcock spectrum yields the $S(p)$ given in Table 6.13.

i Post K-edges, 568–10 000 eV

The summed atomic cross sections of Henke *et al*. are fitted to two 4-term polynomials, one between 568–2042.4 eV, the other from 2042.4–10 000 eV. The coefficients of the polynomials are given in Table 6.14, the corresponding $S(p)$ obtained by analytic integration in Table 6.13. This region contributes slightly more than the two oxygen K-shell electrons to $S(0)$, ~64% to $S(+1)$ and ~47% to $S(+2)$.

j Post K-edges, 10^4–10^5 eV

The calculated atomic cross sections of carbon and oxygen (Chantler, 1995) are summed, and used to evaluate $S(p)$. The contribution of four hydrogen atoms is insignificant. The major impact of this interval is on $S(+2)$, which accrues ~29% of its value.

6.6.3 The analysis

Jhanwar and Meath fitted the refractive-index measurements of Ramaswamy (1936) for methanol to extract a refractive index at infinite wavelength, or $S(-2) =$ 5.485 Ry units. Ramaswamy presented his data for a hypothetical 760 mm and 25.0 °C, i.e. the compressibility was measured, and the results reduced to those which would have been obtained had the gas been ideal (Watson and Ramaswamy, 1936). Our fitting of Ramaswamy's data to the Cauchy expansion yields $\alpha =$ 3.255×10^{-24} cm^3 and $S(-2) = 5.491$ Ry units, slightly higher. Miller (1999) cites $\alpha = 3.32 \times 10^{-24}$ cm^3 from Applequist *et al*. (1972), which is equivalent to $S(-2) = 5.60$ Ry units, but these authors refer their value to $\lambda = 5893$ Å. At $\lambda = \infty$, their value would be close to that deduced here, and by Jhanwar and Meath. The current spectral sum, $S(-2) = 5.5668$ Ry units, is 1.4% higher, but $S(0)$ is only 0.15% above the TRK sum rule. At this point, it is instructive to re-examine the (e,e) data of Burton *et al*., which we eschewed for $h\nu < 30$ eV, but utilized for 30 eV $< h\nu < 150$ eV. Had we used their values for $h\nu < 30$ eV, we would have obtained $S(-2) = 5.189$ Ry units, about 5.4% lower than the expectation value. Olney *et al.* (1997) compute a slightly higher value of $S(-2)$, 5.198 Ry units, from the cross sections of Burton *et al.*, but then re-normalize to their selection of expectation value, $S(-2) = 5.405$, which remains lower than the current choice. Thus, our selection of photoabsorption data for $h\nu < 21.21$ eV not only provides much higher resolution, but also seems justified by the sum rule analysis.

Jhanwar and Meath constrained their optimization procedure to satisfy the $S(-2)$ and $S(0)$ requirements. Their value of $S(-1)$ is about 1% lower than the current spectral sum. This is approximately what we would predict, based on an excess of 1.4% for $S(-2)$ and very nearly the expectation value for $S(0)$.

Our spectral sum for $S(+2)$ is 55 719.9 Ry units, about 2% larger than the 'expectation' value based on the sum of electron charge densities at the nuclei from Hartree–Fock calculations by Fraga *et al.* (1976). An almost identical value can be inferred from the calculations of Bunge *et al.* (1993). The spectral sum obtained by Jhanwar and Meath is very nearly the same as the current one.

The current spectral sum for $S(+1)$ is about 1.5% larger than that of Jhanwar and Meath. Here the additional information available currently for the oscillator strengths in the K-shell regions may play a role. Both spectral sums are substantially higher than 286.0 Ry units based on atomic additivity.

To evaluate $S_i(-1)$, we proceed as follows. From IP-11.8 eV, the photoionization cross sections of Person and Nicole (1970) are directly applicable. Between 11.8–19.5 eV, we combine the quantum yield of ionization from Burton *et al.* with the photoabsorption cross sections of Person and Nicole (1974). Above 19.5 eV, we accept the surmise of Burton *et al.* that the quantum yield is unity. Summing these various components, we arrive at $S_i(-1) = 6.180$, or 6.092, if we correct for the 1% excess assumed for $S(-1)$. This compares very favorably to the directly measured $M_i^2 = S_i(-1) = 6.22 \pm 0.18$ given by Rieke and Prepejchal (1972).

6.7 Benzene (C$_6$H$_6$)

6.7.1 Preamble

Benzene (C$_6$H$_6$) is the prototypical aromatic hydrocarbon, with planar D_{6h} symmetry. The ground state electron configuration is

$$(1a_\text{g})^2(1e_{1\text{u}})^4(1e_{2\text{g}})^4(1b_{1\text{u}})^2(2a_{1\text{g}})^2(2e_{1\text{u}})^4(2e_{2\text{g}})^4(3a_{1\text{g}})^2(2b_{1\text{u}})^2$$

$$(1b_{2\text{u}})^2(3e_{1\text{u}})^4(1a_{2\text{u}})^2(3e_{2\text{g}})^4(1e_{1\text{g}})^4, \ \tilde{X}^1A_{1\text{g}}$$

The four deepest orbitals correspond to the carbon K-shell, with ionization energy of 290.3 eV (Davis and Shirley, 1974). Baltzer *et al.* (1997) list the vertical ionization potentials (VIP) of the outer valence orbitals (in eV) as $\tilde{X}^2E_{1\text{g}}$, 9.45; $\tilde{A}^2E_{2\text{g}}$, 11.7; $\tilde{B}^2A_{2\text{u}}$, 12.3; $\tilde{C}^2E_{1\text{u}}$, $\sim$14.0; $\tilde{D}^2B_{2\text{u}}$, 14.78, $\tilde{E}^2B_{1\text{u}}$, 15.77; and $\tilde{F}^2A_{1\text{g}}$, 17.04. Although the separation between inner and outer valence orbitals is somewhat arbitrary, the successive deeper orbitals are $\tilde{G}^2 E_{2g}$, 19.1; $\tilde{H}^2 E_{1u}$, 22.6; and $\tilde{I}^2 A_{1g}$, 25.9 eV, from Baltzer *et al.* and also Bieri and Åsbrink (1980).

Vibronic structure observed in the photoelectron spectrum of the $\tilde{X}^2 E_{1g}$, and $\tilde{A}^2 E_{2g}$, bands has been interpreted in terms of dynamic Jahn–Teller effects. Distortion from the hexagonal, planar D_{6h} symmetry can accompany Jahn–Teller effects in degenerate states. Lindner *et al.* (1996) have examined this possibility in the electronic ground state, $\tilde{X}^2 E_{1g}$, of $C_6H_6{}^+$. From their high-resolution spectrum, they conclude that 'the cation is definitely distorted by linear Jahn–Teller coupling' to a C–C–C angle of 118.1 degrees, but the wells in the pseudo-rotation potential are only 8 cm^{-1}, whereas the zero-point energy of the Jahn–Teller active ν_6 vibration is 413 cm^{-1}, and hence the $C_6H_6{}^+$ ground state is 'necessarily viewed in D_{6h} symmetry', i.e. effectively planar and hexagonal. The excited $\tilde{A}^2 E_{2g}$ state has also been examined for Jahn–Teller, pseudo-Jahn–Teller and Herzberg–Teller vibronic coupling by Goode *et al.* (1997). (Note that these authors denote $^2 E_{2g}$ as the $\tilde{B}$ state.)

Structure in the photoabsorption spectrum below the IP involves excitation to the valence states $^1 B_{2u}$ and $^1 B_{1u}$ which are electric-dipole forbidden, but derive oscillator strength from vibronic interaction, as well as a strong, dipole-allowed transition to $^1 E_{1u}$. In addition, there are a number of sharp Rydberg series converging to the first IP. At higher energies, broader Rydberg bands are observed converging to $\tilde{A}^2 E_{2g}$, $\tilde{C}^2 E_{1u}$, $\tilde{F}^2 A_{1g}$, and possibly $\tilde{B}^2 A_{2u}$.

A sum-rule analysis has been performed recently by Kumar and Meath (1992). These authors noted data available to them below 70 eV, and made selections in their optimization procedure, but for $h\nu > 70$ eV they opted for mixture rules, particularly $df/dE(C_6H_6) = 3\, df/dE\,(C_2H_2)$. Although there was (and continues to be) sparse experimental photoabsorption information between $\sim$35 eV and the carbon K-edge, they apparently chose to ignore available data between 280–320 eV, i.e. Akimov *et al.* (1985) and Hitchcock (1989), shown by Piancastelli *et al.* (1989). Very recently, Rennie *et al.* (2000) obtained higher-resolution absolute photoabsorption spectra from 284.5–820 eV. Also quite recently, Rennie *et al.* (1998) have presented absolute photoabsorption spectra from IP–35 eV which we shall compare with some earlier work. These authors also present their own sum rule analysis.

6.7.2 The data

The adiabatic ionization potential (AIP) of benzene has been given as $74\,556.57(5)$ cm$^{-1} \equiv 9.243\,836\,(6)$ eV by Neuhauser *et al.* (1997). This verifies, with additional precision, the value $9.243\,76(6)$ eV obtained by Nemeth *et al.* (1993) and supersedes AIP $(C_6H_6) = 9.243\,64(5)$ found by Chewter *et al.* (1987).

a   *The discrete spectrum and transitions below the IP*

Kumar and Meath chose the data of Koch and Otto (1976) for the region 5.44–7.70 eV. Here, we select the photoabsorption data of Pantos *et al.* (1978), because comparison with the data of other workers at higher energies reveals that the Koch/Otto cross sections are systematically lower. Pantos *et al.* provide explicit escillator strengths for the transitions to $^1B_{2u}$, $^1B_{1u}$, $^1E_{1u}$ and some Rydberg transitions. They also offer a total oscillator strength below the IP, $f = 1.24$, which is slightly larger than the sum of the individual transitions. We infer that the difference corresponds to overlapping Rydberg transitions just below the IP, and supplement these transitions accordingly. These f values, and the other $S(p)$, are detailed in Table 6.15.

b   The autoionization region and beyond: IP–35 eV

In Fig. 6.16, three photoabsorption data sets are compared – Koch/Otto and Rennie *et al.*, both using synchrotron radiation and extending to 35 eV, and Person and Nicole (1974), using a helium continuum light source which terminates at 21.2 eV. The overall shapes are similar, but the cross sections of Koch and Otto are distinctly lower than the other two, approximately 15% below Rennie *et al.* and 20% below Person and Nicole. This observation prompted us to eschew the Koch/Otto data in the sub-ionization region. Here, we tentatively select the measurements of Rennie *et al.* because they are more current, and extend to 35 eV. For clarity, only selected points from Rennie *et al.* are plotted. The graphical data presented by those authors were electronically scanned and digitized. Values of $S(p)$ were computed by fine-mesh trapezoidal integration, and are recorded in Table 6.15.

c   35.0–91.5 eV; 91.5–284.5 eV

The domain between 35 eV and the K-edge of carbon is largely unexplored. No structure is anticipated here, since the deepest inner valence IP has been transcended at $\sim$26 eV. Kumar and Meath negotiate the interval 35.0–70.0 eV using data of Kilcoyne *et al.* (1986), but these are calculational results. Some experimental information is given by Gluskin *et al.* (1981) between 90–220 eV, but we obtain the same contribution to oscillator strength in this region using summed atomic cross sections of Henke *et al.* (1993). Rennie *et al.* linearly interpolate from their highest energy (35 eV) to summed atomic cross sections beginning at 49.6 eV from an earlier compilation of Henke *et al.* (1982). In our experience with other hydrocarbons, atomic additivity becomes a reasonable approximation above 90 eV, but not as low as 50 eV. The approach adopted here is to interpolate on a log-log plot between the cross section of Rennie *et al.* at 35.0 eV and the summed atomic cross section of Henke *et al.* (1993) at 91.5 eV. From 91.5–284.5 eV, the summed atomic cross sections are used. Both segments are fitted by four-term polynomials. The coefficients are given in Table 6.16, and the contributions to $S(p)$ in Table 6.15.

d   The carbon K-edge region, 284.5–350 eV

Rennie *et al.* (2000) find that their absolute cross sections are in quantitative agreement with earlier electron energy loss spectra (EELS) displayed by Piancastelli

Table 6.15 Spectral sums, and comparison with expectation values for C_6H_6[a]

Energy, eV	$S(-2)$	$S(-1)$	$S(0)$	$S(+1)$	$S(+2)$
4.90 ($^1A_{1g} \rightarrow {}^1B_{2u}$)[b]	0.0100	0.0036	0.0013	0.0005	0.0002
6.19 ($\rightarrow {}^1B_{1u}$)[b]	0.4348	0.1978	0.090	0.0410	0.0186
6.96 ($\rightarrow {}^1E_{1u}$)[b]	3.6418	1.8630	0.953	0.4875	0.2494
6.93 ($\rightarrow$ Rydberg)[b]	0.2891	0.1472	0.075	0.0382	0.0195
(9.0)[c]	(0.2758)	(0.1825)	(0.1207)	(0.0798)	(0.0528)
$\sum$ to IP	4.6515	2.3941	1.24	0.6470	0.3405
IP–35.0[d]	11.8153	14.8124	20.4760	31.1243	51.5977
35.0–91.5[e]	0.5190	1.7697	6.4057	24.8689	104.115
91.5–284.5[f]	0.0205	0.1878	1.8519	20.0017	238.558
284.5–350[g]	0.0070	0.1602	3.6608	83.9167	1 930.158
350–1740[f]	0.0054	0.1919	7.5296	343.7866	19 001.248
1740–10 000[f]	–	0.0023	0.4203	90.7990	24 157.357
10^4–10^{5}[h]	–	–	0.0096	11.7462	19 770.24
10^5–10^{6}[i]	–	–	–	0.5400	8 557.32
10^6–10^{7}[i]	–	–	–	0.0186	2 912.62
10^7–10^{8}[i]	–	–	–	0.0006	943.26
10^8–10^{9}[i]	–	–	–	–	300.56
10^9–∞[i]	–	–	–	–	139.12
Total	17.0187	19.5184	41.5939	607.450	78 106.5
Expectation values	17.008[j]		42.0		76 908.8[k]
Other values	(16.948)[l]	19.61[l]	(42.0)[l]	620.2[l]	81 480[l]
	16.555[d]	19.546[d]	40.474[d]		–
	16.556[d]	19.208[d]	41.129[d]		–

[a]In Ry units.

[b]Pantos *et al.* (1978).

[c]Oscillator strength added to make Table II of ref. b conform to their stated total oscillator strength below the IP, 1.24.

[d]Rennie *et al.* (1998).

[e]Log-log interpolation. See text.

[f]Summed atomic cross sections from Henke *et al.* (1993).

[g]Rennie *et al.* (2000).

[h]Atomic additivity, using calculated atomic cross sections from Chantler (1995).

[i]Assuming hydrogen-like behavior, bare carbon atom with screening, K-shell only (Betthe and Salpeter (1977), and atomic additivity. See text.

[j]Current evaluation. See Sect. 6.7.3, text.

[k]Using atomic additivity and Hartree–Fock calculations for atomic carbon, from Fraga *et al.* (1976), Bunge *et al.* (1993).

[l]Kumar and Meath (1992).

et al., but 20–30% higher than photoabsorption cross sections presented by Akimov *et al.* Their results in this energy range find some support from atomic additivity. At 350 eV, well above the carbon K-edge (290.42 eV, Lunnell *et al.*, 1978), they report $\sigma = 3.843$ Mb, compared to 3.896 Mb calculated by a fit to atomic cross sections ($\times 6$) given by Henke *et al.* (1993). Hence, the data presented in Figs. 1–3 by Rennie *et al.* have been scanned and digitized, and their contributions to $S(p)$ are recorded in Table 6.15. However, their tabular data deviate by a relatively larger

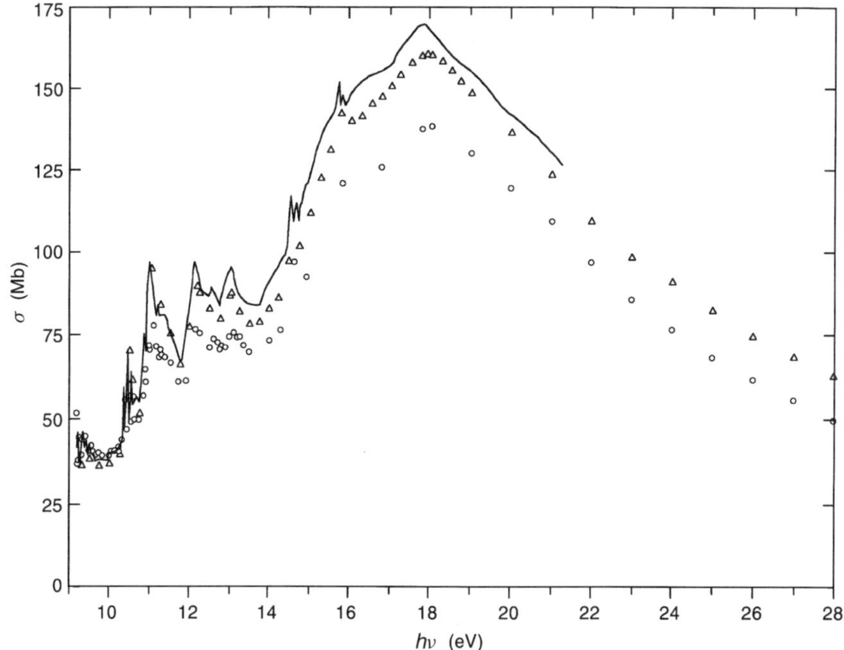

Fig. 6.16 Absolute photoabsorption spectrum of C_6H_6, 9–28 eV. —— Person and Nicole (1974); o Koch and Otto (1976); △ Rennie *et al.* (1998)

Table 6.16 Coefficients of the polynomial $df/dE = ay^2 + by^3 + cy^4 + dy^5$ fitted to data at various energies[a]

Energy range, eV	a	b	c	d
35.0–91.5	12.2207	455.2784	−1507.67	2201.212
91.5–284.5	−17.3124	2017.378	−22956.7	89659.24
350–1740	63.45481	41788.6	−744315	2407277
1740–10000	−12.4692	52985.36	−998930	−10323934

[a] df/dE in Ry units, $y = B/E$, $B = IP = 9.24384$ eV.

percentage as energy increases and the cross section declines, resulting in a value double that of atomic additivity at 800 eV. Consequently, we choose to terminate the Rennie data at 350 eV.

e Post K-edge: 350–10000 eV

We return to the summed atomic cross sections of Henke *et al.* The domain is partitioned into two intervals, 350–1740 eV and 1740–10000 eV. Each is fitted by regression to a four-term polynomial. The coefficients are assembled in Table 6.16.

$$f \quad 10^4 - 10^5 \, eV$$

Calculated atomic cross sections (Chantler, 1995) are used to evaluate $S(p)$.

6.7.3 The analysis

Ramaswamy (1936) measured the refractive index of benzene vapor at several wavelengths, and recorded them for a hypothetical ideal gas at 1 atm and 25.0°C. His data were reproduced by Landolt-Börnstein (1962). Alms *et al.* (1975) cite Landolt-Börnstein as a source, but appear to be using other wavelengths than those given there, in analyzing the wavelength-dependent polarizability. They arrive at a static polarizability (α) of $10.00 \times 10^{-24} \, cm^3$, equivalent to $S(-2) = 16.87$ Ry units. Kumar and Meath use the data of Ramaswamy to arrive at molar refractivity as a function of wavelength, from which they deduce $S(-2) = 16.948$ Ry units. They also calculate molar refractivities at the wavelengths used by Alms *et al.*, and the agreement is very good. Their value corresponds to a static polarizability of $10.045 \times 10^{-24} \, cm^3$. Our fitting of Ramaswamy's data to a Cauchy expansion results in $\alpha = 10.081 \times 10^{-24} \, cm^3$, and $S(-2) = 17.0008$ Ry units. The more recent literature contains calculated values and single-wavelength measurements for liquid benzene, but no extensive vapor measurements. In their sum-rule analysis, Rennie *et al.* cite Bridge and Buckingham (1966) as their source for an experimental value of $\alpha = 10.4 \times 10^{-24} \, cm^3$. Bridge and Buckingham did not measure α, but instead obtained it from an earlier source. Furthermore, their value refers to $\lambda = 6328 \, Å$ (in good agreement with the calculated value of Kumar and Meath at that wavelength) but not to the static polarizability.

The current spectral sum, $S(-2) = 17.0187$ Ry units, is <0.1% above the expectation value. However, the spectral sum for $S(0)$, 41.5939, is 1% below the required TRK value. Comparison with Kumar and Meath reveals that our partial values of $S(0)$ are lower between IP–35.0 eV by 0.4, 283.8–500.0 eV by 0.3 and 500.0–1000 eV by 0.2. In each instance, our spectral sums are based on more current experimental data. Hence, the deficit in $S(0)$ is not readily attributable to a specific spectral range. Part of this deficit may be the result of scanning error in reading the data of Rennie *et al.*, where our partial $S(0)$ are lower by 0.2 than their analysis of their own data, although closer agreement is obtained with their partial $S(-1)$ and $S(-2)$. Their total $S(-2)$ and $S(0)$ are lower than our spectral sums, largely because they have not supplemented the sub-ionization values of Pantos *et al.* just below the IP.

Our spectral sum for $S(-1)$ is 0.5% lower than that of Kumar and Meath. Since they have constrained $S(-2)$ and $S(0)$ to the expectation values in their fitting procedure, their value, $S(-1) = 19.61$ Ry units, may be preferred to our value, $S(-1) = 19.518$ Ry units.

The expectation value for $S(+2)$ is taken as the sum of corresponding expectation values for the constituent atoms, based on Hartree–Fock charge densities at the nuclei. Identical values for this quantity are obtained from Fraga *et al.* (1976) and Bunge *et al.* (1993). Our spectral sum for this quantity, 78 106.5 Ry units, is 1.5% higher than the 'expectation' value. Kumar and Meath obtain a value of $S(+2)$ which is 5.9% higher.

Simple atomic additivity yields $S(+1) = 582.7$ Ry units (Fraga *et al.*, 1976). The true value is expected to be larger. The current spectral sum is 607.5 Ry, while Kumar and Meath obtain 620.2 Ry units. The present result is likely to be closer to the correct value, because the high $S(+2)$ obtained by Kumar and Meath should tend to generate a high $S(+1)$.

To compute $S_i(-1)$, one must utilize the quantum yield of ionization. This quantity, as obtained by Rennie *et al.*, is lower than that given by Person (1965) between IP–11.64 eV, but higher than recorded by Yoshino *et al.* (1973) between 11.8–20.66 eV. For $h\nu \geq 20.66$ eV, the quantum yield is taken to be unity. On balance, the photoionization cross sections from Rennie *et al.* will yield higher values than those inferred from a combination of Person and Yoshino *et al.*, because the photoabsorption cross sections are higher for $h\nu > 11.8$ eV than for $h\nu < 11.8$ eV. Using the photoionization cross sections of Rennie *et al.* from IP–20.66 eV, and the sources in Table 6.15 for $h\nu > 20.66$ eV, we obtain $S_i(-1) = 15.804$. Rennie *et al.* offer two values, 15.986 and 15.648, the higher value associated with the choice of lower K-edge cross sections, which seems unlikely. Nevertheless, the results are in fair agreement, and distinctly lower than the directly measured value, $M_i^2 = S_i(-1) = 17.54 \pm 0.37$ found by Rieke and Prepejchal (1972).

6.8 Buckminsterfullerene (C_{60})

6.8.1 Preamble

Most of the species treated in this monograph are either permanent gases or vapors at ambient temperature, and hence are amenable to absolute photo-absorption measurements by the Beer–Lambert law. Exceptions include atomic nitrogen, oxygen and chlorine, which we describe as transient species, and the alkali elements lithium and sodium, which are characteristic of high-temperature species. Heat pipes (Fung *et al.*, 2000), combined with vapor pressure measurements, have been used for some atoms that require high-temperature for their generation, and recently von dem Borne *et al.* (1995) described an apparatus that uses an interferometer to measure the anomalous dispersion in the vicinity of a resonance line 'quasi-simultaneous' with the absorption measurement, in effect normalizing to a known optical oscillator strength. Cubaynes *et al.* (1998) have commented recently that experimentally determined absolute photoabsorption cross sections for metal vapors are limited to lithium, sodium and barium, and even here the uncertainty is $\pm 25\%$. (Fung *et al.* (2000) claim $\pm 14\%$ for magnesium.) It will be noted that our current sum rule analyses of lithium and sodium have introduced calculated values. For the transient species N and Cl, the relative cross sections were normalized to 1/2 the molecular cross section at sufficiently high energy. Atomic oxygen, generated by electric discharge in molecular oxygen, was calibrated by relating the diminution of a molecular signal upon striking the discharge to the appearance of an atomic signal. None of the above strategies can be considered a general solution to the problem of establishing absolute cross sections for non-permanent gases or vapors. However, a relative spectrum can be realized for either high-temperature vapors or transient species by maintaining constant generating conditions. Pseudo-absolute values for atoms are often arrived at by normalizing to an ab initio calculated value at some energy. Advanced atomic calculations (RPA-random phase approximation; MBPT-many body perturbation theory; R-matrix) have demonstrated accuracy to perhaps $\pm 10\%$. With molecules, there are greater computational difficulties, and consequently larger uncertainties. The purpose of this section is to illustrate the utility of sum rule analysis for molecules, when measured absolute cross sections differ by almost a factor 5.

For buckminsterfullerene (C_{60}), most of the published absolute photoabsorption measurements have been made in the visible and ultraviolet regions. There are three prominent bands, at 3.8, 4.9 and 5.96 eV, which have been identified (Weiss, 1993) as electric-dipole allowed, $^1A_g \rightarrow {}^1T_{1u}$ transitions in I_h symmetry. Table 6.17 compares peak cross sections, as reported by several investigators. Smith (1996) has shown that most of the variation (nearly a factor 5) can be attributed to differing vapor pressures assumed by the researchers. The challenge is to determine which cross section (and by implication, which vapor pressure) is most likely to be correct.

6.8.2   The data

The adiabatic IP of C_{60} has been measured by photoionization mass spectrometry to be 7.57 ± 0.01 eV (Yoo *et al.*, 1992), $7.58^{+0.04}_{-0.02}$ eV (de Vries *et al.*, 1992) and 7.54 ± 0.04 eV (Hertel *et al.*, 1992).

a   Transitions below the IP

Both Smith and Coheur *et al.* (1996) have reviewed earlier vapor-pressure measurements of C_{60}, and both have chosen the data of Piacente *et al.* (1995). (We note from Table 6.17 that their selection leads to much lower cross sections

Table 6.17   Maximum cross sections (Mb) for the lowest dipole-allowed transitions in C_{60}

Energy, eV	3.8	4.9	5.96	Ref.
	420	–	–	Brady and Beiting (1992)
	334	1310	1837	Dai *et al.* (1994)
	490	1900	2400	Gong *et al.* (1994)
	478	1893	2468	Gong *et al.* (1996)
	112	420	550	Smith (1996)
	113	442	–	Coheur *et al.* (1996)
	287	1130	1429	Yasumatsu *et al.* (1996)

than those given by other investigators.) To support their choice, both Smith and Coheur *et al.* have compared their gas phase photoabsorption cross sections to measurements on C_{60} in solutions of n-hexane, by use of the relation

$$A_s/A_g = (n^2 + 2)^2/9n$$

sometimes called the Chako factor (Chako, 1934; Polo and Wilson, 1955; Linder and Abdulnur, 1971). Here, A_s is the integrated absorption coefficient in solution, A_g that in the gas phase, and n is the index of refraction of the solvent at the wavelength of the absorption measurement. This expression is an approximation. Linder and Abdulnur (1971) show that there are higher-order terms. However, the measurements in solution can be performed with known concentrations and known path lengths. Both Smith and Coheur *et al.* find good agreement between their absolute intensities in the gas phase and solution spectra, using the Chako factor, if they employ the vapor pressures of Piacente *et al.* Their respective data are plotted in Fig. 6.17. The data of Coheur *et al.* extend to 5.08 eV, while those of Smith continue to 6.05 eV. In their region of overlap, they are seen to be in quite good agreement. We stress once again (since this is fundamental to the subsequent analysis) that earlier measurements proffered much higher cross sections.

b   IP–11.37 eV

Whereas Smith and Coheur *et al.* used cells with well-defined lengths, Yasumatsu *et al.* (1996) used a windowless cell, and a different selection of vapor pressures. Their relative cross sections track those of Smith and Coheur *et al.* quite

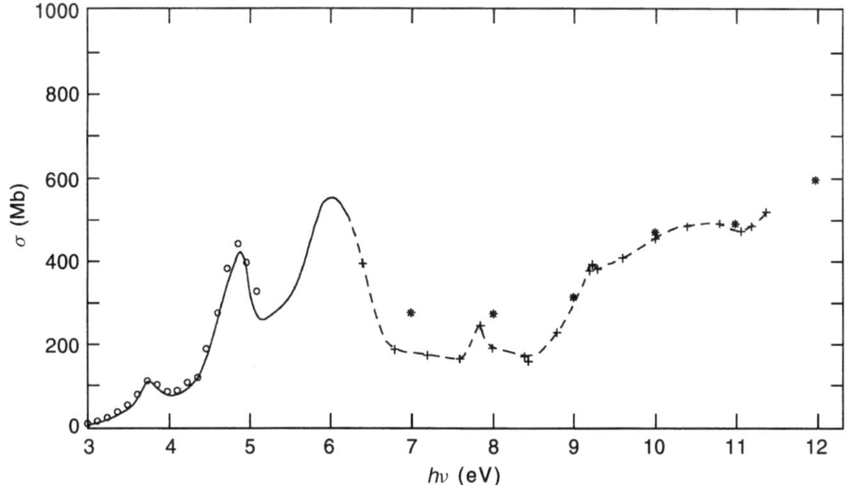

Fig. 6.17 Absolute photoabsorption spectrum of C_{60}, 1st trial, 3–12 eV. —— Smith (1996); ○ Coheur *et al.* (1996); + Yasumatsu *et al.* (1996), renormalized; * Keller and Coplan (1992), normalized

satisfactorily; their absolute values are larger by an average factor of 2.42. Their data, renormalized by this factor, are also plotted in Fig. 6.17, thereby extending our absolute photoabsorption spectrum to 11.37 eV.

c 11.37–27 eV

Relative photoionization cross sections have been measured from the IP to 40.8 eV (Yoo *et al*., 1992) and ~35 eV (Hertel *et al*., 1992). However, these data cannot be directly normalized to photoabsorption measurements, because of the omnipresence (for molecules) of the quantum yield of ionization, which is energy dependent. We shall examine these data in a different context. Quite recently, Jaensch and Kamke (2000a) made their own vapor pressure measurements, and then determined the absolute photoabsorption cross section of C_{60} between 11–25 eV (Jaensch and Kamke, 2000b). In this region (sometimes called the plasma resonance), the absolute cross section attains its maximum; a large fraction of the total oscillator strength is contained therein. They connected their data at 11 eV to the lower-energy measurements of Yasumatsu *et al*. This entailed lowering the absolute values of Yasumatsu *et al*. by nearly a factor 7, considerably larger than the factor 2.42 based on Smith's data. Their spectrum is displayed in Fig. 6.18.

Prior to the availability of the Jaensch/Kamke measurements, the only alternative source for this spectral region was inelastic energy loss spectroscopy (EELS). Such measurements can be transformed into a pseudo-photoabsorption spectrum if the incident electrons have much higher energy than the energy loss, and if the measurements are performed at very small scattering angles (Bethe, 1933; Inokuti, 1971). Two such measurements, with the required transformation, have been published (Keller and Coplan, 1992; Burose *et al*., 1993c). Their pseudo-photoabsorption spectra both display the low energy peaks at ~4.9 and ~6.0 eV, and the much larger band with a maximum near 20 eV, but in detail they are drastically different. This can best be appreciated by reference to Fig. 6.18. Here, we have reproduced the spectrum of Fig. 6.17 from 3–11.37 eV. All of the data consistently display a shoulder between ~10.5–11.5 eV. We normalize the relative spectra of Keller and Coplan, and Burose *et al*., in this region to the adjusted spectrum of Yasumatsu *et al*. It is now evident that the spectrum of Burose *et al*. rises more steeply than the spectra of Keller and Coplan, and of Jaensch and Kamke. Also, its maximum is at ~18 eV, whereas the others peak at ~22 eV. The strikingly similar shapes of the spectral distributions of Keller and Coplan, and of Jaensch and Kamke, arrived at by quite different methods, leads us to conclude their spectra are favored over that of Burose *et al*. Henceforth, we shall ignore the latter, and concentrate on the relative merit of the absolute intensities deduced for Keller and Coplan, and those measured by Jaensch and Kamke.

d Photoionization, 7.6–40.8 eV

In order to estimate the absolute cross section between ~25 eV and ~90 eV (where atomic additivity begins to be reliable) we turn to photoionization data. As discussed in Chap. 3, the quantum yield of ionization is typically low at the

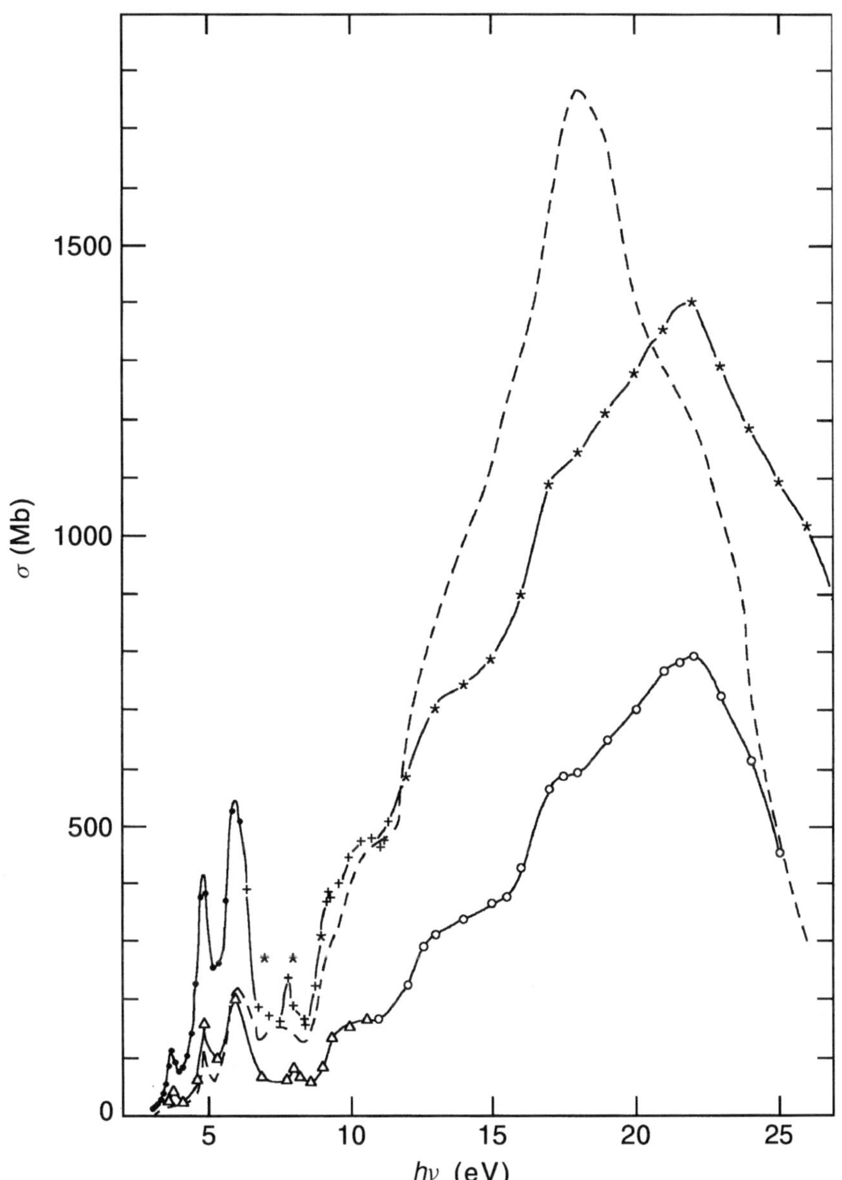

Fig. 6.18 Alternative absolute photoabsorption spectra of C_{60}, 3–27 eV. • Smith (1996); + Yasumatsu *et al.* (1996), renormalized; * Keller and Coplan (1992), normalized; - - - Burose *et al.* (1993), normalized at 11 eV; ○ Jaensch and Kamke (2000b); △ Yasumatsu *et al.* (1996), as renormalized by Jaensch and Kamke (2000b)

IP, and grows to unity at $\sim 20\,\mathrm{eV}$ for many large molecules. Earlier (Berkowitz, 1999) we tried to fit the relative ion yields of Yoo *et al*. and Hertel *et al*. to the normalized absolute cross sections of Keller and Coplan sketched in Fig. 6.18. The ion yields of Yoo *et al*. had the expected behavior when normalized at $\sim 22\,\mathrm{eV}$, i.e., the photoionization cross section was lower than photoabsorption on the lower-energy side, and approximately tracked photoabsorption on the high-energy side. By contrast, the ion yield spectrum of Hertel *et al*., when necessarily constrained not to exceed photoabsorption at low energy, was much lower than photoabsorption at high energy. Effectively, the width of the 'plasma resonance' in the ion yield curve was narrower than that in pseudo-photoabsorption.

In Fig. 6.19, we adopt the same fitting strategy to the new absolute photoabsorption spectrum of Jaensch and Kamke. Similar behavior is found – the ion yield curve of Hertel *et al*. tracks photoabsorption down to $15\,\mathrm{eV}$, but is much lower than photoabsorption for $h v > 21\,\mathrm{eV}$. The ion yield curve of Yoo *et al*. translates to $\sigma_\mathrm{i} < \sigma_\mathrm{a}$ below $20\,\mathrm{eV}$ but it is unrealistically higher than photoabsorption for the 1 or 2 points above $20\,\mathrm{eV}$. (Here we have supplemented the ion yield of $C_{60}{}^{+}$ with the contribution of $C_{60}{}^{++}$.) The data of Jaensch and Kamke begin to display scatter above $24\,\mathrm{eV}$, and terminate at $26\,\mathrm{eV}$. In order to extrapolate their data to higher energies, we link their point at $24\,\mathrm{eV}$ to the isolated points of Yoo *et al*., thus retaining some measure of smoothness and extending the measurements to $40.8\,\mathrm{eV}$. This somewhat arbitrary construction has the effect of enhancing the contribution of the Jaensch/Kamke data to the oscillator strength, but as the sum rule analysis will show, it is nonetheless too small. It neglects the possible

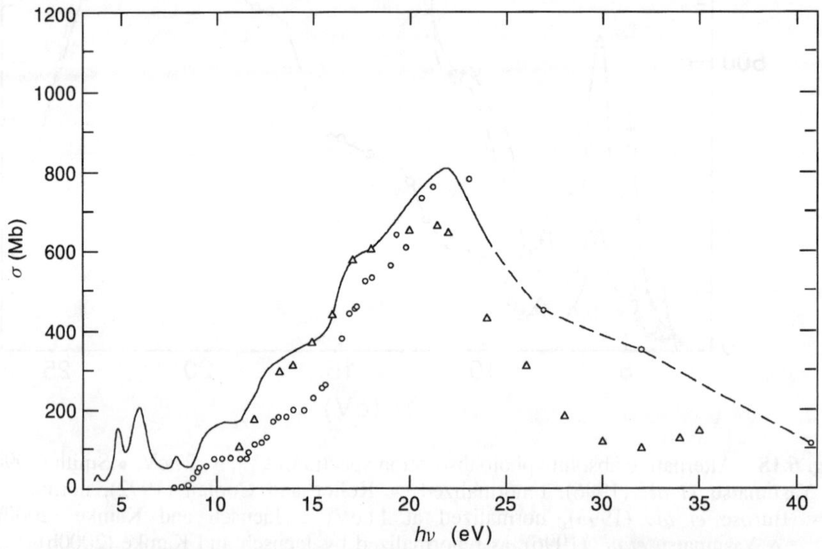

Fig. 6.19   Absolute photoabsorption spectrum of C_{60}, 2nd trial, 3–41 eV. ● Jaensch and Kamke (2000b); ○ Yoo *et al*. (1992), normalized to Jaensch and Kamke (2000b); △ Hertel *et al*. (1992), normalized to Jaensch and Kamke (2000b)

Table 6.18 Spectral sums, and comparison with expectation values for C_{60}[a]

Energy, eV	$S(-2)$	$S(-1)$	$S(0)$	$S(+1)$	$S(+2)$
3.1–7.6 (IP)	61.9	24.4	9.96	4.2	1.8
7.6–40.8	116.2	151.9	230.7	399.6	768.8
Sub-total	178.1	176.3	240.66	403.8	770.6
Adjusted[b]	138.9	137.4	187.6	314.8	600.8
(Sub-total)[c]	76.5	83.9	120.8	20.7	391.5
40.8–280[d]	2.45	10.7	55.2	355.7	3 073
(40.8–280)[e]	2.06	9.6	51.2	350.5	3 005
280–320[f]	0.05	1.2	25.8	568.7	12 560
320–1740[g]	0.07	2.4	87	3 726	197 100
1740–10 000[g]	–	–	4.2	908	241 570
10^4–10^5	–	–	0.1	117.5	197 770
10^5–10^6	–	–	–	5.4	85 570
10^6–∞	–	–	–	0.2	42 960
Total	180.7	190.6	413	6085	781 374
Adjusted[b]	141.5	151.7	360	5997	781 200
(Total)[h]	78.7	97.1	289.1	5882	780 927
Expectation values	~135[i]	–	360	–	–
Additivity	–	205.4[j]	–	5807[j]	768 768[j,k]

[a]In Ry units.
[b]Reduction by 22%, 3.1–40.8 eV, to match TRK sum. See text.
[c]Jaensch and Kamke (2000b), extrapolated to 40.8 eV.
[d]Henke *et al*. (1993) + additivity, extrapolated to 40.8 eV.
[e]Henke *et al*. (1993) + additivity, extrapolated to 40.8 eV based on Jaensch and Kamke (2000b).
[f]Krummacher *et al*. (1993); Itchkawitz *et al*. (1995) normalized.
[g]Henke *et al*. (1993) + additivity.
[h]Utilizing Jaensch and Kamke (2000b).
[i]From refractive index, dielectric constant and deflection expts. See text.
[j]Fraga *et al*. (1976) and atomic additivity.
[k]Bunge *et al*. (1993) and atomic additivity.

subsidiary peak at 35 eV in the data of Hertel *et al*., and discussed by Lambin *et al*. (1992). In Table 6.18, we list the contributions to $S(p)$ based on the values given by Jaensch and Kamke from the onset of absorption to 24 eV, and our extension of their data to 40.8 eV. Also recorded is the earlier result (Berkowitz, 1999), based on the data of Smith, Yasumatsu *et al*. and Keller and Coplan.

e 40.8–280 eV

Figure 6.20 presents estimates of the photoabsorption cross section in this domain. At 40.8 eV, two values are shown, depending upon alternative normalizations of the ion yield of Yoo *et al*. The other points are based on atomic additivity, i.e., $60 \times \sigma(C)$, where $\sigma(C)$ is taken from Henke *et al*. (1993). Other examples in this monograph indicate that this is a rather accurate representation of hydrocarbons from ~90 eV to the carbon K-edge. Below 90 eV molecular effects can be anticipated. Here, we simply connect the alternative values at 40.8 eV by smooth interpolation to the summed atomic cross sections at high energy. The

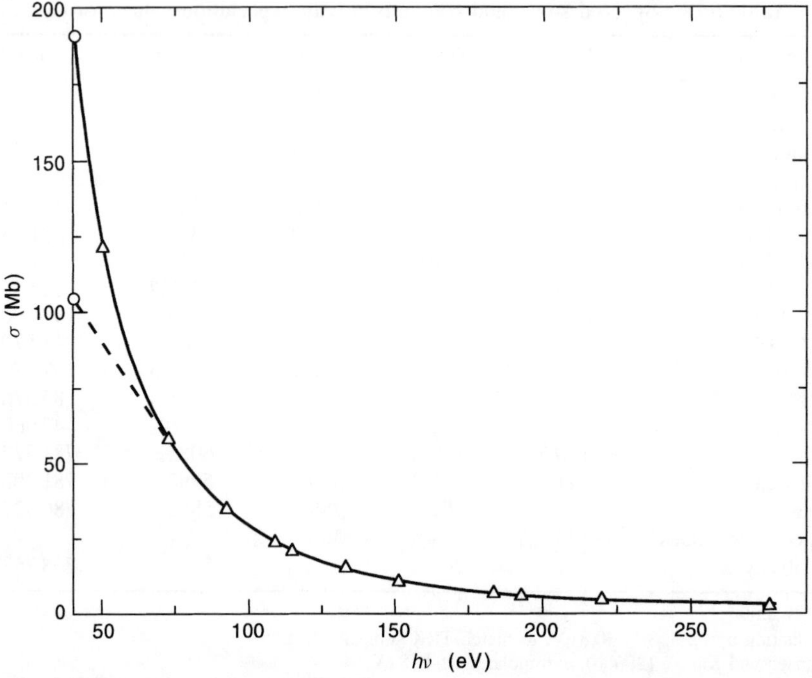

Fig. 6.20 Estimated absolute photoabsorption spectrum of C_{60}, 40–280 eV. ○ Yoo *et al.*
(1992), alternate normalizations; △ $\sigma(C_{60}) = 60 \times \sigma(C)$, from Henke *et al.* (1993)

corresponding contributions to $S(p)$ are recorded in Table 6.18. Although the
alternative estimates of the cross section at 40.8 eV differ by nearly a factor 2,
the differences in $S(p)$ are relatively small.

f Carbon K-edge, 280–320 eV

Krummacher *et al.* (1993) have obtained a gas-phase absorption spectrum, in
relative intensity units, between ~284–289 eV. Itchkawitz *et al.* (1995) have
presented a spectrum from a 'pristine C_{60} film' which is similar, but extends to
308 eV. Hitchcock and Mancini (1994) discuss normalization of such spectra to
absolute cross sections. A commonly used procedure is to normalize at ~25 eV
above the K-edge, which in this case would occur at ~315 eV. Hence, some slight
interpolation is necessary with the available data. Figure 6.21 is a composite of
the data of Krummacher *et al.* (1993) and Itchkawitz *et al.* (1995), appropriately normalized, together with higher-energy points from summed atomic cross
sections.

g 320–10 000 eV

We fit the summed atomic cross sections of Henke *et al.* to two polynominals, spanning the ranges 320–1740 eV and 1740–10 000 eV by regression. The
coefficients of these polynominals are given in Table 6.19.

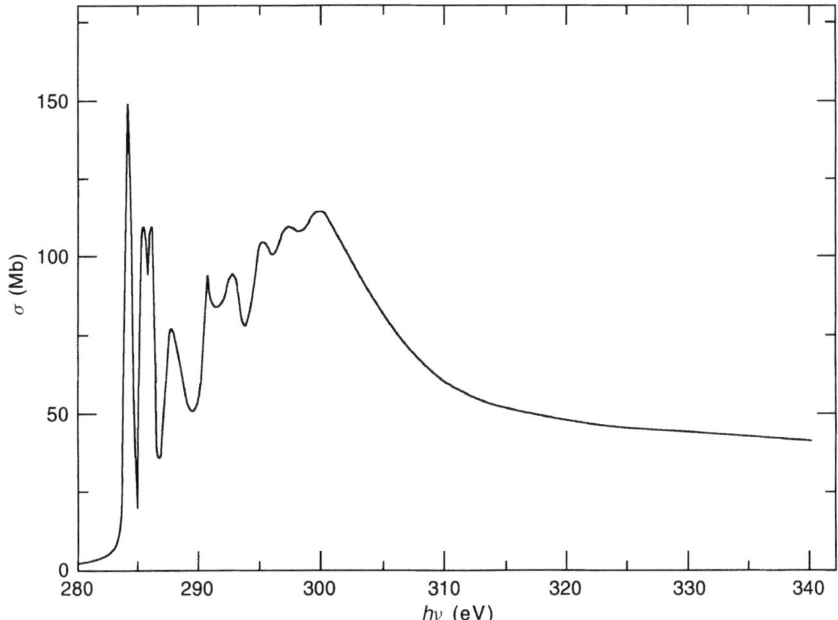

Fig. 6.21 Absolute photoabsorption spectrum of C_{60}, K-edge region. Data of Krummacher *et al.* (1993); Itchkawitz *et al.* (1995), normalized to atomic additivity

Table 6.19 Coefficients of the polynomial $df/dE = ay^2 + by^3 + cy^4 + dy^5$ C_{60}, 320–10 000 eV[a]

Energy range, eV	a	b	c	d
320–1740	0.657 093	13.436 96	−7.525 18	0.705 243
1740–10 000	−0.126 6	17.142 34	−10.298	−3.391 32

[a]df/dE in Ry units, $y = B/E$, B = IP (K-edge) = 290.1 eV; $\sigma(C_{60})$ assumed as $60\,\sigma(C)$, where $\sigma(C)$ is taken from Henke *et al.* (1993).

h 10^4 eV–∞

The calculated and tabulated atomic carbon cross sections of Chantler (1995) ($\times 60$) are utilized between 10^4–10^5 eV. Above 10^5 eV, the atomic carbon cross section is approximated by the Bethe–Salpeter equation (Bethe and Salpeter, 1977).

6.8.3   The analysis

Until recently, the experimental estimates of the electric dipole polarizability of C_{60} were based on measurements of the refractive index n or the dielectric constant $\varepsilon(\varepsilon = n^2)$ of thin films of C_{60}. Ren *et al.* (1991), Eklund (1992) and

Kafafi *et al.* (1992) reported $n = 1.90$, 1.96 and 2.0, respectively. The molar polarization P is given by the Lorenz–Lorentz equation

$$P = [(n^2 - 1)/(n^2 + 2)]M/\rho$$

where M is the molecular weight, and ρ is the density. The latter can be estimated from the lattice constant ($a = 14.17\,\text{Å}$) of an fcc unit cell, with 4 molecules/unit cell, giving 1.406×10^{21} molecules/cm^3, or $\rho = 1.68_3$ g/cm^3. Thus, the three alternative values of n yield $P = 199$, 208 and 214 cm^3/mol, or $\alpha = 79.0$, 82.6 and 84.9×10^{-24} cm^3.

Later, Eklund *et al.* (1995) described an apparently more precise measurement of the dielectric constant of thin-film C_{60} as a function of wavelength. Extrapolating to infinite wavelength, they obtained $\varepsilon_0 = 4.08 \pm 0.05$, or $\alpha = 85 \times 10^{-24}$ cm^3. They presented evidence for infrared lattice vibrations contributing $\approx 2 \times 10^{-24}$ cm^3, and hence a net molecular polarizability of $\approx 83 \times 10^{-24}$ cm^3. Quite recently, Antoine *et al.* (1999) have made a direct measurement of the gas-phase polarizability by deflecting a beam of C_{60} in an inhomogeneous electric field. They obtained $\alpha = 76.5 \pm 8 \times 10^{-24}$ cm^3. This gas-phase measurement is preferred, since it eliminates possible solid-state contributions, but unfortunately it retains a 10% uncertainty. (At this writing, Ballard *et al.* (2000) have determined $\alpha = 79 \pm 4 \times 10^{-24}$ cm^3 for C_{60} vapor at $\lambda = 1.064\,\mu$m, based on the light force experienced by the species in a standing-wave light field. This incident light energy is below the lowest electronic excitation, but above the vibrational (infrared) excitations. Applying a Cauchy expansion, and an estimate of $S(-4)$ from the data sources cited in Table 6.18, we estimate a static dipole polarizability $\alpha = 77 \pm 4 \times 10^{-24}$ cm^3 for the Ballard data, essentially the same result as that obtained by Antoine *et al.*).

Numerous calculations have been performed to evaluate the dipole polarizability of a C_{60} molecule. Westin *et al.* (1996) and Norman *et al.* (1997) summarize several of them. Among the more extensive ab initio calculations, Fowler *et al.* (1990) obtained 65.4×10^{-24} cm^3 as a lower bound. Other significant results include 78.8×10^{-24} cm^3 (Weiss *et al.*, 1993, RPA), 82.7–83.0×10^{-24} cm^3 (Quong and Pederson (1992); Pederson and Quong (1992), local density approximations) and 80.6–82.5×10^{-24} cm^3 (van Gisbergen *et al.* (1997), time-dependent density functional theory). Norman *et al.* used cubic response theory and RPA to obtain 85.8×10^{-24} cm^3, but later (Jonsson *et al.*, 1998) corrected this to 75.3×10^{-24} cm^3. Thus, the calculated polarizabilities (given equal weight) are circumscribed by the values $79 \pm 4 \times 10^{-24}$ cm^3, close to the measured values. For our first test of absolute cross sections by sum rule analysis, we take $\alpha \approx 80 \times 10^{-24}$ cm^3 as a rough average of calculated and experimental results, or $S(-2) \approx 135$ Ry units.

In Table 6.18, we note that the spectral sum for $S(-2)$ based on overlapping the data of Smith, Yasumatsu *et al.* and Keller and Coplan is 180.7, while that stemming from the data of Jaensch and Kamke is 78.7. Furthermore, almost all of the difference occurs between 3.1–40.8 eV. The nature of the construction in this region, in both cases, should enable us to correct each data set by simple scaling (to first approximation). Thus, an enhancement by a factor 1.74 in the Jaensch/Kamke data, or a reduction by a factor 0.74 in the other case, would achieve the desired result.

For $S(0)$, the Smith-based construction yields 413, that originating from the Jaensch/Kamke data is 289, and the TRK value is 360. Applying the same correction factors used for $S(-2)$ yields 351.2 (Smith-based) and 378 (Jaensch/Kamke). The Smith-based data is slightly over-corrected, the Jaensch/Kamke data more so. The latter behavior can be attributed to the manner of extrapolation of the Jaensch/Kamke data above 24 eV (recall Fig. 6.19), where we abruptly switch to the higher values of Yoo *et al.* Between 24–40.8 eV, the contribution to $S(-2)$ is relatively less important than that to $S(0)$. For the Smith-based data, the TRK sum is achieved by applying a reduction factor of 0.78, only slightly smaller than the correction to $S(-2)$.

We had previously remarked about the similar shapes of the spectra of Keller and Coplan, and Jaensch and Kamke (Sect. 6.8.2.c and Fig. 6.18). More detailed analysis reveals that the spectrum resulting from inelastic electron scattering is somewhat broader than that from photoabsorption in the vicinity of the 'plasma resonance'. This could partially result from the differing resolutions (0.6 eV, electrons; 0.13 eV, photons), but may have other causes. Thus, if we adjust the EELS data of Fig. 6.18 by the factor 0.78, the cross section at its maximum (22 eV) is 1100 Mb, only 37.5% larger than that obtained by Jaensch and Kamke. However, at 11 eV they differ by a factor 2.2. The correction factor given by the $S(-2)$ sum rule, 1.74 is a rough average.

The results of the $S(-2)$ and $S(0)$ sum rule analysis favor the absolute cross sections based on the low-energy values of Smith, and of Coheur *et al.*, requiring a reduction of only 22%. This deviation may be within the uncertainty of vapor pressure measurements and Chako factor corrections. Also, the overlapping of the cross sections of Yasumatsu *et al.*, and of Keller and Coplan, can only be approximate, and the latter spectrum has a large weight. When we recall the variation in absolute cross sections of almost a factor 5 given in Table 6.17, this discrepancy of 22% certainly supports the calibrations of Smith, and Couheur *et al.* Their cross sections are the lowest ones in Table 6.17. The absolute cross sections of Jaensch and Kamke are still lower, as we have seen. Furthermore, their juncture with the data of Yasumatsu *et al.*, and renormalization of the latter, results in cross sections for the ultraviolet peaks that are lower than those of Smith by a factor ~ 2.7 (see Fig. 6.18).

In recent correspondence, Kamke (2000) has disclosed to this author a careful reassessment of possible errors in the vapor pressure and absolute cross section measurements of Jaensch and Kamke. Their revised cross sections are increased by 15.3%, with an uncertainty of ± 5%. Thus, their maximum cross section is 922 ± 46 Mb. If we replace our trial value ($\alpha \simeq 80 \times 10^{-24}$ cm^3) with the experimental gas phase value ($\alpha = 76.5 \pm 8 \times 10^{-24}$ cm^3), the expectation value of $S(-2)$ becomes 129 ± 14 Ry units. The 5% reduction in $S(-2)$ imposes a corresponding

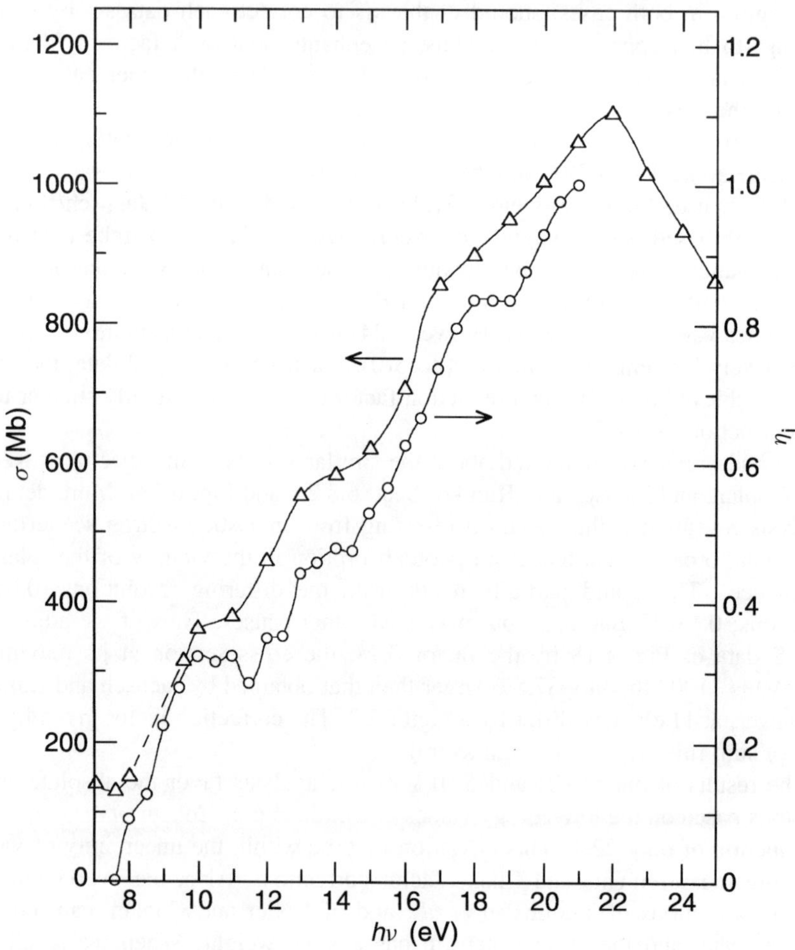

Fig. 6.22 Adjusted absolute photoabsorption spectrum of C_{60}, 7–25 eV, and estimated quantum yield of ionization (σ_i). △ Sum rule adjusted cross section, using Smith (1996), Jaensch and Kamke (2000b); Keller and Coplan (1992); ○ η_i from Yoo *et al.* (1992) normalized at 21 eV to sum-rule-adjusted cross section

decrease in the cross sections based on the data of Smith, Yasumatsu *et al.* and Keller and Coplan, from a maximum of 1100 Mb (Fig. 6.22) to 1050 ±110 Mb. A discrepancy of 13% persists, but the two values now overlap, within their combined error bars. A reduction factor of 0.75 (instead of 0.78) applied to the Smith-based values (3.1–40.8 eV) in Table 6.18 results in $S(-2) = 138$ (cf. expectation value $= 129 \pm 14$ Ry units) and $S(0) = 353$ (cf. TRK value $= 360$). These revised numbers have not been incorporated in Table 6.18, partly to avoid confusion, and also because the corrected values provided by Kamke await publication. The discrepancy between the directly measured cross sections of Smith (Table 6.17) and the extrapolated values of Jaensch and Kamke is reduced from a factor 2.7 to $\sim$2.0.

In order to provide some further insight into the differing breadths of the EELS spectrum (Keller and Coplan, 1992) and the photoabsorption curve (Jaensch and Kamke, 2000b), and also the suspected subsidiary peak at $\sim$35 eV (Lambin *et al.*, 1992), we examine the predictions of ab initio calculations for this domain. Alasia *et al.* (1994) performed a rather extensive calculation, using the local density approximation (LDA) plus RPA equations. They present their results as $S(E)$ versus E, where $S(E)$ is a strength function which may be written

$$S(E) = 2a_0^3 \left(\frac{1}{E(\text{Ry})} \right) \mathrm{d}f/\mathrm{d}E\,(\text{Ry}).$$

Upon converting to our conventional σ versus E plot, we find fairly good agreement with our selected data in absolute magnitude. The maximum cross section ($\sim$1320 Mb, cf. $\sim$1050 $\pm$110 Mb) is shifted to lower energy ($\sim$20.4 eV, cf. 22 eV). The ultraviolet peak (7 eV, cf. 6 eV) has a cross section of $\sim$378 Mb (cf. 430 Mb). The width appears to fall between the EELS and photoabsorption widths. There is a weak subsidiary peak at $\sim$31.5 eV, and the cross section remains quite high between 30–35 eV, but we are cautioned (Alasia *et al.*, 1994) that this region is not very accurate because of the energy cutoff at 40 eV used for the particle-hole basis states. Hansen *et al.* (1995) performed a tight-binding linear response calculation, which they acknowledged to be less extensive than that of Alasia *et al.*. Their calculated maximum cross section ($\sim$2600 Mb) is double that of Alasia *et al.*, and the peak, correspondingly narrower, cannot be considered a reliable basis for distinguishing between experiments.

Westin *et al.* have also used an LDA linear response formalism with an adjustable screening parameter to calculate the oscillator strength distribution up to 40 eV. Their spectrum is presented in relative units, but the corresponding width is slightly larger than that of Keller and Coplan. Gerchikov *et al.* (1997) have calculated the differential inelastic electron scattering cross section and found fairly good agreement with Keller and Coplan. In summary, the available calculations appear to favor the broader distribution, but they lack the accuracy to be conclusive.

The spectral sum for $S(+2)$ is dominated by oscillator strengths at the K-edge and beyond, and consequently is insensitive to the disputed cross sections at lower energies. The spectral sum (781 200 Ry units) is just 1.6% larger than the 'expectation value', taken as 60 times $S(+2)$ for atomic carbon (Fraga *et al.*, 1976; Bunge *et al.*, 1993).

Both $S(-1)$ and $S(+1)$ involve correlation terms, and consequently are expected to deviate from atomic additivity. The adjusted spectral sum for $S(-1)$, 151.7 Ry units, is approximately 74% of the value (Fraga *et al.*, 1976) based on atomic additivity, while the spectral sum for $S(+1)$ is $\sim$3.3% larger than the summed (Fraga *et al.*, 1976; Bunge *et al.*, 1993) atomic value. The fractional decrease in $S(-1)$ and increase in $S(+1)$ are very nearly the same as those obtained for benzene (see Sect. 6.7).

It is useful to estimate $S_i(-1)$, the ionized component of $S(-1)$, since it can serve as a benchmark for high-energy electron impact ionization cross sections (vide infra). In Fig. 6.22, we have plotted the adjusted absolute photoabsorption cross sections in the 7–25 eV region, based on the sum rule analysis. The photoion yield curve of Yoo *et al.* has been normalized to this curve, assuming that the quantum yield of ionization (η_i) is unity for $h\nu \geq 21$ eV. With this assumption, we compute $S_i(-1) = 102 \pm 10$ Ry units.

Itoh *et al.* (1999) have recently reported the absolute ionization cross sections of C_{60} by electron impact between 0.4–5.0 keV. We have analyzed their data in the form of a Fano plot (see Sect. 3.3). Treated in this fashion, their cross sections exhibit significant scatter. Nevertheless, a slope corresponding to $M_i^2 = 100 \pm 15$ seems a fair representation. The value of M_i^2 is fortuitously close to $S_i(-1)$ obtained from the photoabsorption/photoionization spectrum. (See Berkowitz, 1999, ref. 39 for a more detailed discussion.)

The quantum yield of ionization inferred from normalization of the data of Yoo *et al.* is also shown in Fig. 6.22. It displays some structure, superimposed on a monotonic ascent from ~ 0 at 7.6 eV to 1.0 at 21.0 eV.

6.9   Sulfur Hexafluoride (SF$_6$)

6.9.1   Preamble

For the purposes of accurate sum-rule analysis, sulfur hexafluoride (SF$_6$) presents some problems. Although it has high (octahedral, O_h) symmetry, it is a fairly large molecule containing a second-row atom, and hence involves 16 molecular orbitals. Electron escape from the central atom is impeded by a cage of electronegative atoms. This barrier effectively creates a double well potential – an inner well, with favorable transition probabilities to low-lying, unoccupied valence states, and an outer well, with weak transition probabilities to Rydberg states. Excitation to inner-well states above an asymptotic ionization threshold leads to quasi-discrete states, called shape resonances. Such behavior is manifested not just by S(1s), S(2s) and S(2p) excitations, but also F(1s) transitions. Interest in the detailed nature of these resonances has spurred numerous investigations of SF$_6$, including photoabsorption, photoelectron spectroscopy, electron momentum spectroscopy, and of course, theoretical studies. However, apart from the recent measurements of Holland *et al.* (1992) covering the range IP–29.5 eV, absolute cross section data are largely limited to studies of 1979 or earlier vintage. Kumar *et al.* (1985) performed a sum-rule analysis with data available to them, and Holland *et al.* produced their own sum-rule analysis.

In O_h symmetry, and within the independent particle approximation, the sequence of occupied molecular orbitals in the electronic ground state of SF$_6$ may be written

$$(1a_{1g})^2(2a_{1g})^2(1t_{1u})^6(1e_g)^4(3a_{1g})^2(2t_{1u})^6$$

$$-(4a_{1g})^2(3t_{1u})^6(2e_g)^4, (5a_{1g})^2(4t_{1u})^6$$

$$-(1t_{2g})^6(3e_g)^4(1t_{2u})^6(5t_{1u})^6(1t_{1g})^6, {}^1\tilde{A}_{1g}$$

The lower tier of 5 molecular orbitals, from $1t_{1g}$ to $1t_{2g}$, are largely combinations of F(2p) atomic orbitals, according to Weigold $et\ al.$ (1991) and Yang $et\ al.$ (1998b). The vertical ionization potentials (VIP) from these orbitals have been measured several times, most recently with high resolution by Holland $et\ al.$ (1995b). They give (in eV) 15.67 ($\tilde{X}^2T_{1g}$); 16.95 ($\tilde{A}^2T_{1u}$); 17.2 ($\tilde{B}^2T_{2u}$); 18.35 ($\tilde{C}^2E_g$); and 19.68 ($\tilde{D}^2T_{2g}$). The two next deeper orbitals, $4t_{1u}$ and $5a_{1g}$, involve primarily combinations of F(2p) and S(3s) orbitals. The photoelectron spectrum corresponding to ($\tilde{E}^2T_{1u}$) has a well-resolved vibrational progression, suggesting electron removal from a bonding orbital, with VIP $= 22.53$ eV, while that for the ($\tilde{F}^2A_{1g}$) state (VIP $= 26.82$ eV) has weak vibrational structure, though described as mostly bonding with some anti-bonding character (Weigold $et\ al.$, 1991).

Completing the second tier are the orbitals $2e_g$, $3t_{1u}$ and $4a_{1g}$, which have compositions dominated by F(2s). The $2e_g$ orbital is described as F(2s) non-bonding, with a binding energy of ~ 39.5 eV, while $3t_{1u}$ has some F(2s) $+$ S(3p) bonding character (VIP ~ 41.0 eV) and $4a_{1g}$ is found to have F(2s) (some F2(p)) $+$S(3s) bonding character, and VIP $= 44.1$ eV.

The molecular orbitals in the upper tier comprise the atomic core orbitals. The $2t_{1u}$ orbital consists of S($2p_{3/2}$) (IP $= 180.27$ eV) and S($2p_{1/2}$) (IP $= 181.48$ eV) (Hudson $et\ al.$, 1993), while $3a_{1g}$ is S(2s), with IP $= 244.7$ eV (Keshi-Rahkonen, 1976). The six F(1s) orbitals become $1e_g$, $1t_{1u}$ and $2a_{1g}$, with a binding energy of 695.04 eV, while the S(1s) orbital in O_h symmetry is a_{1g}, with a binding energy of 2490.1 eV (Jolly $et\ al.$, 1984).

Absolute photoabsorption cross sections for SF_6 are sparse or inconsistent in several energy intervals. The recent data of Holland $et\ al.$ (1992) from IP–29.5 eV, stated to be accurate to $\leq 3\%$, solidify an important spectral region. Their data lie considerably lower than those of Blechschmidt $et\ al.$ (1972) in the region of overlap, and the latter are a prime source for the sub-ionization region, which implies potential error there. As noted by Holland $et\ al.$, relatively few experiments have been performed at $h\nu > 40$ eV. Zimkina and collaborators have provided several of these measurements, in the vicinity of the S(2s), S(2p), S(1s) and F(1s) edges (Zimkina and Fomikov, 1967; Zimkina and Vinogradov, 1971; Vinogradov and Zimkina, 1972). Unfortunately, comparison with other sources, where possible, indicates that the cross sections of Zimkina and collaborators are lower by 15–20%. We shall ultimately appeal to the sum rules for judging the preferred selection of data.

6.9.2 The data

The uppermost occupied orbital of SF_6 is $1t_{1g}$. The high-resolution photoelectron spectrum (PES) corresponding to formation of ($\tilde{X}^2T_{1g}$) (see, e.g. Holland $et\ al.$,

1995b) consists of a broad band, with VIP = 15.67 eV, and an onset of $\sim$15.3 eV. In a threshold PES study by Yencha *et al*. (1997) this onset has been reduced to 15.116 eV. The photoionization mass spectrum (PIMS) of SF_6 reveals that SF_5^+ is already produced at these lower energies, with no evidence of SF_6^+ (Berkowitz, 1979). The implication is that the electronic ground state of SF_6^+, as accessed in the Franck–Condon region, is unstable. Tichy *et al*. (1987; 1990) examined a number of charge-transfer experiments of the generic type

$$MH^+ + SF_6 \rightarrow SF_5^+ + HF + M$$

with $MH^+ = HCl^+$, HBr^+, CH_5^+. When a charge-transfer reaction is observed to proceed rapidly, the usual assumption is that the reaction is exothermic, i.e., $\Delta H_r < 0$. From such studies, they initially concluded (Tichy *et al*., 1987) that the appearance potential (AP) of SF_5^+ from SF_6 was 13.98 $\pm$0.03 eV, and later (Tichy *et al*., 1990) that AP $(SF_5^+/SF_6) < 13.78$ eV.

More recently, Evans *et al*. (1997) photoionized SF_6 near the onset of observable signal (15.311 and 15.508 eV) and measured the kinetic energy of the SF_5^+ fragment. Noting that conservation of momentum requires most of the kinetic energy release to occur in the neutral F partner, they concluded that >1.2 eV of kinetic energy was liberated, and hence that the (unobservable) threshold for SF_5^+ from SF_6 was 14.11 $\pm$0.08 eV, about 0.33 eV larger than the upper limit from charge transfer. Recent ab initio calculations (Irikura, 1995; Bauschlicher and Ricca, 1998) appeared to favor the photoionization result. Irikura (1995) has made the interesting suggestion that, rigorously, the criterion for a reaction to proceed is that $\Delta F_r < 0$. In most charge-transfer reactions the number of products and reactants is the same, and the $T\Delta S$ term is frequently ignored. For the generic reaction considered here, there is an additional product, and $T\Delta S$ at 298 K is 0.48 eV. Adding this quantity to the prior upper limit (<13.78 eV) makes it compatible with the photoionization result. In any case, the electronic ground state of SF_6^+ appears to be unbound, and consequently an adiabatic IP, in the usual sense, is undefined. Here, we adopt 14.9 eV (832 Å) for the pragmatic purpose of comparing our oscillator strength sums with those of Holland *et al*. (1992).

a   *Transitions below 14.9 eV*

Holland *et al*. (1992) have discussed the inconsistent photoabsorption and electron energy loss spectra below $\sim$10 eV, and have ignored them in their sum rule analysis. If we were to use the electron energy loss oscillator strengths of Hitchcock and Van der Wiel (1979) between 5–10 eV, it would significantly enhance $S(-2)$, but only slightly affect $S(0)$. We shall see that transitions above 10 eV yield a spectral sum for $S(-2)$ that already exceeds the expectation value. Thus, we assume the lower limit of significant absorption to commence at 10 eV. Kumar *et al*. have made a different judgment, retaining the cross sections of Hitchcock and Van der Wiel below 10 eV, thus increasing $S(-2)$ by $\sim$0.53 Ry units.

Between 10–14.9 eV, the choices involve electron energy loss data of Hitchcock and Van der Wiel and Simpson *et al.* (1966) and photoabsorption measurements by Blechschmidt *et al.* Kumar *et al.* continue with the electron impact values, while Holland *et al.* (1992) choose the photoabsorption results. Our practice has been to utilize photoabsorption data when available. A typical criticism of photoabsorption in low-energy regions is the possibility of saturation at sharp absorption lines. This is not an issue in SF_6.

b   14.9–26.38 eV; 26.38–70.85 eV

Figure 6.23 compares the photoabsorption data of Holland *et al.* (1992) with similar measurements by Lee *et al.* (1977) and with the (e,e) data of Hitchcock and Van der Wiel above 20 eV. The Holland and Lee cross sections are in fairly good agreement up to $\sim$27 eV, where the Holland data display more scatter. The values of Hitchcock and Van der Wiel are consistently higher up to $\sim$56 eV, beyond which they merge with the data of Lee *et al.* For this sum rule analysis, we retain the cross sections of Holland *et al.* from 14.9–26.38 eV, then continue with the data of Lee *et al.* from 26.38–70.85 eV. All of the above data have been scanned from figures. To test the fidelity of data extraction, we have compared our values of $S(0)$, $S(-1)$ and $S(-2)$ with those given by Holland *et al.* across their experimental range. Our values are $\sim$2% larger for $S(-2)$, and $\sim$3% larger for $S(0)$. All the integrations have been performed trapezoidally. The $S(p)$ values are listed in Table 6.20.

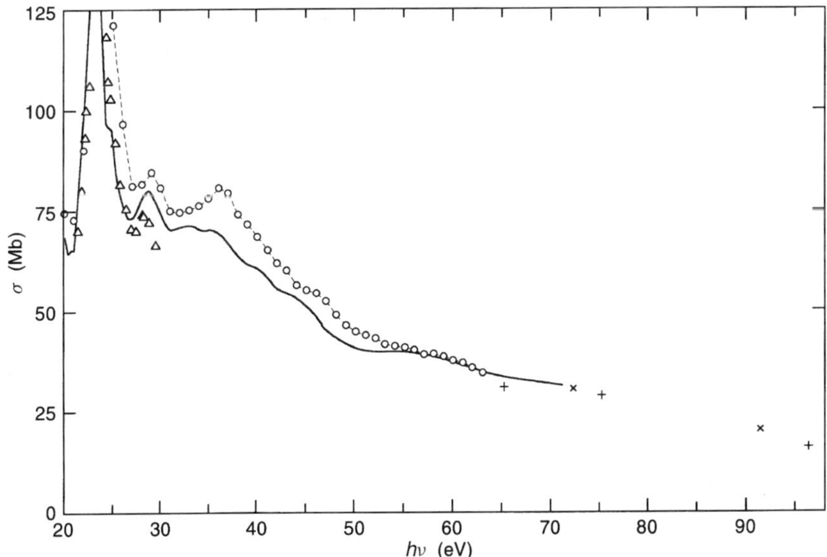

Fig. 6.23   Absolute photoabsorption of SF_6, 20–96 eV. —— Lee *et al.* (1977); o Hitchcock and Van der Wiel (1979); △ Holland *et al.* (1992); + Vinogradov and Zimkina (1972); × Henke *et al.* (1993); + additivity

Table 6.20 Spectral sums, and comparison with expectation values for SF_6[a]

Energy, eV	$S(-2)$	$S(-1)$	$S(0)$	$S(+1)$	$S(+2)$
10.0–14.9[b]	1.1821	1.0675	0.9761	0.9037	0.8469
14.9–26.38[c]	3.9027	5.7971	8.8354	13.7979	22.0344
26.38–70.85[d]	2.3561	6.6600	20.2837	66.8494	237.785
70.85–171.0[e]	0.2468	1.7234	12.7067	99.5576	830.435
172.3–173.4[f]	0.0005	0.0067	0.085	1.0799	13.72
177.4–177.5[g]	–	0.0007	0.0085	0.1109	1.45
180.9[g]	–	–	0.0006	0.0080	0.11
183.4–184.6[g]	0.0012	0.0166	0.225	3.0426	41.14
196.2[f]	0.0015	0.0211	0.304	4.3838	63.22
205.8[h]	0.0003	0.0047	0.0707	1.0687	16.16
241.2[h]	0.0002	0.0028	0.049	0.8687	15.40
171–210[h]	0.0076	0.1063	1.489	20.8482	291.91
210–243[h]	0.0049	0.0817	1.360	22.6405	376.91
243–277[h]	0.0031	0.0588	1.124	21.4792	410.46
277–685[e]	0.0076	0.2044	5.8114	175.7966	5 680.56
689.0[f]	–	0.0017	0.0865	4.3804	221.83
694.7[f]	–	0.0021	0.1087	5.5502	283.39
699.9[f]	–	0.0012	0.0631	3.2460	166.98
713.2[f]	–	0.0007	0.0358	1.8766	98.37
685–735[f]	0.0004	0.0203	1.0631	55.7385	2 923.15
735–2490[e]	0.0017	0.1211	9.5192	827.9096	80 942.5
2486[j]	–	0.0002	0.0349	6.377	1 165.2
2490–2580[j]	–	0.0010	0.1828	34.0787	6 355.2
2580–10 000[e]	–	0.0083	2.2933	706.6848	250 389.3
10^4–10^5[k]	–	0.0001	0.1746	227.4771	408 439.3
10^5–10^6[l]	–	–	0.0012	14.8500	241 330.8
10^6–10^7[l]	–	–	–	0.5773	91 195.1
10^7–10^8[l]	–	–	–	0.0196	30 629.0
10^8–10^9[l]	–	–	–	0.0006	9 876.4
$10^9 \to \infty$[l]	–	–	–	–	4593.0
Total	7.7091	15.9085	66.8923	2321.20	1 136 611.7
	(7.599)[m]	(16.03)[m]	(69.4)[m]	(2353.5)[m]	(1 137 138.0)[m]
Expectation values	7.595[n]		70.0		1 121 231.6[o]
	7.510[p]				1 121 265.8[q]
Other values	(7.510)[p]	15.725[p]	(70.0)[p]	2360.0[p]	1 126 400.0[p]
	7.65[c]	15.58[c]	63.47[c]		

[a]In Ry units.

[b]Blechschmidt *et al.* (1972).

[c]Holland *et al.* (1992).

[d]Lee *et al.* (1977).

[e]Summed atomic cross sections from Henke *et al.* (1993).

[f]From normalized spectrum of Hudson *et al.* (1993). See text.

[g]Turci *et al.* (1995).

[h]Zimkina and Fomichev (1967).

[i]Zimkina and Vinogradov (1971).

[j]LaVilla (1972).

[k]Atomic additivity. Calculated atomic cross sections from Chantler (1995).

[l]Atomic additivity. Calculated atomic cross sections, K-shells only, using hydrogenic equation with screening from Bethe and Salpeter (1977).

[m]Suggested adjustments. See text.

[n]Pack (1982).

[o]Fraga *et al.* (1976).

[p]Kumar *et al.* (1985).

[q]Bunge *et al.* (1993).

c 70.85–171 eV

Vinogradov and Zimkina present tabulated points at selected X-ray lines which support the view that this energy region is structureless until the approach to the S(2p) edge. In Fig. 6.23, some of these points are shown, and compared with summed atomic cross sections obtained from Henke *et al.* (1993). The latter are distinctly larger, and derive some support from their smooth juncture with the values of Lee *et al.* at lower energies. We shall find a similar pattern at higher energies (above the F(1s) edge, 705–775 eV), where direct measurements by LaVilla (1972) are about 20–25% larger than those of Zimkina and Vinogradov (1971). Thus, we opt to traverse this energy interval with the summed Henke cross sections, fitted by regression to a 4-term polynomial. The coefficients of the polynomial are given in Table 6.21. The relative merit of this choice can be assessed subsequently by sum rule analysis. The contribution of this region to $S(0)$ is substantial – 12.7 using Henke *et al.* (1993), 10.7 using Vinogradov and Zimkina (1972).

d S(2p), S(2s) edges and beyond: 171–277 eV

A photoabsorption spectrum with high resolution and low statistical uncertainty in the vicinity of the S(2p) edge (170–215 eV) has been presented by Hudson *et al.* (1993). Unfortunately, it provides only relative intensities. However, their relative oscillator strengths offer some assistance.

The four lowest-lying unoccupied valence orbitals are $6a_{1g}$, $6t_{1u}$, $2t_{2g}$, $4e_g$. Since S(2p) has odd parity ($2t_{1u}$ in O_h symmetry), one can anticipate three optically allowed transitions to these valence orbitals. Zimkina and Fomichev (1967) found three prominent bands in the region immediately below and above the S(2p) edge, denoted as a_{1g} (172.27–173.44 eV), t_{2g} (183.40–184.57 eV) and e_g (196.2 eV). Turci *et al.* (1995) have deduced optical oscillator strengths for these bands from electron energy loss spectroscopy. We initially focus on the t_{2g} band, for which Turci *et al.* find $f = 0.225$ (avg), whereas our reading of Zimkina and Fomichev yields $f = 0.218$. The agreement is encouraging; we standardize on $f(t_{2g}) = 0.225$. The ratio of band areas, t_{2g}/a_{1g}, is 2.76/2.63 (two separate figures, Zimkina and Fomichev) and 2.59 (Hudson *et al.*, 1993). With an average of 2.66, we deduce $f(a_{1g}) = 0.085$, slightly larger than that given by Turci *et al.* Similarly, the ratio of band areas, e_g/t_{2g}, is 1.36 (Zimkina and Fomichev)

Table 6.21 Coefficients of the polynomial $df/dE = ay^2 + by^3 + cy^4 + dy^5$ fitted to data at various energies[a]

Energy range, eV	a	b	c	d
70.85–171.0	74.894 67	178.9066	536.9304	−5152.18
277–685	−43.452 4	12 085.99	−220 820	1 360 455
735–2490	28.607 24	64 437.52	−1 518 986	7 299 618
2580–10 000	6.134 174	154 709.9	−3 520 536	−429 099 974

[a] df/dE in Ry units, $y = B/E$, $B = 14.9$ eV.

and 1.34 (Hudson *et al.*, 1993). With an average of 1.35, $f(e_g) = 0.304$, 50% larger than the value of Turci *et al*. Some weak bands in the interval 170–200 eV, whose oscillator strengths are given by Turci *et al*., are included in Table 6.20 for completeness. Two additional bands, at 205.8 and 241.2 eV (the latter at about the expected energy for $S(2s) \rightarrow t_{1u}$) as observed by Zimkina and Fomichev, are also included. The contribution of the underlying continuum, evaluated in three segments (171–210; 210–243; 243–277 eV), is taken from Zimkina and Vinogradov (1971) and Zimkina and Fomichev (1967). It adds $\sim$4 to $S(0)$, and may be too low by $\sim$20%, as discussed earlier.

e   Between S(2s) and F(1s): 277–685 eV

Vinogradov and Zimkina present tabulated data for SF_6 at nine X-ray lines in this interval. Some of these energies coincide with those used by Henke *et al*. in their tabulation of atomic cross sections. Direct comparison between summed atomic cross sections and the molecular data shows that the former are consistently larger than the latter. At 277.0 eV, σ(summed) $= 4.19$ Mb, σ(mol) $= 3.3$ Mb. In our experience, the individual atomic cross sections are probably more accurate than the earlier molecular values. One may question the validity of atomic additivity, especially if σ(mol) $> \sigma$(summed). However, in this case the reverse is true, and so we opt for σ(summed), especially since comparison of the cross sections of Zimkina and collaborators with other direct measurements (see Sect. 6.9.2.c.) consistently indicates that the Zimkina numbers are lower.

We anticipate no significant structure in this spectral range, and the isolated points of Vinogradov and Zimkina support this view. We traverse this interval by fitting the summed atomic cross sections of Henke *et al*. to our usual 4-term polynomial. The $S(p)$ thereby obtained by analytical integration appear in Table 6.20. The major effect of choosing σ(summed) over σ(mol) is to increase $S(0)$ from $\sim$4.86 to 5.81, with lesser relative effect on the other $S(p)$.

f   The F(1s) region, 685–735 eV

Three photoabsorption data sets are available in this domain: Zimkina and Vinogradov (1971), LaVilla (1972) and Hudson *et al*. (1993). The latter are clearly the best resolved, but lack an absolute calibration. The absolute cross sections of LaVilla are approximately 25% higher than those of Zimkina and Vinogradov. The spectrum in this region is composed of four prominent absorption bands surmounting a continuum which increases significantly around the F(1s) edge. The spectral bands have been assigned as a_{1g} (689.0 eV), t_{1u}(694.7 eV), t_{2g}(699.9 eV) and e_g(713.2 eV). The gerade levels involve transitions to the same unoccupied valence levels encountered in the S(2p) spectrum, from the ungerade molecular orbital composed of F(1s) atomic orbitals. The t_{1u} band, nominally forbidden from S(2p), is allowed and prominent from the gerade F(1s) orbitals.

Our modus operandi is to transfer absolute calibrations from either Zimkina and Vinogradov or LaVilla to the cleaner, better-resolved spectrum of Hudson *et al*., and thence to evaluate oscillator strengths by graphical integration. The results from the LaVilla calibration are given in Table 6.20. The alternative normalization yields the expected lower $S(p)$, but only by 0.21 in $S(0)$ since the spectral range is relatively small.

g   Between F(1s) and S(1s): 735–2490 eV

As discussed in Sect. 6.9.2.e., we can compare the measured $\sigma(\mathrm{mol})$ of Vinogradov and Zimkina with $\sigma(\mathrm{summed})$ from Henke *et al.* at energies corresponding to selected X-ray lines. The latter are consistently larger, by $\sim20\%$. We again choose the experimental atomic cross sections of Henke *et al.*, and fit them to a 4-term polynomial. The resulting $S(p)$ obtained by analytical integration differ more markedly from a corresponding treatment of the data of Vinogradov and Zimkina, since this is a substantial spectral range. $S(0)$ differs by ~1.6, $S(+1)$ by ~122 Ry and $S(+2)$ by ~6000 Ry units.

h   The S(1s) region, 2490–2580 eV

The atomic S(1s) orbital becomes $1a_{1g}$ in O_h symmetry. Of the four lowest-energy unoccupied valence orbitals, only t_{1u} has the proper parity for allowed transitions from $1a_{1g}$. Both Zimkina and Vinogradov and LaVilla display an absolute photoabsorption spectrum in this region, and the only prominent band is assigned to the $1a_{1g} \rightarrow t_{1u}$ transition at 2486 eV. We choose the spectrum of LaVilla for graphical integration. The cross sections are 10% larger than those of Zimkina and Vinogradov, but the absolute values of $S(p)$ are small for this short interval.

i   Post sulfur K-edge: 2580–10 000 eV

Here, we turn once again to the summed atomic cross sections of Henke *et al.*, fitted to a 4-term polynomial. From this polynomial, we compute $\sigma(2580\,\mathrm{eV}) = 0.19$ Mb, compared with $\sigma(2580\,\mathrm{eV}) = 0.22$ Mb from the spectrum of LaVilla. The discrepancy is probably within the uncertainty of both determinations, but it might indicate residual structure beyond the K-edge. Significant contributions to $S(0)$, $S(+1)$ and $S(+2)$ accrue in this interval.

j 10^4–10^5 eV

We employ the sum of calculated atomic cross sections of Chantler (1995).

6.9.3   The analysis

The molar polarization of a gas is the sum of electronic, atomic and orientational effects. The latter is a function of the electric dipole moment, which is absent in SF_6. However, the atomic contribution, which depends upon the transition probabilities of the relatively low-frequency vibrations, is substantial. The polarizability of SF_6 is $\sim6.54 \times 10^{-24}$ cm^3 if measurements are made in the microwave region, but only $\sim4.48 \times 10^{-24}$ cm^3 from refractive index measurements in the visible or ultraviolet (Berkowitz, 1979). For comparison with spectral sums involving electronic transitions, only this latter component is relevant. Watson and Ramaswamy (1936) have determined the refractive index of SF_6 at five wavelengths in the visible. Pack (1982) has introduced these values into the conventional Cauchy expansion, and used Stieltjes constraints and Padé approximants to evaluate the electronic contribution to $S(-2)$ at infinite wavelength. He obtained $S(-2) = 7.595 \pm 0.005$ Ry units, equivalent to $\alpha = 4.502 \pm 0.003 \times 10^{-24}$ cm^3. Kumar *et al.* also analyzed the refractive index data of Watson and Ramaswamy, and found it somewhat inconsistent. They adopted $S(-2) = 7.510$, but allowed for 1% error, which almost overlaps Pack's value. Our spectral sum for $S(-2)$ is still larger, by 1.5%. Although this level of agreement is quite good for a complex molecule such as SF_6, it is instructive to examine possible sources of this discrepancy. The major contributions to $S(-2)$ accrue from cross sections between 10–70.85 eV. A reduction of oscillator strength to satisfy $S(-2)$ will also decrease $S(0)$. Our total $S(0)$ is lower than the TRK expectation value by 4.4%. (Had we chosen the cross sections of Zimkina and collaborators, it would have been still lower.) The requirement of lowering $S(-2)$, without seriously diminishing $S(0)$, focuses attention on the lowest-energy range, 10–14.9 eV. Here, we employed the data of Blechschmidt *et al.*, which were found by Holland *et al.* (1992) to lie considerably higher than their data at slightly higher energies. A 10% reduction in the cross sections of Blechschmidt *et al.* below 14.9 eV would satisfy $S(-2)$, without significantly affecting $S(0)$. Kumar *et al.*, using electron energy loss data in this region, arrive at a much lower (35%) contribution.

Kumar *et al.* employ constraints on $S(-2)$ and $S(0)$ in their optimization procedure to determine the distribution of oscillator strengths. Since this ensures agreement with the TRK sum rule, a comparison of their distribution with ours should be instructive. Relative to our oscillator strengths, their $\Delta S(0)$ is larger by ~ 1.0 (27.89–68.88 eV), ~ 1.3 (68.88–164.1 eV) and ~ 2.5 (164.1–300.0 eV). The largest discrepancy occurs in the last segment, where their value depends strongly on Blechschmidt *et al.* and enhances their cross section in the optimization procedure, whereas ours leans heavily on the data of Zimkina and collaborators, which is probably too low by 0.8, according to other comparisons presented here. The 68.88–164.1 eV interval contrasts their use of the electron energy loss data of Hitchcook *et al.*, enhanced by optimization, with our summed atomic cross sections of Henke *et al.*, which we preferred to the lower measured cross sections of Zimkina and Vinogradov. The discrepancy in the interval 27.89–68.88 eV largely stems from our use of photoabsorption data (Lee *et al.*, 1977) and their selection of electron energy loss measurements (Hitchcock and Van der Wiel, 1979). Thus, in two of these regions, the discrepancy results from electron energy loss versus photoabsorption measurements. The use of electron energy loss values is somewhat redundant, since they were initially normalized to the TRK sum rule.

The total $S(0)$ obtained by Holland *et al.* (1992), 63.47, would be enhanced by ~ 0.47 if extended to infinite energy. It is still ~ 3 units lower than the present result, which is ~ 3 units lower than the expectation value.

We take the 'expectation' value of $S(+2)$ to be the sum of atomic $S(+2)$ values, proportional to the sum of atomic charge densities at the nuclei obtained from Hartree–Fock calculations. From Fraga *et al.* (1976) we deduce $S(+2) = 1\,121\,231.6$ Ry units, while the calculations of Bunge *et al.* (1993) lead to $S(+2) = 1\,121\,265.8$ Ry units. The current spectral sum is only 1.4% larger, while that of Kumar *et al.* is even closer, $\sim 0.5\%$ high.

Making a simple atomic additivity estimate for $S(+1)$ yields 2315.3 Ry units (Fraga *et al.*, 1976). Correlation effects usually increase this quantity in the molecule. Our spectral sum, ~ 2321 Ry units, is slightly larger. Making the corrections suggested above for $S(0)$ would increase $S(+1)$ to ~ 2357 Ry units, very nearly the same as that obtained by Kumar *et al.*, 2360 Ry units. However, it would concomitantly increase $S(-1)$ by ~ 0.5 Ry units, placing it $\sim 4\%$ higher than that of Kumar *et al.* The simplest and perhaps most defensible adjustment is to increase $S(0)$ in the 164–300 eV region (where the data of Zimkina and Vinogradov were suspect), ignoring the corrections at lower energy. This would increase $S(-1)$ by 0.12, $S(0)$ by 2.5 and $S(+1)$ by 32.3, bringing their totals to 16.03, 69.4 and 2353.5 Ry units, respectively.

The value of $S_i(-1)$, can be deduced utilizing the current $S(-1)$, and combining the quantum yield of ionization and photoabsorption cross section given by (Holland *et al.*, 1992). We have arbitrarily assumed that the quantum yield is unity below 600 Å (20.66 eV), since the measured value exceeds unity for some shorter wavelengths. Our calculated $S_i(-1)$ is 13.894, while Holland *et al.* obtain 13.82. We are unaware of direct measurements of this quantity. Between IP–20.66 eV, about 68% of absorption leads to ionization.

6.10　Silane (SiH$_4$)

6.10.1　Preamble

Silane, like methane, has tetrahedral symmetry in its neutral ground state. In the independent particle model, the aufbau of molecular orbitals may be written as

$$(1a_1)^2(2a_1)^2(1t_2)^6(3a_1)^2(2t_2)^6,\ \tilde{X}^1A_1$$

Ionization from the triply degenerate uppermost occupied orbital (HOMO) distorts the tetrahedral structure, in accordance with the Jahn–Teller theorem, and analogous to the behavior of methane. However, as will be seen, the distortion is so extreme in this case that the Franck–Condon factors connecting SiH$_4$ and the ionic ground state, which may be characterized as SiH$_2^+$.H$_2$, are barely detectable between 11.0–11.6 eV by photoionization mass spectrometry, and not observed by conventional photoelectron spectroscopy. The photoabsorption spectrum which may be attributed to excitation of HOMO displays broad structure as in the case of CH$_4$. Ionization from $3a_1$ (nominally Si (3s)) occurs at ~ 18 eV; weak oscillations attributable to Rydberg series approaching this IP can be seen in the total photoabsorption cross section. The $1t_2$ orbital, predominantly Si (2p), has an ionization energy corresponding to $(2p_{3/2})^{-1}$ at 107.31 eV, with appreciable pre-edge structure down to $\sim$102 eV. Weak features in the absorption spectrum at $\sim$155 eV may herald the ionization from $2a_1$ (Si(2s)). The K-edge occurs at 1847 eV, with a broad peak observable down to $\sim$1840 eV.

We are unaware of any complete sum rule analysis for SiH_4, but a partial analysis has been given by Cooper et al. (1995c).

6.10.2   The data

In the He I photoelectron spectrum, the first detectable onset of ionization occurs at 11.60 eV (Potts and Price, 1972). This corresponds roughly to the appearance potentials of SiH_2^+ and SiH_3^+ from SiH_4. Early mass spectrometric studies were ambiguous regarding the stability of SiH_4^+, most claiming that it was unobservable. (By contrast CH_4^+ is quite stable.) The identification is complicated by the isotopic structure of Si, which has weak abundances for ^{29}Si ($\sim$4.7%) and ^{30}Si ($\sim$3.1%) in addition to the main component, ^{28}Si ($\sim$92.2%). Consequently, $^{30}SiH_2^+$ and $^{29}SiH_3^+$ occur at the same nominal mass as $^{28}SiH_4^+$. Berkowitz et al. (1987b) accounted for the isotopic structure, and demonstrated that SiH_4^+ did indeed exist, and that its photoionization spectrum exhibited structure down to 11.00 $\pm$0.02 eV. Subsequently, three ab initio calculations (Pople and Curtiss, 1987; Kudo and Nagase, 1988, Frey and Davidson, 1988b) showed that the most stable structure of SiH_4^+ had C_s symmetry, essentially $SiH_2^+ \cdot H_2$, with Si–H bond lengths of approximately 1.46 and 1.9 Å. We take the adiabatic IP of SiH_4 to be 11.00 $\pm$0.02 eV.

a   Transitions below the IP

Absolute photoabsorption spectra have been recorded by Suto and Lee (1986) with a resolution of 0.015 eV and an estimated uncertainty in cross sections of $\pm$10%, and by Itoh et al. (1986) with a resolution of 0.035 eV and a statistical uncertainty in cross sections of 15%. More recently, Cooper et al. presented pseudo-photoabsorption cross sections obtained by inelastic electron scattering (e,e) with a resolution of 0.05 eV and an estimated uncertainty in cross sections of $\pm$5%. These three data sets can be compared in Fig. 6.24. From 8–9.5 eV, they are quite similar. Between 9.5–10.8 eV, the values of Cooper et al. and Itoh et al. are notably larger than those of Suto and Lee. Beyond 10.8 eV, both photoabsorption measurements converge, and plunge more rapidly than the (e,e) determinations. We evaluate the contributions to $S(p)$ by trapezoidal integration of the electronically scanned and digitized data sets. In Table 6.22, we record separately the contributions from Itoh et al. and Cooper et al. The values of Suto and Lee are lower than those of Itoh et al. and will be shown to be less likely.

b   The continuum

b.1   IP–22.0 eV   We are unaware of any direct photoabsorption measurements between 11.6–13.5 eV. Kameta et al. (1991) obtained absolute photoabsorption

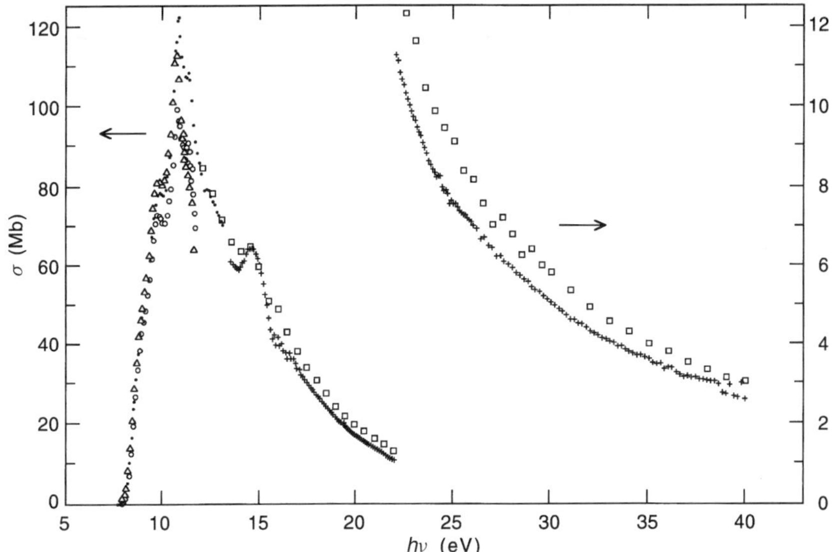

Fig. 6.24 Absolute photoabsorption spectrum of SiH$_4$, 8–40 eV. •, □ Cooper *et al.* (1995c); ○ Suto and Lee (1986); △ Itoh *et al.* (1986); + Kameta *et al.* (1991)

cross sections from 13.5 to 40.0 eV, with a resolution of 0.09 eV and an uncertainty in cross sections of ±10%. In the 13.5–22.0 eV region, their values are somewhat lower than the (e,e) cross sections tabulated by Cooper *et al.*, as can be seen in Fig. 6.24. The 11.6–13.5 eV gap is bridged only by the data of Cooper *et al.* Although Kameta *et al.* state that 'a very good continuation is expected between the present photoabsorption cross sections and those by Suto and Lee', this is not apparent in Fig. 6.24. Indeed, continuity arguments would seem to favor the higher cross sections of Cooper *et al.* In Table 6.22, we register the $S(p)$ from Cooper *et al.* between 11.0–13.5 eV, and then separately the alternative values of Kameta *et al.* and Cooper *et al.* between 13.5–22.0 eV.

b.2 22.0–40.0 eV In Fig. 6.24, we note that the cross sections of Cooper *et al.* lie above those of Kameta *et al.* by 10–15%. Both sets of contributions to $S(p)$ are listed in Table 6.22.

b.3 40.0–101.0 eV We must rely on the (e,e) data of Cooper *et al.* (1995c) in this region, which is shown in Fig. 6.25, together with an isolated point based on summed atomic cross sections from Henke *et al.* (1993). The relative contributions to all the $S(p)$ in this region are not large; the uncertainty in $S(0)$ may amount to 0.1.

b.4 101.0–108.0 eV This is the region exhibiting structure, some of it sharp, largely preceding the Si(2p) edge. Hayes and Brown (1972) obtained an absolute

Table 6.22 Spectral sums and comparison with expectation values for SiH_4[a]

Energy, eV	$S(-2)$	$S(-1)$	$S(0)$	$S(+1)$	$S(+2)$
7.94–11.00[b]	3.2723	2.3661	1.7201	1.2568	0.9229
7.68–11.00[c]	3.3466	2.4211	1.7615	1.2886	0.9475
11.00–11.60[b]	0.6590	0.5465	0.4534	0.3762	0.3122
11.00–11.60[c]	0.8234	0.6831	0.5668	0.4704	0.3905
11.60–13.50[c]	1.5632	1.4314	1.3132	1.2070	1.1116
13.50–22.0[d]	1.8314	2.1514	2.5688	3.1214	3.8635
13.50–22.0[c]	1.9879	2.3464	2.8153	3.4372	4.2736
22.0–40.0[d]	0.2062	0.4163	0.8638	1.8455	4.0623
22.0–40.0[c]	0.2416	0.4847	0.9992	2.1211	4.6403
40.0–101.0[c]	0.0374	0.1463	0.6064	2.6863	12.7744
101.0–108.0[c]	0.0027	0.0205	0.1582	1.2202	93.4158
101.0–108.0[e]	0.0051	0.0393	0.3032	2.3372	18.0208
108.0–194.0[c]	0.0265	0.2790	3.0189	33.4867	380.4728
108.0–194.0[e]	0.0367	0.3837	4.1084	45.1168	507.8112
194–350[c]	0.0086	0.1550	2.8732	54.8122	1 075.99
194–350[f]	0.0083	0.1483	2.7111	50.9037	984.06
350–851.5[f]	0.0013	0.0450	1.5982	60.2484	2 421.63
851.5–1840[f]	0.0001	0.0045	0.3736	32.5023	2 967.82
1840–1872[g]	–	0.0005	0.0671	9.1566	1 248.72
1872–10 000[f]	–	0.0072	1.6794	363.7729	108 728.27
10^4–10^5	–	0.0001	0.0662	86.5324	156 227.58
10^5–∞	–	–	0.0005	5.6369	146 099.24
Totals[h]	8.0393	8.0248	17.8977	658.5792	419 183.3
Totals[i]	7.6210	7.6886	18.4334	666.7004	419 226.2
Expectation values	7.986[j]		18.0		414 178.2[k]
					414 182.3[l]
Other values	7.943[c]				
	8.00[m]	7.915[m]			

[a] In Ry units.
[b] Itoh *et al.* (1986).
[c] Cooper *et al.* (1995c).
[d] Kameta *et al.* (1991).
[e] Hayes and Brown (1972).
[f] Summed atomic cross sections from Henke *et al.* (1993).
[g] Spectrum of Bodeur and Nenner (1986), normalized to summed atomic cross sections at 1872 eV.
[h] $S(p)$ obtained using (e,e) data of ref. c.
[i] $S(p)$ obtained using photoabsorption data of refs. b, d and e.
[j] Obtained from refractive index measurements. Watson and Ramaswamy (1936). See text.
[k] Stoichiometric sum of atomic silicon $S(+2)$ from Fraga *et al.* (1976) and atomic hydrogen $S(+2)$.
[l] Same as ref. k, but atomic silicon $S(+2)$ from Bunge *et al.* (1993).
[m] Olney *et al.* (1997).

photoabsorption spectrum with 0.04 eV resolution with an uncertainty in cross section of ±20%. Somewhat later, Friedrich *et al.* (1979) obtained a similar spectrum, with slightly lower resolution. With this apparent agreement, it is surprising to find that the (e,e) data of Cooper *et al.*, obtained with 0.1 eV resolution, display a cross section that is ∼20% lower at the 103.2 eV peak, and 50% lower at 108 eV.

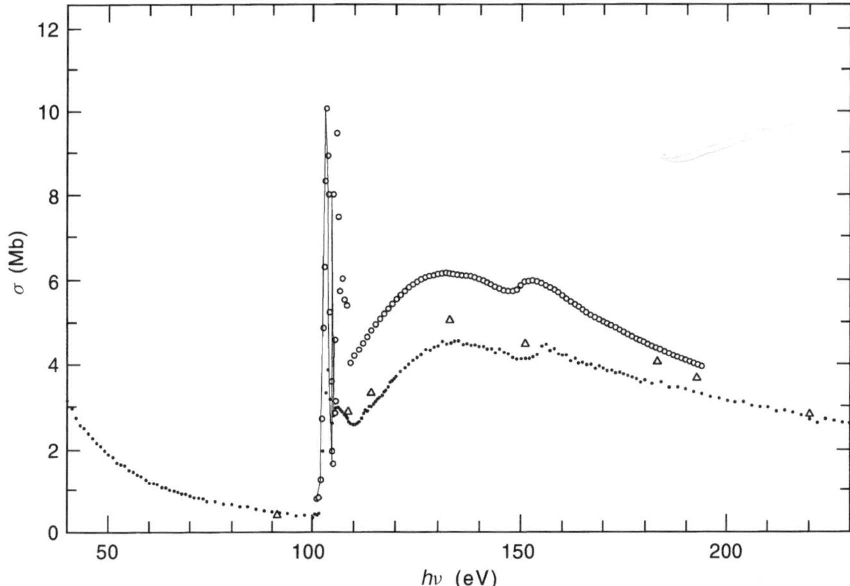

Fig. 6.25 Absolute photoabsorption spectrum of SiH$_4$, 40–230 eV. • Cooper *et al.*
(1995c); ○ Hayes and Brown (1972); △ Henke *et al.* (1993) + additivity

This comparison stands in marked contrast to the valence region, where the (e,e)
data are higher than photoabsorption values.

b.5   108.0–194.0 eV   The comparison between photoabsorption and (e,e) data
in this interval can be visualized in Fig. 6.25, where the photoabsorption values
are seen to be ∼35% larger. Also shown are sparse points using additivity of
atomic cross sections from Henke *et al.* They fall between the photoabsorption
and (e,e) data sets, but for most of the range they lie closer to the (e,e) values.
This suggests, but does not prove, that the latter may be more reliable. The
contributions of photoabsorption and inelastic electron scattering to $S(0)$ differ
by 1.1 units, and to $S(+1)$ by ∼12 Ry units, which is substantial, and will be
judged in the sum rule analysis.

b.6   194.0–350.0 eV   Cooper *et al.* extend their (e,e) data to 350 eV. We
compare with stoichiometrically summed atomic photoabsorption cross sections
from Henke *et al.*, which have been fitted by regression to a 4-term polynomial.
The coefficients of the polynomial are given in Table 6.23. From Table 6.22, we
note that the alternative contributions to $S(p)$ do not differ greatly.

b.7   350–1840 eV   No structure is anticipated in this region, which terminates
just short of the pre-K-edge structure. Lacking other data, we traverse this region

Table 6.23 Coefficients of the polynomial $df/dE = ay^2 + by^3 + cy^4 + dy^5$ fitted to data at various energies[a]

Energy range, eV	a	b	c	d
194.0–851.5	21.2782	6012.946	−108772	685123.3
851.5–1840.0	−32.4784	20680.26	−1458751	41520091
1872–10000	9.528262	140616.1	−6310654	380068525

[a]df/dE in Ry units, $y = B/E$, $B = 11.00$ eV.

using summed atomic cross sections from Henke *et al.*, which we fit to two 4-term polynomials, spanning 350–851.5 and 851.5–1840 eV.

b.8 1840–1872 eV A relative photoabsorption spectrum in the K-edge region was presented by Bodeur and Nenner (1986). There is very likely more pre-edge structure than shown in the spectrum of Bodeur and Nenner, since the total ion yield curve of Shigemasa *et al.* (1990) displays at least two peaks with a photon resolution of 1 eV. Cavell and Sodhi (1979) place the K-edge of SiH_4 at 1847.0 eV. We normalize the spectrum of Bodeur and Nenner at $1847 + 25 = 1872$ eV to the fitted sum of atomic cross sections from Henke *et al.*, which gives $\sigma = 0.2$ Mb at 1872 eV. With this calibration, the spectrum of Bodeur and Nenner is digitized and trapezoidally integrated to provide the $S(p)$ given in Table 6.22.

b.9 1872–10000 eV The summed atomic cross sections of Henke *et al.* are fitted by regression to a 4-term polynomial, whose coefficients can be found in Table 6.23. This region supplies the bulk of the Si (1s) contribution to $S(0)$, as well as major contributions to $S(+1)$ and $S(+2)$.

6.10.3 The analysis

Experimental information providing values of the electric dipole polarizability α of silane is based on early work of Watson and collaborators. Watson *et al.* (1934) measured the dielectric constant, and from that reported a molar polarization of 13.72 cm^3/mol. In the older literature, this quantity was recognized as $P_E + P_A$, where P_E is the electronic contribution, and P_A is called the atomic contribution, but refers to the infrared vibrational excitations. Bishop and Cheung (1982) reported the vibrational contribution to $\alpha(SiH_4)$ to be 0.566×10^{-24} cm^3 and hence $P_A = 1.43$ cm^3/mol, and $P_E = 12.29$ cm^3/mol. Later, Watson and Ramaswamy (1936) measured the refractive index of SiH_4 at five wavelengths. From the functional relationship

$$n - 1 = A/(B - v^2)$$

where n = refractive index, v = light frequency and A, B are constants, they deduced the refractive index at zero frequency, and hence $P_E = 11.95$ cm^3/mol.

This value is 2.8% smaller than that derived from dielectric constant measurements. Since the refractive index measurement, based on interferometric methods, is sensitive only to P_E, it is expected to be more reliable. Dougherty and Spackman (1994), using the same data with a quadratic extrapolation, obtained $\alpha = 4.72_7 \times 10^{24}\,cm^3$, and $P_E = 11.92\,cm^3/mol$. (They also reported the results of high level ab initio calculations, which are 1–2% lower.) Our fitting of the data of Watson and Ramaswamy to a Cauchy expansion yields $P_E(0) = 11.94\,cm^3/mol$, $\alpha = 4.734 \times 10^{-24}\,cm^3$, and $S(-2) = 7.986\,Ry$ units. (We note parenthetically that the value compiled by Miller (1999) $\alpha = 5.44 \times 10^{-24}\,cm^3$, refers to $P_E + P_A$).

In Table 6.22, we note that the photoabsorption cross sections below 40 eV are lower than those forthcoming from the (e,e) measurements, while those between 101–194 eV are higher. Above 40 eV, the contributions to $S(-2)$ are small; below 40 eV, the difference in oscillator strength between the (e,e) data and photoabsorption, $\Delta S(0)$, amounts to ~ 0.5. Thus, we compute two grand sums, one utilizing the (e,e) data of Cooper *et al.* (1995c), and the other employing only photoabsorption data where possible. The (e,e) data yield $S(-2) = 8.0393\,Ry$ units, within 0.67% of the expectation value, whereas the photoabsorption cross sections attain $S(-2) = 7.6210\,Ry$ units, 4.6% too low. Clearly, the inelastic electron scattering data are favored for $E < 40\,eV$. This conclusion supports the earlier observation that there is an awkward discontinuity in the photoabsorption data between 11.6–13.5 eV, whereas the (e,e) cross sections smoothly bridge this gap. It also justifies our neglect of the data of Suto and Lee (1986) in Table 6.22, since they are even lower than those of Itoh *et al.*

For $S(0)$, the (e,e) based result is ~ 0.1 shy of the TRK sum, whereas the photoabsorption data are 0.43 too high. Since we have made the tentative conclusion that the values of Cooper *et al.* are preferred to the photoabsorption data below 40 eV, incorporating them together with the photoabsorption data at higher energies from Hayes and Brown and Friedrich *et al.* would make $S(0)$ almost 1 unit too high. We conclude that the latter cross sections are too high.

Accurate values of the electron charge density at the silicon nucleus are available from Hartree–Fock calculations by Fraga *et al.* (1976) and from Bunge *et al.* (1993). When supplemented by the small hydrogen contributions, they yield values of $S(+2)$ which are only 1.2% lower than the spectral sum.

Simple additivity, using the results of Fraga *et al.*, yields $S(+1) = 656.07\,Ry$ units. Using our preferred (e,e) data, the spectral sum for $S(+1)$ is only 2.5 Ry units higher, which implies that correlation plays only a small role here.

Olney *et al.* (1997) report a value of $S(-1) = 7.915\,Ry$ units, based on the data of Cooper *et al.* Since the latter data terminate at 350 eV, we supplement their value of $S(-1)$ from the higher energy data in Table 6.22, and arrive at $S(-1) = 7.972$, close to our spectral sum of 8.025.

We wish to approximate $S_i(-1)$, the ionized component of $S(-1)$, since it relates to the square of the transition dipole in high-energy electron impact experiments. Kameta *et al.* report quantum yields of ionization, η_i, between

13.5–22.0 eV with stated errors $<\pm5\%$. Earlier values by Cooper *et al.* (1990) deviate significantly from those of Kameta *et al.* Since other aspects of Cooper *et al.* (1990) have been corrected and superceded by Cooper *et al.* (1995c), we consider η_i from Cooper *et al.* (1990) less reliable. To estimate η_i below 13.5 eV, we normalize the relative total ion yield curve of Hayaishi *et al.* (1987) at 16 eV to the absolute photoabsorption cross sections of Cooper *et al.* At 16 eV, Kameta *et al.* find $\eta_i \approx 1.0$. With this normalization, the values of η_i between 13.5–16.0 eV are in fair agreement with Kameta *et al.*, and enable us to extend the quantum yield measurements to 11.87 eV. Below 11.87 eV, the ion yield diminishes to almost zero at 11.6 eV, although a very weak SiH_4^+ signal persists to 11.0 eV. We normalize the summed partial ion yield curves of Berkowitz *et al.* to the total ion yield of Hayaishi *et al.* for this short interval. Where available, the ion yields of Hayaishi *et al.* are preferred over those of Berkowitz *et al.* not only because the total ion yield is given directly, but also because the former utilized a time-of-flight spectrometer, whose collection efficiency is more uniform than that from the quadrupole mass spectrometer used by the latter. We now combine the η_i with photoabsorption cross sections (σ_a) from Cooper *et al.* to determine the photoionization cross sections (σ_i) between 11.6–22.0 eV. Integration yields $\Delta S_i(-1) = 2.8088$ in this interval. When supplemented by $\Delta S(-1) = 1.1428$ between 22.0 eV $\rightarrow \infty$ (see Table 6.22, data of Cooper *et al.*), we arrive at $S_i(-1) = 3.952$. Direct electron impact values for comparison are unknown to us. Between IP–22.0 eV, $\sim74\%$ of total absorption leads to ionization, but only $\sim16\%$ between 11.0–13.0 eV.

7

Aspirations for the Future

This author has had numerous requests for absolute photoabsorption and photoionization cross sections of transient molecular species such as OH, NH_n ($n = 1, 2$) and CH_n ($n = 1-3$). While some data are available in limited spectral regions, they are deemed insufficient for fruitful application of sum rules. For example, experimental absolute photoabsorption cross sections have been reported for the hydroxyl radical below 10.8 eV (Rouse and Engleman, 1973; Nee and Lee, 1985), but only relative photoionization cross sections between the IP (13.02 eV) and 18.23 eV (Cutler *et al.*, 1995, Dehmer, 1984). No experimental polarizability measurements have been cited, and theoretical values (Esposti and Werner, 1990, Paldus and Li, 1996, Karna, 1996) differ by as much as 17%. Clearly, much more experimental information is needed.

In the realm of theory, there are two domains where additional effort would be desirable. The expectation values of $S(-1)$ and $S(+1)$ obtained from calculations with correlated wave functions have been most helpful in assessing absolute cross sections for the noble gases and alkali elements included in this monograph. The sole molecular representative is H_2. Various authors (see, for example, Chipman *et al.*, 1977, Mulder and Meath, 1981) have had limited success in calculating these quantities for a few small molecules, using contemporary calculational capabilities. With the current level of progress in quantum chemistry, it is hoped that more accurate matrix elements for evaluating $S(-1)$ and $S(+1)$ will be forthcoming for at least the smaller molecules. The other area still requiring attention from theorists concerns elaboration of the formulas for the expectation values $S(p)$ to heavier elements ($Z > 18$). Some progress has been made incorporating relativistic velocities of one-electron atoms and approximations for many-electron atoms. (See, for example, Cohen and Leung, 1998; 1999 and references therein.) A more complete theory may require inclusion of higher-multipole interactions of photons with the target, which have terms of the same order as relativistic effects.

References

Adam, M. Y. (1986) *Chem. Phys. Lett.* **128**, 280

Akimov, V. N., Vinogradov, A. S. and Zhadenov, A. V. (1988) *Opt. Spectrosc. (USSR)* **65**, 210

Akimov, V. N., Vinogradov, A. S. and Zimkina, T. M. (1982) *Opt. Spectrosc.* **53**, 548

Akimov, V. N., Vinogradov, A. S., Pavlychev, A. A. and Sivkov, V. N. (1985) *Opt. Spectrosc. (USSR)* **59**, 206

Alasia, F., Broglia, R. A., Roman, H. E., Serra, L., Colò, G. and Pacheco, J. M. (1994) *J. Phys. B* **27**, L643

Alaverdov, V. I. and Podolyak, E. R. (1982) *Opt. Spectrosc. (USSR)* **53**, 665

Aleksandrov, Yu. M., Gruzdov, P. F., Kozlov, M. G., Loginov, A. V., Makhov, V. N., Fedorchuk, R. V. and Yakimenko, M. N. (1983) *Opt. Spectrosc. (USSR)* **54**, 4

Alexander, W. I., Abraham, E. R. I. and Hulet, R. G. (1996) *Phys. Rev. A* **54**, R5

Allison, A. C. and Dalgarno, A. (1970) *Atom. Data* **1**, 289; (1970) *Mol. Phys.* **19**, 567

Allison, D. C. S., Burke, P. G. and Robb, W. D. (1972) *J. Phys. B* **5**, 55

Alms, G. R., Burnham, A. K. and Flygare, W. H. (1975) *J. Chem. Phys.* **63**, 3321

Alpher, R. A. and White, D. R. (1959) *Phys. Fluids* **2**, 153

Amusia, M. Ya., Cherepkov, N., Živanović, Dj. and Radojević, V. (1976) *Phys. Rev. A* **13**, 1466

Amusia, M. Ya., Cherepkov, N. A., Radojević, V. and Živanović, Dj. (1976) *J. Phys. B* **9**, L469

Amusia, Ya. M. (1990) *Atomic Photoeffect* (Plenum Press, New York)

Andersen, J. N., Johansson, U., Nyholm, R., Sorensen, S. L. and Wiklund, M., MaxLab Ann. Rept. No. 180 (1997)

Andersen, J. O. and Veseth, L. (1994) *Phys. Scripta* **50**, 13

Anderson, J. H. B., Osborne, P. J. K. and Glass, I. I. (1967) *Phys. Fluids* **10**, 1848

Anderson, S. M. (1993) *Geophys. Res. Lett.* **20**, 1579

Andersson, K. and Sadlej, A. J. (1992) *Phys. Rev. A* **46**, 2356

Andersson, K., Borowski, P., Fowler, P. W., Malmqvist, P.-Å., Roos, B. O. and Sadlej, A. J. (1992) *Chem. Phys. Lett.* **190**, 367

Angel, G. C. and Samson, J. A. R. (1988) *Phys. Rev. A* **38**, 5578

Antoine, R., Dugourd, Ph., Rayane, D., Benichou, E. and Broyer, M. (1999) *J. Chem. Phys.* **110**, 9771

Applequist, J., Carl, J. R. and Fung, K.-K. (1972) *J. Am. Chem. Soc.* **94**, 2952

Arrighini, G. P., Biondi, F., Guidotti, C., Biagli, A. and Marinelli, F. (1980) *Chem. Phys.* **52**, 133

Asplund, L., Gelius, U., Hedman, S., Helenlund, K., Siegbahn, K. and Siegbahn, P. E. M. (1985) *J. Phys. B* **18**, 1569

Au, J. W. and Brion, C. E. (1997) *Chem. Phys.* **218**, 109

Au, J. W., Cooper, G., Burton, G. R., Olney, T. N. and Brion, C. E. (1993a) *Chem. Phys.* **173**, 241

Au, J. W., Cooper, G., Burton, G. R., Olney, T. N. and Brion, C. E. (1993b) *Chem. Phys.* **173**, 209

Avaldi, L., Dawber, G., Hall, R. I., King, G. C., McConkey, A. G., MacDonald, M. A. and Stefani, G. (1995) *J. Electron Spectrosc.* **71**, 93

Azuma, Y., Berry, H. G., Gemmell, D. S., Suleiman, J., Westerlind, M., Sellin, I. A., Woicik, J. C. and Kirkland, J. P. (1995) *Phys. Rev. A* **51**, 447

Bacis, R., Bouvier, A. J. and Flaud, J. M. (1998) *Spectrochim. Acta A* **54**, 17

Backx, C., Wight, G. R. and van der Wiel, M. J. (1976) *J. Phys. B* **9**, 315

Backx, C., Wight, G. R., Tol, R. R. and van der Wiel, M. J. (1975) *J. Phys. B* **8**, 3007

Bacon, J. A. and Pratt, S. T. (2000) *J. Chem. Phys.* **112**, 4153

Bader, R. F. W., Henneker, W. H. and Cade, P. E. (1967) *J. Chem Phys.* **46**, 3341

Baer, T. and Guyon, P. M. (1986) *J. Chem. Phys.* **85**, 4765

Baer, T., Guyon, P.-M., Nenner, I., Tabche-Fouhaille, A., Botter, R., Ferreira, L. F. A. and Govers, T. R. (1979) *J. Chem. Phys.* **70**, 158

Baig, M. A. and Connerade, J. P. (1984) *J. Phys. B* **17**, 1785

Baig, M. A., Mahmood, M. S., Sommer, K. and Hormes, J. (1994) *J. Phys. B* **27**, 389

Baker, A. D., Baker, C., Brundle, C. R. and Turner, D. W. (1968) *Int. J. Mass Spectrom. Ion Phys.* **1**, 285

Ballard, A., Bonin, K. and Louderback, J. (2000) *J. Chem. Phys.* **113**, 5732

Baltzer, P., Karlsson, L., Lundqvist, M., Wannberg, B., Holland, D. M. P. and Mac-Donald, M. A. (1995) *Chem. Phys.* **195**, 403

Baltzer, P., Karlsson, L., Wannberg, B., Holland, D. M. P., MacDonald, M. A., Hayes, M. A. and Eland, J. H. D. (1998) *Chem. Phys.* **237**, 451

Baltzer, P., Karlsson, L., Wannberg, B., Öhrwall, G., Holland, D. M. P., MacDonald, M. A., Hayes, M. A. and von Niessen, W. (1997) *Chem. Phys.* **224**, 95

Baltzer, P., Wannberg, B., Karlsson, L., Carlsson, Göthe M. and Larsson, M. (1992) *Phys. Rev. A* **45**, 4374

Banna, M. S., Frost, D. C., McDowell, C. A., Noodleman, L. and Wallbank, B. (1977) *Chem. Phys. Lett.* **49**, 213

Banna, M. S., Wallbank, B., Frost, D. C., McDowell, C. A. and Perera, J. S. H. Q. (1978) *J. Chem. Phys.* **68**, 5459

Barrell, H. and Sears, J. E. (1940) *Philos. Trans. Roy. Soc. Lond. A* **238**, 1

Barrientos, C. and Martin, I. (1987) *Can. J. Phys.* **65**, 435

Barrus, D. M., Blake, R. L., Burek, A. J., Chambers, K. C. and Pregenzer, A. L. (1979) *Phys. Rev. A* **20**, 1045

Bass, A. M., Ledford, A. E., Jr. and Laufer, A. H. (1976) *J. Res. Natl. Bur. Stand.* **80A**, 143

Baudais, F. L. and Taylor, J. W. (1980) *J. Electron Spectrosc.* **18**, 85

Bauschlicher, C. W., Jr. and Ricca, A. (1998) *J. Phys. Chem. A* **102**, 4722

Bell, K. L. and Berrington, K. A. (1991) *J. Phys.* **B24**, 933

Bell, K. L. and Kingston, A. E. (1971) *J. Phys. B* **4**, 1308

Bell, K. L. and Kingston, A. E. (1994) *Adv. At. Mol. Opt. Phys.* **32**, 1

Bell, K. L., Burke, P. G., Hibbert, A. and Kingston, A. E. (1989) *J. Phys. B* **22**, 3197

Benzaid, S., Menzel, A., Jimenez-Mier, J., Schaphorst, S. J., Krause, M. O. and Caldwell, C. D. (1996) *Phys. Rev. A* **54**, R2537

Bergeson, S. D., Balakrishnan, A., Baldwin, K. G. H., Lucatorto, T. B., Marangos, J. P., McIlrath, T. J., O'Brian, T. R., Rolston, S. L., Sansonetti, C. J., Wen, J., Westbrook, N., Cheng, C. H. and Eyler, E. E. (1998) *Phys. Rev. Lett.* **80**, 3475

Berkowitz, J. (1971) *Adv. High Temp. Chem.*, vol. **3** (Academic Press, N.Y.)

Berkowitz, J. (1978) *J. Chem. Phys.* **69**, 3044

Berkowitz, J. (1979) *Photoabsorption, Photoionization and Photoelectron Spectroscopy* (Academic Press, New York)

Berkowitz, J. (1997a) *J. Phys. B* **30**, 583

Berkowitz, J. (1997b) *J. Phys. B* **30**, 881

Berkowitz, J. (1999) *J. Chem. Phys.* **111**, 1446

Berkowitz, J. (2000) Preliminary Franck–Condon calculations

Berkowitz, J. and Chupka, W. A. (1969) *J. Chem. Phys.* **51**, 2341

Berkowitz, J. and Eland, J. H. D. (1977) *J. Chem. Phys.* **67**, 2740

Berkowitz, J. and Goodman, G. L. (1979) *J. Chem. Phys.* **71**, 1754

Berkowitz, J. and Greene, J. P. (1984) *J. Chem. Phys.* **81**, 4328

Berkowitz, J. and Spohr, R. (1973) *J. Electron Spectrosc.* **2**, 143

Berkowitz, J., Greene, J. P. and Cho, H. (1987a) *J. Chem. Phys.* **86**, 674

Berkowitz, J., Greene, J. P., Cho, H. and Ruščić, B. (1987b), *J. Chem. Phys.* **86**, 1235

Berkowitz, J., Ruščić, B. and Yoo, R. K. (1992) *Comments At. Mol. Phys.* **28**, 95

Berrah, N., Langer, B., Bozek, J., Gorczyca, T. W., Hemmers, O., Lindle, D. W. and Tosder, O. (1996) *J. Phys. B* **29**, 5351

Bethe, H. (1930) *Ann. Phys. Lpz.* **5**, 325

Bethe, H. (1933) in *Handbuch der Physik*, eds. Geiger, H. and Scheel, K. (Springer, Berlin), vol. **24/1**, p. 273ff

Bethe, H. A. and Salpeter, E. E. (1977) *Quantum Mechanics of One and Two-Electron Atoms* (Plenum, New York), p. 305

Bethke, G. W. (1959) *J. Chem. Phys.* **31**, 662

Bettendorf, M., Peyerimhoff, S. D. and Buenker, R. J. (1982) *Chem. Phys.* **66**, 261

Bhaskar, N. D. and Lurio, E. (1976) *Phys. Rev. A* **13**, 1484

Bianconi, A., Petersen, H., Brown, F. C. and Bachrach, R. Z. (1978) *Phys. Rev. A* **17**, 1907

Biémont, E. Gebarowski, R. and Zeippen, C. J. (1994) *Astron. Astrophys.* **287**, 290

Bieri, G. and Åsbrink, L. (1980) *J. Electron Spectrosc.* **20**, 149

Biernacki, D. T., Colson, S. D. and Eyler, E. E. (1988) *J. Chem. Phys.* **89**, 2599

Birnbaum, G. and Chatterjee, S. K. (1952) *J. Appl. Phys.* **23**, 220

Bishop, D. M. and Cheung, L. M. (1982) *J. Phys. Chem. Ref. Data* **11**, 119

Bishop, D. M. and Pipin, J. (1995) *Chem. Phys. Lett.* **236**, 15

Bizau, J. M. and Wuilleumier, F. J. (1995) *J. Electron Spectrosc.* **71**, 205

Black, G., Sharpless, R. L., Slanger, T. G. and Lorents, D. C. (1975) *J. Chem. Phys.* **62**, 4266

Blechschmidt, D., Haensel, R., Koch, E. E., Nielsen, U. and Sagawa, T. (1972) *Chem. Phys. Lett.* **14**, 33

Bobeldijk, M., van der Zande, W. J. and Kistemaker, P. G. (1994) *Chem. Phys.* **179**, 125

Bodeur, S. and Esteva, J. M. (1985) *Chem. Phys.* **100**, 415

Bodeur, S. and Nenner, I. (1986) *J. Phys.* **47**, C8–79

Bodeur, S., Marechal, J. L., Reynaud, C., Bazin, D. and Nenner, I. (1990) *Z. Phys. D* **17**, 291

Boer, F. P. and Lipscomb, W. N. (1969) *J. Chem. Phys.* **50**, 989

Boschi, R., Murrell, J. N. and Schmidt, W. (1972) *Faraday Disc. Chem. Soc.* **54**, 116

Boyd, A. H. (1964) *Planet Space Sci.* **12**, 729

Brady, B. R. and Beiting, E. J. (1992) *J. Chem. Phys.* **97**, 3855

Brage, T., Fischer, C. F. and Jönsson, P. (1994), *Phys. Rev. A* **49**, 2181

Breinig, M., Chen, M. H., Ice, G. E., Parente, F., Crasemann, B. and Brown, G. S. (1980) *Phys. Rev. A* **22**, 520

Breton, J., Guyon, P. M. and Glass-Maujean, M. (1980) *Phys. Rev. A* **21**, 1909

Bridge, N. J. and Buckingham, A. D. (1966) *Proc. Roy. Soc. Lond. A* **295**, 334

Brion, C. E., Iida, Y. and Thomson, J. P. (1986) *Chem. Phys.* **101**, 449

Brion, J., Chakir, A., Charbonnier, J., Maumont, D., Parisse, C. and Malicet, J. (1998) *J. Atmos. Chem.* **30**, 291

Brown, E. R., Carter, S. L. and Kelly, H. P. (1980) *Phys. Rev. A* **21**, 1237

Brundle, C. R., Neumann, D., Price, W. C., Evans, D., Potts, A. W. and Streets, D. G. (1970) *J. Chem. Phys.* **53**, 705

Bryant, G. P., Jiang, Y., Martin, M. and Grant, E. R. (1994) *J. Chem. Phys.* **101**, 7199

Buenker, R. J., Peyerimhoff, S. D. and Peric, M. (1976) *Chem. Phys. Lett.* **42**, 383

Bunge, C. F., Barrientos, J. A. and Bunge, A. V. (1993) *Atom. Data Nucl. Data Tables* **53**, 113

Burke, V. M. and Lennon, D. J. (1996) Data available at *http://cdsweb.u-strasbg.fr/topbase.html*

Burkholder, J. B. and Talukdar, R. K. (1994) *Geophys. Res. Lett.* **21**, 581

Burns, R. C., Graham, C. and Weller, A. R. M. (1986) *Mol. Phys.* **59**, 41

Burose, A. W., Dresch, T. and Ding, A. M. G. (1993) *Z. Phys. D* **26**, S294

Burrows, J. P., Richter, A., Dehn, A., Deters, B., Hommelmann, S., Voight, S. and Orphal, J. (1999) *J. Quant. Spectr. Rad. Transf.* **61**, 509

Burton, G. R., Chan, W. F., Cooper, G. and Brion, C. E. (1992) *Chem. Phys.* **167**, 349

Burton, G. R., Chan, W. F., Cooper, G. and Brion, C. E. (1993a) *Chem. Phys.* **177**, 217

Burton, G. R., Chan, W. F., Cooper, G., Brion, C. E., Kumar, A. and Meath, W. J. (1993b) *Can. J. Chem.* **71**, 341

Butler, K. and Mendoza, C. (1983) *J. Phys. B* **16**, L707

Butler, K. and Zeippen, C. J. (1991) *J. Physique IV*, **C1**, 1

Cairns, R. B. and Samson, J. A. R. (1966) *J. Opt. Soc. Am.* **56**, 526

Caldwell, C. D., Krause, M. O., Cowan, R. D., Menzel, A., Whitfield, S. B., Hallman., S. Frigo, S. P. and Severson, M. C. (1999) *Phys. Rev. A* **59**, R926

Carlson, R. W. (1974) *J. Chem. Phys.* **60**, 2350

Carlson, R. W., Judge, D. L., Ogawa, M. and Lee, L. C. (1973) *Appl. Optics* **12**, 409

Carlson, T. A., Krause, M. O., Fahlman, A., Keller, P., Taylor, J. W., Whitley, T. and Grimm, F. A. (1983) *J. Chem. Phys.* **79**, 2157

Carravetta, V., Gel'mukhanov, F. Kh., Ågren, H., Sundin, S., Osborne, S. J., Naves de Brito, A., Björneholm, O., Ausmees, A. and Svensson, S. (1997) *Phys. Rev. A* **56**, 4665

Carravetta, V., Luo, Y. and Ågren, H. (1993), *Chem. Phys.* **174**, 141

Carroll, P. K. and Collins, C. P. (1969) *Can. J. Phys.* **47**, 563

Carroll, P. K., Collins, C. P. and Yoshino, K. (1970) *J. Phys. B* **3**, L127

Carroll, P. K., Huffman, R. E., Larrabee, J. C. and Tanaka, Y. (1966) *Astrophys. J.* **146**, 553

Cavell, R. G. and Sodhi, R. (1979) *J. Electron Spectrosc.* **15**, 45

Celotta, R. J., Mielczarek, S. R. and Kuyatt, C. E. (1974) *Chem. Phys. Lett.* **24**, 428

Chako, N. Q. (1934) *J. Chem. Phys.* **2**, 644

Chan, W. F., Cooper, G. and Brion, C. E. (1992c) *Chem. Phys.* **168**, 375

Chan, W. F., Cooper, G. and Brion, C. E. (1993b) *Chem. Phys.* **170**, 99

Chan, W. F., Cooper, G. and Brion, C. E. (1993c) *Chem. Phys.* **170**, 123

Chan, W. F., Cooper, G. and Brion, C. E. (1993d) *Chem. Phys.* **170**, 111

Chan, W. F., Cooper, G. and Brion, C. E. (1993e) *Chem. Phys.* **178**, 387

Chan, W. F., Cooper, G. and Brion, C. E. (1993f) *Chem. Phys.* **178**, 401

Chan, W. F., Cooper, G. and Brion, C. E. (1994) *Chem. Phys.* **180**, 77

Chan, W. F., Cooper, G., Guo, X. and Brion, C. E. (1992a) *Phys. Rev. A* **45**, 1420

Chan, W. F., Cooper, G., Guo, X., Burton, G. R. and Brion, C. E. (1992b) *Phys. Rev. A* **46**, 149

Chan, W. F., Cooper, G., Sodhi, R. N. S. and Brion, C. E. (1993a) *Chem. Phys.* **170**, 81

Chang, J.-J. and Kelly, H. P. (1975) *Phys. Rev. A* **12**, 92

Chang, T. N. (1975) *J. Phys. B* **8**, 743

Chantler, C. T. (1995) *J. Phys. Chem. Ref. Data* **24**, 71

Chen, C.-T. and Robicheaux, F. (1994) *Phys. Rev. A* **50**, 3968

Chen, F. Judge, D. L., Wu, C. Y. R., Caldwell, J., White, H. P. and Wagener, R. (1991) *J. Geophys. Res.* **96**, 17519

Chen, F. Z. and Wu, C. Y. R. (1999) *J. Phys. B* **32**, 3283

Cherepkov, N. A., Chernysheva, L. V., Radojević, V. and Pavlin, I (1974) *Can. J. Phys.* **52**, 349

Chewter, L. A., Sander, M., Müller-Dethlefs, K. and Schlag, E. W. (1987) *J. Chem. Phys.* **86**, 4737

Child, M. S. (1997) *Philos. Trans. Roy. Soc. Lond. A* **355**, 1623

Child, M. S. and Jungen, Ch. (1990) *J. Chem. Phys.* **93**, 7756

Child, M. S., Gilbert, R. D. and Jungen, Ch. (1991) *Laser Chem.* **11**, 253

Chipman, D. M., Kirtman, B. and Palke, W. E. (1977) *J. Chem. Phys.* **67**, 2236

Christiansen, O., Gauss, J. and Stanton, J. F. (1999) *Chem. Phys. Lett.* **305**, 147

Chung, K. T. (1997) *Phys. Rev. A* **56**, R3330

Chung, Y. M., Lee, E. M., Masuoka, T. and Samson, J. A. R. (1993) *J. Chem. Phys.* **99**, 885

Chupka, W. A. and Berkowitz, J. (1969) *J. Chem. Phys.* **51**, 4244

Chupka, W. A. and Berkowitz, J. (1971) *J. Chem. Phys.* **54**, 4256

Chupka, W. A., Miller, P. J. and Eyler, E. E. (1988) *J. Chem. Phys.* **88**, 3032

Clark, I. D. and Frost, D. C. (1967) *J. Am. Chem. Soc.* **89**, 244

Clark, L. B. and Simpson, W. T. (1965) *J. Chem. Phys.* **43**, 3666

Clyne, M. A. A. and Nip, W. S. (1977) *J. Chem. Soc. Faraday. Trans.* II **73**, 161

Codling, K., Hamley, J. R. and West, J. B. (1977) *J. Phys. B* **10**, 2797

Codling, K., Madden, R. P. and Ederer, D. L. (1967) *Phys. Rev.* **155**, 26

Cohen, S. M. and Leung, P. T. (1998) *Phys. Rev. A* **57**, 4994

Cohen, S. M. and Leung, P. T. (1999) *Phys. Rev. A* **59**, 4847

Coheur, P. F., Carleer, M. and Colin, R. (1996) *J. Phys. B* **29**, 4987

Cole, B. E. and Dexter, R. N., (1978) *J. Phys. B* **11**, 1011

Colmont, J.-M., Demaison, J. and Cosléou, J. (1995) *J. Mol. Spectrosc.* **171**, 453

Cook (1970) in *Recent Developments in Mass Spectroscopy*, Ogata, K. and Hayakawa, T., eds. (University Park, Baltimore), p. 761

Cook, G. R. (1968) *Trans. Am. Geophys. Union* **49**, 736A

Cook, G. R., Metzger, P. H. and Ogawa, M. (1965) *Can. J. Phys.* **43**, 1706

Cooper, G., Burton, G. R. and Brion, C. E. (1995a) *J. Electron Spectrosc.* **73**, 139

Cooper, G., Burton, G. R., Chan, W. F. and Brion, C. E. (1995c) *Chem. Phys.* **196**, 293

Cooper, G., Ibuki, T. and Brion, C. E. (1990) *Chem. Phys.* **140**, 33

Cooper, G., Ibuki, T., Iida, Y. and Brion, C. E. (1988) *Chem. Phys.* **125**, 307

Cooper, G., Olney, T. N. and Brion, C. E. (1995b) *Chem. Phys.* **194**, 175

Cooper, G., Zarate, E. B., Jones, R. K. and Brion, C. E. (1991) *Chem. Phys.* **150**, 237,251

Cooper, J. W. (1974) *Phys. Rev. A* **9**, 2236

Cooper, J. W. (1996) *Radiat. Phys. Chem.* **47**, 927

Cossart-Magos, C., Jungen, M. and Launay, F. (1987) *Mol. Phys.* **61**, 1077

Coulon, R., Montixi, G. and Occelli, R. (1981) *Can. J. Phys.* **59**, 1555

Coville, M. and Thomas, T. D. (1995) *J. Electron Spectrosc.* **71**, 21

Crasemann, B., Koblas, P. E., Wang, T.-C., Birdseye, H. E. and Cheng, M. H. (1974) *Phys. Rev. A* **9**, 1143

Cubaynes, D., Voky, L., Wuilleumier, F. J., Rouvellou, B., Hibbert, A., Faucher, P., Bizau, J.-M., Journel, L., Saraph, H. E. and Bely-Dubau, F. (1998) *Phys. Rev. A* **57**, 4432

Cubric, D., Willis, A. A., Comer, J. and Hammond, P. (1996) *J. Phys. B* **29**, 4151

Cuthbertson, C. and Cuthbertson, M. (1909) *Proc. Roy. Soc. Lond. A* **83**, 171

Cuthbertson, C. and Cuthbertson, M. (1913a) *Philos. Trans. Roy. Soc. (Lond.) A* **213**, 1

Cuthbertson, C. and Cuthbertson., M. (1913b) *Proc. Roy. Soc. Lond. A* **89**, 361

Cuthbertson, C. and Cuthbertson, M. (1936) *Proc. Roy. Soc. Lond. A* **155**, 213

Cutler, J. N., He, Z. X. and Samson, J. A. R. (1995) *J. Phys. B* **28**, 4577

Dai, S., MacToth, L. Del Cul, G. D. and Metcalf, D. H. (1994) *J. Chem. Phys.* **101**, 4470

Dalgarno, A. and Lynn, N. (1957) *Proc. Phys. Soc. Lond. A* **70**, 802

Dalgarno, A. and Sadeghpour, H. R. (1994) *Comments At. Mol. Phys.* **30**, 143

Dalgarno, A. L. and Stewart, A. L. (1960) *Proc. Phys. Soc. Lond.* **76**, 49

Dasgupta, A. and Bhatia, K. (1985) *Phys. Rev. A* **31**, 759

Daviel, S., Iida, Y., Carnovale, F. and Brion, C. E. (1984) *Chem. Phys.* **83**, 319

Davis, D. W. and Shirley, D. A. (1974) *J. Electron Spectrosc.* **3**, 157

de Reilhac, L. and Damany, N. (1971) *J. Phys. (Paris)* **32**, C4, 32

de Reilhac, L. and Damany, N. (1977) *J. Quant. Spectr. Rad. Transf.* **18**, 121

de Vries, J., Steger, H., Kamke, B., Menzel, C., Weisser, B., Kamke, W. and Hertel, I. V. (1992) *Chem. Phys. Lett.* **188**, 159

Dehmer, P. M. (1984) *Chem. Phys. Lett.* **110**, 79

Dehmer, P. M. and Chupka, W. A., (1976) *J. Chem Phys.* **65**, 2243

Dehmer, P. M. and Holland, D. M. P. (1991) *J. Chem. Phys.* **94**, 3302

Dehmer, P. M., Berkowitz, J. and Chupka, W. A. (1973) *J. Chem. Phys.* **59**, 5777

Dehmer, P. M., Berkowitz, J. and Chupka, W. A. (1974) *J. Chem. Phys.* **60**, 2676

Dehmer, P. M., Luken, W. L. and Chupka, W. A. (1977) *J. Chem. Phys.* **67**, 195

Denne, D. R. (1970) *J. Phys. D* **3**, 1392

Deslattes, R. D., LaVilla, R. E., Cowan, P. L. and Henins, A. (1983) *Phys. Rev. A* **27**, 923

Dickinson, H., Rolland, D. and Softley, T. P. (1997) *Philos. Trans. Roy. Soc. Lond. A* **355**, 1585

Diercksen, G. H. F. and Langhoff, P. H. (1987) *Chem. Phys.* **112**, 227

Diercksen, G. H. F., Kraemer, W. P., Rescigno, T. N., Bender, C. F., McKoy, B. V, Langhoff, S. R. and Langhoff, P. W. (1982) *J. Chem. Phys.* **76**, 1043

Dillon, M. A. and Inokuti, M. (1981) *J. Chem. Phys.* **74**, 6271

Ditchburn, R. W., Jutsum, P. J. and Marr, G. V. (1953) *Proc. Roy. Soc. Lond.* **A219**, 89

Dixit, S.N., Lynch, D. L., McKoy, B. V. and Hazi, A. U. (1989) *Phys. Rev. A* **40**, 1700

Doering, J. P., Gulcicek, E. E. and Vaughan, S. O. (1985) *J. Geophys. Res.* **90**, 5279

Domke, M., Xue, C., Puschmann, A., Mandel, T., Shirley, D. A. and Kaindl, G. (1990) *Chem. Phys. Lett.* **173**, 122

Dougherty, J. and Spackman, M. A. (1994) *Mol. Phys.* **82**, 193

Drake, G. W. F., (1996) in *Atomic, Molecular and Optical Physics Handbook*, Drake, G. W. F., ed. (Amer. Inst. of Phys., New York), Chap. 11

Drescher, M., Brockhinke, A., Böwering, N., Heinzmann, U. and Lefebvre-Brion, H. (1993) *J. Chem. Phys.* **99**, 2300

Dressler, K. and Wolniewicz, L. (1985) *J. Chem. Phys.* **82**, 4720

Dunn, A. F. (1964) *Can. J. Phys.* **42**, 1489

Dutuit, O., Tabche-Fouhaille, A., Nenner, I., Frohlich, H. and Guyon, P. M. (1985) *J. Chem. Phys.* **83**, 584

Dyck, M. K., Olney, T. N., Cooper, G. and Brion, C. E. (1995) *19th Intl. Conf. on the Physics of Electronic and Atomic Collisions*, Whistler, B. C., Book of Abstracts, p. 71

Dyke, J. M., Golob, L., Jonathan, N., Morris, A. and Okuda, M. (1974) *J. Chem. Soc. Faraday. Trans.* 2 **70**, 1828

Ederer, D. L. and Tomboulian, D. H. (1964) *Phys. Rev. A* **133**, 1525

Ederer, D. L., Lucatorto, T. and Madan, R. P (1970), *Phys. Rev. Lett.* **35**, 1537

Edqvist, O., Åsbrink, L. and Lindholm, E. (1971) *Z. Naturforsch. A* **26**, 1407

Edqvist, O., Lindholm, E., Selin, L. E., Åsbrink, L., Kuyatt, C. E., Mielczarek, S. R., Simpson, J. A. and Fischer-Hjalmars, I. (1970) *Phys. Scripta* **1**, 172

Edvardsson, D., Baltzer, P., Karlsson, L., Lundqvist, M. and Wannberg, B. (1995) *J. Electron Spectrosc.* **73**, 105

Edvardsson, D., Baltzer, P., Karlsson, L., Wannberg, B., Holland, D. M. P., Shaw, D. A. and Rennie, E. E. (1999) *J. Phys. B* **32**, 2583

Eggarter, E. (1975) *J. Chem. Phys.* **62**, 833

Eidelsberg, M., Rostas, F., Breton, J. and Thieblemont, B. (1992) *J. Chem. Phys.* **96**, 5585

Eikema, K. S. E., Ubachs, W., Vassen, W. and Hogervorst, W. (1997) *Phys. Rev.* **A55**, 1866

Eklund, P. (1992) *Bull. Am. Phys. Soc.* **37**, 191

Eklund, P. C., Rao, A. M., Wang, Y., Zhou, P., Wang, K.-A., Holden, J. M., Dresselhaus, M. S. and Dresselhaus, G. (1995) *Thin Solid Films* **257**, 211

Ekstrom, C. R., Schmiedmayer, J., Chapman, M. S., Hammond, T. D. and Pritchard, D. E. (1995) *Phys. Rev. A* **51**, 3883

Epprecht, G. W. (1950) *Z. Angew. Math. Physik* **1**, 138

Erickson, K. E. (1962) *J. Opt. Soc. Am.* **52**, 777

Eriksson, K. B. S. (1958) *Arkiv Fysik* **13**, 303

Eriksson, K. B. S. (1986) *Phys. Scripta* **34**, 211

Eriksson, K. B. S. and Isberg, H. B. S. (1963) *Arkiv Fysik* **24**, 549

Erman, P., Brzozowski, J. and Smith, W. H. (1974) *Astrophys. J.* **192**, 59

Erman, P., Hatherly, P. A., Karawajczyk, A., Köble, U., Rachlew-Källne, E., Stankiewicz, M. and Yoshiki Franzén, K. (1996) *J. Phys. B* **29**, 1501

Erman, P., Karawajczyk, A., Rachlew-Källne, E. and Stromholm, C. (1995) *J. Chem. Phys.* **102**, 3064

Erman, P., Karawajczyk, A., Rachlew-Källne, E., Stankiewicz, M., Yoshiki Franzén, K. Sannes, P. and Veseth, L. (1997) *Chem. Phys. Lett.* **273**, 239

Esposti, A. D. and Werner, H.-J. (1990) *J. Chem. Phys.* **93**, 3351

Esteva, J. M., Gauthé, B., Dhez, P. and Karnatak, R. C. (1983) *J. Phys. B* **16**, L263

Evans, M., Ng, C.Y., Hsu, C.-W. and Heimann, P. (1997) *J. Chem. Phys.* **106**, 978

Fang, T. K. and Chung, K. T. (2001) *Phys. Rev.* **A63**, 020702 (R)

Fano, U. (1954) *Phys. Rev.* **95**, 1198

Fano, U. (1961) *Phys. Rev.* **124**, 1866

Fano, U. and Cooper, J. W. (1965) *Phys. Rev.* **137**, A1364

Fano, U. and Cooper, J. W. (1968) *Rev. Mod. Phys.* **40**, 441

Federman, S. R., Menningen, K. L., Lee, W. and Stoll, J. B. (1997) *Astrophys. J.* **477**, L61

Feng, R., Cooper, G. and Brion, C. E. (1999a) *Chem. Phys.* **244**, 127

Feng, R., Cooper, G., Burton, G. R., Brion, C. E. and Avaldi, L. (1999b) *Chem. Phys.* **240**, 371; (1999b) *erratum Chem. Phys.* **248**, 293

Field, R. W., Benoist d'Azy, O., Lavollée, M., Lopez-Delgado, R. and Tramer, A. (1983) *J. Chem. Phys.* **78**, 2838

Filippov, A. and Prokofjew, W. (1928) *Z. Phys.* **56**, 458

Filippov, A. N. (1932) *Zh. Eksp. Teor. Fiz.* **2**, 24

Fischer, C. F., Jönsson, P. and Godefroid, M. (1998) *Phys. Rev. A* **57**, 1753

Flannery, M. R., Tai, H. and Albritton, D. L. (1977) *Atom. Data Nucl. Data Tables* **20**, 563

Fock, J. H., Gürtler, P. and Koch, E. E. (1980) *Chem. Phys.* **47**, 87

Fowler, P. W., Lazzeretti, P. and Zanasi, R. (1990) *Chem. Phys. Lett.* **165**, 79

Fraga, S., Karwowski, J. and Saxena, K. M. S. (1976) *Handbook of Atomic Data*, (Elsevier Science Publishing Company, Amsterdam), p. 329

Fredin, S., Gauyacq, D., Horani, M., Jungen, Ch., Lefevre, G. and Masnou-Seeuws, F. (1987) *Mol. Phys.* **60**, 825

Freund, H.-J., Kossmann, H. and Schmidt, V. (1986) *Chem. Phys. Lett.* **123**, 463

Frey, R. F. and Davidson, E. R. (1988a) *J. Chem. Phys.* **88**, 1775

Frey, R. F. and Davidson, E. R. (1988b) *J. Chem. Phys.* **89**, 4227

Friedrich, H., Sonntag, B., Rabe, P., Butscher, W. and Schwarz, W. H. E. (1979) *Chem. Phys. Lett.* **64**, 360

Frivold, O. E., Hassel, O. and Rustad, U. S. (1937) *Phys. Z.* **38**, 191

Frohlich, H. (private communication) has pointedly noted the error bars in Frohlich and Glass-Maujean (1990) (up to 30%). He feels that the dip ($\sim$0.3 in η_i at $\sim$785 Å) may be more accurately represented by the EELS data (Daviel *et al.*, 1984), where it is $\sim$0.1, more like their absolute fluorescence

Frohlich, H. and Glass-Maujean, M. (1990) *Phys. Rev. A* **42**, 1396

Frohlich, H., Guyon, P. M. and Glass-Mauhean, M. (1991) *Phys. Rev. A* **44**, 1791

Frost, D. C., Lee, S. T. and McDowell, C. A. (1974) *Chem. Phys. Lett.* **24**, 149

Frost, D. C., McDowell, C. A. and Vroom, D. A. (1967a) *Can. J. Chem.* **45**, 1343

Frost, D. C., McDowell, C. A., Sandhu, J. S. and Vroom, D. A. (1967b) *J. Chem. Phys.* **46**, 2008

Fung, H. S., Chu, C. C., Hsu, S. J., Wu, H. H. and Yih, T. S. (2000) *Rev. Sci. Instrum* **71**, 1564

Gallusser, R. and Dressler, K. (1982) *J. Chem. Phys.* **76**, 4311

Garcia, J. D. (1966) *Phys. Rev.* **147**, 66

Gardner, A., Lynch, M., Stewart, D. T. and Watson, W. S. (1973) *J. Phys. B* **6**, L262

Gaupp, A., Kuske, P. and Andrä, H. J. (1982) *Phys. Rev. A* **26**, 3351

Gedat, E., Püttner, R., Domke, M. and Kaindl, G. (1993) *J. Chem. Phys.* **109**, 4471

Gejo, T., Okada, K. and Ibuki, T. (1997) *Chem. Phys. Lett.* **277**, 497

Gejo, T., Okada, K., Ibuki, T. and Saito, N. (1999) *J. Phys. Chem. A* **103**, 4598

Gerchikov, L. G., Solov'yov, A. V., Connerade, J.-P. and Greiner, W. (1997) *J. Phys. B* **30**, 4133

Gibson, N. D. and Risley, J. S. (1995) *Phys. Rev. A* **52**, 4451

Gibson, S. T., Gies, H. P. F., Blake, A. J., McCoy, D. G. and Rogers, P. J. (1983) *J. Quant. Spectr. Rad. Transf.* **30**, 385

Gibson, S. T., Greene, J. P., Ruščić, B. and Berkowitz, J. (1986) *J. Phys. B* **19**, 2825

Giusti-Suzor, A. and Jungen, Ch. (1984) *J. Chem. Phys.* **80**, 986

Glass-Maujean, M. (1984) *Atom. Data Nucl. Data Tables* **30**, 301

Glass-Maujean, M., Breton, J. and Guyon, P. M. (1984) *Chem. Phys. Lett.* **112**, 25

Glass-Maujean, M., Breton, J. and Guyon, P. M. (1985) *J. Chem. Phys.* **83**, 1468

Glass-Maujean, M., Breton, J. and Guyon, P. M. (1987) *Z. Phys. D* **5**, 189

Glownia, J. H., Riley, S.J., Colson, S. D. and Niemann, G. C. (1980) *J. Chem. Phys.* **73**, 1296

Gluskin, E. S., Krasnoperova, A. A. and Mazalov, L. N. (1977) *Zhur. Strukt. Khim.* **18**, 665

Gluskin, E. S., Rapatskii, L. A., Raitsimring, A. M. and Yu. Tsvetkov, D. (1981) *High Energy Chem.* **15**, 71

Goebel, D., Hohm, U. and Kerl, K. (1995) *J. Mol. Struct.* **349**, 253

Goebel, D., Hohm, U., Kerl, K., Trümper, U. and Maroulis, G. (1994) *J. Phys. Chem.* **98**, 13123

Goldman, A., Bonomo, F. S., Williams, W. J., Murcray, D. G. and Snider, D. E. (1975) *J. Quant. Spectr. Rad. Transf.* **15**, 107

Golomb, D., Watanabe, K. and Marmo, F. F. (1962) *J. Chem. Phys.* **36**, 958

Gong, Q., Sun, Y., Huang, Z., Zhou, X., Gu, Z. and Qiang, D. (1996) *J. Phys. B* **29**, 4981

Gong, Q., Sun, Y., Huang, Z., Zhu, X., Gu, Z. N. and Qiang, D. (1994) *J. Phys. B* **27**, L199

Goode, J. G., Hofstein, J. D. and Johnson, P. M. (1997) *J. Chem. Phys.* **107**, 1703

Grimm, F. A., Whitley, T. A., Keller, P. R. and Taylor, J. W. (1991) *Chem. Phys.* **154**, 303

Guest, M. F. and Saunders, V. R. (1975) *Mol. Phys.* **29**, 873

Gürtler, P., Saile, V. and Koch, E. E. (1977a) *Chem. Phys. Lett.* **48**, 245

Gürtler, P., Saile, V. and Koch, E. E. (1977b) *Chem. Phys. Lett.* **51**, 386

Guyon, P. M., Baer, T. and Nenner, I. (1983) *J. Chem. Phys.* **78**, 3665

Habenicht, W., Reiser, G. and Müller-Dethlefs, K. (1991), *J. Chem. Phys.* **95**, 4809

Haddad, G. N. and Samson, J. A. R. (1986) *J. Chem. Phys.* **84**, 6623

Hamdy, H., He, Z. X. and Samson, J. A. R. (1991) *J. Phys. B* **24**, 4803

Hammond, B. L. and Rice, J. E. (1992) *J. Chem. Phys.* **97**, 1138

Han, J. C., Ye, C., Suto, M. and Lee, L. C. (1989) *J. Chem. Phys.* **90**, 4000

Hansen, M. S., Pacheco, J. M. and Onida, G. (1995) *Z. Phys. D* **35**, 141

Hardis, J. E., Ferrett, T. A., Southworth, S. H., Parr, A. C., Roy, P., Dehmer, J. L., Dehmer, P. M. and Chupka, W. A. (1988) *J. Chem. Phys.* **89**, 812

Harth, K., Raab, M. and Hotop, H. (1987) *Z. Phys. D* **7**, 213

Hasson, V. and Nicholls, R. W. (1971) *J. Phys. B* **4**, 1769

Hattori, H., Hikosaka, Y., Hikida, T. and Mitsuke, K. (1997) *J. Chem. Phys.* **106**, 4902

Hayaishi, T., Iwata, S., Sasanuma, M., Ishiguro, E., Morioka, Y., Iida, Y. and Nakamura, M. (1982) *J. Phys. B* **15**, 79

Hayaishi, T., Koizumi, T., Matsuo, T., Nagata, T., Sato, Y., Shibata, H. and Yagishita, A. (1987) *Chem. Phys.* **116**, 151

Hayes, W. and Brown, F. C. (1972) *Phys. Rev. A* **6**, 21

Henke, B. L., Gullikson, E. M. and Davis, J. C. (1993) *Atom. Data Nucl. Data Tables* **54**, 181

Henke, B. L., Lee, P., Tanaka, T. J., Shimabukuro, R. L. and Fujikawa, B. K. (1982) *Atom. Data Nucl. Data Tables* **27**, 1

Henry, R. J. W. (1967) *Planet. Space Sci.* **15**, 1747

Hepburn, J. W. (1997) *J. Chem. Phys.* **107**, 7106

Hertel, I. V., Steger, H., de Vries, J., Weisser, B., Menzel, C., Kamke, B. and Kamke, W. (1992) *Phys. Rev. Lett.* **68**, 784

Hibbert, A., Biémont, E. Godefroid, M. and Vaeck, N (1991) *J. Phys. B* **24**, 3943

Hibbert, A., Le Dourneuf, M. and Lan, V. K. (1977) *J. Phys. B* **10**, 1015

Hillier, I. H. and Saunders, V. R. (1971) *Mol. Phys.* **22**, 193

Hino, K., Ishihara, T., Shimizu, F., Toshima, N. and McGuire, J. H. (1993) *Phys. Rev. A* **48**, 1271

Hirschfelder, J. O., Brown, W. B. and Epstein, S. T. (1964) *Adv. Quantum. Chem.* **1**, 255

Hitchcock, A. P., Data presented in Piacente *et al.* (1989)

Hitchcock, A. P. (1990) *Phys. Scripta T* **31**, 159

Hitchcock, A. P. and Brion, C. E. (1980) *J. Phys. B* **13**, 3269

Hitchcock, A. P. and Mancini, D. C. (1994) *J. Electron Spectrosc.* **67**, 1

Hitchcock, A. P. and Van der Wiel, M. J. (1979) *J. Phys. B* **12**, 2153

Hitchcock, A. P., Wen, A. T. and Ruhl, E. (1990) *J. Electron Spectrose.* **51**, 653

Hodgeson, J. A., Sibert, E. E. and Curl, R. F. Jr. (1963) *J. Phys. Chem.* **67**, 2833

Hohm, H. (1994) *Mol. Phys.* **81**, 157

Hohm, U. (1993) *Mol. Phys.* **78**, 929

Hohm, U. and Kerl, K. (1990) *Mol. Phys.* **69**, 803, 819

Hölemann, P. and Goldschmidt, H. (1934) *Z. Phys. Chem. B* **24**, 199

Holland, D. M. P. and West, J. B. (1987a) *J. Phys. B* **20**, 1479

Holland, D. M. P. and West, J. B. (1987b) *Z. Phys. D* **4**, 367

Holland, D. M. P., MacDonald, M. A. and Hayes, M. A. (1990) *Chem. Phys.* **142**, 291

Holland, D. M. P., MacDonald, M. A., Baltzer, P., Karlsson, L., Lundqvist, M., Wannberg, B. and von Niessen, W. (1995b) *Chem. Phys.* **192**, 333

Holland, D. M. P., MacDonald, M. A., Hayes, M. A., Baltzer, P., Karlsson, L., Lundqvist, M., Wannberg, B. and von Niessen, W. (1994) *Chem. Phys.* **188**, 317

Holland, D. M. P., MacDonald, M. A., Hayes, M. A., Karlsson, L. and Wannberg, B. *J. Electron Spectrosc.* **104**, 243 (1999); originally *J. Electron Spectrosc.* **97**, 253 (1998)

Holland, D. M. P., Shaw, D. A. and Hayes, M. A. (1995a) *Chem. Phys.* **201**, 299

Holland, D. M. P., Shaw, D. A., Hayes, M. A., Shpinkova, L. G., Rennie, E. E., Karlsson, L., Baltzer, P. and Wannberg, B (1997) *Chem. Phys.* **219**, 91

Holland, D. M. P., Shaw, D. A., Hopkirk, A., MacDonald, M. A. and McSweeney, S. M. (1992) *J. Phys. B* **25**, 4823

Holland, D. M. P., Shaw, D. A., McSweeney, S. M., MacDonald, M. A., Hopkirk, A. and Hayes, M. A. (1993) *Chem. Phys.* **173**, 315

Hsu, C.-W., Evans, M., Heimann, P., Lu, K. T. and Ng, C. Y. (1996) *J. Chem. Phys.* **105**, 3950

Hubbell, J. H. (1969) Natl. Stand. Ref. Data Ser. NSRDS-NBS 29 (U.S. Govt. Print. Off., Washington, D.C.)

Hubbell, J. H. (1977) *Radiat. Res.* **70**, 58

Huber, K. P. (1961) *Helv. Phys. Acta* **34**, 929

Huber, K. P. and Jungen, Ch. (1990) *J. Chem. Phys.* **92**, 850

Hudson, E., Shirley, D. A., Domke, M., Remmers, G. and Kaindl, G. (1994) *Phys. Rev. A* **49**, 161

Hudson, E., Shirley, D. A., Domke, M., Remmers, G., Puschmann, A., Mandel, T., Xue, C. and Kaindl, G. (1993) *Phys. Rev. A* **47**, 361

Hudson, R. D. and Carter, V. L. (1965) *Phys. Rev.* **137**, A1648

Hudson, R. D. and Carter, V. L. (1967) *J. Opt. Soc. Am.* **57**, 651

Hudson, R. D. and Carter, V. L. (1968) *J. Opt. Soc. Am.* **58**, 430

Huebner, R. H., Celotta, R. J., Mielczarek, S. R. and Kuyatt, C. E. (1975a) *J. Chem. Phys.* **63**, 241

Huebner, R. H., Celotta, R. J., Mielczarek, S. R. and Kuyatt, C. E. (1975b) *J. Chem. Phys.* **63**, 4490

Huffman, R. E., Larrabee, J. C. and Tanaka, Y. (1964) *J. Chem. Phys.* **40**, 2261

Ibuki, T., Cooper, G. and Brion, C. E. (1989) *Chem. Phys.* **129**, 295

Ibuki, T., Horie, Y., Kamiuchi, A., Morimoto, Y., Tinone, M. C. K., Tanaka, K. and Honma, K. (1995) *J. Chem. Phys.* **102**, 5301

Ibuki, T., Koizumi, H., Yoshimi, T., Morita, M., Arai, S., Hironaka, K., Shinsaka, K., Hatano, Y., Yagishita, Y. and Ito, K. (1985) *Chem. Phys. Lett.* **119**, 327

Iida, Y., Carnovale, F., Daviel, S. and Brion, C. E. (1986) *Chem. Phys.* **105**, 211

Inokuti, M. (1971) *Rev. Mod. Phys.* **43**, 297

Irikura, K. K. (1995) *J. Chem. Phys*, **102**, 5357

Isenberg, E. M., Carter, S. L., Kelly, H. P. and Salomonson, S. (1985) *Phys. Rev. A* **32**, 1472

Ishiguro, E., Sasanuma, M., Masuko, H., Morioka, Y. and Nakamura, M. (1978) *J. Phys. B* **11**, 993

Ishii, I. and Hitchcock, A. P. (1988) *J. Electron Spectrosc.* **46**, 55

Ishii, I., McLaren, R., Hitchcock, A. P. and Robin, M. B. (1987) *J. Chem. Phys.* **87**, 4344

Ishii, I., McLaren, R., Hitchcock, A. P., Jordan, K. D., Choi, Y. and Robin, M. B. (1988) *Can. J. Chem.* **66**, 2104

Ishihara, T., Hino, K. and McGuire, J. H. (1991) *Phys. Rev. A* **44**, R6890

Itchkawitz, B. S., Long, J. P., Schedel-Niedrig, T., Kabler, M. N., Bradshaw, A. M., Schlögl, R. and Hunter, W. R. (1995) *Chem. Phys. Lett.* **243**, 211

Ito, K., Ueda, K., Namioka, T., Yoshino, K. and Morioka, Y. (1988) *J. Opt. Soc. Am. B* **5**, 2006

Itoh, A., Tsuchida, H., Miyabe, K., Majima, T. and Imanishi, N. (1999) *J. Phys. B* **32**, 277

Itoh, U., Toyoshima, Y., Onuki, H., Washida, N. and Ibuki, T. (1986) *J. Chem. Phys.* **85**, 4867

Jaensch, R. and Kamke, W. (2000a) *Mol. Mater.* **13**, 163

Jaensch, R. and Kamke, W. (2000b) *Mol. Mater.* **13**, 143

Jenouvrier, A., Coquart, B. and Mérienne, M. F. (1996), *J. Atmos. Chem.* **25**, 21

Jhanwar, B. L. and Meath, W. J. (1982) *Chem. Phys.* **67**, 185

Jhanwar, B. L. and Meath, W. J. (1984) *Can. J. Chem.* **62**, 373

Jhanwar, B. L., Meath, W. J. and MacDonald, J. C. F. (1981) *Can. J. Phys.* **59**, 185

Jhanwar, B. L., Meath, W. J. and MacDonald, J. C. F. (1983) *Can. J. Phys.* **61**, 1027

Jochims, H. W., Baumgartel, H. and Leach, S. (1996) *Astron. Astrophys.* **314**, 1003

Jolly, A., Lemaire, J. L., Belle-Oudry, D., Edwards, S., Malmasson, D., Vient, A. and Rostas, F. (1997) *J. Phys. B* **30**, 4315

Jolly, W. L., Bomben, K. D. and Eyermann, C. Y. (1984) *Atom. Data Nucl. Data Tables* **31**, 433

Jonsson, D., Norman, P., Ruud, K., Ågren, H. and Helgaker, T. (1998) *J. Chem. Phys.* **109**, 572

Jönsson, P., Ynnerman, A., Fischer, C. F., Godefroid, M. R. and Olsen, J. (1996) *Phys. Rev. A* **53**, 4021

Jungen, C. and Raoult, M. (1981) *Faraday Disc. Chem. Soc.* **71**, 253

Kabir, P. K. and Salpeter, E. E. (1957) *Phys. Rev.* **108**, 1256

Kafafi, Z. H., Lindle, J. R., Pong, R. G. S., Bartoli, F. J., Lingg, L. J. and Milliken, J. (1992) *Chem. Phys. Lett.* **188**, 492

Kagann, R. H. (1982) *J. Mol. Spectr.* **95**, 297

Kameta, K., Machida, S., Kitajima, M., Ukai, M., Kouchi, N., Hatano, Y. and Ito, K. (1996) *J. Electron Spectrosc.* **79**, 391

Kameta, K., Ukai, M., Chiba, R., Nagano, K., Kouchi, N., Hatano, Y. and Tanaka, K. (1991) *J. Chem. Phys.* **95**, 1456

Kamke, W. (2000) Private Communication

Kaplan, I. G. and Markin, A. P. (1973) *Zh. Eksp. Teor. Fiz.* **64**, 424

Kaplan, I. G. and Markin, A. P. (1975) *Dokl. Akad. Nauk. SSSR* **223**, 1172

Karawajczyk, A., Erman, P., Rachlew-Källne, E., Riusi Riu, J., Stankiewicz, M., Yoshiki Franzén, K. and Veseth, L. (2000) *Phys. Rev. A* **61**, 032718

Karlsson, L., Matsson, L., Jadrny, R., Bergmark, T. and Siegbahn, K. (1976) *Phys. Scripta* **14**, 230

Karna, S. P. (1996) *J. Chem. Phys.* **104**, 6590

Kassimi, N. E. and Thakkar, A. J. (1994) *Phys. Rev. A* **50**, 2948

Katayama, D. H., Huffman, R. E. and O'Bryan, C. L. (1973) *J. Chem. Phys.* **59**, 4309

Katsumata, S., Shiromaru, H. and Kimura, T. (1984) *Bull. Chem. Soc. Japan* **57**, 1784

Kaufman, V. and Minnhagen, L. (1972) *J. Opt. Soc. Am.* **62**, 92

Keller, J. W. and Coplan, M. A. (1992) *Chem. Phys. Lett.* **193**, 89

Kelly, H. P. (1969) *Phys. Rev.* **182**, 84

Kempgens, B., Itchkawitz, B. S., Randall, K. J., Feldhaus, J., Bradshaw, A. M., Köppel, H., Gadea, F. X., Nordfors, D., Schirmer, J. and Cederbaum, L. S. (1995) *Chem. Phys. Lett.* **246**, 347

Kempgens, B., Kivimäki, A., Itchkawitz, B. S., Köppe, H.M., Schmidbauer, M., Neeb, M., Maier, K., Feldhaus, J. and Bradshaw, A. M. (1998) *J. Electron Spectrosc.* **93**, 39

Kempgens, B., Kivimäki, A., Köppe, H. M., Neeb, M., Bradshaw, A. M. and Feldhaus, J. (1997b) *J. Chem. Phys.* **107**, 4219

Kempgens, B., Kivimäki, A., Neeb, M., Köppe, H. M., Bradshaw, A. M. and Feldhaus, J. (1996) *J. Phys. B* **29**, 5389

Kempgens, B., Maier, K., Kivimäki, A., Köppe, H. M., Neeb, M., Piancastelli, M. N., Hergenhahn, U. and Bradshaw, A. M. (1997) *J. Phys. B* **30**, L741

Keski-Rahkonen, O. and Krause, M. O. (1976) *J. Electron Spectrosc.* **9**, 371

Kharchenko, P., Babb, J. F. and Dalgarno, A. (1997) *Phys. Rev. A* **55**, 3566

Kheifets, A. S. and Bray, I. (1998a) *Phys. Rev. A* **57**, 2590

Kheifets, A. S. and Bray, I. (1998b) *Phys. Rev. A* **58**, 4501

Kilcoyne, D. A. L., Nordholm, S. and Hush, N. S. (1986) *Chem. Phys.* **107**, 255

Killgoar, P. C., Jr., Leroi, G. E., Chupka, W. A. and Berkowitz, J. (1973) *J. Chem. Phys.* **59**, 1370

Kim, K. and King, W. T. (1979) *J. Mol. Struct.* **57**, 201

Kim, Y.-K., Naon, M. and Cornille, M. (1973), Argonne National Laboratory, RER Div., Annual Rept. ANL-8060, Part I, pp. 14–23

Kimman, J., Verschuur, J. W. J., Lavollee, M., van Linden, H. B., van den Heuvell and van der Wiel, M. J. (1986) *J. Phys. B* **19**, 3909

Kimura, K., Katsumata, S., Achiba, Y., Yamazaki, T. and Iwata, S. (1981) *Handbook of He I Photoelectron Spectra of Fundamental Organic Molecules* (Japan Scientific Societies Press, Tokyo)

King, F. W. (1989) *Phys. Rev. A* **40**, 1735

Kirby, K. and Cooper, D. L. (1989) *J. Chem. Phys.* **90**, 4895

Kirouac, S. and Bose, T. K. (1973) *J. Chem. Phys.* **59**, 3043

Kivimäki, A., Neeb, M., Kempgens, B., Köppe, H. M. and Bradshaw, A. M. (1996) *J. Phys. B* **29**, 2701

Kivimäki, A., Neeb, M., Kempgens, B., Köppe, H. M., Maier, K. and Bradshaw, A. M. (1997) *J. Phys. B* **30**, 4279

Knowles, P. J., Rosmus, P. and Werner, H.-J. (1988) *Chem. Phys. Lett.* **146**, 230

Kobayashi, R., Koch, H., Jørgensen, P. and Lee, T. J. (1993) *Chem. Phys. Lett.* **211**, 94

Kobayashi, T., Sasagane, K., Aiga, F. and Yamaguchi, K. (1999) *J. Chem. Phys.* **110**, 11720

Koch, A. and Peyerimhoff, S. D. (1992) *Chem. Phys. Lett.* **195**, 104

Koch, E. E. and Otto, A. (1976) *Int. J. Rad. Phys. Chem.* **8**, 113

Koch, E. E. and Skibowski, M. (1971) *Chem. Phys. Lett.* **9**, 429

Koizumi, H. (1991) *J. Chem. Phys.* **95**, 5846

Kolos, W. and Wolniewicz, L. (1964) *J. Chem. Phys.* **41**, 3663

Kolos, W. and Wolniewicz, L. (1965) *J. Chem. Phys.* **43**, 2429

Kolos, W. and Wolniewicz, L. (1967) *J. Chem. Phys.* **46**, 1426

Kondratenko, A., Mazalov, L. N. and Neiman, K. M. (1980) *Opt. Spectrosc.* **49**, 266

Kong, W. and Hepburn, J. W. (1994) *Can. J. Phys.* **72**, 1284

Kong, W., Rodgers, D., Hepburn, J. W., Wang, K. and McKoy, V. (1993) *J. Chem. Phys.* **99**, 3159

Kornberg, M. A. and Miraglia, J. E. (1993) *Phys. Rev. A* **48**, 3714

Köster, C. and Grotemeyer, J. (1992) *Org. Mass Spectrom.* **27**, 463

Kosugi, N., Adachi, J., Shigemasa, E. and Yagishita, A. (1992) *J. Chem. Phys.* **97**, 8842

Krasnoperova, A. A., Gluskin, E. S., Mazalov, L. N. and Kochubel, V. A. (1976) *J. Struct. Chem.* **17**, 947

Krummacher, S., Biermann, M., Neeb, M., Liebsch, A. and Eberhardt, W. (1993) *Phys. Rev. B* **48**, 8424

Kudo, T. and Nagase, S. (1988) *Chem. Phys.* **122**, 233

Kumar, A. and Meath, W. J. (1985a) *Can. J. Chem.* **63**, 1616

Kumar, A. and Meath, W. J. (1985b) *Can. J. Phys.* **63**, 417

Kumar, A. and Meath, W. J. (1992) *Mol. Phys.* **75**, 311

Kumar, A., Fairley, G. R. G. and Meath, W. J. (1985) *J. Chem. Phys.* **83**, 70

Kumar, A., Meath, W. J., Bündgen, P. and Thakkar, A. J. (1996) *J. Chem. Phys.* **105**, 4927

Kurucz, R. L. and Peytremann, E. (1975) *Smithsonian Astrophys. Obs. Spec. Rept.* 362

Kutzner, M., Kelly, H. P., Larson, D. J. and Altun, Z. (1988) *Phys. Rev. A* **38**, 5107

Ladenburg, R. and Wolfsohn, G. (1932) *Z. Phys.* **79**, 42

Lagerqvist, A. and Miescher, E. (1958) *Helv. Phys. Acta* **31**, 221

Lambin, Ph., Lucas, A. A. and Vigneron, J. P. (1992) *Phys. Rev. B* **46**, 1794

Landolt-Bornstein Tables, (1962), Vol. **II**, Part 8, *Optical Constants*, (Springer-Verlag, Berlin), p. 6–8 81

Langer, B., Berrah, N., Wehlitz, R., Gorczyca, T. W., Bozek, J. and Farhat, A. (1997) *J. Phys. B* **30**, 593

Langhoff, P. W. and Karplus, M. (1969) *J. Opt. Soc. Am.* **59**, 863

Larsén, T. (1938) *Zeits. Phys.* **111**, 391

Lassettre, E. N. and Skerbele, A. (1971) *J. Chem. Phys.* **54**, 1597

Laughlin, C. (1995) *J. Phys. B* **28**, L701

LaVilla, R. E. (1972) *J. Chem. Phys.* **57**, 899

LaVilla, R. E. (1979) *Phys. Rev. A* **19**, 1999

LaVilla, R. E., Mehlman, G. and Salomon, E. B. (1981) *J. Phys. B* **14**, L1

Lavin, C., Velasco, A. M. and Martin, I. (1997) *Astron. Astrophys.* **328**, 426

Le Dourneuf, M., Lan, V. K. and Zeippen, C. J. (1979) *J. Phys. B* **12**, 2449

Lee, L. C. and Suto, M. (1984) *J. Chem. Phys.* **80**, 4718

Lee, L. C. and Suto, M. (1986) *Chem. Phys.* **110**, 161

Lee, L. C., Carlson, R. W., Judge, D. L. and Ogawa, M. (1973) *J. Quant. Spectr. Rad. Transf.* **13**, 1023

Lee, L. C., Phillips, E. and Judge, D. L. (1977) *J. Chem. Phys.* **67**, 1237

Lee, L. C., Wang, X. and Suto, M. (1987) *J. Chem. Phys.* **86**, 4353

Lee, S. S. and Oka, T. (1991) *J. Chem. Phys.* **94**, 1698

LeFèvre, R. J. W., Ross, I. G. and Smythe, B. M. (1950) *J. Chem. Soc.* part 1, 276

Lefebvre-Brion, H. and Keller, F. (1989) *J. Chem. Phys.* **90**, 7176

Lefebvre-Brion, H., Dehmer, P. M. and Chupka, W. A. (1988) *J. Chem. Phys.* **88**, 811

Lehmann, J. K., Simon, D. and Splitt, R. (1987) *Z. Phys. Chem. (Leipzig)* **268**, 209

Letzelter, C., Eidelsberg, M., Rostas, F., Breton, J. and Thieblemont, B. (1987) *Chem. Phys.* **114**, 273

Lewis, B. R. (1974) *J. Quant. Spectr. Rad. Transf.* **14**, 537

Lewis, B. R. and Carver, J. H. (1983) *J. Quant. Spectr. Rad. Transf.* **30**, 297

Lewis, B. R., Gibson, S. T., Emami, M. and Carver, J. W. (1988a) *J. Quant. Spectr. Rad. Transf.* **40**, 1

Lewis, B. R., Gibson., S. T., Emami, M. and Carver, J. W. (1988b) *J. Quant. Spectr. Rad. Transf.* **40**, 469

Leyh, B., Delwiche, J., Hubin-Franskin, M.-J. and Nenner, I. (1987) *Chem. Phys.* **115**, 243

Lias, S. G., Bartmess, J. E., Liebman, J. F., Holmes, J. L., Levin, R. D. and Mallard, W. G. (1988) *J. Phys. Chem. Ref. Data* **17**, Suppl. No. 1

Lias, S. G., NIST Website (Feb. 2000). http://webbook.nist.gov/chemistry/

Ligtenberg, R. C. G., van der Burgt, P. J. M., Renwick, S. P., Westerveld, W. B. and Risley, J. S. (1994) *Phys. Rev. A* **49**, 2363

Linder, B. and Abdulnur, S. (1971) *J. Chem. Phys.* **54**, 1807

Lindgård, A. and Nielsen, S. E. (1977) *Atom. Data Nucl. Data Tables* **19**, 533

Lindner, R., Müller-Dethlefs, K., Wedum, E., Haber, K. and Grant, E. R., *Science* (1996) **271**, 1698

Lisini, A. (1992) in *AIP Conf. Proc. No. 258, Synchrotron Radiation and Dynamic Phenomena*, Beswick, A., ed. (AIP, New York)

Locht, R., Leyh, B., Denzer, W., Hagenow, G. and Baumgartel, H. (1991) *Chem. Phys.* **155**, 407

Lukirskii, A. P. and Zimkina, T. M. (1963) *Bull. Acad. Sci. USSR, Phys. Ser.* **27**, 808

Lunell, S., Svensson, S., Malmqvist, P. Å., Gelius, U., Basilier, E. and Siegbahn, K. (1978) *Chem. Phys. Lett.* **54**, 420

Lynch, D., Lee, M.-T., Lucchese, R. R. and McKoy, V., (1984) *J. Chem. Phys.* **80**, 1907

Ma, Y., Chen, C. T., Meigs, G., Randall, K. and Sette, F. (1991) *Phys. Rev. A* **44**, 1848

Mack, K. M. and Muenter, J. S. (1977) *J. Chem. Phys.* **66**, 5278

MacNeil, K. A. G. and Dixon, R. N. (1977) *J. Electron Spectrosc.* **11**, 315

Madden, R. P., Ederer, D. L. and Codling, K. (1969) *Phys. Rev.* **177**, 136

Malathy Devi, V., Das, P. P., Bano, A., Rao, K. N., Flaud, J.-M., Camy-Peyret, C. and Chevillard, J.-P. (1981) *J. Mol. Spectrosc.* **88**, 251

Malathy Devi, V., Fridovich, B., Jones, G. D., Snyder, D. G. S., Das, P. P., Flaud, J.-M., Camy-Peyret, C. and Rao, K. N. (1982) *J. Mol. Spectrosc.* **93**, 179

Manatt, S. L. and Lane, A. L. (1993) *J. Quant. Spectrose. Rad. Transf.* **50**, 267

Marmo, F. F. (1953) *J. Opt. Soc. Am.* **43**, 1186

Maroulis, G. (1991) *J. Chem. Phys.* **94**, 1182

Maroulis, G. (1994) *J. Chem. Phys.* **101**, 4949

Maroulis, G. (1998a) *J. Chem. Phys.* **108**, 5432

Maroulis, G. (1998b) *Chem. Phys. Lett.* **289**, 403

Maroulis, G. and Thakkar, A. J. (1989) *Chem. Phys. Lett.* **156**, 87

Marr, G. V. and Creek, D. M. (1968) *Proc. Roy. Soc. A* **304**, 233

Marr, G. V. and West, J. B. (1976) *Atom. Data Nucl. Data Tables* **18**, 505

Martin, G. A. and Wiese, W. L. (1976) *Phys. Rev. A* **13**, 699 (1976); *J. Phys. Chem. Ref. Data* **5**, 537

Martin, I. and Barrientos, C. (1987) *Can. J. Phys.* **64**, 435

Martin, W. C. (1980) *J. Opt. Soc. Am.* **70**, 784

Martin, W. C. (1984) *Phys. Rev. A* **29**, 1883

Maryott, A. A. and Buckley, F. (1953) *Natl. Bur. Stand. Circ.* 537, (U. S. Govt. Print. Off.)

Mason, N. J. and Pathak, S. K. (1997) *Contemp. Phys.* **38**, 289

Mason, N. J., Gingell, J. M., Davies, J. A., Zhao, H., Walker, I. C. and Siggel, M. R. F. (1996) *J. Phys. B* **29**, 3075

Masuko, H., Morioka, Y., Nakamura, M., Ishiguro, E. and Sasanuma, M. (1979) *Can. J. Phys.* **57**, 745

Matsui, H. and Grant, E. R. (1996) *J. Chem. Phys.* **104**, 42

Matsunaga, F. M. and Watanabe, K. (1967) *Science of Light* **16**, 31; *ibid* **16**, 191

Mauersberger, K., Hanson, D., Barnes, J. and Morton, J. (1987) *J. Geophys. Res. D* **92**, 8480

Mazalov, L. N., Bertenev, V. M., Sadovskii, A. P. and Guzhavina, T. I. (1992) *J. Struct. Chem.* **13**, 799

McEachran, R. P. and Cohen, M. (1983) *J. Phys. B* **16**, 3125

McLaren, R., Clark, S. A. C., Ishii, I. and Hitchcock, A. P. (1987) *Phys. Rev. A* **36**, 1683

Meath, W. J. and Kumar, A. (1990) *Int. J. Quantum Chem. Symp.* **24**, 501

Meerts, W. L., Stolte, S. and Dymanus, A. (1977) *Chem. Phys.* **19**, 467

Mehlman, G., Cooper, J. W. and Salomon, E. B. (1982) *Phys. Rev. A* **25**, 2113

Mehlman, G., Ederer, D. L. and Salomon, E. B. (1978b) *J. Chem. Phys.* **68**, 1862

Mehlman, G., Ederer, D. L., Salomon, E. B. and Cooper, J. W. (1978a) *J. Phys. B* **11**, L689

Mellinger, A., Vidal, C. R. and Jungen, Ch. (1996) *J. Chem. Phys.* **104**, 8913

Mérienne, M. F., Jenouvrier, A. and Coquart, B. (1995) *J. Atmos. Chem.* **20**, 281

Merkt, F. and Softley, T. P. (1992) *Phys. Rev. A* **46**, 302

Merkt, F., MacKenzie, S. R., Rednall, R. J. and Softley, T. P. (1993) *J. Chem. Phys.* **99**, 8430

Merkt, F., Signorell, R., Palm, H., Osterwalder, A. and Sommavilla, M. (1998) *Mol. Phys.* **95**, 1045

Metzger, P. H. and Cook, G. R. (1964) *J. Chem. Phys.* **41**, 642

Metzger, P. H., Cook, G. R. and Ogawa, M. (1967) *Can. J. Phys.* **45**, 203

Meyer, K. W. and Greene, C. H. (1994) *Phys. Rev. A* **50**, R3573

Miescher, E. (1976) *Can. J. Phys.* **54**, 2074

Miescher, E. and Huber, K. P. (1976) *Intern. Rev. of Science, Phys. Chem. Ser. 2*, **Vol. 3**, *Spectroscopy*, Ramsay, D. A., ed. (Butterworths, London), pp. 37–43

Miller, P. J., Chupka, W. A. and Eland, J. H. D. (1988) *Chem. Phys.* **122**, 395

Miller, T. M. (1999) in *CRC Handbook of Chemistry and Physics*, 80th edn., 1999–2000, Lide, D. R. ed. (CRC Press, Boca Raton)

Miller, W. F. and Platzman, R. L. (1957) *Proc. Phys. Soc. (Lond) A* **70**, 299

Minnhagen, L. (1971) *J. Opt. Soc. Am.* **61**, 1257

Minnhagen, L. (1973) *J. Opt. Soc. Am.* **63**, 1185

Mitsuke, K., Hikosaka, Y., Hikida, T. and Hattori, H. (1996) *J. Electron Spectrosc.* **79**, 395

Molina, L. T. and Molina, M. J. (1986) *J. Geophys. Res. D* **91**, 14501

Molof, R. W., Schwartz, H. L., Miller, T. M. and Bederson, B. (1974) *Phys. Rev. A* **10**, 1131

Monfils, A. (1968) *J. Mol. Spectrosc.* **25**, 513

Moore, C. E. (1971) *Atomic Energy Levels, NSRDS-NBS 35*, vol. **1** (U.S. Govt. Print. Off., Washington, D.C.)

Morino, Y. and Tanimoto, M. (1984) *Can. J. Phys.* **62**, 1315

Morioka, Y., Masuko, H., Nakamura, M., Sasanuma, M. and Ishiguro, E. (1978) *Can. J. Phys.* **56**, 962

Morton, D. C. (1991) *Astrophys. J. Suppl. Ser.* **77**, 119

Morton, D. C. and Noreau, L. (1994) *Astrophys. J. Suppl.* **95**, 301

Moseley, J. T., Ozenne, J. B. and Cosby, P. C. (1981) *J. Chem. Phys.* **74**, 337

Mulder, F. and Meath, W. J. (1981) *Mol. Phys.* **42**, 629

Müller, T (1996) private communication

Mulliken, R. S. (1959) *Tetrahedron* **5**, 253

Myer, J. A. and Samson, J. A. R. (1970) *J. Chem. Phys.* **52**, 266

Nagournet, W., Happer, W. and Lurio, A. (1978) *Phys. Rev. A* **17**, 1394

Nahar, S. N., (1998) *Phys. Rev.* **A58**, 3766

Nahar, S. N. and Pradhan, A. K. (1997) *Astrophys. J. Suppl. Ser.* **111**, 339

Nakata, R. S., Watanabe, K. and Matsunaga, F. M. (1965) *Science of Light* **14**, 54

Nakayama, T., Kitamura, M. Y. and Watanabe, K. (1959) *J. Chem. Phys.* **30**, 1180

Namioka, T. (1964) *J. Chem. Phys.* **41**, 2141

Naon, M., Cornille, M. and Kim, Y.-K. (1975) *J. Phys. B* **8**, 864

Narayan, B. (1972) *Mol. Phys.* **23**, 281

Narayana, B. and Price, W. C. (1972) *J. Phys. B* **5**, 1784

Natali, S., Kuyatt, C. E. and Mielczarek, S. R. (1979) unpublished, cited in Berkowitz (1979a)

Nee, J. B. and Lee, L. C. (1984) *J. Chem. Phys.* **81**, 31

Nee, J. B., Suto, M. and Lee, L. C. (1985) *Chem. Phys.* **98**, 147

Nee, J. B., Suto, M. and Lee, L. C. (1986) *J. Chem. Phys.* **85**, 719

Nemeth, G. I., Selzle, H. L. and Schlag, E. W. (1993) *Chem. Phys. Lett.* **215**, 151

Nesbet, R. K. (1977) *Phys. Rev. A* **16**, 1

Neuhauser, R. G., Siglow, K. and Neusser, H. J. (1997) *J. Chem. Phys.* **106**, 896

Newbound, K. B. (1949) *J. Opt. Soc. Am.* **39**, 835

Newell, A. C. and Baird, R. C. (1965) *J. Appl. Phys.* **36**, 3751

Nicholson, A. J. C. (1965) *J. Chem. Phys.* **43**, 1171

Nielson, G. C., Parker, G. A. and Pack, R. T. (1976) *J. Chem. Phys.* **64**, 2055

Ninomiya, K., Ishiguro, E., Iwata, S., Mikuni, A. and Sasaki, T. (1981) *J. Phys. B* **14**, 1777

Nishikawa, S. and Watanabe, T. (1973) *Chem. Phys. Lett.* **22**, 590

Nishimura, H. and Tawara, H. (1994) *J. Phys. B* **27**, 2063

Nordfors, D., Nilsson, A., Mårtensson, N. Svensson, S., Gelius, U. and Ågren, H. (1991) *J. Electron Spectrosc.* **56**, 117

Norman, P., Luo, Y., Jonsson, D. and Ågren, H. (1997) *J. Chem. Phys.* **106**, 8788

Oates, C. W., Julienne, P. S., Lett, P. D., Phillips, W. D., Tiesinga, E. and Williams, C. J. (1996) *Europhys. Lett.* 35, **85**

Ogawa, M. and Cook, G. R. (1958a) *J. Chem. Phys.* **28**, 173

Ogawa, M. and Cook, G. R. (1958b) *J. Chem. Phys.* **28**, 747

Ogawa, S. and Ogawa, M. (1975) *Can. J. Phys.* **53**, 1845

Ojha, P. C. and Hibbert, A. (1990) *Phys. Scripta* **42**, 424

Olney, T. N., Cann, N. N., Cooper, G. and Brion, C. E. (1997) *Chem. Phys.* **223**, 59

Ono, Y., Osuch, E. A. and Ng, C. Y. (1982) *J. Chem. Phys.* **76**, 3905

Orcutt, R. H. and Cole, R. H. (1967) *J. Chem. Phys.* **46**, 697

Owens, J. C. (1967) *Appl. Opt.* **6**, 51

Pack, R. T. (1974) *J. Chem. Phys.* **61** 2091

Pack, R. T. (1982) *J. Phys. Chem.* **86**, 2794

Padial, N., Csanak, G., McKoy, B. V. and Langhoff, P. W. (1981) *J. Chem. Phys.* **74**, 4581

Page, R. H., Larkin, R. J., Shen, Y. R. and Lee, Y. T. (1988) *J. Chem. Phys.* **88**, 2249

Paldus, J. and Li, X. (1996) *Can. J. Chem.* **74**, 918

Pantos, E., Philis, J. and Bolovinos, A. (1978) *J. Mol. Spectrosc.* **72**, 36

Parker, G. A. and Pack, R. T. (1976) *J. Chem. Phys.* **64**, 2010

Pazur, R. J., Kumar, A., Thuraisingham, R. A. and Meath, W. J. (1988) *Can. J. Chem.* **66**, 615

Peach, G., Seraph, H. E. and Seaton, M. J. (1988) *J. Phys. B* **21**, 3669

Peck, E. R. and Khanna, B. N. (1966) *J. Opt. Soc. Am.* **56**, 1059

Pederson, M. R. and Quong, A. A. (1992) *Phys. Rev. B* **46**, 13584

Pekeris, C. L. (1954) *Phys. Rev.* **115**, 1216

Perrin, A., Mandin, J.-Y., Camy-Peyret, C., Flaud, J.-M., Chevillard, J.-P. and Guelachvili, G. (1984) *J. Mol. Spectrosc.* **103**, 417

Person, J. C. (1965) *J. Chem. Phys.* **43**, 2553

Person, J. C. and Nicole, P. P. (1968) *J. Chem. Phys.* **49**, 5421

Person, J. C. and Nicole, P. P. (1970) *J. Chem. Phys.* **53**, 1767; Tabular data are available in Ann. Rept., Radiological Phys. Div., Argonne Natl. Lab., ANL-7760, Part 1 (1970)

Person, J. C. and Nicole, P. P. (1974) Argonne Natl. Lab., RER Div. Ann. Rept. ANL-75-3, Part I

Person, J. C. and Nicole, P. P. (1977), Argonne Natl. Lab., RER Div. Ann. Rept. ANL-77-65, Part I, p. 19

Person, J. C. and Nicole, P. P., Argonne Natl. Lab., RER Div. Ann. Rept. ANL-78-65, Part I, p. 33

Person, J. C., Fowler, D. E. and Nicole, P. P., Argonne Natl. Lab. RER Div. Ann. Rept. ANL-75-60, Part I, p. 30

Persson, W. (1971) *Phys. Scripta* **3**, 133

Phillips, E., Lee, L. C. and Judge, D. L. (1977) *J. Quant. Spectr. Rad. Transf.* **18**, 309

Piacente, V., Gigli, G., Scardala, P., Giustini, A. and Ferro, D. (1995) *J. Phys. Chem.* **99**, 14052

Piancastelli, M. N., Ferrett, T. A., Lindle, D. W., Medhurst, L. J., Heimann, P. A., Liu, S. H. and Shirley, D. A. (1989) *J. Chem. Phys.* **90**, 3004

Pipin, J. and Bishop, D. M. (1992) *Phys. Rev. A* **45**, 2736

Platzman, R. L. (1960) *J. Phys. et le Radium* **21**, 853

Platzman, R. L. (1962) *Radiat. Res.* **17**, 419

Platzman, R. L. (1962) *Vortex* **23**, 372

Polo, S. R. and Wilson, M. K. (1955) *J. Chem. Phys.* **23**, 2376

Pople, J. A. and Curtiss, L. A. (1987) *J. Phys. Chem.* **91**, 155

Potts, A. W. and Price, W. C. (1972) *Proc. Roy. Soc. Lond. A* **326**, 165

Pradhan, A. K. (1978) *J. Phys. B* **11**, L729

Pratt, S. T., Dehmer, P. M. and Dehmer, J. L. (1993) *J. Chem. Phys.* **99**, 6233

Preston, K. F. and Barr, R. F. (1971) *J. Chem. Phys.* **54**, 3347

Quong, A. A. and Pederson, M. R. (1992) *Phys. Rev. B* **46**, 12906

Rabalais, J. W., Karlsson, L., Werme, L. O., Bergmark, T. and Siegbahn, K. (1973) *J. Chem. Phys.* **58**, 3370

Rabalais, J. W., McDonald, J. M., Scherr, V. and McGlynn, S. P. (1971) *Chem. Rev.* **71**, 73

Radler, K. and Berkowitz, J. (1979) *J. Chem. Phys.* **70**, 221

Radziemski, Jr., L. J. and Kaufman, V. (1969) *J. Opt. Soc. Am.* **59**, 424

Radziemski, Jr., L. J. and Kaufman, V. (1974) *J. Opt. Soc. Am.* **64**, 366

Ramaswamy, K. L. (1936) *Proc. Indian Acad. Sci. A* **4**, 675

Raoult, M. and Jungen, C. (1981) *J. Chem. Phys.* **74**, 3388

Raymonda, J. W. and Simpson, W. T. (1967) *J. Chem. Phys.* **47**, 430

Reinartz, J. M. L. J., Meerts, W. L. and Dymanus, A. (1978) *Chem. Phys.* **31**, 19

Reinsch, E.-A. and Meyer, W. (1976) *Phys. Rev. A* **14**, 915

Reiser, G., Habenicht, W. and Müller-Dethlefs, K. (1993) *J. Chem. Phys.* **98**, 8462

Reiser, G., Habenicht, W., Müller-Dethlefs, K. and Schlag, E. (1988) *Chem. Phys. Lett.* **152**, 119

Remmers, G., Domke, M., Puschmann, A., Mandel, T., Kaindl, G., Hudson, E. and Shirley, D. A. (1993) *Chem. Phys. Lett.* **214**, 241

Ren, S. L., Wang, Y., Rao, A. M., McRae, E., Holden, J. M., Hager, T., Wang, K., Lee, W.-T., Ni, H. F., Selegue, J. and Eklund, P. C. (1991) *Appl. Phys. Lett.* **59**, 2678

Rennie, E. E., Johnson, C. A. F., Parker, J. E., Holland, D. M. P., Shaw, D. A. and Hayes, M. A. (1998) *Chem. Phys.* **229**, 107

Rennie, E. E., Kempgens, B., Köppe, H. M., Hergenhahn, U., Feldhaus, J., Itchkawitz, B. S., Kilcoyne, A. L. D., Kivimäki, A., Maier, K., Piancastelli, M. N., Polcik, M., Rüdel, A. and Bradshaw, A. M. (2000) *J. Chem. Phys.* **113**, 7362

Reutt, J. E., Wang, L. S., Lee, Y. T. and Shirley, D. A. (1986a) *J. Chem. Phys.* **85**, 6928

Reutt, J. E., Wang, L. S., Pollard, J. E., Trevor, D. J., Lee, Y. T. and Shirley, D. A. (1986b) *J. Chem. Phys.* **84**, 3022

Reynaud, C., Gaveau, M.-A., Bisson, K., Millié, P., Nenner, I., Bodeur, S., Archirel, P. and Lévy, B. (1996) *J. Phys. B* **29**, 5403

Rice, J. E., Taylor, P. R., Lee, T. J. and Almlöf, J. (1991), *J. Chem. Phys.* **94**, 4972

Richards, J. A. and Larkins, F. P. (1986) *J. Phys. B* **19**, 1945

Rieke, F. F. and Prepejchal, W. (1972) *Phys. Rev. A* **6**, 1507

Robicheaux, F. and Greene, C. H. (1992) *Phys. Rev. A* **46**, 3821

Robin, M. B. and Kuebler, N. A. (1972) *J. Electron Spectrosc.* **1**, 13

Roos, B. and Siegbahn, P. (1971) *Theor. Chim. Acta* **21**, 368

Rothenberg, S. and Davidson, E. R. (1967) *J. Mol. Spectrosc.* **22**, 1

Rouse, P. E. and Engleman, Jr., R. (1973) *J. Quant. Spectr. Rad. Transf.* **13**, 1503

Rudd, M. E., DuBois, R. D., Toburen, L. H., Ratcliffe, C. A. and Goffe, T. F. (1983) *Phys. Rev. A* **28**, 3244

Rudd, M. E., Goffe, T. V., DuBois, R. D. and Toburen, L. H. (1985b) *Phys. Rev. A* **31**, 492

Rudd, M. E., Kim, Y.-K., Madison, D. H. and Gallagher, J. W. (1985a) *Rev. Mod. Phys.* **57**, 965

Rühl, E., Price, S. D. and Leach, S. (1989) *J. Phys. Chem.* **93**, 6312

Ruščić, B. and Berkowitz, J. (1983) *Phys. Rev. Lett.* **50**, 675

Ruščić, B., Greene, J. P. and Berkowitz, J. (1984) *J. Phys. B* **17**, 1503

Russell, A. J. and Spackman, M. A. (1995) *Mol. Phys.* **84**, 1239

Russell, A. J. and Spackman, M. A. (1996) *Mol. Phys.* **88**, 1109

Russell, A. J. and Spackman, M. A. (1997) *Mol. Phys.* **90**, 251

Rustgi, O. P. (1964) *J. Opt. Soc. Am.* **54**, 464

Saethre, L. J., Siggel, M. R. F. and Thomas, T. D. (1989) *J. Electron Spectrosc.* **49**, 119

Saha, H. P. (1993) *Phys. Rev. A* **47**, 2865

Saha, H. P. and Caldwell, C. D. (1991) *Phys. Rev. A* **44**, 5642

Saha, H. P., Froese-Fischer, C. and Langhoff, P. W. (1988) *Phys. Rev. A* **38**, 1279

Salhi-Benachenhou, N., Engels, B., Huang, M.-B. and Lunell, S. (1998) *Chem. Phys.* **236**, 53

Salpeter, E. E. and Zaidi, M. H. (1962) *Phys. Rev.* **125**, 248

Samson, J. A. R. (1966) in *Adv. Atomic Molec. Phys.*, Bates, D. R. and Estermann, I., eds. (Academic Press, N. Y), vol. **2**, p. 177

Samson, J. A. R. and Angel, G. C. (1990) *Phys. Rev. A* **42**, 1307

Samson, J. A. R. and Gardner, J. L. (1976) *J. Electron Spectrosc.* **8**, 35

Samson, J. A. R. and Haddad, G., cited by Gallagher, J. W., Brion, C. E., Samson, J. A. R. and Langhoff, P. W. (1988) *J. Phys. Chem. Ref. Data* **17**, 9

Samson, J. A. R. and Haddad, G. N. (1994) *J. Opt. Soc. Am. B* **11**, 277

Samson, J. A. R. and Pareek, P. N. (1985) *Phys. Rev. A* **31**, 1470

Samson, J. A. R. and Yin, L. (1989) *J. Opt. Soc. Am. B* **6**, 2326

Samson, J. A. R., Gardner, J. L. and Starace, A. F. (1975) *Phys. Rev A* **12**, 1459

Samson, J. A. R., Haddad, G. N. and Kilcoyne, L. D. (1987b) *J. Chem. Phys.* **87**, 6416

Samson, J. A. R., Haddad, G. N., Masuoka, T., Pareek, P. N. and Kilcoyne, D. A. L. (1989) *J. Chem. Phys.* **90**, 6925

Samson, J. A. R., He, Z. X., Yin, L. and Haddad, G. N. (1994) *J. Phys. B* **27**, 887

Samson, J. A. R., Lyn, L., Haddad, G. N. and Angel, G. C. (1991) *J. Physique 4, Coll.* C1–99

Samson, J. A. R., Masuoka, T., Pareek, P. N. and Angel, G. C. (1987a) *J. Chem. Phys.* **86**, 6128

Samson, J. A. R., Rayburn, G. H. and Pareek, P. N. (1982) *J. Chem. Phys.* **76**, 393

Samson, J. A. R., Shefer, Y. and Angel, G. C. (1986) *Phys. Rev. Lett.* **56**, 2020

Sasanuma, M., Morioka, Y., Ishiguro, E. and Nakamura, M. (1974) *J. Chem. Phys.* **60**, 327

Saxon, R. P. (1973) *Phys. Rev. A* **8**, 839

Schaphorst, S. J., Whitfield, S. B., Saha, H. P., Caldwell, C. D. and Azuma, Y. (1993) *Phys. Rev.* **A47**, 3007

Schectman, R. M., Federman, S. R., Beideck, D. J. and Ellis, D. G. (1993) *Astrophys. J.* **406**, 735

Schiff, B., Pekeris, C. L. and Accad, Y. (1971) *Phys. Rev. A* **4**, 885

Schirmer, J., Trofimov, A. B., Randall, K. J., Feldhaus, J., Bradshaw, A. M., Ma, Y., Chen, C. T. and Sette, F. (1993) *Phys. Rev. A* **47**, 1136

Schlag, E. W. and Levine, R. D. (1992) *J. Phys. Chem.* **96**, 10608

Schlag, E. W., Grotemeyer, J. and Levine, R. D. (1992) *Chem. Phys. Lett.* **190**, 521

Schmidbauer, M., Kilcoyne, A. L. D., Köppe, H.-M. Feldhaus, J. and Bradshaw, A. M. (1992) *Chem. Phys. Lett.* **199**, 119

Schoen, R. I. (1962) *J. Chem. Phys.* **37**, 2032

Schram, B. L., de Heer, F. J., van der Wiel, M. J. and Kistemaker, J. (1965) *Physica* **31**, 94

Schulz, K., Kaindl, G., Domke, M., Bozek, J. D., Heimann, P. A., Schlachter, A. S. and Rost, J. M. (1996) *Phys. Rev. Lett.* **77**, 3086

Schulz, R. W. (1938) *Zeits. Phys.* **109**, 517

Schutten, J., de Heer, F. J., Moustafa, H. R., Boerboom, A. J. H. and Kistemaker, J. (1966) *J. Chem. Phys.* **44**, 3924

Schwab, J. J. and Anderson, J. G. (1982) *J. Quant. Spectr. Rad. Transf.* **27**, 445

Schram, B. L., van der Wiel, M. J., de Heer, F. J. and Moustafa, H. R. (1966) *J. Chem. Phys.* **44**, 49

Semenov, R. I. and Strugach, B. A. (1968) *Opt. Spectrosc (USSR)* **24**, 258

Shafizadeh, N., Fillion, J.-H., Gauyacq, D. and Couris, S. (1997) *Philos. Trans. Roy. Soc. Lond. A* **355**, 1637

Sham, T. K., Yang, B. X., Kirz, J. and Tse, J. S. (1989) *Phys. Rev. A* **40**, 652

Shapiro, M. (1977) *Chem. Phys. Lett.* **46**, 442

Shaw, D. A., Holland, D. M. P., Hayes, M. A., MacDonald, M. A., Hopkirk, A. and McSweeney, S. M. (1995) *Chem. Phys.* **198**, 381

Shaw, D. A., Holland, D. M. P., MacDonald, M. A., Hopkirk, A., Hayes, M. A. and McSweeney, S. M. (1992a) *Chem. Phys.* **166**, 379

Shaw, D. A., Holland, D. M. P., MacDonald, M. A., Hopkirk, A., Hayes, M. A. and McSweeney, S. M. (1992b) *Chem. Phys.* **163**, 387

Shaw, D. A., Holland, D. M. P., MacDonald, M. A., Shpinkova, L. G., Hayes, M. A. and Rennie, E. E. (1997) *J. Phys. B* **30**, 905

Shigemasa, E., Hayaishi, T, Sasaki, T. and Yagishita, A. (1993) *Phys. Rev. A* **47**, 1824

Shigemasa, E., Ueda, K., Sato, Y., Yagishita, A., Maezawa, H., Sasaki, T., Ukai, M. and Hayaishi, T. (1990) *Phys. Scripta* **41**, 60

Shiner, D., Gilligan, J. M., Cook, B. M. and Lichten, W. (1993) *Phys. Rev. A* **47**, 4042

Siegbahn, K., Nordling, C., Johansson, G., Hedman, J., Hedén, P. F., Hamrin, K., Gelius, U. and Bergmark, T. (1969) *ESCA Applied to Free Molecules* (North-Holland, Amsterdam)

Signorell, R. and Merkt, F. (1999) *J. Chem. Phys.* **110**, 2309

Simpson, J. A., Kuyatt, C. E. and Mielczarek, S. R. (1966) *J. Chem. Phys.* **44**, 4403

Sivkov, V. N., Akimov, V. N. and Vinogradov, A. S. (1987) *Opt. Spectrosc. (USSR)* **63**, 162

Sivkov, V. N., Akimov, V. N., Vinogradov, A. S. and Zimkina, T. M. (1984) *Opt. Spectrosc. (USSR)* **57**, 160

Sivkov, V. N., Akimov, V. N., Vinogradov, A. S. and Zimkina, T. M. (1986) *Opt. Spectrosc. (USSR)* **60**, 194

Slanger, T. G. and Black, G. (1978) *J. Chem. Phys.* **68**, 1844

Smith, A. L. (1970) *Philos. Trans. Roy. Soc. Lond. A* **268**, 169

Smith, A. L. (1996) *J. Phys. B* **29**, 4975

Smith, P. L., Stark, G., Yoshino, K. and Ito, K. (1994) *Astrophys. J.* **431**, L143

Smith, P. L., Yoshino, K. and Parkinson, W. H. (1991) *J. Geophys. Res.* **96**, 17529

Sokell, E., Wills, A. A. and Comer, J. (1996) *J. Phys. B* **29**, 3417

Sokell, E., Wills, A. A., Comer, J. and Hammond, P. (1997) *J. Phys. B* **30**, 2635

Sorokin, A. A., Shmaenok, L. A., Bobashev, S. V., Mobus, B. and Ulm, G. (1998) *Phys. Rev. A* **58**, 2900

Sorokin, A. A., Shmaenok, L. A., Bobashev, S. V., Mobus, B., Richter, M. and Ulm, G. (2000) *Phys. Rev. A* **61**, 022723

Sosa, C. and Ferris, K. F. (1990) *Mater. Res. Soc. Symp. Proc.* **193**, 131

Southworth, S. H., Ferrett, T. A., Hardis, J. E., Parr, A. C. and Dehmer, J. L. (2000) *Physics Essays* **13**, 290

Spelsberg, D. and Meyer, W. (1998) *J. Chem. Phys.* **108**, 1532

Staib, A. and Domcke, W. (1991) *J. Chem. Phys.* **94**, 5402

Stark, G., Lewis, B. R., Gibson, S. T. and England, J. P. (1998) *Astrophys. J.* **505**, 452

Stark, G., Lewis, B. R., Gibson, S. T. and England, J. P. (1999) *Astrophys. J.* **520**, 732

Stark, G., Yoshino, K., Smith, P. L., Ito, K. and Parkinson, W. H. (1991) *Astrophys. J.* **369**, 574

Stewart, A. L. (1978) *J. Phys. B* **11**, 2449

Stewart, R. F. (1975) *Mol. Phys.* **29**, 1577

Stiehler, J. and Hinze, J. (1995) *J. Phys. B* **28**, 4055

Stoll, E. (1922) *Ann. Phys.* **69**, 81

Stolow, A. and Lee, Y. T. (1993) *J. Chem. Phys.* **98**, 2066

Stolte, W. C., Lu, Y., Samson, J. A. R., Hemmers, O., Hansen, D. L., Whitfield, S. B., Wang, H., Glans, P. and Lindle, D. W. (1997) *J. Phys. B* **30**, 4489

Strobel, A., Fischer, I., Straecker, J., Nieder-Schatteburg, G., Müller-Dethlefs, K. and Bondybey, V. E. (1992) *J. Chem. Phys.* **97**, 2332

Sukhorukov, V. L., Khoperskii, A. N., Petrov, I. D., Yavna, V. A. and Demekhin, V. F. (1987) *J. Physique* **48**, 1677

Suto, M. and Lee, L. C. (1983) *J. Chem. Phys.* **78**, 4515

Suto, M. and Lee, L. C. (1984) *J. Chem. Phys.* **80**, 4824

Suto, M. and Lee, L. C. (1986) *J. Chem. Phys.* **84**, 1160

Syage, J. A., Cohen, R. B. and Steadman, J. (1992) *J. Chem. Phys.* **97**, 6072

Sze, K.-H., Brion, C. E., Tong, X.-M. and Li, J.-M. (1987) *Chem. Phys.* **115**, 433

Takezawa, S. (1970) *J. Chem. Phys.* **52**, 2575

Tanaka, Y. (1942) *Sci. Pap. Inst. Phys. Chem. Res. Tokyo* **39**, 456

Tanaka, Y., Inn, E. C. Y. and Watanabe, K. (1953) *J. Chem. Phys.* **21**, 1651

Tang, J.-Z. and Shimamura, I. (1995) *Phys. Rev. A* **52**, R3413

Taniguchi, N., Takahashi, K., Matsui, Y., Dylewski, S. M., Geiser, J. D. and Houston, P. L. (1999) *J. Chem. Phys.* **111**, 6350

Taylor, K. T. and Burke, P. G. (1976) *J. Phys. B* **9**, L353

Terwilliger, D. T. and Smith, A. L. (1975) *J. Chem. Phys.* **63**, 1008

Thomas, D., Coville, M., Thissen, R. and Morin, P. (1992) *Synchr. Rad. News* **5**(3), 9

Thomas, G. F. and Meath, W. J. (1977) *Mol. Phys.* **34**, 113

Thomas, T. D. and Shaw, R. W. (1974) *J. Electron Spectr.* **5**, 1081

Thoss, M. and Domcke, W. (1997) *J. Chem. Phys.* **106**, 3174

Tichy, M., Jahavery, G., Twiddy, N. D. and Ferguson, E. E. (1987) *Int. J. Mass Spectrom. Ion Proc.* **79**, 231

Tichy, M., Jahavery, G., Twiddy, N. D. and Ferguson, E. E. (1990) *Int. J. Mass Spectrom. Ion Proc.* **97**, 211

Tilford, S. G. and Simmons, J. D. (1972) *J. Phys. Chem. Ref. Data* **1**, 147

Tonkyn, R. G., Wiedmann, R., Grant, E. R. and White, M. G. (1991) *J. Chem. Phys.* **95**, 7033

Tonkyn, R. G., Wiedmann, R. T. and White, M. G. (1992) *J. Chem. Phys.* **96**, 3696

Tsurubuchi, S., Watanabe, K. and Arikawa, T. (1990) *J. Phys. Soc. Japan* **59**, 497

Tuilier, W. H., Laporte, D. and Esteva, J. M. (1982) *Phys. Rev. A* **26**, 372

Turci, C. C., Francis, J. T., Tyliszczak, T., de Souza, G. G. B. and Hitchcock, A. P. (1995) *Phys. Rev. A* **52**, 4678

Turner, D. W., Baker, C., Baker, A. D. and Brundle, C. R. (1970) *Molecular Photoelectron Spectroscopy* (Wiley-Interscience, London)

Ueda, K., Okunishi, M., Chiba, H., Shimizu, Y., Ohmori, K., Sato, Y., Shigemasa, E. and Kosugi, N. (1995) *Chem. Phys. Lett.* **236**, 311

Ukai, M., Kameta, K., Chiba, R., Nagano, K., Kouchi, N., Shinsaka, K., Hatano, Y., Umemoto, H., Ito, Y. and Tanaka, K. (1991) *J. Chem. Phys.* **95**, 4142

Ukai, M., Kameta, K., Kouchi, N., Nagano, N., Hatano, Y. and Tanaka, K. (1992b) *J. Chem. Phys.* **97**, 2835

Ukai, M., Kameta, K., Machida, S., Kouchi, N., Hatano, Y. and Tanaka, K. (1994) *J. Chem. Phys.* **101**, 5473

Ukai, M., Kouchi, N., Kameta, K., Terazawa, N., Chikahiro, Y., Hatano, Y. and Tanaka, K. (1992a) *Chem. Phys. Lett.* **195**, 298

Vaida, V., McCarthy, M. I., Engelking, P. C., Rosmus, P., Werner, H. J. and Botschwina, P. (1987) *J. Chem. Phys.* **86**, 6669

van der Meer, W. J., Butselaar, R. J. and de Lange, C. A. (1986) *Austr. J. Phys.* **39**, 779

van der Meer, W. J., van der Meulen, P., Volmer, M. and de Lange, C. A. (1988) *Chem. Phys.* **126**, 385

van der Meulen, P., Krause, M. O., Caldwell, C. D., Whitfield, S. B. and de Lange, C. A. (1992) *Phys. Rev. A* **46**, 2468

van Gisbergen, S. J. A., Snijders, J. G. and Baerends, E. J. (1997) *Phys. Rev. Lett.* **78**, 3097

Vandaele, A. C., Hermans, C., Simon, P. C., Carleer, M., Colin, R., Fally, S., Mérienne, M. F., Jenouvrier, A. and Coquart, B. (1998) *J. Quant. Spectr. Rad. Transf.* **59**, 171

Velchev, I., Hogervorst, W. and Ubachs, W. (1999) *J. Phys. B* **32**, L 511

Verner, D. A., Barthel, P. D. and Tytler, D. (1994) *Astron. Astrophys. Suppl. Ser.* **108**, 309

Verstraete, L., Leger, A., d'Hendecourt, L., Dutuit, O. and Defourneau, D. (1990) *Astron. Astrophys.* **237**, 436

Victor, G. A. and Dalgarno, A. (1969) *J. Chem. Phys.* **50**, 2535

Vinogradov, A. S. and Zimkina, T. M. (1971) *Opt. Spectrosc. (USSR)* **31**, 364

Vinogradov, A. S. and Zimkina, T. M. (1972) *Opt. Spectrosc. (USSR)* **32**, 17

Volz, U. and Schmoranzer, H. (1996) *Phys. Scripta* **T65**, 48

Volz, U., Majerus, N., Liebel, H., Schmitt, A. and Schmoranzer, H. (1996) *Phys. Rev. Lett.* **76**, 2862

von dem Borne, A., Federmann, F., Lee, M. K. and Sonntag, B. (1995) *J. Phys. B* **28**, 2591

Vondrácek, M., Prince, K. C., Karvonen, J., Coreno, M., Camilloni, R., Avaldi, L. and de Simone, M. (1999) *Synchr. Rad. News* **12**, 27

Vrakking, M. J. J., Lee, Y. T., Gilbert, R. D. and Child, M. S. (1993) *J. Chem. Phys.* **98**, 1902

Wüst, J. and Reindel, H. (1934) *Z. Phys. Chem. B* **24**, 155

Wang, H.-T., Felps, W. S. and McGlynn, S. P. (1977) *J. Chem. Phys.* **67**, 2614

Wang, L., Lee, Y. T. and Shirley, D. A. (1987) *J. Chem. Phys.* **87**, 2489

Wang, Z.-W. and Chung, K. T. (1994) *J. Phys. B* **27**, 855

Warneck, P., Marmo, F. F. and Sullivan, J. O. (1964) *J. Chem. Phys.* **40**, 1132

Watanabe, K. (1954) *J. Chem. Phys.* **22**, 1564

Watanabe, K. and Jursa, A. S. (1964) *J. Chem. Phys.* **41**, 1650

Watanabe, K. and Sood, S. P. (1965) *Science of Light* **14**, 36

Watanabe, K., Inn, E. C. Y. and Zelikoff, M., (1953a) *J. Chem. Phys.* **21** 1026

Watanabe, K., Matsunaga, F. M. and Sakai, H. (1967) *Appl. Optics* **6**, 391

Watanabe, K., Zelikoff, M. and Inn, E. C. Y. (1953b), Air Force Cambridge Technical Report AFCRC-TR-53-23, p. 65

Watson, H. E. and Ramaswamy, K. L. (1936) *Proc. Roy. Soc. Lond. A* **156**, 144

Watson, H. E., Rao, G. G. and Ramaswamy, K. L. (1934) *Proc. Roy. Soc. Lond. A* **143**, 558

Watson, W. C., Stewart, D. T., Gardner, A. B. and Lynch, M. J. (1975) *Planet. Space Sci.* **23**, 384

Watson, W. S. (1972) *J. Phys. B* **5**, 2292

Weigold, E., Zheng, Y. and von Niessen, W. (1991) *Chem. Phys.* **150**, 405

Weiss, A. W. (1963) *Astrophys. J.* **138**, 1262

Weiss, A. W. (1992) *Can. J. Chem.* **70**, 456

Weiss, H., Ahlrichs, R. and Häser, M. (1993) *J. Chem. Phys.* **99**, 1262

Weiss, M. J., Berkowitz, J. and Appelman, E. H. (1977) *J. Chem. Phys.* **66**, 2049

Weissler, G. L. (1956) in *Handbuch der Physik* (Springer-Verlag, Berlin), vol. 21, p. 318

Wells, M. C. and Lucchese, R. R. (1999) *J. Chem. Phys.* **111**, 6290

Werner, H.-J. and Meyer, W. (1976) *Phys. Rev. A* **13**, 13

West, J. B. and Marr, G. V. (1976) *Proc. Roy. Soc. Lond. A* **349**, 397

West, J. B., Hayes, M. A., Parr, A. C., Hardis, J. E., Southworth, S. H., Ferrett, T. A., Dehmer, J. L., Hu, X.-M. and Marr, G. V. (1990) *Phys. Scripta* **41**, 487

West, J. B., Hayes, M. A., Siggel, M. R. F., Dehmer, J. L., Dehmer, P. M., Parr, A. C. and Hardis, J. E. (1996) *J. Chem. Phys.* **104**, 3923

Westerveld, W. B., Mulder, Th. F. A. and van Eck, J. (1979) *J. Quant. Spectr. Rad. Transf.* **21**, 553

Westin, E., Rosén, A., Te Velde, G. and Baerends, E. J. (1996) *J. Phys. B* **29**, 5087

Wettlaufer, D. E. and Glass, I. I. (1972) *Phys. Fluids* **15**, 2065

White, M. G., Leroi, G. E., Ho, M.-H. and Poliakoff, E. D. (1987) *J. Chem. Phys.* **87**, 6653

Wiedmann, R. T. and White, M. G. (1992) *SPIE Proc. Ser.* **1638**, 273

Wiedmann, R. T., Grant, E. R., Tonkyn, R. G. and White, M. G. (1991) *J. Chem. Phys.* **95**, 746

Wiedmann, R. T., White, M. G., Lefebvre-Brion, H. and Cossart-Magos, C. (1995) *J. Chem. Phys.* **103**, 10417

Wiese, W. L. and Martin, G. A. (1980) *Wavelengths and Transition Probabilities for Atoms and Ions. Part II. Transition Probabilities*, NSRDS-NBS 68

Wiese, W. L., Fuhr, J. R. and Deters, T. M. (1996) *J. Phys. Chem. Ref. Data Monograph* No. 7

Wiese, W. L., Smith, M. W. and Glennon, B. M. (1966) *Atomic Transition Probabilities*, Natl. Inst. Stand. Tech. U.S. Circ. No. 4 (U.S. Govt. Print. Off., Wash., D.C.) Vol. **I**

Wiese, W. L., Smith, W. W. and Miles, B. M. (1969) Atomic Transition Probabilities, NSRDS-NBS 22, Vol. **II**

Wight, G. R. and Brion, C. E. (1974a) *J. Electron Spectrosc.* **4**, 313

Wight, G. R. and Brion, C. E. (1974b) *J. Electron Spectrosc.* **3**, 191

Wight, G. R. and Brion, C. E. (1974c) *J. Electron Spectrosc.* **4**, 25

Williams, B. A. and Cool, T. A. (1991) *J. Chem. Phys.* **94**, 6358

Williams, J. H., Schwingel, C. H. and Winning, C. H. (1936) *J. Am. Chem. Soc.* **58**, 197

Wolff, H. W., Radler, K., Sonntag, B. and Haensel, R. (1972) *Z. Phys.* **257**, 353

Wolniewicz, L. (1966) *J. Chem. Phys.* **45**, 515

Wolniewicz, L. (1969) *J. Chem. Phys.* **51**, 2002

Wolniewicz, L. (1975) *Chem. Phys. Lett.* **31**, 248

Wolniewicz, L. (1993) *J. Chem. Phys.* **99**, 1851

Woon, D. E. and Dunning, T. J. Jr., (1994) *J. Chem. Phys.* **100**, 2975

Wu, C. Y. R. and Chen, F. Z. (1998) *J. Quant. Spectr. Rad. Transf.* **60**, 17

Wu, C. Y. R. and Judge, D. L. (1985) *J. Chem. Phys.* **82**, 4495

Wu, C. Y. R., Chen, F. Z., Judge, D. L., Hus, X.-M. and Caldwell, J. (1999) *J. Chem. Phys.* **110**, 267

Wu, C. Y. R., Chien, T. S., Liu, G. S., Judge, D. L. and Caldwell, J. J. (1989) *J. Chem. Phys.* **91**, 272

Wu, J. Z., Whitfield, S. B., Caldwell, C. D., Krause, M. O., van der Meulen, P. and Fahlman, A. (1990) *Phys. Rev. A* **42**, 1350

Wu, S. L., *et al.*, unpublished but cited by Zhong *et al.* (1997)

Wuilleumier, F. (1965) *J. Physique* **26**, 776

Wuilleumier, F. (1971) *J. Physique* **32**, C4-88

Xia, T. J., Chien, T. S., Wu, C. Y. R. and Judge, D. L. (1991) *J. Quant. Spectr. Rad. Transf.* **45**, 77

Yamada, C., Kanamori, H. and Hirota, E. (1985) *J. Chem. Phys.* **83**, 552

Yan, L., Ågren, H., Carravetta, V., Vahtras, O., Karlsson, L., Wannberg, B., Holland, D. M. P. and MacDonald, M. A. (1998b) *J. Electron Spectrosc.* **94**, 163

Yan, M., Sadeghpour, H. R. and Dalgarno, A. (1998a) *Astrophys. J.* **496**, 1044

Yan, Z.-C. and Drake, G. W. F. (1995) *Phys. Rev. A* **52**, R4316

Yasuike, T. and Yabushita, S. (2000) *Chem. Phys. Lett.* **316**, 257

Yasumatsu, H., Kondow, T., Kitagawa, H., Tabayashi, K. and Shobatake, K. (1996) *J. Chem. Phys.* **104**, 899

Yavna, V. A., Khoperskii, A. N., Demekhina, L. A. and Sukhorukov, V. L. (1986b) *Opt. Spectrosc. (USSR)* **61**, 273

Yavna, V. A., Petrov, I. D., Demekhina, L. A., Khoperskii, A. N. and Sukhorukov, V. L. (1986a) *Opt. Spectrosc. (USSR)* **61**, 552

Yencha, A. J., Thompson, D. B., Cormack, A. C., Cooper, D. R., Zubek, M., Bolognesi, P. and King, G. C. (1997) *Chem. Phys.* **216**, 227

Yoo, R. K., Ruščić, B. and Berkowitz, J. (1992) *J. Chem. Phys.* **96**, 911

Yoshimine, M., Tanaka, K., Tatewaki, H., Obara, S., Sadaki, F. and Ohno, K. (1976) *J. Chem. Phys.* **64**, 2254

Yoshino, K. (1970) *J. Opt. Soc. Am.* **60**, 1220

Yoshino, K. (1998) Private communication, see Internet: http://cfa-www.harvard.edu /ampdata/VUV_H2O/

Yoshino, K., Esmond, J. R. and Parkinson, W. H. (1997) *Chem. Phys.* **221**, 169

Yoshino, K., Esmond, J. R., Freeman, D. E. and Parkinson, W. R. (1993) *J. Geophys. Res. D* **98**, 5205

Yoshino, K., Esmond, J. R., Parkinson, W. H., Ito, K. and Matsui, T. (1996a) *Chem. Phys.* **211**, 387

Yoshino, K., Esmond, J. R., Sun, Y., Parkinson, W. H., Ito, K. and Matsui, T. (1996b) *J. Quant. Spectr. Rad. Transf.* **55**, 53

Yoshino, K., Tanaka, Y., Carroll, P. K. and Mitchell, P. (1975) *J. Mol. Spectrosc.* **54**, 87

Yoshino, K., private communication, cited by Stahel, D., Leoni, M. and Dressler, K. (1983) *J. Chem. Phys.* **79**, 254

Yoshino, M., Takeuchi, J. and Suzuki, H. (1973) *J. Phys. Soc. Jpn* **34**, 1039

Younglove, B. A. (1972) *J. Res. Natl. Bur. Stand.* **76A**, 37

Zahn, C. T. (1933) *Phys. Z.* **34**, 461

Zeiss, G. D. and Meath, W. J. (1975) *Mol. Phys.* **30**, 161

Zeiss, G. D. and Meath, W. J. (1977) *Mol. Phys.* **33**, 1155

Zeiss, G. D., Meath, W. J., MacDonald, J. C. F. and Dawson, D. J. (1977) *Can. J. Phys.* **55**, 2080

Zeiss, G. D., Meath, W. J., MacDonald, J. C. F. and Dawson, D. J. (1980) *Mol. Phys.* **39**, 1055

Zelikoff, M. and Watanabe, K. (1953) *J. Opt. Soc. Am.* **43**, 756

Zelikoff, M., Watanabe, K. and Inn, E. C. Y. (1953) *J. Chem. Phys.* **21**, 1643

Zhadenov, A. V., Akimov, V. N. and Vinogradov, A. S. (1987) *Opt. Spectrosc (USSR)* **62**, 204

Zhang, W., Sze, K. H., Brion, C. E., Tong, X. M. and Li, J. M. (1990) *Chem. Phys.* **140**, 265

Zhong, Z. P., Feng, R. F., Zu, K. Z., Wu, S. L., Zhu, L. F., Zhang, X. J., Ji, Q. and Shi, Q. C. (1997) *Phys. Rev. A* **55**, 1799

Zhu, Y.-F. and Gordon, R. J. (1990) *J. Chem. Phys.* **92**, 2897

Zhu, Y.-F., Grant, E. R. and Lefebvre-Brion, H. (1993) *J. Chem. Phys.* **99**, 2287

Zimkina, T. M. and Fomichev, V. A. (1967) *Soviet Physics-Doklady* **11**, 726

Zimkina, T. M. and Vinogradov, A. S. (1971) *J. Phys. (Paris)* **32**, C4–3

Zipf, E. C. and McLaughlin, R. W. (1978) *Planet. Space Sci.* **26**, 449

Zubek, M., King, G. C. and Rutter, P. M. (1988) *J. Phys. B* **21**, 3585

Index

acetylene (C_2H_2) 252–60
  absolute photoabsorption spectrum
      255, 256
  adiabatic ionization potential 252–3
  autoionization region 253–4
  carbon K-shell 252, 258
  contributions to S(p) of transitions
      below IP 254
  electron configuration 252
  oscillator strength distribution below IP
      253
  polynomial fit to data 258
  post K-edge 258
  quantum yields 110–11, 259–60
  spectral sums and comparison with
      expectation values 257
  static polarizability 258
  sum-rule analysis 258–60
ammonia (NH_3) 237–45
  absolute photoabsorption spectrum
      239, 243
  adiabatic ionization potential 238
  autoionization region 241
  contributions to S(p) of transitions
      below IP 240–1
  discrete spectrum and transitions below
      IP 238–41
  electronic ground state 237
  nitrogen K-edge 238, 243, 244
  photoionization cross sections 245
  polynomial fit to data 244
  post K-edge 244
  predissociation 245
  refractivity 244
  quantum yield 108–10, 245
  spectral sums and comparison with
      expectation values 242
  sum-rule analysis 244–5

argon 82–94
  absolute photoabsorption spectrum 91
  continuum 87–92
  contributions of discrete spectrum to
      S(p) sums 88
  dielectric constant 93
  discrete spectrum 82–7
  ionization potential 82
  polynomial fit to data 90
  refractivity 93
  resonances 89
  spectral sums, and comparison with
      expectation values 89–90
  sum-rule analysis 93–4
atomic chlorine 66–82
  absolute photoabsorption spectrum 78,
      80
  continuum 76–80
  contributions of $(3p)^{-1}$ spectrum to S(p)
      sums 69–70
  electronic ground state 66
  ionization potential 66
  polynomial fit to data 78
  spectral sums, and comparison with
      theoretical sums 77
  static electric dipole polarizability 81
  sum-rule analysis 81–2
  valence shell spectrum 66–76
atomic hydrogen 8
  expectation values for S(p) 8
  ionization potential 8
  oscillator strength distribution in the
      continuum 8
  oscillator strengths, Lyman series 8
atomic nitrogen 27–35
  absolute photoabsorption
      (photoionization) spectrum
      29–30

atomic nitrogen (*Continued*)
  determinization of static electric dipole
      polarizability   34
  discrete spectrum   27–9
  ionization potential   27
  polynomial fit to data   33
  resonances   31
  spectral sums, and comparison with
      expectation values   32
  sum-rule analysis   33–5
atomic oxygen   35–43
  autoionization peaks   37–9
  continuum   36–42
  contribution to sum rules from
      autoionizing transitions   38
  contribution to sum rules of discrete
      transitions   36
  discrete spectrum   35–6
  ionization potential   35
  polynomial fit to data   41
  spectral sums, and comparison with
      expectation values   40
  static electric dipole polarizability   42
  sum-rule analysis   42–3
atoms   8–94

Beer-Lambert law   1, 3, 44, 140, 261,
      287
benzene (C_6H_6)   281–7
  absolute photoabsorption spectrum
      285
  adiabatic ionization potential (AIP)
      282
  autoionization region and beyond   283
  carbon K-edge region   283–5
  carbon K-shell   281
  discrete spectrum and transitions below
      IP   282–3
  ground state electron configuration
      281
  polynomial fit to data   285
  post K-edge   285–6
  refractivity   286
  quantum yield   112, 287
  spectral sums and comparison with
      expectation values   284
  sum-rule analysis   286–7
  vibrionic structure   282
Bethe-Salpeter equation   2

buckminsterfullerene (C_{60})   287–300
  absolute photoabsorption spectrum
      289, 292, 294, 295, 298
  adiabatic IP   288
  alternative absolute photoabsorption
      spectra   291
  carbon K-edge   294
  electric dipole polarizability   295, 296
  maximum cross sections (Mb) for
      lowest dipole-allowed transitions
      288
  photoabsorption measurements   290
  photoionization   290–3
  polynomial fit to data   295
  pseudo-photoabsorption spectrum   290
  quantum yield   298, 300
  spectral sums and comparison with
      expectation values   293
  sum-rule analysis   295–300
  transitions below IP   288–9

C_2H_2 *see* acetylene
C_2H_4 *see* ethylene
C_2H_6 *see* ethane
C_3H_8, quantum yield   117
C_6H_6 *see* benzene
$C_{10}H_8$, quantum yield   116, 118
$C_{14}H_{10}$, quantum yield   116, 118
$C_{18}D_{12}$, quantum yield   116
$C_{24}H_{12}$, quantum yield   116
carbon dioxide (CO_2)   189–97
  absolute photoabsorption spectrum
      193–4
  adiabatic ionization potential   190
  autoionization region and beyond
      191–3
  continuum   193–4
  discrete spectrum and transitions below
      IP   190–1
  electric dipole polarizability   196
  inter-edge continuum   195
  molecular orbital structure   190
  near carbon K-edge   194–5
  near oxygen K-edge   195
  photoabsorption cross sections
      189–90
  polynomial fit to data   194
  post K-edge continuum   195
  quantum yield   104–5, 197

spectral sums, and comparison with
expectation values 192
sum-rule analysis 196–7
carbon monoxide (CO) 156–66
ab initio calculations 163
absolute photoabsorption spectrum
157, 161, 162
adiabatic ionization potential 158
autoionization region 157, 158, 165
carbon K-edge 159
carbon K-edge region 161–2
continuum 158–60
discrete oscillator strengths 163
discrete spectrum and transitions below
IP 158
inter-edge continuum 162
oxygen K-edge 162
oxygen K-edge region 162–3
post K-edge continuum 163
predissociation 156
polynomial fit to data 161
quantum yield 100, 165
refractive index 163
spectral sums, and comparison with
expectation values 160
sub-ionization oscillator strengths 163,
166
sum-rule analysis 163–6
C-C bond 267
C-C double bond 260
C-C triple bond 252
CH_3OH *see* methanol
CH_4 *see* methane
chlorine *see* atomic chlorine

Debye equation 212, 235
diatomic molecules 125–80
competitive processes 98–103
direct dissociation 96
direct ionization 96

EELS 222, 225, 227, 233, 283, 290,
297, 299
electric dipole-electric quadrupole (E1-E2)
5
electric dipole-magnetic dipole (E1-M1)
5
electron synchrotrons 1
electronic autoionization 97

ethane (C_2H_6) 267–74
absolute photoabsorption spectrum
269, 271–2
adiabatic ionization potential 268
autoionization 269
carbon K-edge region 271
discrete spectrum and transitions below
IP 268–9
inelastic electron scattering 268
photoabsorption measurements 268
polynomial fit to data 272
quantum yield 116, 273
spectral sums and comparison with
expectation values 270
sum-rule analysis 272–4
ethylene (C_2H_4) 260–7
absolute photoabsorption spectrum
263, 264, 267
adiabatic ionization potential 261
autoionization region and beyond
261–3
carbon K-edge 264, 265
continuum 263–4
discrete spectrum and transitions below
IP 261
electronic ground state 260
oscillator strength distribution 260,
262
polynomial fit to data 265
post K-edge 265
refractivity 265
quantum yield 111–12, 266–7
spectral sums and comparison with
expectation values 262
sum-rule analysis 265–7

Franck-Condon factors 96, 98, 309
Franck-Condon region 302
Franck-Condon span 100

'gross' ionization cross section 122

H_2O molecule *see* water (H_2O)
Hartree-Fock atomic and molecular charge
densities at the nucleus 146,
155
HC1 *see* hydrogen chloride

helium 8–17
  absolute photoabsorption spectrum 11, 12, 14
  continuum 10–17
  discrete spectrum 9–10
  excess oscillator strength 10
  ionization potential 9
  polynomial fit to data 15
  resonances 10
  spectral sums and comparison with expectation values 11
  sum-rule analysis 17
hydrogen *see* atomic hydrogen; molecular hydrogen
hydrogen chloride (HC1) 175–80
  absolute photoabsorption spectrum 176
  adiabatic ionization potential 177
  continuum 177–9
  discrete spectrum, and transitions below IP 177
  electronic ground state 175
  L_{III} edge 176
  polynomial fit to data 179
  quantum yield 100, 102, 180
  spectral sums and comparison with expectation values 178
  static electric dipole polarizability 179
  sum-rule analysis 179–80
  valence shell orbital sequence 101
hydrogen sulfide (H_2S) 214–21
  absolute photoabsorption spectrum 219
  adiabatic IP 215
  continuum 217–20
  discrete region 215–16
  polynomial fit to data 219
  refractivity 220
  S(2s) edge 218
  spectral sums and comparison with expectation values 216
  sulfur K-edge 219
  sulfur K-edge region 219–20
  sulfur $L_{II,III}$ edges 217
  sum-rule analysis 220–1
hyperspherical coordinates with close-coupling (HSCC) 13

internal conversion 96
ionization, quantum yield of *see* quantum yield of ionization

Jahn-Teller interaction 246, 282, 309

Lamb shift 4
lithium 18–26
  absolute photoabsorption spectrum 23, 24
  determination of expectation values of S(-1), S(+1) and S(+2) 25–6
  discrete spectrum 18–19
  discrete spectrum oscillator strengths 19
  energies and oscillator strengths of resonances 22
  ionization continuum 19–23
  ionization potential 18
  polynomial fit to data 20
  spectral sums, and comparison with expectation values 21
  static electric dipole polarizability 24
  sum-rule analysis 23–5
Lorenz-Lorentz equation 296
Lyman bands *see* molecular hydrogen
Lyman series 8

methane (CH_4) 246–52
  absolute photoabsorption spectrum 246, 250
  adiabatic ionization potential 247
  carbon K-edge region 249–50
  continuum 248–9
  electronic configuration 24
  molar refractivity 251
  oscillator strength distribution below IP 247–8
  photoionization cross section 252
  polynomial fit to data 250
  post K-edge 250–1
  quantum yield 114
  spectral sums and comparison with expectation values 249
  sum-rule analysis 251–2
methanol (CH_3OH) 274–81
  absolute photoabsorption spectrum 275, 277, 278

adiabatic ionization potential (AIP)
275
carbon K-edge 279
discrete spectrum and transitions below
IP 275
inter-edge continuum 279
orbital sequence 274
oxygen K-edge 279–80
polynomial fit to data 278
post K-edges 280
refractive index 280
spectral sums and comparison with
expectation values 276
sum-rule analysis 280–1
molecular hydrogen (H_2) 125–40
autoionization 130–1
autoionization oscillator strengths
132–3
determination of expectation values of
$S(-1)$, $S(+1)$ and $S(+2)$
137–40
discrete spectrum and transitions below
IP 126–8
dissociation continuum above IP 129
dissociation continuum below IP 129
ionization continuum 133–4
ionization continuum and transitions
above IP 129–35
Lyman bands 126–7
oscillator strengths attributed to
predissociation above IP 131
polynomial fit to data 134
predissociation above IP 129
quantum yield 98
spectral sums, and comparison with
expectation values 130
sum-rule analysis 136–7
underlying ionization continuum 131
Werner bands 126–7
molecular nitrogen (N_2) 140–7
adiabatic ionization potential 141
autoionization region 144
continuum 144–5
discrete spectrum and transitions below
IP 141–4
electric dipole polarizability 146
K-edge structure 145
polynomial fit to data 144

post K-edge continuum 145
quantum yield 98, 99, 147
resonances preceding K-edge 145
spectral sums, and comparison with
expectation values 143
sum-rule analysis 145–7
molecular oxygen (O_2) 147–56
absolute photoabsorption spectrum
149, 150, 154
adiabatic ionization potential 148
autoionization region 150–1
autoionization structure 147
continuum 151–3
K-edge region 153
photoionization, ZEKE 97
polynomial fit to data 153
post K-edge continuum 153
quantum yield yield 98–100, 156
Schumann-Runge bands 147–9
spectral sums, and comparison with
expectation values 152
sub-ionization region 148–50
sub-ionization $S(p)$ 151
sum-rule analysis 153–6
multi-channel quantum defect theory
(MQDT) 97, 102

neon 43–55
absolute photoabsorption spectrum 55
continuum 50–5
contributions from discrete spectrum to
$S(p)$ sums 49
discrete spectrum 43–50
ionization potential 43
oscillator strengths for discrete
transitions 45–6
polynomial fit to data 52
resonances 50–3
spectral sums, and comparison with
expectation values 51
sum-rule analysis 53–5
nitric oxide (NO) 166–75
absolute photoabsorption spectrum
172, 173
adiabatic IP 167–8
autoionization 167, 169, 171, 175
continuum 171
discrete spectrum and transitions below
IP 168–9

nitric oxide (NO) (*Continued*)
  inter-edge region 172
  nitrogen K-edge structure 171–2
  oxygen K-edge structure 173–4
  polynomial fit to data 171, 173
  post oxygen K-edge 174
  quantum yield 100, 175
  spectral sums, and comparison with
      expectation values 170
  sum-rule analysis 174–5
nitrogen *see* atomic nitrogen; molecular
      nitrogen
nitrogen dioxide (NO$_2$) 206–14
  absolute photoabsorption spectrum
      207, 209, 211
  continuum 210–12
  dielectric constant 212
  discrete spectrum, and transitions below
      IP 207–10
  inelastic electron scattering
      measurements 207
  inter-edge region 211–12
  nitrogen K-edge 210
  oxygen K-edge 212
  polynomial fit to data 211
  post K-edge region 212
  quantum yield 107, 214
  refractive index 213
  spectral sums and comparison with
      expectation values 208
  sum-rule analysis 212–14
nitrous oxide (N$_2$O) 197–206
  absolute photoabsorption spectrum
      202, 203, 204
  adiabatic ionization potential 198
  autoionization region 199–200
  continuum 200–2
  contributions to S(p) of transitions
      below IP 199
  discrete spectrum and transitions below
      IP 198–9
  inter-edge region 202–3
  nitrogen K-edge region 202
  oxygen K-edge region 204
  photoionization 97
  polynomial fit to data 203
  post K-edges 205
  quantum yield 105–7, 206

spectral sums, and comparison with
      expectation values 201
  sum-rule analysis 205–6

oxygen *see* atomic oxygen; molecular
      oxygen
ozone (O$_3$) 228–36
  absolute photoabsorption spectrum
      229, 230
  adiabatic ionization potential 230–1
  Chappuis band 228, 231
  Hartley band 229, 231
  K-shell region 230, 234
  photoabsorption cross section 229,
      231, 233
  photoionization spectrum 236
  polynomial fit to data 234
  spectral sums, and comparison with
      expectation values 232
  static electric dipole polarizability 234
  sum-rule analysis 234–6
  thermochemical cycle 231
  Wulf band 228–9, 231

photoelectron spectroscopy 5
photoionization mass spectrometry 5
Poincaré recurrence time 119
polyaromatic hydrocarbons (PAHs)
      116–17
  quantum yield 116
polyatomic molecules 237–316
  general observations 114–21
  quantum yield 108–21
predissociation 96
pulsed field ionization, zero kinetic energy
      (PFI-ZEKE)
  photoelectron spectroscopy *see* ZEKE
      measurements

quantum yield and M$_i^2$ 121–4
quantum yield of ionization 95–125

radiationless transition 96
random phase approximation (RPA) 10
Reference Table 6
Renner-Teller splitting 191
R-matrix 97
Rydberg series 95, 191, 242, 261, 309
Rydberg states 141, 142, 168, 237, 260

Rydberg transitions 219
Rydberg Units 6,8

Schumann-Runge bands *see* molecular
    oxygen
shape resonance 96
silane (SiH_4) 309–16
  absolute photoabsorption spectrum
    311, 313
    adiabatic ionization potential 310
    aufbau of molecular orbitals 309
    continuum 310–14
    dielectric constant 314–15
    K-edge region 314
    photoelectron spectrum 310
    polynomial fit to data 314
    quantum yield 115, 315–16
    refractive index 314–15
    spectral sums and comparison with
      expectation values 312
    sum-rule analysis 314–16
    transitions below IP 310
sodium 55–56
  absolute photoabsorption spectrum 59,
    63
    autoionizing resonances 61–2
    continuum 57–64
    discrete spectrum 56–7
    ionization potential 55
    oscillator strengths 56, 57
    photoabsorption cross sections 60
    polynomial fit to data 63
    resonances around K-edge 63–4
    spectral sums, and comparison with
      expectation values 58
    static electric dipole polarizability 65
    sum-rule analysis 65–6
squared dipole moment M_i^2 121–4
  from photoionization and charged
    particle ionization 123
Stark effect 5
sulfur dioxide (SO_2) 221–8
    adiabatic IP 222
    autoionization 224
    continuum 224–6
    dielectric constant 226–7
    discrete spectrum and transitions below
      IP 222–3

    electron configuration in ground state
      221
    lowest energy unoccupied orbitals 221
    photoionization cross sections 228
    polynomial fit to data 225
    refractive index 226–7
    spectral sums and comparisons with
      expectation values 223
    sum rule analysis 226–8
sulfur hexafluoride (SF_6) 300–9
  absolute photoabsorption spectrum
    303
    charge-transfer experiments 302
    F(1s) region 306–7
    high-resolution photoelectron spectrum
      (PES) 301–2
    photoabsorption cross sections,
      14.9–70.85 eV 303
    photoabsorption cross sections,
      70.85–277 eV 305–6
    photoabsorption between S(2s) and
      F(1s): 277–685 eV 306
    photoabsorption between F(1s) and
      S(1s) 307
    photoionization mass spectrum 302
    polynomial fit to data 305
    post sulfur K-edge 307
    quantum yield 112–13, 309
    refractive index 307–8
    S(1s) region 307
    sequence of occupied molecular orbitals
      in electronic ground state
      300–1
    spectral sums, and comparison with
      expectation values 304
    sum-rule analysis 307–309
    transitions below 14.9 eV 302–3

Thomas-Reiche-Kuhn (TRK) sum rule 2,
    6, 33, 43, 93, 122, 126, 140,
    163, 188, 196, 205, 213, 237,
    247, 251, 261, 297, 308, 315
Thomas-Reiche-Kuhn value 25, 180, 220
threshold photoelectron spectrum (TPES)
    116–17, 119
triatomic molecules 181–236
    quantum yield 103–7

vibrational autoionization 97

water (H_2O)   181–9
  ab initio calculations of polarizability
      188
  absolute photoabsorption cross section,
      UV-VUV   181
  absolute photoabsorption spectrum
      183, 184
  adiabatic ionization potential   183
  autoionization region   184–5
  continuum   185
  contributions to S(p) below IP   185
  discrete spectrum and transitions below
      IP   184
  electron energy loss spectrum   182
  higher pressure experiments   188
  oxygen K-edge region   185–7
  polynomial fit to data   186
  post K-edge   187
  quantum yield   103–4, 189
  refractivity measurements   187–8
  spectral sums and comparison with
      expectation values   186
  sum-rule analysis   187–9
  valence orbital sequence   103
Werner bands *see* molecular hydrogen

ZEKE measurements   5, 97, 158, 167,
      198, 207, 215, 238, 253, 275